AF606427

Gmelin Handbuch der Anorganischen Chemie

Achte völlig neu bearbeitete Auflage

8th Edition

Volumes published on "Rare Earth Elements" (Syst.-No. 39)

* Completely or ° in part in English

Gmelin Handbuch der Anorganischen Chemie

Achte völlig neu bearbeitete Auflage

BEGRÜNDET VON Leopold Gmelin

ACHTE AUFLAGE BEGONNEN im Auftrag der Deutschen Chemischen Gesellschaft von R. J. Meyer

FORTGEFÜHRT VON E. H. E. Pietsch und A. Kotowski
Margot Becke-Goehring

HERAUSGEGEBEN VOM Gmelin-Institut für Anorganische Chemie der Max-Planck-Gesellschaft zur Förderung der Wissenschaften
Direktor: Ekkehard Fluck

Springer-Verlag
Berlin · Heidelberg · New York 1981

Gmelin Handbook of Inorganic Chemistry

8th Edition

Sc, Y, La–Lu
RARE EARTH ELEMENTS

Part D 3

Coordination Compounds (Continuation)

With 42 illustrations

AUTHORS

Edward R. Birnbaum, Department of Chemistry, New Mexico State University, Las Cruces, New Mexico, USA

John H. Forsberg, Department of Chemistry, Monsanto Hall, Saint Louis University, Saint Louis, Missouri, USA

Yizhak Marcus, Department of Inorganic and Analytical Chemistry, The Hebrew University of Jerusalem, Jerusalem, Israel

FORMULA INDEX

Helga Köttelwesch, Gmelin-Institut, Frankfurt am Main

EDITORS

Therald Moeller, Department of Chemistry, Arizona State University, Tempe, Arizona, USA

Edith Schleitzer-Rust, Gmelin-Institut, Frankfurt am Main

System Number 39

Springer-Verlag
Berlin · Heidelberg · New York 1981

LITERATURE CLOSING DATE: UP TO MID 1979
IN MANY CASES MORE RECENT DATA HAVE BEEN CONSIDERED

Die vierte bis siebente Auflage dieses Werkes erschien im Verlag von Carl Winter's Universitätsbuchhandlung in Heidelberg

Library of Congress Catalog Card Number: Agr 25-1383

ISBN 3-540-93432-4 Springer-Verlag, Berlin · Heidelberg · New York
ISBN 0-387-93432-4 Springer-Verlag, New York · Heidelberg · Berlin

LN-Druck Lübeck

Preface

Coordination compounds of scandium, yttrium, and the lanthanides are described in series D of the "Rare Earth Elements". Complexes with nitrogen donor and with oxygen donor ligands are described in volumes D 1 and D 2. This volume, volume D 3, continues the discussions of complexes with oxygen donor ligands.

Chapter 1 deals with complexes derived from water. Reference is made to salt hydrates described in volumes of series B and C. Coordination numbers of terpositive rare earth ions toward water molecules in aqueous solutions are discussed in detail in terms of interpretations of physical data.

Chapters 2, 3 and 4 deal successively with the preparation and properties of complexes derived from alcohols and alcoholates, aromatic hydroxy compounds, aldehydes and monoketones. Where Ce^{IV} complexes have been reported for a specific ligand, they are described after the M^{III} complexes. Solution studies are described, when data are available.

Chapter 5 describes complexes with 1,3-diketones and other polyketones. The discovery in 1963 of laser action in rare earth diketonates prompted investigations of their fluorescence spectra and led to isolation and characterization of the first eight-coordinate rare earth complexes. A review of the luminescence properties is presented as an introduction to Chapter 5. Following the initial communication in 1969 several hundred publications have appeared demonstrating the ability of certain tris(diketonato) chelates to act as NMR shift reagents. Since the applications of lanthanide shift reagents are well documented in several extensive review articles, the pertinent sections of this chapter concentrate mainly on the preparation, characterization, and physicochemical properties of the shift reagent adducts, and includes only those NMR studies most closely associated with structural aspects and magnetic properties of the complexes. Chromatographic separation of the rare earths in the gas phase as the volatile, thermally stable tris(dipivaloylmethanato)- and tris(heptafluorodimethyloctanedionato) chelates has been also demonstrated.

Chapter 6 describes complexes with quinones and triphenylmethane dyes. Most of these complexes have been studied in solution, and only a limited number of complexes have been isolated and characterized. The use of triphenylmethane dyes as reagents for the quantitative determination of rare earth ions in solution is well documented.

Chapter 7 covers complexes with ethers and O-heterocycles, including crown ethers, pyranones, derivatives of coumarin, hydroxyflavones, and hematein. Emphasis is substantially on syntheses. Where available, data on crystal and molecular structures and on solution properties are included.

A formula index at the end of this volume lists all ligands and their empirical molecular formulas.

Tempe, Arizona, USA
Frankfurt/Main
December 1980

Therald Moeller
Edith Schleitzer-Rust

Table of Contents

Page

Page

1 Complexes with Water

Y. Marcus
Department of Inorganic and Analytical Chemistry, The Hebrew University of Jerusalem
Jerusalem, Israel

1.1 General

Much of the earlier work concerning the coordination of water with Sc, Y, and the lanthanide ions in solution has been rather speculative, the speculations resting on the coordination of water to the metal ion in crystalline compounds. Moeller et al. [1] summarize this phase of the work, stating that the compositions of coordination compounds (solid and in solution) were consistent with the assignment of the coordination number six to Y and the rare earth metal ions. Although this coordination number has been assumed by many authors, there was an increasingly impressive array of evidence that the true coordination number was larger than six [1, 2].

The evidence for coordination numbers for water > 6 includes the structures of the crystalline salt hydrates $M(BrO_3)_3 \cdot 9H_2O$ [3] and $M(C_2H_5OSO_3)_3 \cdot 9H_2O$ [4], which form isomorphous enneahydrate crystals along the whole series of the rare earths (M = La to Lu) [5 to 9] each with all nine water molecules coordinated to the metal ion. It includes also, however, structures such as $MCl_3 \cdot 7H_2O$ (M = La to Pr) and $MCl_3 \cdot 6H_2O$ (M = Nd to Lu) [10 to 14], $M(NO_3)_3 \cdot 6H_2O$ (M = La to Lu) [15], and $M_2(SO_4)_3 \cdot 9H_2O$ (M = La to Pr) and $M_2(SO_4)_3 \cdot 8H_2O$ (M = Nd to Lu) [16 to 18], where only a part of the coordination shell is made up of the oxygen atoms of the water molecules, while the rest is made up of chlorine atoms or oxygen atoms from the anions. The structures of these and other crystal hydrates are discussed in detail in the Gmelin Handbook (Sc, Y, La to Lu), see "Seltenerdelemente" Volumes C 2, 1974 (the nitrates), C 4 (the chlorides, to be published), C 6, 1978 (the bromates), and C 8, 1981 (the sulfates and ethylsulfates), and will not be treated here. The structure [3] of the inner coordination shell of crystalline $Nd(BrO_3)_3 \cdot 9H_2O$, i.e., that of $Nd(H_2O)_9^{3+}$, is shown in Fig. 1-2, p. 8, for reference purposes only.

This chapter discusses the aqua-complexes formed in solution from the following aspects: (a) the coordination number, i.e., the primary hydration number, of the ions Sc^{3+}, Y^{3+}, and the lanthanides La^{3+} to Lu^{3+}; (b) the structures of these hydrated ions; (c) the rate at which H_2O is released from the hydration shell, as a measure of the tightness of its binding; (d) some comments on the thermodynamic data for the Gibbs free energy, enthalpy, and entropy of hydration, as a measure of the strength of the coordination of the water; (e) some comments on the bonding of the water to the metal ions.

1.2 Coordination Numbers of the Aqua-Ions

Hydration numbers are definable only operationally, i.e., by the method according to which they are determined. Consequently, they are generally more sensitive to the method of determination than to the ion, the hydration of which they should describe. Fortunately, however, the first hydration shell can be described rather definitely by the requirement, that the oxygen atoms of the water molecules be in direct contact with the metal ion. Hence so called "primary hydration numbers" are more or less well defined descriptors of the number of water molecules in the first coordination shell of the metal ions in dilute aqueous solutions, devoid of coordinating ligands other than water. In the following, "primary hydration number" will be taken as synonymous with "coordination number" in such dilute aqueous solutions.

For some time a controversy raged, concerning the question, whether the coordination number stays constant or changes, abruptly or over few neighboring elements, as the atomic number Z varies from 57 (La) to 71 (Lu). The suggestion that the primary hydration number changes from a higher one (presumably 9) for La^{3+}, Ce^{3+}, Pr^{3+}, and Nd^{3+} to a lower one (presumably 8) for Tb^{3+}, Dy^{3+}, Ho^{3+}, Er^{3+}, Tm^{3+}, Yb^{3+}, and Lu^{3+} has been made in a number of papers, amongst the first ones and the most influential of which being that of Spedding et al. [19]. However, other authors [20 to 25] concluded that their data could very well be interpreted with, or even required for their interpretation, a constant primary hydration number along the series. These suggestions and conclusions were all based on indirect evidence, and must yield to recent results from more nearly direct determinations.

1.2.1 Data from X-Ray Scattering

The most direct values of primary hydration numbers are available from the analysis of atomic radial distribution functions (ARDF's), obtained from low angle X-ray diffraction by aqueous solutions of the rare earth metal halides. The results obtained to date by this method are summarized in Table 1/1.

Table 1/1

Information Obtained from the First and Some of the Subsequent Peaks in the ARDF's of Rare Earth Halides in Aqueous Solutions.

salt	concentration mol/kg H_2O	first peak coord. No.	first peak $d_{M\text{-}O}$, in Å	peak for M-X $d_{M\text{-}X}$, in Å	Ref.
$LaCl_3$	1.74, 2.10, 2.67	8.0	2.48	4.7	[26, 27]
	2.0, 3.81	9.13	2.58	5.00	[28]
$LaBr_3$	2.66	8.0	2.48	4.8	[29]
$PrCl_3$	3.80	9.22	2.54	4.97	[28]
$NdCl_3$	0.56, 0.85, 1.73	8.0	2.41	4.9	[30, 31]
	3.37	8.90	2.51	4.89	[28]
$SmCl_3$	1.43	9.9	2.42	4.5	[32]
$GdCl_3$	0.85, 2.66	8.0	2.37	4.8	[31, 33]
	1.43	9.9	2.40	4.5	[32]
$TbCl_3$	3.49	8.18	2.41	4.82	[34]
$DyCl_3$	3.29	7.93	2.40	4.85	[34]
$ErCl_3$	3.54	8.19	2.37	4.79	[34]
	0.95, 1.37, 3.05	6.5	2.3	4.6	[35]
ErI_3	1.33	6.3	2.3	5.2	[35]
$TmCl_3$	3.63	8.12	2.36	4.79	[34]
$LuCl_3$	3.61	7.97	2.34	4.78	[34]

The ARDF is $4\pi r^2\rho(r)$, where r is the distance from a given (metal) atom and $\rho(r)$ the density of atoms at the distance r (**Fig. 1-1**). Its units are e^2/Å (e is an electron). The area under the curve from suitably chosen r_1 to r_2, divided by the area (in e^2) expected per single neighbor, is the number of nearest neighbors of the atom at the origin. The starting distance, r_1 may be taken as zero, or as the distance where $\rho(r)$ starts practically to differ from zero. The determination of the final distance, r_2, is rather a matter of judgement, and the accuracy of the coordination number depends heavily on its proper choice. It is unfortunate that the results published by various authors do not agree. Thus the ARDF for aqueous $ErCl_3$, published already in 1960 [35] by Brady, differs in the positions and the shapes of the peaks from that given by Habenschuss, Spedding in 1979 [34], and the ARDF for aqueous $LaCl_3$ of Smith, Wertz [26] differs in these respects from the one of Habenschuss, Spedding [28].

Fig. 1-1

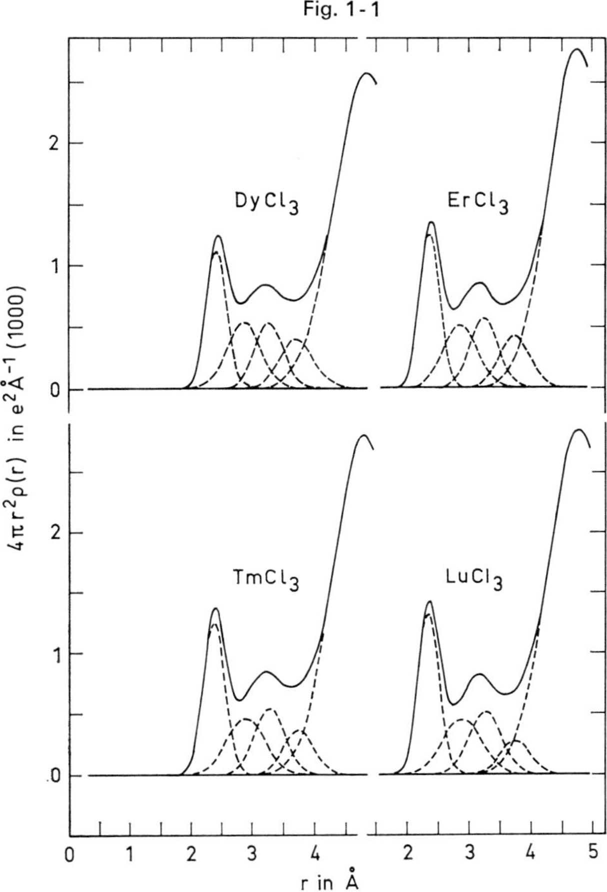

Atomic radial distribution functions of aqueous Dy, Er, Tm, and Lu chlorides [34]. The dashed curves represent the resolution of the experimental curve into 5 Gaussian components.

The work of Brady [35] is now superceded. (Brady himself considered that rather than his direct result of 6.5 for the first coordination shell, "within the experimental error the number is six". The distance to the first peak is given with only one significant decimal figure, rather than the two or three given in the later studies.) A coordination number of eight (8.1 ± 0.1) as a mean for five elements for the heavy rare earth aquo-complexes seems to be well established by the painstaking work of Habenschuss, Spedding [34]. With this agrees also the coordination number of eight (8.0 ± 0.2) given by Steele, Wertz [31, 33] for Gd^{3+}, but not the discordant result 9.9 given by Ryss et al. [32] for this ion, as well as for Sm^{3+}. A real discrepancy is noted for the light rare earth metal ions, where Habenschuss, Spedding [28] give a coordination number of nine (9.1 ± 0.1 as a mean for three elements), whereas Wertz et al. [26, 27, 29 to 31] report a coordination number of eight.

No dependence on the anion was discerned in those studies where two different salts of a given metal were compared (aqueous $LaCl_3$ [26, 27] and $LaBr_3$ [29], aqueous $ErCl_3$ and ErI_3 [35]). Comparative data obtained and interpreted in the same way by the same author should be valid, in this respect, even if the quantitative conclusion is in error, as it evidently is for the results with Er^{3+}.

1.2.2 Data from NMR Peak Areas

Another way to obtain primary hydration numbers in a direct manner is the evaluation of the peak areas obtained in proton NMR measurements. These are made on solutions in a mixed aqueous organic solvent, capable of being cooled to such a low temperature that a peak corresponding to the bound water molecules is observed [36 to 38], separate from that for bulk water. A suitable solvent is deuterated acetone, d^6-CH_3COCH_3, capable of being cooled down to −100°C (and down to −120°C when mixed with Freon −12, CCl_2F_2, if it is necessary to lower the temperature further) in order to slow down the exchange rate of the water (see section 1.4, p. 9) sufficiently for a separate peak to be observed. The acetone is said not to coordinate to the metal ion in these solutions in the presence of a sufficient amount of water, but the low dielectric constant in these solutions and the low temperature promote ion pairing. Nitrate ions are known to enter the first coordination shell of the lanthanides, and no convincing evidence was presented in these studies that perchlorate ions do not do so too, in the relatively water-poor solutions. This explains the low primary hydration numbers obtained in perchlorate solutions, shown in Table 1/2. Another explanation [37] is that only some of the water molecules in the first coordination shell exchange sufficiently slowly to give a separate peak, whereas the others, still in the first coordination shell but bound a bit more loosely, exchange too rapidly, and are included with the signal for the bulk water. Inspection of Fig. 1-2 shows that in the crystalline enneahydrate six water molecules are equivalent, but differ from the other three. That inner-sphere ion association decreases the primary hydration numbers appreciably, under the conditions of the experiments, is demonstrated by the results for nitrate salts, or where nitric acid was added to the perchlorate salts [36 to 38]. The hydration numbers recorded in Table 1/2, even for the perchlorate salts, must therefore be considered as lower limits.

Table 1/2

Hydration Numbers Observed at <−95°C in d^6-Acetone Containing Some Water[a)] by Proton NMR.

cation	perchlorate		nitrate	
Sc^{3+}	[b),c)]		3.7 to 4.1	[36]
Y^{3+}	[c)]		2.2 to 2.6	[36]
La^{3+}	6.0 to 6.4	[37]		
Ce^{3+}	4.5 to 5.5	[37]		
Pr^{3+}	≈6	[38]		
Nd^{3+}	≈6	[38]		
Tb^{3+}			≈2	[38]
Dy^{3+}			≈3	[38]
Ho^{3+}	≈3	[38]		
Er^{3+}	4.2 to 6.4	[38]	1 to 2	[37, 38]
Tm^{3+}			≈2	[38]
Yb^{3+}	3.6 to 4.2[d)]	[38]	1.6 to 2.0	[38]
Lu^{3+}	5.8 to 6.1	[38]	2.2[d)]	[38]

[a)] Unless otherwise stated, mole ratio of water to metal ion in solution is (6 to 15) : 1.—[b)] No separate signals for bound and bulk water observable down to −100°C [39].—[c)] In the presence of 1 M $HClO_4$ additional to the metal nitrate, hydration numbers of 5.1 for Sc^{3+} and ≈2.4 for Y^{3+} are observed [39].—[d)] In the presence of a large excess of water (>50 times the metal ion concentration), the hydration number 8 is observed for Yb^{3+} [40] and 6 for Lu^{3+} [108].

1.2.3 Data from Compressibilities

Another method, that is purported by its users to give primary hydration numbers, is the compressibility method. According to Passinsky [41] the central ion with the primary water of hydration are incompressible, because of the strong electrical field of the ion. The large pressure exerted by this field is greater than the limit of ca. 6.8×10^8 Pa, above which water is practically no longer compressible [42]. The electrostriction caused by this field-induced pressure produces thus an incompressible hydrated core, and the number of water molecules affected in this way per ion is identified with its primary hydration number. This is given as $h^{\infty} = -\varnothing\varkappa^{\infty}/\varkappa^{\circ}_{H_2O}V^{\circ}_{H_2O}$, where $\varnothing\varkappa^{\infty}$ is the apparent molar isothermal compressibility, extrapolated to infinite dilution, and $\varkappa^{\circ}_{H_2O}$ and $V^{\circ}_{H_2O}$ are the compressibility and molar volume of pure water [43]. Experimental data on compressibilities pertain to electrolytes rather than to ions, but are additive with respect to ionic contributions at infinite dilution. If the contribution of one ion is fixed arbitrarily, those of the others are obtained [43]. The compressibilities involved should be the isothermal ones, but adiabatic compressibilities, obtained from ultrasound velocities, are used in their stead, if not all the data required for the conversion (expansibilities, molar volumes and heat capacities) are available [44 to 48]. The results from several authors are shown in Table 1/3. No cognizance of the hydration of the anions is made in the work of Jezowska-Trzebiatowska et al. [44 to 47] so that the hydration number obtained for the whole electrolyte is ascribed to the rare earth metal cation only. However, the differences between chloride, nitrate, and perchlorate data [46] show the effect of the anion. According to Padova's calculation [49] from the compressibility data of Spedding, Atkinson [50], ca. 2.7 should be subtracted for each chloride ion and ca. 1.3 for each nitrate ion from the hydration number of the salt, to give that for the metal ion, recorded in Table 1/3 in the column marked – [43, 49]. Janenas [42] proposed a method differing somewhat from Passinsky's [41, 43], and obtained with it hydration numbers larger by 4 to 5 units. The fluctuations in the hydration numbers reported by Voleisiene et al. [48] have not been explained. In general, it is seen that the primary hydration numbers obtained from the compressibilities are larger than the coordination numbers obtained from the ARDF's, and should represent upper limits. A slight trend of these hydration numbers to increase with Z, the atomic number of the elements, can be discerned. Apparently these hydration numbers comprise some of the secondary solvation, which may, indeed, increase as the size of the metal ions decreases with increasing Z, and the strength of the electrical field around them increases.

Table 1/3

Primary Hydration Numbers Obtained from Compressibilities.

cation	anions and references								
	ClO_4^- [44]	ClO_4^- [45]	Cl^- [46]	NO_3^- [46]	ClO_4^- [46]	NO_3^- [47]	NO_3 [42]	– [43, 49]	NO_3^- [48]
Y^{3+}							15.59		
La^{3+}	12.7					10.4	12.75	11	11.6
Ce^{3+}						*)	12.36	13	15.6
Pr^{3+}	12.9		15.7	10.1	12.0	10.1			9.1
Nd^{3+}	12.5		17.1	10.9	12.8	10.9		12	8.8
Sm^{3+}	11.05					12.0			16.8
Eu^{3+}	10.45		*)	*)	*)	*)			
Gd^{3+}	11.0					12.0			9.5
Tb^{3+}	11.7	11.74				13.0			
Dy^{3+}	11.95	12.01				12.8			17.5
Ho^{3+}	12.5	12.71				12.8			
Er^{3+}	12.6	12.39				12.2		12	16.4
Tm^{3+}	12.8	12.93				*)			
Yb^{3+}	13.0	12.98						13	

*) Compressibilities varied with concentration in a manner that precluded the calculation of hydration numbers.

A method related to the compressibility method involves the viscosity of the solution. In the variant proposed by Padova [51], the Jones-Dole B coefficient, related to the hydrated volume of the ion by $B = 2.5\ V^{\circ}_{hydr}$, is related in turn to V^{∞}_{salt}, the partial molar volume at infinite dilution: $V^{\circ}_{hydr} = V^{\infty}_{salt} + h^{\infty} V^{\infty}_{H_2O}$, where $V^{\infty}_{H_2O}$ is the partial molar volume of water, 18.60 cm^3/mol. From the B coefficients and the partial molar volumes given by Spedding et al. [19, 52], and on assigning the value $h^{\infty}_{Cl^-} = 1$, the following hydration numbers are obtained: La^{3+} 8.5, Pr^{3+} 9.5, Nd^{3+} 9.0, Sm^{3+} 9.3, Tb^{3+} 10.2, Dy^{3+} 11.0, Ho^{3+} 11.0, and Er^{3+} 11.0. Janenas [42] obtained by a somewhat different method the following numbers for the nitrate salts: $Y(NO_3)_3$ 15.0, $La(NO_3)_3$ 12.6, $Ce(NO_3)_3$ 12.6. The primary hydration numbers of the cations are obtained by subtracting three times $h^{\infty}_{NO_3^-}$, but the numerical value of the latter quantity is not given.

1.2.4 Data from Thermodynamic and Other Measurements

Many other properties of the rare earth metal salt solutions have been utilized to give hydration numbers, but only for a few can it be claimed that they yield the primary hydration numbers. For example, the partial molar volumes yield numbers in the range 26 to 29 [53], which clearly include secondary hydration. The activity coefficients, on the other hand, yield the values La^{3+} 7.5, Ce^{3+} 7.5, Eu^{3+} 8.1, Y^{3+} 8.8, and Sc^{3+} 9.5 [54], which seem to correspond more or less to the primary hydration numbers. A hydration number of 6 is ascribed [55] to Pr^{3+} by an unstated method from apparent molar volume data for dilute nitrate and perchlorate solutions. Density data are also used [56] to ascribe hydration numbers of 9.4 to the cations in 1.43 m $SmCl_3$ and $GdCl_3$ (a hydration number of 6.0 is assigned to the chloride ion).

Interesting in this connection are the attempts to calculate Gibbs free energies, enthalpies, or entropies of hydration of the rare earth metal cations by means of models, and compare the results with experimental values, based on an appropriate extrathermodynamic assumption for one ion. The models used for these calculations involve a contribution from the bonding of the water molecules in the inner coordination shell—expressed by means of a hydration number—as well as contributions from further water molecules, which need be of no concern here. The Gibbs free energies of hydration were fitted best with the hydration number 6 for Sc^{3+} and 8 for Y^{3+} and La^{3+} [57]. A somewhat different model was used for fitting simultaneously Gibbs free energies and enthalpies of hydration, with the hydration number 5.6 common to all the cations from La^{3+} to Lu^{3+} [58]. A yet different model was used [59] to fit entropies of hydration, with the hydration numbers 7.92 for Sc^{3+}, 8.09 for Y^{3+}, 7.59 for La^{3+}, values for the light lanthanides falling gradually to 7.34 for Sm^{3+}, rising again for the heavy lanthanides to 7.94 for Yb^{3+} and to 8.41 for Lu^{3+}. In all these studies the "hydration numbers" are treated as adjustable parameters, without being ascribed the physically meaningful function of denoting the number of nearest neighbors the metal ion has in dilute aqueous solutions.

Other papers, in which the relation of thermodynamic and transport properties of aqueous solutions of rare earth metal salts to their hydration is discussed, are satisfied with noting the S-shape of plots of the properties versus the atomic number (Z), the crystalline ionic radius (r_c), or its reciprocal ($1/r_c$). The latter two quantities vary monotonically with Z, so that S-shaped plots would be obtained in all three representations. The properties discussed by Spedding et al. include the partial molar volumes [19, 60, 61], expansivities [62, 63], heats of dilution [64 to 66], entropies [67], heat capacities [68, 69], activity coefficients [70 to 72], conductivities [73], and viscosities [74]. Similar observations of an S-shaped plot were made by Bertha, Choppin [75] concerning entropies of hydration. A U-shaped plot was found by Mioduski, Siekierski [76] for the Gibbs free energy of exchange of hydrated rare earth metal ions with rare earth metal ethylsulfates. The interpretation of the nonmonotonic dependence of these quantities on Z (or on r_c or $1/r_c$) favored by these authors is the following. The primary hydration number, and with it the size of the hydrated ion, has a constant value (presumably 9) for the light lanthanides La^{3+} to Nd^{3+}, has a different, lower, constant value (presumably 8) for the heavy lanthanides (Tb^{3+} to Lu^{3+}), and has a fractional value, or corresponds to an equilibrium between two species ($Ln(H_2O)_9^{3+} \rightleftharpoons Ln(H_2O)_8^{3+} + H_2O$) for the intermediate elements, Sm^{3+} to Gd^{3+} [19]. Since alternative interpretations of the S-shape of the curves have been presented (e.g., [22, 52]), in terms of structural changes of the inner hydration shell at a constant hydration number, and in view of the direct determinations now available from ARDF's (Table 1/1), no further reliance on this early suggestion [19] of a varying primary hydration number should be made.

1.2.5 Data from Spectroscopic Measurements

A still different approach to the evaluation of the primary hydration numbers is the comparison of spectroscopic properties of a lanthanide ion in aqueous solutions, which are sensitive to the field exerted by the immediate environment of the ion on its electronic orbitals, with those of well defined crystalline hydrates. The visible spectrum of Nd^{3+} in dilute aqueous solutions was found [77] to resemble that of ennea-hydrated neodymium in crystalline $Nd(BrO_3)_3 \cdot 9H_2O$, whereas other compounds have different band splitting patterns. Absorption bands of Nd^{3+} are quite sensitive to the environment (in contrast to those of Ho^{3+} and Er^{3+}), and while in dilute aqueous solution the spectrum of $NdCl_3$ resembles closely that of crystalline $Nd(BrO_3)_3 \cdot 9H_2O$, as the concentration increases the spectrum resembles more those of $NdCl_3 \cdot 6H_2O$ and $Nd_2(SO_4)_3 \cdot 8H_2O$, where the neodymium is said to be octa-coordinated [78, 79]. The spectrum of Er^{3+} in dilute aqueous solutions resembles that of the octa-coordinated crystalline $ErCl_3 \cdot 6H_2O$ (the inner shell is $Er(H_2O)_6Cl_2^+$) [79], while that of Eu^{3+} in solutions differs from both those of the crystalline nona-coordinated bromate and octa-coordinated chloride [79].

More quantitative information can be obtained if the rate of quenching of the fluorescence of europium or terbium by the protons of the water of hydration is measured. If an anhydrous solution of $Eu(NO_3)_3$ in acetone is prepared by dissolving the hexahydrate in acetone and drying the solution with zeolite NaA, the fluorescence life time is rather long [80]. The europium is coordinated with three nitrate ions and several acetone molecules, the protons of which are so far from the europium, that they do not contribute to the radiationless transitions that limit the lifetime. When water is gradually added, it displaces the acetone molecules first, up to three moles water per mole of europium, and only the fourth mole of water displaces nitrate anions. The proximity of the protons shortens the lifetime of the fluorescence, the more, the more water is present. Consecutive stability constants of the aqua-complexes have been calculated. Up to 0.32 mol water/dm^3 the species: $Eu(NO_3)_3(C_3H_6O)_n$; $Eu(NO_3)_3H_2O(C_3H_6O)_{n-1}$, lg $K_1 = 3.10 \pm 0.16$; $Eu(NO_3)_3(H_2O)_2(C_3H_6O)_{n-2}$, lg $K_2 = 1.69 \pm 0.05$; $Eu(NO_3)_3(H_2O)_3(C_3H_6O)_{n-3}$, lg $K_3 = 0.90 \pm 0.08$, coexist in equilibrium. At higher water concentrations $Eu(NO_3)_2(H_2O)_4(C_3H_6O)^+_{n-3}$, lg $K_4 = 0.13 \pm 0.07$, is formed [80].

A different way of using the fluorescence life times has been proposed by Horrocks et al. [81, 82]. For each of the ions Eu^{3+} and Tb^{3+}, the rate of quenching of the fluorescence excited by a dye laser in their solutions in H_2O and in D_2O was compared with that in crystalline compounds, containing definite numbers of light or heavy water molecules of hydration. The difference in the rate constants with H_2O and D_2O, plotted against the hydration number of the crystalline compounds, yielded a straight line, from which the number for the aqua-ions in solution can be inter- or extrapolated. It is assumed that the efficiency of the radiationless quenching by protons is independent of the state of the matter. The resulting primary hydration numbers for Eu^{3+} and Tb^{3+} aqua-complexes in dilute aqueous solutions are 9.6 and 9.0, respectively [81, 82].

The chemical shifts observed in NMR spectra of lanthanide ions in $H_2{}^{17}O$ should depend on the number of water molecules in the inner coordination shell, directly exposed to the paramagnetic lanthanide ions. Measurements in erbium nitrate solutions containing an excess of lithium nitrate [83] gave shifts that were compatible with the assignment of 8 water molecules to the aqua-ion of Er^{3+} in dilute aqueous solutions. The linear dependence of the chemical shift in rare earth metal perchlorate solutions on the salt concentration over a wide range (≈ 0.1 to 2.2 m) was taken as evidence of the presence of a single hydrate species independent of the concentration [20] for Pr^{3+}, Nd^{3+}, Gd^{3+}, Tb^{3+}, Dy^{3+}, Ho^{3+}, Er^{3+}, Tm^{3+}, and Yb^{3+}. What this species is could not be evaluated from this type of experiment. Proton chemical shifts in dilute $Gd(ClO_4)_3$ solutions in 0.1 mol/dm^3 $HClO_4$ [84] were compatible with both coordination numbers 8 and 9, giving 16 and 18 protons in a coordination shell around the gadolinium atom. The experimentally found number 19.6 or 19.9 should be somewhat larger than the calculated ones, but not as much as 6 or 7 units, resulting for the hydration number 6 also considered [85].

1.3 Structure of the Hydrate Coordination Shell

Much less can be said about the structure of the aqua-complex, i.e., of the first coordination shell of the hydrated ions Sc^{3+}, Y^{3+}, and La^{3+} to Lu^{3+} in aqueous solutions, than about the coordination number.

The assignment of a coordination number of 6 to the hydrated Sc^{3+} ion in solution [36, 57] seems to be noncontroversial. There is also no evidence that the first coordination shell of water molecules does not have the expected octahedral structure.

An octahedral structure for the water molecules, as proposed by Brady [35] for Er^{3+} from X-ray diffraction data, is not compatible with the more recent evidence obtained by this method, as discussed in Section 2.1. In fact, it was rather early found by Freed [86] from the visible absorption spectrum of aqueous Eu^{3+}, that this ion cannot be located in a highly symmetrical environment. The number of lines into which the $f \rightarrow f$ absorption bands split is too large to be accounted for by a cubic or other centrosymmetric local symmetry ([86], but see [87]). This qualitative, negative conclusion was confirmed also more recently for aqueous Ce^{3+}, Nd^{3+}, Er^{3+}, and Yb^{3+} by electron paramagnetic resonance (EPR) measurements on frozen solutions of their salts [88].

In a more positive sense, the structures of the aqua-complexes of Nd^{3+} [77, 79] and Er^{3+} [79] have been likened to those of the solids with coordination numbers 9 and 8, respectively. The enneahydrate, $Nd(H_2O)_9^{3+}$, thus has the trigonal prismatic structure shown in **Fig. 1-2** with symmetry D_{3h} [79, 88], although the presence of some extra bands in the spectrum point to an even lower symmetry, due to some deformation [79]. The, presumed, octahydrate, $Er(H_2O)_8^{3+}$, has been assigned the structure of a square antiprism.

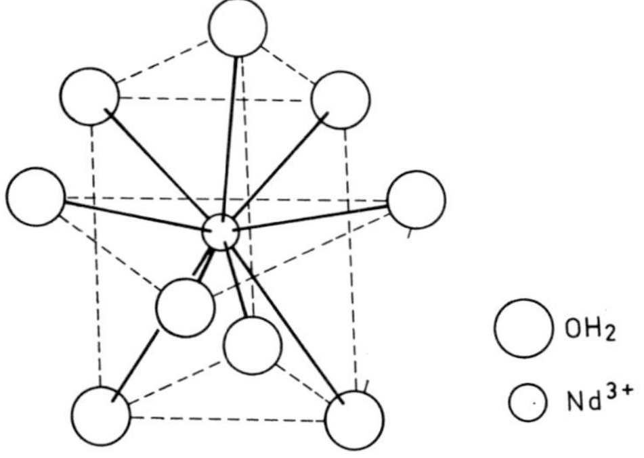

Fig. 1-2

The structure of the first coordination shell of $Nd(H_2O)_9^{3+}$ in $Nd(BrO_3)_3 \cdot 9\,H_2O$ crystals [3].

That the structure is different for the light and the heavy lanthanides is indicated by results from several techniques. Differential Raman spectroscopy in 5% D_2O in water, compared to H_2O solutions is stated by Gutierrez et al. [89] to show that La^{3+} affects the water molecules in a manner differing from that, in which do Gd^{3+} and Lu^{3+}.

The equilibrium constants for the coprecipitation of the various lanthanide ions M^{3+} with ethylsulfates of La, Nd, Gd, or Er led Mioduski, Siekierski [72] to speculations concerning the structure of the aqueous ions. The symmetry D_{3h} was assigned to enneahydrated ions from La^{3+} to Pm^{3+}, the symmetry in a distorted face centered trigonal prism, with "partial" coordination of the ninth water molecule in the inner coordination sphere, was assigned to the ions from Pm^{3+} to Gd^{3+}, and that of a square antiprism, D_{4d}, to the octahydrated ions from Gd^{3+} to Lu^{3+}.

A change in the structure of the species as the atomic number Z increases and the radius decreases from La^{3+} to Lu^{3+} has been invoked also by others in support of the contention that the coordination number remains constant throughout the series (e.g., [22]). This possibility was conceded by Spedding, Pikal [52], who were influential proponents of the view that the coordination number does change. They noted that the break in the viscosity versus Z curve may be due to a change in the shape factor rather than in the volume of the hydrated ions.

The relaxation of the quadrupolar line broadening of ^{139}La NMR in aqueous chloride, nitrate, perchlorate, and sulfate solutions has been interpreted [90] as signifying a transition from the less symmetrical $La(H_2O)_9^{3+}$ to the more symmetrical (cubic?) $La(H_2O)_8^{3+}$, accompanied by the loss of a water molecule as the solutions become more concentrated. This interpretation has, however, been critisized [91], on the grounds that the observed band narrowing is an artifact due to over- and under-compensation in the extrapolation made [90] to zero modulation amplitude, and to a partial saturation at the high frequencies that had to be employed for the dilute solutions. The NMR results on 0.1 to 1.7 mol/dm³ $^{139}LaCl_3$ solutions show no effects other than the expected dependency on the viscosity, i.e., no change in species as the concentration is increased [91].

The most direct evidence on the structure of the aqua-complexes comes from the interpretation of the X-ray ARDF's undertaken by Wertz et al. [26, 27, 30, 31, 33]. The octahydrate $La(H_2O)_8^{3+}$ is assigned cubic symmetry, in order to account for the locations and shapes of all the features of the ARDF [26, 27]. A square antiprism, on the other hand, is definitely unable to account for them. Chloride ions, that associate as outer sphere ion-pairs with the $La(H_2O)_8^{3+}$ ions, may be located on either the faces or the edges of the cubes. The structures of $Nd(H_2O)_8^{3+}$ and $Gd(H_2O)_8^{3+}$ have been termed pseudocubic, although in the latter case this structure is not completely consistent with the ARDF. Again, outer sphere ion-paired chloride ions take up octahedral positions (i.e., at the faces of the cube). In the presence of a large excess of HCl (10 mol/dm³) the chloride ions penetrate the inner sphere, to give $NdCl(H_2O)_7^{2+}$ and $GdCl_2(H_2O)_6^{+}$ of the cubic structure [30, 31, 33].

1.4 The Rate of Release of Water from the Inner Coordination Shell

The bonding of the water in the inner coordination shell of the lanthanide ions can be characterized by the rate at which it is released on exchange, either with other water molecules, or with ligands which come in from the outer coordination sphere or from the bulk of the solution. The rate of exchange, although very fast (residence times of the order of 10 ns), is yet not sufficiently fast to be diffusion controlled, and is definitely controlled by the kinetics of water elimination. Most of the evidence on this rate has been obtained from the absorption or the relaxation of ultrasonic energy, but some has been derived from NMR measurements, as shown in Table 1/4, p. 10.

The NMR data are based on the transverse line width of the ^{17}O signal, $1/T_2$ [96, 97], on the longitudinal line width of the ^{17}O signal, $1/T_1$ [97], on ^1H relaxation rates [98], in conjunction with ESR data [97]. They require for their interpretation the assumption of a coordination number, and slightly different rate constants are obtained with coordination numbers 8 and 9 [96]. The results may also be described by means of different equations, leading again to somewhat different rate constants, hence the values given [96] are considered as lower limits only. However, the interpretation of the NMR data describes the self exchange of the water, and does not depend on a mechanism that involves the exchange of water molecules with incoming ligands.

The ultrasound absorption data depend on the assumption that the exchange proceeds according to the following two-step mechanism

$$Ln(H_2O)_n^{3+} + A^- \underset{k_{-1}}{\overset{k_1}{\rightleftharpoons}} Ln(H_2O)_n^{3+}A^-;\ K_1 = k_1/k_{-1} \tag{1}$$

$$Ln(H_2O)_n^{3+}A^- \underset{k_{-2}}{\overset{k_2}{\rightleftharpoons}} [Ln(H_2O)_{n-1}A]^{2+} + H_2O;\ K_2 = k_2/k_{-2} \tag{2}$$

The first equilibrium, producing the outer sphere ion-pair, is considered to be very fast, and the overall rate is controlled by the rate constant k_2 for the release of the water from the inner sphere [92 to 95, 99 to 101]. The incoming ion should, therefore, have no influence on the rate constant. The overall forward rate constant of the exchange k_f, observed as a peak in the ultrasonic intensity frequency dependence, is related to the forward rate of the step (2), k_2, as

$$k_f = k_2 \cdot K_1/(1 + k_2/k_{-1}) \tag{3}$$

where k_{-1} is the rate of dissociation of the outer sphere ion-pair, and K_1 is its formation equilibrium constant:

$$K_1 = K_o/(1+K_2) \tag{4}$$

where K_o is the observed overall association constant, and K_2 is the formation equilibrium constant of the inner sphere complex. Combination of eqs. (3) and (4) leads to

$$k_2 = k_f \cdot K_o^{-1}(1+k_2/k_{-1})(1+K_2) \tag{5}$$

Only if $k_2 \ll k_{-1}$ and $K_2 \ll 1$ can the rate constant k_2 for the release of the water be obtained from the observable quantities k_f and K_o, since k_{-1} and K_2 are not directly measurable. Hence all the rate constants quoted in Table 1/4 are in fact only lower limits [94].

Table 1/4
Rate Constants for Release of Water Molecules from the Inner Sphere, lg k (k in s^{-1}).

method	ultra-son.	ultra-son.	ultra-son.	NMR	NMR, ESR	NMR	ultra-son.	ultra-son.	ultra-son.
incoming anion	SO_4^{2-}	NO_3^-	NO_3^-	b)	b)	b)	ClO_4^-	NO_3^-	NO_3^-
Ref.	[92, 93]	[94]	[95]	[96]	[97]	[98]	[99]	[100]	[101]
Sc^{3+}						$\gg$5.8 d)			
La^{3+}	8.32	≧8.09							
Ce^{3+}	8.52	≧8.35							
Pr^{3+}	8.64	≧8.36							
Nd^{3+}	8.72	≧8.46 a)	8.75	≧7.9				8.26	8.18
Sm^{3+}	8.87								
Eu^{3+}	8.82			≧7.9					
Gd^{3+}	8.83	≧8.19 a)	8.69		8.95				
Tb^{3+}	8.72			>7.32					
Dy^{3+}	8.62			>7.15		>8.0			
Ho^{3+}	8.45			>6.97					
Er^{3+}	8.28			>6.73			8.08 c)	7.78	
Tm^{3+}	8.15			>6.52					
Yb^{3+}	7.9	≧8.20							
Lu^{3+}	7.8								

a) Recalculated from data in ref. [95]; b) in the presence of ClO_4^- ions; c) in 50wt% methanol; d) At −20°C [110].

There is, however, no general agreement that the association inevitably proceeds according to the two step mechanism (1) + (2). The ultrasonic absorption data for aqueous Y^{3+} nitrate and perchlorate solutions [102] have been interpreted in terms of a rate controlled by the association with nitrate ions, rather than by the elimination of a water molecule. The rate constant has therefore the units $dm^3 \cdot mol^{-1} \cdot s^{-1}$ (lg k = 8.0) and is not comparable with the zero order rate constants (units s^{-1}) in Table 1/4. It is comparable, however, with the rate constants obtained by fast chemical kinetic methods (temperature jump) for the association of Y^{3+} and the lanthanide ions with ligands such as murexide [103, 104] and oxalate [105]. These, in turn, are first- order rate constants (in $dm^3 \cdot mol^{-1} \cdot s^{-1}$) that have been interpreted as due to the rate- controlling step of the release of a molecule of water, reaction (2), in an SN1-type mechanism. The data are shown in Table 1/5.

Table 1/5

Rate Constants for Association with Murexide at 12°C [103, 104] and with Oxalate at 25°C [105], Involving the Release of Water molecules, lg k (k in $dm^3 \cdot mol^{-1} \cdot s^{-1}$).

metal ion	Sc^{3+}	Y^{3+}	La^{3+}	Ce^{3+}	Pr^{3+}	Nd^{3+}	Sm^{3+}	Eu^{3+}
lg k (murexide)	7.68	7.11	7.93	7.98	7.93	7.97	7.98	7.91
lg k (oxalate)			7.90			7.93		7.89
metal ion	Gd^{3+}	Tb^{3+}	Dy^{3+}	Ho^{3+}	Er^{3+}	Tm^{3+}	Yb^{3+}	Lu^{3+}
lg k (murexide)	7.72	7.48	7.23	7.15	7.00	7.04	7.04	7.11
lg k (oxalate)	7.76	7.38	7.11	7.00	6.80	6.80		

The great similarity of these rate constants for ligands as different as murexide and oxalate is a strong basis for accepting that the rate is controlled by the rate of release of water (i.e., k_2), rather than by the rate of association.

1.5 The Strength of the Bonding of the Water

Very little can be said about the strength of the bonding of the ions Sc^{3+}, Y^{3+}, and the lanthanides La^{3+} to Lu^{3+} with water, if one is concerned only with the water as a ligand in the first coordination shell. Equilibrium constants for the consecutive exchange of three acetone molecules and a nitrate ion from $Eu(NO_3)_3(C_3H_6O)_n$ in acetone solutions have been obtained by noting the rate of radiationless relaxation of fluorescence of the europium in the solutions [80] (Section 1.2.5). In so far as the rate of the release of the water from the inner coordination sphere is a measure of the strength of the bonding, as it is at least in a relative sense, the data in the preceding section should be consulted.

The strength of the bonding of the water in the inner coordination shell is, of course, also reflected in the total Gibbs free energy and enthalpy of the hydration. These quantities, as well as the corresponding entropy, which include also the interactions with water molecules in the second, outer, coordination shell and with further water molecules, have been examined in terms of models based on electrostatics [57 to 59] (see Section 1.2.4, p. 7). The thermodynamic data themselves have been presented in "Seltenerdelemente" B 7, 1979, pp. 5/6.

The Gibbs free energy and enthalpy changes for the association of the gaseous Sc^{3+}, Y^{3+} and La^{3+} ions with 6 water molecules to form the gaseous hexa-coordinated octahedral ion hydrates have been calculated [109]. The values are shown in Table 1/6. They constitute 116% for Sc^{3+}, 96% for Y^{3+} and 71% for La^{3+} of the total, experimental, standard molar hydration Gibbs free energies and epthalpies. Further interactions make up the necessary differences.

Table 1/6

Gibbs Free Energy and Enthalpy Changes for Hydration of Gaseous Ions to Form Gaseous Hexahydrates [109].

ion	hydration No.	$-\Delta G°$ (in kJ/mol)	$-\Delta H°$ (in kJ/mol)
Sc^{3+}	6	4445	4716
Y^{3+}	6	3405	3675
La^{3+}	6	2132	2394

These values give an indication of the strength of the bonding of the water, which depends, of course, on the assumptions on which the calculations are based. For the other lanthanides the contribution of this step to the total calculated values was not specified [58].

1.6 The Bonding of the Water Molecules to the Metal Ions

Most of the data on the aqua-complexes of Sc^{3+}, Y^{3+}, and the lanthanides La^{3+} to Lu^{3+} are compatible with the assumption that the bonding is purely electrostatic in nature. Electrostatic models [57 to 59, 109] were found to account well for the thermodynamic quantities for hydration. There is only slight evidence that the electrons from the oxygen atoms of the water molecules participate to some extent in the orbitals of the metal ions. The chemical shift of ^{17}O NMR signals in aqueous rare earth perchlorate solutions has been interpreted [106] in terms of indirect contact hyperfine interaction of unpaired 4f electrons with the O nucleus. Very weak covalent bonds were said to be formed between the oxygen 2s and the metal 6s orbitals [106, 107].

References to 1:

[1] T. Moeller, D. F. Martin, L. C. Thompson, R. Ferrús, G. R. Feistel, W. J. Randall (Chem. Rev. **65** [1965] 1/50). — [2] J. L. Hoard, G. S. Smith, M. Lind (in: S. Kirschner, Advances in the Chemistry of the Coordination Compounds, New York 1961, pp. 296/302). — [3] L. Helmholz (J. Am. Chem. Soc. **61** [1939] 1544/50). — [4] J. A. A. Ketelaar (Physica **4** [1937] 619/30). — [5] I. Mayer, A. Glasner (J. Inorg. Nucl. Chem. **29** [1967] 1605/9).

[6] S. K. Sikka (Acta Cryst. A **25** [1969] 621/6). — [7] D. R. Fitzwater, R. E. Rundle (Z. Krist. **112** [1959] 362/74). — [8] C. R. Hubbard, C. O. Quicksall, R. A. Jacobson (Acta Cryst. B **30** [1974] 2613/9). — [9] J. Albertson, I. Elding (Acta Cryst. B **33** [1977] 1460/9). — [10] J. Pabst (Am. J. Sci. [5] **22** [1931] 426/9).

[11] V. I. Ivanov (Kristallografiya **13** [1968] 905/7; Soviet Phys.-Cryst. **13** [1969] 786/8). — [12] M. Marezio, H. A. Plettinger, W. H. Zachariasen (Acta Cryst. **14** [1961] 234/6). — [13] N. K. Bel'skii, Yu. T. Strukhkov (Kristallografiya **10** [1965] 21/8; Soviet Phys.-Cryst. **10** [1965] 15/20). — [14] V. V. Babakin, R. F. Klevstova, L. P. Solov'eva (Zh. Strukt. Khim. **15** [1974] 820/30; J. Struct. Chem. [USSR] **15** [1974] 723/32). — [15] B. Ericksson, L. O. Larsson (J. Chem. Soc. Chem. Commun. **1978** 616/7).

[16] E. B. Hund, R. E. Rundle, A. J. Stosick (Acta Cryst. **7** [1954] 106/9). — [17] A. Dereigne, G. Pannetier (Bull. Soc. Chim. France **1968** 174/80). — [18] J. H. Burns, R. D. Baybarz (Inorg. Chem. **11** [1972] 2233/7). — [19] F. H. Spedding, M. J. Pikal, B. O. Ayres (J. Phys. Chem. **70** [1966] 2440/9). — [20] J. Reuben, D. Fiat (J. Chem. Phys. **51** [1969] 4909/17).

[21] G. Geier, U. Karlen, A. v. Zelewsky (Helv. Chim. Acta **52** [1969] 1967/75). — [22] G. Geier, U. Karlen (Helv. Chim. Acta **54** [1971] 135/53). — [23] G. Anderegg, F. Wenk (Helv. Chim. Acta **54** [1971] 216/29). — [24] I. Grenthe, G. Hessler, H. Ots (Acta Chem. Scand. **27** [1973] 2543/53). — [25] I. Dellien (Acta Chem. Scand. **27** [1973] 733/42).

[26] L. S. Smith, D. L. Wertz (J. Am. Chem. Soc. **97** [1975] 2365/8). — [27] L. S. Smith, D. C. McCain, D. L. Wertz (J. Am. Chem. Soc. **98** [1976] 5125/8). — [28] A. Habenschuss, F. H. Spedding (J. Chem. Phys. **70** [1979] 3758/63). — [29] L. S. Smith, D. L. Wertz (J. Inorg. Nucl. Chem. **39** [1977] 95/8). — [30] M. L. Steele, D. L. Wertz (Inorg. Chem. **16** [1977] 1225/8).

[31] M. L. Steele (Diss. Univ. Southern Mississippi 1976). — [32] A. I. Ryss, M. K. Levositskaya, I. M. Shapovalov (Deposited Doc. VINITI-856-76 [1976]; C.A. **89** [1978] No. 95116). — [33] M. L. Steele, D. L. Wertz (J. Am. Chem. Soc. **98** [1976] 4424/8). — [34] A. Habenschuss, F. H. Spedding (J. Chem. Phys. **70** [1979] 2797/806). — [35] G. W. Brady (J. Chem. Phys. **33** [1960] 1079/82).

[36] A. Fratiello, V. M. Nishida, R. E. Lee, R. E. Schuster (J. Chem. Phys. **50** [1969] 3624/8). — [37] A. Fratiello, V. Kubo, S. Peak, B. Sanchez, R. E. Schuster (Inorg. Chem. **10** [1971] 2552/7). — [38] A. Fratiello, V. Kubo, G. A. Vidulich (Inorg. Chem. **12** [1973] 2066/71). — [39] A. Fratiello, R. E. Lee, R. E. Schuster (Inorg. Chem. **9** [1970] 391/2). — [40] V. A. Shcherbakov, O. G. Golubovskaya (Zh. Neorgan. Khim. **21** [1976] 314/6; Russ. J. Inorg. Chem. **21** [1976] 168/70).

[41] A. Passinsky (Acta Physicochim. URSS **8** [1938] 385/418; C.A. **1939** 4494). — [42] V. Janenas (Zh. Fiz. Khim. **52** [1978] 1462/5; Russ. J. Phys. Chem. **52** [1978] 839/40). — [43] J. Padova (Bull. Res. Council Israel A **10** [1961] 63/4). — [44] B. Jezowska-Trzebiatowska, S. Ernst, J. Legendziewicz, G. Oczko (Bull. Acad. Polon. Sci. Ser. Sci. Chim. **26** [1978] 805/11). — [45] B. Jezowska-Trzebiatowska, S. Ernst, J. Legendziewicz, G. Oczko (Chem. Phys. Letters **60** [1978] 19/21).

[46] B. Jezowska-Trzebiatowska, S. Ernst, J. Legendziewicz, G. Oczko (Bull. Acad. Polon. Sci. Ser. Sci. Chim. **24** [1976] 997/1002). — [47] B. Jezowska-Trzebiatowska, S. Ernst, J. Legendziewicz, G. Oczko (Bull. Acad. Polon. Sci. Ser. Sci. Chim. **25** [1977] 649/54). — [48] B. Voleisiene, A. Voleisis, E. Jaronis (Nauchn. Tr. Vyssh. Uchebn. Zavedenii Litov.SSR Ul'trazvuk **1976** No. 8, pp. 16/21; C.A. **87** [1977] No. 123539). — [49] J. Padova (J. Chem. Phys. **40** [1964] 691/4). — [50] F. H. Spedding, G. Atkinson (in: J. W. Hamer, Structure of Electrolyte Solutions, Wiley, New York 1959, pp. 333).

[51] J. Padova (J. Phys. Chem. **75** [1967] 2347/8). — [52] F. H. Spedding, M. J. Pikal (J. Phys. Chem. **70** [1966] 2430/40). — [53] J. Padova (J. Chem. Phys. **39** [1963] 1552/7). — [54] E. Glueckauf (Trans. Faraday Soc. **51** [1955] 1235/44). — [55] V. Jedinakova (Collection Czech. Chem. Commun. **42** [1977] 1285/9).

[56] J. M. Shapovalov, A. I. Ryss (Deposited Doc. VINITI-2592-76 [1976]; C.A. **90** [1979] No. 13042). — [57] W. E. Morf, W. Simon (Helv. Chim. Acta **54** [1971] 794/810). — [58] S. Goldman, L. R. Morss (Can. J. Chem. **53** [1975] 2695/700). — [59] P. R. Tremaine, S. Goldman (J. Phys. Chem. **82** [1978] 2317/21). — [60] F. H. Spedding, V. W. Saeger, K.A. Gray, P.K. Boneau, M. A. Brown, C. W. DeKock, J. I. Baker, L. E. Shiers, H. O. Weber, A. Habenschuss (J. Chem. Eng. Data **20** [1975] 72/81).

[61] F. H. Spedding, L. E. Shiers, M. A. Brown, J. L. Derer, D. L. Swanson, A. Habenschuss (J. Chem. Eng. Data **20** [1975] 81/8). — [62] A. Habenschuss, F. H. Spedding (J. Chem. Eng. Data **21** [1976] 95/113). — [63] W. M. Gildseth, A. Habenschuss, F. H. Spedding (J. Chem. Eng. Data **22** [1977] 58/70). — [64] F. H. Spedding, J. L. Derer, M. A. Mohs, J. A. Rard (J. Chem. Eng. Data **21** [1976] 474/83). — [65] F. H. Spedding, M. A. Mohs, J. L. Derer, A. Habenschuss (J. Chem. Eng. Data **22** [1977] 142/53).

[66] F. H. Spedding, C. W. DeKock, G. W. Pepple, A. Habenschuss (J. Chem. Eng. Data **22** [1977] 58/70). — [67] F. H. Spedding, J. A. Rard, A. Habenschuss (J. Phys. Chem. **81** [1977] 1069/74). — [68] F. H. Spedding, J. P. Walters, J. L. Baker (J. Chem. Eng. Data **20** [1975] 88/93). — [69] F. H. Spedding, K. C. Jones (J. Phys. Chem. **70** [1966] 2450/5). — [70] F. H. Spedding, H. O. Weber, V. W. Saeger, H. H. Petheram, J. A. Rard, A. Habenschuss (J. Chem. Eng. Data **21** [1976] 341/60).

[71] J. A. Rard, H. O. Weber, F. H. Spedding (J. Chem. Eng. Data **22** [1977] 187/201). — [72] J. A. Rard, J. L. Shiers, D. J. Heiser, F. H. Spedding (J. Chem. Eng. Data **22** [1977] 337/47). — [73] F. H. Spedding, J. A. Rard, V. W. Saeger (J. Chem. Eng. Data **19** [1974] 373/7). — [74] F. H. Spedding, D. Witte, L. E. Shiers, J. A. Rard (J. Chem. Eng. Data **19** [1974] 369/73). — [75] S. L. Bertha, G. R. Choppin (Inorg. Chem. **8** [1969] 613/7).

[76] T. Mioduski, S. Siekierski (J. Inorg. Nucl. Chem. **37** [1975] 1647/51). — [77] P. Krumholz (Spectrochim. Acta **10** [1958] 274/80). — [78] D. G. Karraker (Inorg. Chem. **7** [1968] 473/9). — [79] N. K. Davidenko, L. N. Lugina (Zh. Neorgan. Khim. **13** [1968] 980/7; Russ. J. Inorg. Chem. **13** [1968] 512/7). — [80] V. P. Gruzdev, V. L. Ermolaev (Zh. Neorgan. Khim. **20** [1975] 2650/5; Russ. J. Inorg. Chem. **20** [1975] 1467/70).

[81] W. De W. Horrocks, G. F. Schmidt, D. R. Sudnick, C. Kittrell, R. A. Bernheim (J. Am. Chem. Soc. **99** [1977] 2378/80). — [82] W. De W. Horrocks, D. R. Sudnick (J. Am. Chem. Soc. **101** [1979] 334/40). — [83] I. Abrahamer, Y. Marcus (Inorg. Chem. **6** [1967] 2103/6). — [84] L. O. Morgan, A. W. Nolde (J. Chem. Phys. **31** [1959] 365/8). — [85] L. O. Morgan (J. Chem. Phys. **38** [1963] 2788/9).

[86] S. Freed (Rev. Mod. Phys. **14** [1942] 105/11). — [87] L. N. Lugina, N. K. Davidenko (Koord. Khim. **3** [1977] 193/9; Soviet J. Coord. Chem. **3** [1977] 142/8). — [88] W. B. Mims, J. L. Davis (J. Chem. Phys. **65** [1976] 3266/74). — [89] L. Gutierrez, W. C. Mundy, F. H. Spedding (J. Chem. Phys. **61** [1974] 1953/9). — [90] K. Nakamura, K. Kawamura (Bull. Chem. Soc. Japan **44** [1971] 330/4).

[91] J. Reuben (J. Phys. Chem. **79** [1975] 2154/7). — [92] N. Purdie, C. A. Vincent (Trans. Faraday Soc. **63** [1967] 2745/57). — [93] D. P. Fay, D. Litchinsky, N. Purdie (J. Phys. Chem. **73** [1969] 544/52). — [94] G. S. Darbani, F. Fittipaldi, S. Petrucci, P. Hemmes (Acustica **25** [1971] 125/38). — [95] R. Garnsey, D. W. Ebdon (J. Am. Chem. Soc. **91** [1969] 50/6).

[96] J. Reuben, D. Fiat (J. Chem. Phys. **51** [1969] 4918/72). — [97] R. S. Marianelli (Univ. California, Berkeley 1966; Diss. Abstr. B **28** [1967] 85). — [98] J. Granot, D. Fiat (J. Magn. Resonance **19** [1975] 372/81). — [99] H. B. Silber (J. Phys. Chem. **78** [1974] 1940/4). — [100] H. B. Silber, N. Scheinin, G. Atkinson, J. J. Grecsek (J. Chem. Soc. Faraday Trans. I **68** [1972] 1200/12).

[101] H. B. Silber, J. Fowler (J. Phys. Chem. **80** [1976] 1451/6). — [102] H. C. Wang, P. Hemmes (J. Phys. Chem. **78** [1974] 261/5). — [103] G. Geier (Ber. Bunsenges. Physik. Chem. **69** [1965] 617/25). — [104] G. Geier (Helv. Chim. Acta **51** [1968] 94/105). — [105] A. J. Graffeo, J. C. Bear (J. Inorg. Nucl. Chem. **30** [1968] 1577/84).

[106] W. B. Lewis, J. A. Jackson, J. F. Lemons, H. Taube (J. Chem. Phys. **36** [1962] 694/701). — [107] G. R. Choppin (Pure Appl. Chem. **27** [1971] 23/41). — [108] V. A. Kvun, B. N. Chernyasov (Koord. Khim. **5** [1979] 53/9). — [109] S. Goldman, R. G. Bates (J. Am. Chem. Soc. **94** [1972] 1476/84). — [110] J. W. Neely (UCRL-20580 [1971] 1/72).

2 Alcoholates and Complexes with Alcohols

Edward R. Birnbaum
Department of Chemistry, New Mexico State University
Las Cruces, New Mexico, USA

2.1 General Considerations

Only a limited number of isolatable alcohol adducts of the rare earths have been reported in the literature. A somewhat larger number of reports of alkoxide derivatives are found. The neutral and anionic complexes are discussed separately here. There is also a considerable body of literature discussing the cerium(IV) oxidation of alcohols. It is not possible to treat the extensive literature on the redox reactions between Ce^{IV} and alcohols within the scope of this chapter. For kinetic data and for stability constants, which have been determined in the course of oxidation studies, see the following references: methanol [1 to 3], ethanol [3, 5], 1-propanol [9], 2-propanol [3, 4, 6, 18], allyl alcohol [7, 9], butanol [3, 15], sec butyl alcohol [4, 8, 11, 15], isobutyl alcohol [8], tert butyl alcohol [3, 8, 15], 3-buten-1-ol [3], 2,2-dimethyl-1-propanol = neopentyl alcohol [3, 8], hexanol [3], 5-hexen-1-ol [3], 3-heptanol [3], nonanol [3], 2-methoxyethanol [3], 3-ethoxypropanol [3], cycloalkyl derivatives of methanol [3], 2-cyclohexylethanol [3], cyclopentanol [3], cyclohexanol [3, 9, 10, 12], trans-2-methoxy cyclohexanol [12], 1-methyl cyclohexanol [3], cycloheptanol and -octanol [3], benzyl alcohol [3, 9, 19, 20], 3-, and 4-nitrobenzyl alcohol [3], 2-, 3-, and 4-chloro-benzyl alcohols [3], 4-methylbenzyl alcohol [3], 1- and 2-phenylethanol [3], 4-phenylbutanol [3], diphenylmethanol (= benzhydrol) [3], 2,3-butanediol, glycerol, and other diols [3, 12 to 17]. The rate constants of formation and dissociation depend on the nature of the alcohol: primary > secondary > ternary. In addition, there is an increase in the rate with acidity [8].

Alcohol adducts of complexes containing other ligands are described with the complexes of that ligand. For example, see alcohol solvates of tris(acetylacetonato) complexes on p. 95.

References to 2.1:

[1] S. S. Muhammed, K. V. Rao (Bull. Chem. Soc. Japan **36** [1963] 943/9). — [2] S. S. Muhammed, K. V. Rao (Bull. Chem. Soc. Japan **36** [1963] 949/53). — [3] L. B. Young, W. S. Trahanovsky (J. Am. Chem. Soc. **91** [1969] 5060/8). — [4] B. Sethuram, S. S. Muhammed (Acta Chim. Acad. Sci. Hung. **46** [1965] 115/24). — [5] M. Ardon (J. Chem. Soc. **1957** 1811/5).

[6] B. Sethuram, S. S. Muhammed (Acta Chim. Acad. Sci. Hung. **46** [1965] 125/36). — [7] B. Sethuram (Current Sci. [India] **35** [1966] 254/5). — [8] G. Boivin, M. Zador (Can. J. Chem. **48** [1970] 3053/8). — [9] M. Santappa, B. Sethuram (Proc. Indian Acad. Sci. A **67** [1968] 78/89). — [10] J. S. Litter (J. Chem. Soc. **1959** 4135/6).

[11] D. L. Mathur, G. V. Bakore (Bull. Chem. Soc. Japan **44** [1971] 2600/2). — [12] H. L. Hintz, D. C. Johnson (J. Org. Chem. **32** [1967] 556/64). — [13] F. R. Duke, R. F. Bremer (J. Am. Chem. Soc. **73** [1951] 5179/81). — [14] F. R. Duke, A. A. Forist (J. Am. Chem. Soc. **71** [1949] 2790/2). — [15] H. G. Offner, D. A. Skoog (Anal. Chem. **37** [1965] 1018/21).

[16] G. G. Guilbault, W. H. McCurdy (J. Phys. Chem. **67** [1963] 283/5). — [17] A. Prakash, R. N. Mehrotra, R. C. Kapoor (J. Chem. Soc. Dalton Trans. **1979** 205/10). — [18] D. L. Mathur, G. V. Bakore (J. Indian Chem. Soc. **48** [1971] 363/5). — [19] D. Paquette, M. Zador (Can. J. Chem. **46** [1968] 3507/10). — [20] V. P. Kudesia (Cienc. Cult. [Sao Paulo] **23** [1971] 221/4).

2.2 With Methanol CH_3OH

2.2.1 Scandium Compounds

Methanol adducts of scandium chloride have been prepared in two ways. The first method involved maintaining samples of the anhydrous salt and methanol in separate cups in a desiccator containing a small amount of P_4O_{10} to remove moisture. After evacuation of the desiccator

(≈150 Torr), the weight of the scandium chloride cup was measured as a function of time. The solid material continued to absorb methanol until the solid adduct formed eventually dissolved, at which time the methanol was removed. After continued evacuation, the tetrakis adduct was formed, $ScCl_3 \cdot 4\,CH_3OH$. Upon reduction of the pressure in the desiccator from 150 to 80 mm Hg, in stages, the tris and bis adducts were isolated, $ScCl_3 \cdot n\,CH_3OH$, where n = 3 and 2. The bromide salt treated in the same way yielded $ScBr_3 \cdot n\,CH_3OH$, where n = 4 and 5 [1].

The second method involved stirring the anhydrous chloride salt in methanol. The solid phase left corresponded to the tris adduct after drying. After prolonged standing in a desiccator, a very hygroscopic bis adduct formed. The solubility of scandium chloride in methanol is 45.5 mg/100 mg solvent, very close to a 1:6 molar ratio, at 25°C [2].

References to 2.2.1:

[1] F. Petrů, B. Hajek, F. Jost (Croat. Chem. Acta **29** [1957] 457/60). — [2] E. M. Kirmse (Z. Chem. [Leipzig] **1** [1961] 332/4).

2.2.2 Lanthanide Compounds

2.2.2.1 Chlorides

Adducts of methanol with lanthanide salts have been prepared in several ways. Each hydrated lanthanide(III) chloride (4.4 g) was dissolved in dry methanol (34.0 g). The reaction was exothermic. After heating at reflux for 2 h, cooling of the clear solutions resulted in glassy needle-like crystals. After drying under reduced pressure at room temperature the tris adducts were obtained with the formulation $MCl_3 \cdot 3\,CH_3OH$ where M = La, Ce [1, 3, 6], Pr, Nd [22], and Sm [4]. Distillation of a benzene/water/methanol azeotrope and a benzene/methanol azeotrope from the hydrated cerium(III) chloride salt reportedly yielded the tris adduct, $CeCl_3 \cdot 3\,CH_3OH$, after concentration and cooling [6]. However, the hydrated lanthanum [1, 5], praseodymium, neodymium [22], and samarium [4] chloride salts yielded the $MCl_3 \cdot 2\,CH_3OH \cdot H_2O$ adducts without the distillation of the azeotropes to remove water. The tetrakis adducts of each of the lanthanide chloride salts could be prepared starting from the dried, free-flowing hydrated salt by dissolving 4 g of the appropriate salt in a solution (65 ml) of methanol (30%) and 2,2-dimethoxypropane (70%). After evaporation on a steam bath to ≈ $^1/_3$ the volume, an equal volume of 2,2-dimethoxypropane was added, resulting in the formation of two layers, the heavier of which contained the product. The crystals that resulted were pulverized in the mother liquor. Evaporation at 90 to 100 Torr then removed excess methanol, 2,2-dimethoxypropane, and acetone. The evaporation was stopped after the "first visual dryness" was obtained. Longer drying times resulted in less than four methanol molecules coordinated to each metal ion. Crystals formed in 1 to 2 h for the lighter lanthanides and in up to 60 h for the heavier metal ions. Tetrakis adducts, $MCl_3 \cdot 4\,CH_3OH$, were obtained for M = La, Pr, Nd, Gd, Tb, Ho [2]. The crystals were very hygroscopic. Continued evacuation of the tetrakis adducts at 28 Torr resulted in conversion to the tris adducts [2]. A similar method, using trimethyl orthoformate as the dehydrating agent, resulted in the formation of the tetrakis adducts, $MCl_3 \cdot 4\,CH_3OH$ (where M = La, Nd, Sm, Gd, Dy, Er, Yb, and Y), all of which are isomorphous [7]. The classification of the dysprosium adduct is not reported.

The melting point of $PrCl_3 \cdot 4\,CH_3OH$ occurs over the range 97 to 100°C, whereas that of $NdCl_3 \cdot 3\,CH_3OH$ occurs over the range 110 to 118°C [9]. Thermal decomposition of the praseodymium adduct resulted in loss of methanol molecules at 154 to 156°C, 180 to 182°C, and 262°C while a similar thermal curve for the neodymium adduct showed a loss of methanol at 176 to 178°C, 209°C, and 256°C [9]. The reflectance spectrum of the tris praseodymium adduct, $PrCl_3 \cdot 3\,CH_3OH$, is also reported [17].

An X-ray diffraction study of a solution of $LaCl_3$ in methanol has been reported (1.95 molal, d = 1.15 g/ml). The species existing in solution is suggested to be the dimer $[La_2Cl_4(CH_3OH)_{10}]^{2+}$, where the molecular structure consists of two edge-linked octahedra. Two of the chloride ions act as bridging ligands, and two chloride ions act as terminal ligands. The remaining coordinating positions are filled by the oxygen atoms of the methanol molecules [23]. The model is based on scattering peaks appearing at 1.48, 2.48, 2.95, 3.3, 3.9, 4.6, 5.8, and 7.7 Å from the lanthanum ion. The ^{139}La

and ^{35}Cl NMR spectra of this solution have also been reported, with broad peaks appearing at 293 ± 5 ppm and 167 ± 10 ppm, respectively, compared to the aqueous solution of $LaCl_3$ [23]. Presumably rapid chloride exchange is causing the chemicall nonequivalent chloride ions to appear as one peak in the NMR spectrum.

The solubility of the chloride salt of a rare earth in a particular alcohol is one measure of the strength of complex interaction. For solubilities see [7 to 9]. The compositions of the solid phases found from the solubility data are $PrCl_3 \cdot 3.5\,CH_3OH$ and $NdCl_3 \cdot 3\,CH_3OH$ [9]. Conductivity data as a function of concentration in methanol are reported for lanthanum [1], cerium(III) [1], praseodymium [22], and samarium chloride [4]. The values (in $\Omega^{-1} \cdot cm^2 \cdot mol^{-1}$) for the four salts are 67.25 (at 3.5 mmol), 66.9 (at 2.07 mmol), 72.5 (infinite dilution), and 278.3 (at 60 mmol), respectively. The conductivity vs. concentration plots are linear for the praseodymium and samarium salts and extrapolate to 308 $\Omega^{-1} \cdot cm^2 \cdot mol^{-1}$ for $SmCl_3$ at infinite dilution [4]. Stability constants determined spectrally are $\lg K_2 = 1.52 \pm 0.32$ and $\lg K_3 = 1.15 \pm 0.10$ for neodymium ($\lg K_1$ not determined) [10] and $K_1 = 1.7$ for didymium [11]. Both sets of values were determined in an H_2O/methanol mixture. Extinction coefficients for the solvates $[Nd(H_2O)_2(CH_3OH)_4]^{3+}$ and $[Nd(H_2O)_4(CH_3OH)_2]^{3+}$ were reported to be 4.28×10^2 and 2.17×10^2, respectively, compared to 1.72×10^3 for the aqua ion [10]. Additional visible spectral data are reported for the chloride salts of Nd [11, 12, 22] and Er and Gd [12]. The inner-sphere coordination of La^{3+} changes considerably with time in very concentrated $LaCl_3$/CH_3OH solution. Initially the average solute species is $La(CH_3OH)_8^{3+}$; about 1 year after the preparation of the solution, the average La^{3+} species has ≈4 Cl and ≈4 CH_3OH nearest neighbors, consistent with solute association. The atomic radial distribution function indicates the presence of μ-dichloro bridges between adjacent $LaCl_3$ units in the associated solute species after the solution has aged for 1 year. The presence of μ-dichloro bridges in the highly associated solute species is supported by the large peak at 4.7 Å due to nonbonded La···La and Cl···Cl atom-pair distances across the bridge units. The inner-sphere La-O and La-Cl bond distances are 2.48 and 2.95 Å, respectively, with uncertaities of ≈0.02 and 0.05 Å. The ion-pair La···Cl distance is ≈5.0 Å in solutions where chlorides are found in the second coordination sphere [24].

2.2.2.2 Methanolates

Lanthanum [13], gadolinium [21], and erbium [21] methanolates were prepared by the slow addition of each chloride salt dissolved in methanol (≈18 g/50 ml) to a solution of lithium methanolate prepared by dissolving lithium metal in methanol (12.2 g/250 ml). The resulting voluminous white precipitates, $M(OCH_3)_3$, were removed by filtering, extracted with 500 ml of methanol, and dried under reduced pressure. The praseodymium and neodymium derivatives were also prepared in this way [23]. A similar method, starting with anhydrous salts, was used to prepare the praseodymium and neodymium methanolates $M(OCH_3)_3$ [14]. Molecular weight measurements in benzene (ebullioscopical method) showed the compounds to be monomeric. The methanolate of ytterbium could also be prepared by the addition of methanol (40 ml) to a solution of ytterbium 2-propanolate in benzene (1.66 g/20 ml). After removal of the excess solvent by distillation and drying in vacuo (37°C/0.5 mm), white microcrystals of $Yb(OCH_3)_3 \cdot 0.5\,CH_3OH$ were obtained, which after heating in vacuo (50 to 55°C/0.5 mm) yielded $Yb(OCH_3)_3$. This ytterbium methanolate decomposed above 200°C [15]. A similar method, utilizing alcohol interchange with the 2-propanolate salt, resulted in the preparation of the samarium [18], gadolinium [19, 21], erbium [19], holmium [20], and ytterbium [19] methanolates. The tetrakis compound of cerium(IV), $Ce(OCH_3)_4$, has also been prepared by passing ammonia through a red solution of the complex $(C_5H_6N)_2CeCl_6$ in a methanol/benzene mixture (11 g/47 g/43 g, respectively). This product contained some impurity which reduced the $Ce:OCH_3$ ratio to 1:3.3 instead of 1:4. However, a second method involving the addition of methanol (130 g) to a solution of cerium(IV) 2-propanolate, $Ce(C_3H_7O)_4 \cdot C_3H_8O$, in benzene (4.8 g/50 g) resulted in the immediate precipitation of a yellow powder. After heating at reflux for 3 h, the supernatant liquid was decanted, additional methanol was added (120 g), and the treatment was repeated. Yellow crystals of $Ce(OCH_3)_4$ were obtained after drying at 0.1 mm and room temperature [16]. The cerium(IV) methanolate is insoluble in boiling methyl alcohol, benzene, or pyridine and is nonvolatile [16]. The gadolinium derivative is sparingly soluble in methanol, benzene, and ether [21].

The reaction of holmium aluminium 2-propanolate, $Ho[Al\{OCH(CH_3)_2\}_4]_3$, with methanol in benzene yielded three different products depending upon the amount of methanol used. A 1:6 molar ratio gave $Ho[Al_3(OCH_3)_3\{OCH(CH_3)_2\}_9]$, a 1:9 ratio gave $Ho[Al_3(OCH_3)_6\{OCH(CH_3)_2\}_6]$, and excess methanol gave the insoluble $Ho[Al(OCH_3)_4]_3$ [20]. The other two complex derivatives are soluble in benzene.

References to 2.2.2:

[1] S. N. Misra, T. N. Misra, R. C. Mehrotra (J. Inorg. Nucl. Chem. **27** [1965] 105/13). — [2] L. L. Quill, G. L. Clink (Inorg. Chem. **6** [1967] 1433/5). — [3] U. D. Tripathi, J. M. Batwara, R. C. Mehrotra (J. Chem. Soc. A **1967** 991/2). — [4] B. S. Sankhla, R. N. Kapoor (Australian J. Chem. **20** [1967] 383/6). — [5] R. J. Myers, M. Koss (Chem. Ber. **35** [1902] 2622).

[6] M. A. Saad, H. T. Soliman (J. Chem. U.A.R. **5** [1962] 29/39). — [7] A. Merbach, M.-N. Pitteloud, P. Jaccard (Helv. Chim. Acta. **55** [1972] 44/52). — [8] F. R. Hartley, A. W. Wylie (J. Chem. Soc. **1962** 679/81). — [9] Z. I. Grigorovich (Zh. Neorgan. Khim. **8** [1963] 986/9; Russ. J. Inorg. Chem. **8** [1963] 507/9). — [10] E. D. Romanenko, N. A. Kostromina, T. V. Ternovaya (Zh. Neorgan. Khim. **12** [1967] 700/6; Russ. J. Inorg. Chem. **12** [1967] 365/9).

[11] J. Bjerrum, C. Klixbüll Jørgensen (Acta. Chem. Scand. **7** [1953] 951/5). — [12] I. I. Antipova-Karataeva, Yu. I. Kutsenko (Zh. Neorgan. Khim. **9** [1964] 615/22; Russ. J. Inorg. Chem. **9** [1964] 341/5). — [13] D. C. Bradley, M. M. Factor (Chem. Ind. [London] **1958** 1332). — [14] S. N. Misra, T. N. Misra, R. C. Mehrotra (Australian J. Chem. **21** [1968] 797/800). — [15] U. D. Tripathi, J. M. Batwara, R. C. Mehrotra (J. Chem. Soc. A **1967** 991/2).

[16] D. C. Bradley, A. K. Chatterjee, W. Wardlaw (J. Chem. Soc. **1956** 2260/4). — [17] B. Jezowska-Trzebiatowska, K. Bukietynska, B. Keller (Bull. Acad. Polon. Sci. Ser. Sci. Chim. **25** [1977] 159/63). — [18] S. Sankhla, S. N. Misra, R. N. Kapoor (Chem. Ind. [London] **1965** 382/3). — [19] J. M. Batwara, U. D. Tripathi, R. K. Mehrotra, R. C. Mehrotra (Chem. Ind. [London] **1966** 1379). — [20] A. Mehrotra, R. C. Mehrotra (Indian J. Chem. **10** [1972] 532/5).

[21] R. C. Mehrotra, J. M. Batwara (Inorg. Chem. **9** [1970] 2505/10). — [22] R. C. Mehrotra, T. N. Misra, S. N. Misra (J. Indian Chem. Soc. **42** [1965] 351/8). — [23] L. S. Smith, Jr., D. C. McCain, D. L. Wertz (J. Am. Chem. Soc. **98** [1976] 5125/8). — [24] D. L. Wertz, S. Finch, S. Terry (Inorg. Chem. **18** [1979] 1590/3).

2.3 With Ethanol CH_3CH_2OH (= C_2H_5OH)

2.3.1 Scandium Compounds

Tris and bis ethanol adducts of scandium chloride, as well as the tetrakis and tris adducts of scandium bromide, were prepared by the desiccator method described for methanol [1]. Similarly the tris and bis adducts were also prepared by the second method discussed in that section. The solubility of scandium chloride in ethanol is 37.3 mg/100 mg solvent at 25°C. X-ray powder diffraction data are reported for the bis adduct [2].

References to 2.3.1:

[1] F. Petrů, B. Hajek, F. Jost (Croat. Chem. Acta **29** [1957] 457/60). — [2] E. M. Kirmse (Z. Chem. [Leipzig] **1** [1961] 332/4).

2.3.2 Lanthanide Compounds

2.3.2.1 Chlorides

Adducts of ethanol with the lanthanide chloride salts were prepared in the same manner as the methanol adducts. The tris adducts were also isolated with the formulation, $MCl_3 \cdot 3C_2H_5OH$ where M = La, Ce [1, 2, 5, 11], Pr, Nd [20], and Sm [3]. As with methanol, the azeotropic method, starting with the hydrated cerium(III) salt, yielded the tris adduct [5], whereas without utilizing the azeotrope, the $MCl_3 \cdot 2C_2H_5OH \cdot H_2O$ (M = La [1, 2], Pr, Nd [20], and Sm [3]) adducts were produced. The dehydration of hydrated rare earth chloride salts with triethyl orthoformate has resulted in the preparation of adducts with the formulation, $MCl_3 \cdot 3C_2H_5OH$ where M = La, Nd, Sm, Gd, and Yb, and $MCl_3 \cdot 4C_2H_5OH$ where M = Dy, Er, and Y [6]. The tetrakis adducts are isomorphous except for M = Y, while the tris adducts form a second isomorphous series except for M = Y, which is also unique. The classification of the dysprosium adduct is not reported [6]. The melting point of $PrCl_3 \cdot 3C_2H_5OH$ occurs over the range 104 to 110°C, whereas that of $NdCl_3 \cdot 2C_2H_5OH$ occurs over the range 125 to 128°C [8]. Thermal decomposition of the praseodymium adduct revealed a loss of ethanol molecules at 155 to 159°C, 178 to 180°C, and 253°C, whereas a similar thermal curve for the neodymium adduct showed a loss of ethanol at 162 and 185°C [8]. The reflectance spectra of the praseodymium and neodymium adducts, $MCl_3 \cdot 3C_2H_5OH$, are also reported [13].

Solubilities of the lanthanide chlorides in ethanol are reported in [6 to 8]. The compositions of the solid phases found from the solubility data are $CeCl_{2.98} \cdot 3.11C_2H_5OH$ [7], $PrCl_3 \cdot 2.5C_2H_5OH$ [8], and $NdCl_3 \cdot 2C_2H_5OH$ [8]. Conductivity measurements for $LaCl_3$ in ethanol suggest a limiting molecular conductance of 50.55 for the 1:1 electrolyte species, corresponding to $LaCl_2^+$, and a value of 70.57 for the 1:2 electrolyte species, $LaCl_2^+$, both values in $\Omega^{-1} \cdot cm^2 \cdot mol^{-1}$. Equivalent conductance as a function of concentration is reportet [9]. Typical conductivity values in ethanol are 29.1 ($LaCl_3$ at 3.16 mM) [1], 29.2 ($CeCl_3$ at 2.249 mM) [1], and 64.63 ($SmCl_3$ at 18.4 mM) [3]. The infinite dilution values of the praseodymium and samarium salts are 28.0 [20] and $65 \Omega^{-1} \cdot cm^2 \cdot mol^{-1}$ [3], respectively. The molecular weights of the anhydrous chlorides, determined ebullioscopically, show similar evidence of dissociation. Values for the lanthanum salt are: 208 (0.0686 M), 196 (0.1496 M), and 242 (0.1995 M) compared to 245.3 (calculated); whereas for the cerium salt the values are: 206 (0.098 M), 220 (0.186 M), and 214 (0.2599 M) compared to 246.51 (calculated) [1]. Spectral data in the visible region are reported for the chloride salts of Nd, Er, and Gd [10]. The absorption spectrum of the neodymium adduct in ethanol solution is also reported [20].

2.3.2.2 Thiocyanates

Ethanol adducts of samarium thiocyanate were obtained by mixing ethanol solutions of samarium thiocyanate and anhydrous MCNS (M = K, Rb, and Cs) in a 1:1 molar ratio and allowing the adducts to crystallize in a desiccator over P_4O_{10} [12]. Three adducts were isolated, $KSm(CNS)_4 \cdot 4C_2H_5OH$, $RbSm(CNS)_4 \cdot 5C_2H_5OH$, and $CsSm(CNS)_4 \cdot 6C_2H_5OH$. These adducts lost the solvate molecules at 115, 95, and 90°C for the potassium, rubidium, and caesium double salts, respectively. Decomposition of the completely desolvated material occurred over the ranges, 200 to 280, 250 to 380, and 350 to 440°C, respectively.

2.3.2.3 Ethanolates

Ytterbium ethanolate was prepared by the addition of ethanol (40 ml) to a benzene solution of ytterbium 2-propanolate (1.08 g/20 ml). The white crystals of $Yb(OC_2H_5)_3$ which separated were isolated by distillation of the solvents followed by drying in vacuo (36°C/0.5 mm) [2]. The preparations of samarium [15], gadolinium [16, 19], erbium [16], holmium [17], and ytterbium [16, 18] ethanolates by alcohol interchange with the respective 2-propanolates have also been reported, in analogy with the methanolates. The ethanolate of cerium(IV), $Ce(OC_2H_5)_4$, was prepared either by passing ammonia through an ethanol/benzene mixture containing $(C_5H_6N)_2CeCl_6$, or by successive treatment of the 2-propanolate with ethanol in analogy with the preparation of the methanolate. The compound is insoluble in benzene [14]. The praseodymium and neodymium ethanolates were prepared by the direct reaction of each chloride salt with lithium ethanolate in the dried alcohol, as described earlier for the methanolates [21]. The reflectance spectrum of $Pr(OC_2H_5)_3$ is also reported [13].

The reaction of excess ethanol with holmium aluminium 2-propanolate, $Ho[Al\{OCH(CH_3)_2\}_4]_3$, gave $Ho[Al(OC_2H_5)_4]_3$ after heating at reflux for 4 h, evaporating excess solvent, and drying in a vacuum [17].

References to 2.3.2:

[1] S. N. Misra, T. N. Misra, R. C. Mehrotra (J. Inorg. Nucl. Chem. **27** [1965] 105/13). — [2] U. D. Tripathi, J. M. Batwara, R. C. Mehrotra (J. Chem. Soc. A **1967** 991/2). — [3] B. S. Sankhla, R. N. Kapoor (Australian J. Chem. **20** [1967] 383/6). — [4] R. J. Meyer, M. Koss (Chem. Ber. **35** [1902] 2622). — [5] M. A. Saad, H. T. Soliman (J. Chem. U.A.R. **5** [1962] 29/39).

[6] A. Merbach, M.-N. Pitteloud, P. Jaccard (Helv. Chim. Acta **55** [1972] 44/52). — [7] F. R. Hartley, A. W. Wylie (J. Chem. Soc. **1962** 679/81). — [8] Z. I. Grigorovich (Zh. Neorgan. Khim. **8** [1963] 986/9; Russ. J. Inorg. Chem. **8** [1963] 507/9). — [9] A. M. El-Aggan, D. C. Bradley, W. Wardlaw (J. Chem. Soc. **1958** 2092/9). — [10] I. I. Antipova-Karataeva, Yu. I. Kutsenko (Zh. Neorgan. Khim. **9** [1964] 615/22; Russ. J. Inorg. Chem. **9** [1964] 341/5).

[11] Z. A. Sheka, E. E. Kriss (Zh. Neorgan. Khim. **4** [1959] 1809/13; Russ. J. Inorg. Chem. **4** [1959] 816/8). — [12] A. M. Golub, A. N. Borshch (Ukr. Khim. Zh. **32** [1966] 923/5; Soviet Progr. Chem. **32** [1966] 685/6). — [13] B. Jezowska-Trzebiatowska, K. Bukietynska, B. Keller (Bull. Acad. Polon. Sci. Ser. Sci. Chim. **25** [1977] 159/63). — [14] D. C. Bradley, A. K. Chatterjee, W. Wardlaw (J. Chem. Soc. **1956** 2260/4). — [15] S. Sankhla, S. N. Misra, R. N. Kapoor (Chem. Ind. [London] **1965** 382/3).

[16] J. M. Batwara, U. D. Tripathi, R. K. Mehrotra, R. C. Mehrotra (Chem. Ind. [London] **1966** 1379). — [17] A. Mehrotra, R. C. Mehrotra (Indian J. Chem. **10** [1972] 532/5). — [18] U. D. Tripathi, J. M. Batwara, R. C. Mehrotra (J. Chem. Soc. A **1967** 991/2). — [19] R. C. Mehrotra, J. M. Batwara (Inorg. Chem. **9** [1970] 2505/10). — [20] R. C. Mehrotra, T. N. Misra, S. N. Misra (J. Indian Chem. Soc. **42** [1965] 351/8).

2.4 With 1-Propanol $CH_3CH_2CH_2OH$ (= C_3H_8OH)

2.4.1 Scandium Compounds

Tris and bis 1-propanol adducts of scandium chloride were prepared by the desiccator method described for methanol [1], see p. 15. Similarly, the tetrakis and bis adducts were also prepared by the second method discussed in that section. The solubility of scandium chloride in 1-propanol is 26.1 mg/100 mg solvent at 25°C [2].

References to 2.4.1:

[1] F. Petrů, B. Hajek, F. Jost (Croat. Chem. Acta. **29** [1957] 457/60). — [2] E. M. Kirmse (Z. Chem. [Leipzig] **1** [1961] 332/4).

2.4.2 Lanthanide Compounds

2.4.2.1 Chlorides

The adducts of n-propanol with cerium(III) [2] and samarium [1] chlorides were prepared in the same manner as for the methanol adducts. The tris adducts were isolated with the formulation $MCl_3 \cdot 3\,n\text{-}C_3H_7OH$. The hydrated samarium chloride salt yielded $SmCl_3 \cdot H_2O \cdot 2\,n\text{-}C_3H_7OH$, rather than the tris adduct [1]. As with methanol, the azeotropic method yielded the tris adduct for the cerium(III) chloride salt [2].

The solubilities of the cerium(III) [3], praseodymium, and neodymium salts [4] in n-propanol have been reported. The compositions of the solid phases found from the solubility data are $CeCl_{2.98} \cdot 3.10\,n\text{-}C_3H_7OH$ [3], $PrCl_3 \cdot 2\,n\text{-}C_3H_7OH$ [4], and $NdCl_3 \cdot 2\,n\text{-}C_3H_7OH$ [4]. The melting point of $PrCl_3 \cdot 2\,n\text{-}C_3H_7OH$ occurs over the range 108 to 113°C, whereas that of the analogous neodymium adduct occurs over the range 120 to 125°C [4]. Thermal decomposition of the praseodymium adduct reveals a loss of propanol molecules at 159, 179, and 261°C, whereas a similar thermal curve for the neodymium adduct shows a loss of propanol at 166 and 184°C [4]. A preliminary analysis of reflectance and absorption data of the Pr and Nd adducts, $MCl_3 \cdot 3\,n\text{-}C_3H_7OH$, suggests that the metal ions have C_{3v} symmetry [5].

2.4.2.2 1-Propanolates

Samarium 1-propanolate, $Sm(OC_3H_7)_3$, was prepared by alcohol interchange of the 2-propanolate [7], as were the Gd, Er, and Yb derivatives [8, 9]. The 1-propanolate of cerium(IV), $Ce(OC_3H_7)_4$, was prepared by azeotropic distillation of the cerium(IV) 2-propanolate and a mixture of 1-propanol and benzene. Removal of the remaining solvent by evaporation yielded the 1-propanolate [6]. The cerium derivative is nonvolatile and has a molecular weight of 1620 g/mol (molecular complexity = 4.30) in benzene and 1295 g/mol (molecular complexity = 3.44) in toluene [6]. The molecular weights were determined ebullioscopically.

References to 2.4.2:

[1] B. S. Sankhla, R. N. Kapoor (Australian J. Chem. **20** [1967] 383/6). — [2] M. A. Saad, H. T. Soliman (J. Chem. U.A.R. **5** [1962] 29/39). — [3] F. R. Hartley, A. W. Wylie (J. Chem. Soc. **1962** 679/81). — [4] Z. I. Grigorovich (Zh. Neorgan. Khim. **8** [1963] 986/7; Russ. J. Inorg. Chem. **8** [1963] 507/9). — [5] B. Jezowska-Trzebiatowska, K. Bukietynska, B. Keller (Bull. Acad. Polon. Sci. Ser. Sci. Chim. **25** [1977] 159/63).

[6] D. C. Bradley, A. K. Chatterjee, W. Wardlaw (J. Chem. Soc. **1956** 2260/4). — [7] S. Sankhla, S. N. Misra, R. N. Kapoor (Chem. Ind. [London] **1965** 382/3). — [8] J. M. Batwara, U. D. Tripathi, R. K. Mehrotra, R. C. Mehrotra (Chem. Ind. [London] **1966** 1379). — [9] R. C. Mehrotra, J. M. Batwara (Inorg. Chem. **9** [1970] 2505/10).

2.5 With 2-Propanol $(CH_3)_2CHOH$ (= Isopropyl Alcohol = i-C_3H_7OH)

2.5.1 Chlorides

Adducts of 2-propanol with the lanthanide chloride salts were prepared in the same manner as the methanol and ethanol adducts. The tris adducts were isolated with the formulation $MCl_3 \cdot 3\,i\text{-}C_3H_7OH$ where M = La, Ce [1, 2, 5], Pr, Nd [24], Sm [3], and Ho [9]. The Pr and Nd derivatives could be recrystallized from 2-propanol [24]. The tris adducts of Gd and Er could also be prepared by passing dry HCl through a solution of the 2-propanolates in 2-propanol [22]. Bis adducts containing water, $MCl_3 \cdot H_2O \cdot 2\ i\text{-}C_3H_7OH$ (with M = La [1], Pr, Nd [24], and Sm [3]) were reported if the hydrated chloride salts were used as the starting materials. Other procedures, starting with the hydrated lanthanide chloride salts, used distillation of a benzene/2-propanol/water azeotrope, followed by evaporation to dryness and recrystallization from 2-Propanol to prepare the tris adducts, $MCl_3 \cdot 3\ i\text{-}C_3H_7OH$ where M = La [4], Ce [5], Pr and Nd [24]. 2-Propanol adducts of rare earth chloride salts have also been prepared by treating the hydrated salts with trimethyl orthoformate, and 2-propanol followed by crystallization. The tris adducts are reported with the formulation $MCl_3 \cdot 3\,i\text{-}C_3H_7OH$, with M = La, Nd, Sm, Gd, Dy, Er, Yb, and Y [6]. X-ray powder diffraction data show that the La, Nd, Sm, and Gd adducts form one isomorphous series whereas the Er, Yb, and Y adducts form a second series. Classification of the dysprosium adduct was not reported [6]. A monokis adduct of 2-propanol with scandium chloride was prepared by the desiccator method described for methanol [31].

Solubilities of some MCl_3 salts in 2-propanol are given in [6 to 8]. The composition of the solid phase found from the solubility data of the cerium salt in 2-propanol is $CeCl_{3.01} \cdot 2.2\ i\text{-}C_3H_7OH$ [7]. Conductivity data for the samarium chloride salt in 2-propanol are reported over a wide concentration range, with a value of 16 $\Omega^{-1} \cdot cm^2 \cdot mol^{-1}$ for the equivalent conductance at infinite dilution [3].

2.5.2 2-Propanolates

The rare earth 2-propanolates are the alcoholates discussed most often in the literature. They are frequently used as starting points for the syntheses of other alcoholates via an alcoholate interchange reaction. The only method reported in detail for the preparation of the yttrium and lanthanide tris(2-propanolates) involved the reaction of turnings of the metals (5 g) with 2-propanol (300 ml) in the presence of a small amount of $HgCl_2$ (10^{-4} mol/mol of metal) as a catalyst. The resulting mixtures were heated at reflux at 82°C for 24 h, cooled, and filtered, after which the crude product was purified by recrystallization from hot 2-propanol or by vacuum sublimation. For some of the larger metal ions, a mixture of $HgCl_2$ and $Hg(CH_3COO)_2$ or HgI_2 was used as the catalyst. In each case the tris (2-propanolate) was produced, $M(i\text{-}OC_3H_7)_3$, with M = Y, La to Lu [9]. A second more generally followed method follows the procedure of allowing the 2-propanol solvate of the lanthanide chloride salt, $MCl_3 \cdot 3i\text{-}C_3H_7OH$, to react with sodium 2-propanolate in 2-propanol or in a 2-propanol/benzene mixture as a solvent. Heating at reflux for several hours usually leads to the tris (2-propanolate) as the product. The presence of water in the starting chloride solvate may explain the occasionally poor results reported for this method. Using this method, the 2-propanolates of lanthanum [10, 11], praseodymium [11, 13], neodymium [11, 13], samarium [12], gadolinium [14, 22], erbium [14, 22], holmium [15], and ytterbium [14, 16] have been prepared. A third method, using the procedure just described but starting with the anhydrous trichloroacetate lanthanide salts instead of the chloride salts, was reported for the preparation of the 2-propanolates of praseodymium, neodymium, and samarium [19]. A fourth method, reported for the preparation of yttrium 2-propanolate involved rapidly stirring a solution containing lithium 2-propanolate (24.3 g), 150 ml of 2-propanol, yttrium trichloride (43.1 g), and tetrahydrofuran (25 ml). The mixture was allowed to react at 45°C for 3 h, after which the solvents were distilled off under reduced pressure until the product was nearly dry. The solid was treated with benzene, the suspension filtered, and the filtrate evaporated to dryness. The compound was recrystallized from 2-propanol to remove the lithium chloride contaminent [17].

The tetrakis(2-propanolate) of cerium(IV) was prepared by mixing 2-propanol (41.5 g), ammonia in benzene (167.5 g), and dipyridinium hexachlorocerate(IV), $(C_5H_6N)_2CeCl_6$. After removal of excess ammonia, a yellow solid, $Ce(i\text{-}OC_3H_7)_4 \cdot C_5H_5N$, was obtained by evaporation of the filtrate. The pyridine adduct could be recovered after recrystallization from pyridine and drying at 0.1 mm at room temperature. After recrystallization of the pyridine adduct 4 times from 2-propanol and drying at room temperature at reduced pressure, an alcoholate was obtained, $Ce(i\text{-}OC_3H_7)_4 \cdot i\text{-}C_3H_7OH$ [18]. The alcohol could be removed at 100°C/0.05 mm to yield $Ce(i\text{-}OC_3H_7)_4$, which has a molecular weight of 1179 g/mol (molecular complexity = 3.13) in benzene and 1045 g/mol (molecular complexity = 2.78) in toluene, as determined ebullioscopically [20]. The compound is also reported to be dimeric in 2-propanol [20].

The crystal structure of a neodymium 2-propanolate has been reported, showing the solid to contain hexameric units and one chloride ion and having the formulation $Nd_6(i\text{-}OC_3H_7)_{17}Cl$ [21]. This compound was prepared by the addition of a benzene/2-propanol solution of sodium 2-propanolate to a gently refluxing solution of neodymium trichloride in isopropanol. The suspension was heated at reflux for 12 h, centrifuged, and filtered. The residue was washed with benzene twice and the combined extracts evaporated to dryness in vacuo. The blue solid was crystallized from toluene (−15°C) as light blue needles which did not melt or decompose when heated to 300°C in a sealed capillary. The crystals are monoclinic with space group $P2_1/n\text{-}C_{2h}^5$ (No. 14) and cell dimensions a = 24.52(2), b = 22.60(2), c = 14.22(1) Å, β = 101.05(5)°. The calculated density is 1.636 g/cm³ with 4 molecules in the unit cell and a cell volume of 7736 Å³. The R factor for 2327 reflections was 0.076. The six neodymium ions lie at the corners of a regular trigonal prism with the single chloride ion at the center of the prism, 3.05(1) Å away from the metal ions. There are three types of Nd-O bonds. Each Nd atom is bonded singly to a 2-propanolate ligand with an average Nd-O distance of 2.05(2) Å. A bridging 2-propanolate ligand bonds two Nd atoms at each edge of the prism to give 18 Nd-O bonds that average 2.36(4) Å. In addition there are two 2-propanolate ligands centered on the trigonal faces which are bonded to three Nd atoms to give Nd-O bonds that average 2.45(5) Å. The molecular structure thus contains 6 terminal 2-propanolate ligands, 9 edge-bridging ligands, and 2 face-bridging ligands, resulting in a severely distorted octahedral geometry about each neodymium ion. The coordination polyhedron is shown in **Fig. 2-1** and the important bond distances in Table 2/1 [21]. The implication of this work is that all the methods of synthesis reported earlier may have produced complex polymeric units containing one or more chloride ions, since the chloride salts were invariably used as the starting material. The magnetic moment, μ_{eff}, is 3.22 μ_B per Nd ion. The magnetic susceptibility obeys the Curie law over the range 7.5 to 100 K ($\chi = C_M/(T+\Theta)$), where $C_M = 1.293$ and $\Theta = -2.8$ K) [21].

Fig. 2-1

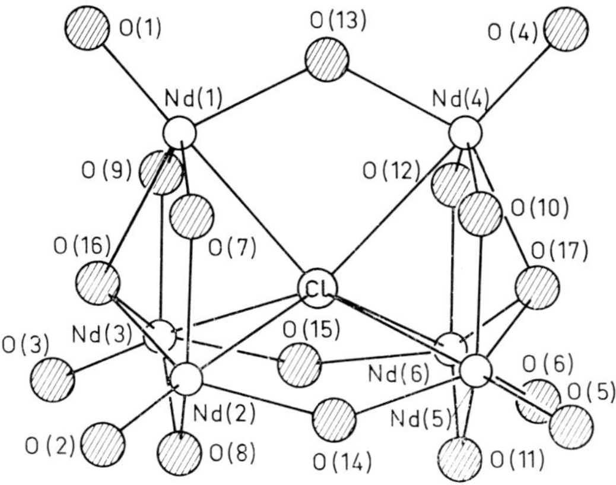

Coordination polyhedron for $Nd_6(i\text{-}OC_3H_7)_{17}Cl$.

The 2-propanolates appear to be polymeric in the solid state, but there is at least one report that the derivatives are monomeric in benzene [11]. All the compounds sublime below ≈200°C (Table 2/2) and decompose via loss of an olefin molecule above that temperature. The infrared spectra of all the lanthanide 2-propanolates are reported, but no analyses are made [17, 23]. Similarly, the UV and visible absorption spectra of these compounds are reported, but the spectra are not used to characterize the compounds [19, 23]. The nephelauxetic parameter, β, is calculated to be 0.9979 for the praseodymium compound dissolved in benzene [19]. A brief discussion of the spectra of the diamagnetic complexes and the mass spectra of the series indicates that these compounds are at least partially polymeric in benzene as well as in the solid [23].

Table 2/1
Interatomic Distances for the $Nd_6(i\text{-}OC_3H_7)_{17}Cl$ Coordination Polyhedron.

atoms	distance in Å	atoms	distance in Å	atoms	distance in Å
Cl-Nd(1)	3.05(2)	O(8) -Nd(3)	2.31(4)	O(16)-Nd(1)	2.49(3)
Cl-Nd(2)	3.07(2)	O(9) -Nd(3)	2.38(3)	O(16)-Nd(2)	2.51(3)
Cl-Nd(3)	3.05(2)	O(9) -Nd(1)	2.37(4)	O(16)-Nd(3)	2.38(3)
Cl-Nd(4)	3.04(2)	O(10)-Nd(4)	2.39(4)	O(17)-Nd(4)	2.44(3)
Cl-Nd(5)	3.05(2)	O(10)-Nd(5)	2.30(4)	O(17)-Nd(5)	2.44(3)
Cl-Nd(6)	3.05(2)	O(11)-Nd(5)	2.36(4)	O(17)-Nd(6)	2.43(3)
O(1)-Nd(1)	2.03(4)	O(11)-Nd(6)	2.33(3)	Nd(1)-Nd(2)	3.81(1)
O(2)-Nd(2)	2.07(4)	O(12)-Nd(6)	2.43(3)	Nd(1)-Nd(3)	3.80(1)
O(3)-Nd(3)	2.04(4)	O(12)-Nd(4)	2.27(3)	Nd(2)-Nd(3)	3.82(1)
O(4)-Nd(4)	2.07(4)	O(13)-Nd(1)	2.36(3)	Nd(4)-Nd(5)	3.81(1)
O(5)-Nd(5)	2.02(4)	O(13)-Nd(4)	2.37(3)	Nd(4)-Nd(6)	3.80(1)
O(6)-Nd(6)	2.05(4)	O(14)-Nd(2)	2.34(3)	Nd(5)-Nd(6)	3.81(1)
O(7)-Nd(1)	2.38(4)	O(14)-Nd(5)	2.38(3)	Nd(1)-Nd(4)	4.22(1)
O(7)-Nd(2)	2.42(4)	O(15)-Nd(3)	2.36(4)	Nd(2)-Nd(5)	4.23(1)
O(8)-Nd(2)	2.43(4)	O(15)-Nd(6)	2.34(4)	Nd(3)-Nd(6)	4.24(1)

Table 2/2
Sublimation Temperatures of the Rare Earth Tris (2-propanolates) and Ce^{IV}Tetrakis(2-propanolates) *).

M	sublimation point in °C/Torr	color	M	sublimation point in °C/Torr	color
Y	200/0.10	white	Tb	190/0.10	white
Ce^{IV}	160 to 70/0.05	—	Dy	190/0.17	light yellow
Pr	175/0.04	green	Ho	195/0.18	peach
Nd	—	blue	Er	195/0.35	pink
Sm	180/0.04**)	light yellow	Tm	185/0.06	light green
Eu	—	orange	Yb	195/0.02	white
Gd	200/0.15	white	Lu	198/0.10	white

*) Data from reference [23], except for Ce^{IV} from reference [20].—**) Sublimation point of Sm derivative also reported as 280°C/1 mm [12].

Double Salts

Double salts of the rare earths with 2-propanol have been obtained by three methods. The first involves the reaction of the rare earth chloride tris(2-propanolates) with potassium aluminium 2-propanolate. The solid reactants were mixed together in stoichiometric amounts, and 2-propanol was added. After being heated at reflux for ≈1 h, the mixtures were left to stand overnight, and the KCl precipitates were removed by filtering. After removing the excess alcohol from the filtrates under reduced pressure, the alcoholates were obtained, $M[Al(i\text{-}OC_3H_7)_4]_3$ with M = Sc, Y [26], La, Pr [25, 26], Ce, Nd, Sm, Gd, Dy, Ho [15, 26], Er, Yb, and Lu [26]. In the case of the cerium(III) derivative, a nitrogen atmosphere was used to prevent oxidation and the final product distilled under reduced pressure, leaving KCl behind [26]. A second method involved the reaction of the lanthanide 2-propanolates with aluminium 2-propanolate dissolved in 2-propanol. After being heated at reflux for 4 to 5 h, the same products, $M[Al(i\text{-}OC_3H_7)_4]_3$ with M = Gd, Ho, and Er, were obtained [26]. If excess aluminium 2-propanolate is used in the reaction, distillation at ≈70 to 80°C/0.1 mm will remove it, leaving the double salts which can be distilled at ≈180 to 200°C/0.1 mm. If excess lanthanide 2-propanolates are used, the double salts can be distilled from the simple 2-propanolates. A third method involves the reaction of lanthanide chlorides with potassium and aluminium chloride dissolved in 2-propanol to produce $M[Al(i\text{-}OC_3H_7)_4]_3$, KCl and H_2 gas, with M = Gd, Ho, and Er [26]. The first method described has also been used to prepare the analogous gallium double salt $M[Ga(i\text{-}OC_3H_7)_4]_3$ with M = La, Pr, Nd, Sm, Gd, Ho, and Er, starting with the chloride salts and potassium gallium 2-propanolate dissolved in 2-propanol [27]. The second method, the reaction of the lanthanide 2-propanolate with gallium 2-propanolate in 2-propanol, also results in the same product for erbium and gadolinium [27]. The boiling points of these derivatives are listed in Table 2/3.

Table 2/3
Boiling Points (b.p.) and Molecular Weights (m.w.) of the Double Salts, $M[M'(i\text{-}OC_3H_7)_4]_3$.

M	M'	b. p. in °C/Torr	m. w. in g/mol	Ref.
Sc	Al	145/0.5	837 (835)	[26]
Y	Al	192/0.5	981 (879)	[26]
La	Al	208/0.5	1030 (929)	[26]
Ce	Al	200/0.5	—	[26]
Pr	Al	192/0.3	969 (931)**)	[26]

Table 2/3 (Continued)

M	M′	b. p. in °C/Torr	m. w. in g/mol*)	Ref.
Nd	Al	186/0.3	853 (934)	[26]
Sm	Al	203/0.5	983 (940)	[26]
Gd	Al	195/0.3	957 (947)	[26]
Dy	Al	203/0.5	997 (952)	[26]
Ho	Al	180/0.2	956 (954)	[26]
Er	Al	187/0.2	948 (957)	[26]
Yb	Al	200/0.5	1048 (963)	[26]
Lu	Al	190/0.3	981 (965)	[26]
La	Ga	139/0.6	921 (1057)**)	[27]
Pr	Ga	123/1.0	893 (1059)**)	[27]
Nd	Ga	165/0.7	952 (1062)**)	[27]
Sm	Ga	140/0.4	980 (1068)**)	[27]
Gd	Ga	145/0.05	1078 (1075)	[27]
Ho	Ga	165/0.7	1087 (1083)**)	[27]
Yb	Ga	—	1077 (1090)**)	[27]

*) Found (calc) in boiling benzene; **) in 2-propanol.

The molecular structure assigned to these derivatives is that of $Al[Al(i\text{-}OC_3H_7)_4]_3$, where the central octahedrally coordinated aluminium ion is replaced by a rare earth ion. The 6 oxygen atoms from the 2-propanolate groups act as bridges to the remaining aluminium ions. Each aluminium ion has two additional nonbridging 2-propanolate groups bound to it, resulting in a coordination number of 4 for the aluminium ion. The ^1H NMR spectrum of the lanthanum derivative consists of a single methyl doublet centered at 1.48 ppm from tetramethylsilane, suggesting rapid exchange between the terminal and bridging 2-propanolate groups even at −60°C [28]. The other proton signals in the spectrum are not discussed. The scandium derivative in CCl_4 (at 100 MHz) has τ values for the methyl doublets at 8.52, 8.68, 8.88, and 8.91 ppm. The first two values are assigned to bridging 2-propanolate groups and the latter two values to terminal groups. In each case J = 6 Hz [29]. The mass spectra of several of these compounds have been determined and their fragmentation patterns reported. In each case the molecular ion for the tetrameric compound is found [28, 30].

References to 2.5:

[1] S. N. Misra, T. N. Misra, R. C. Mehrotra (J. Inorg. Nucl. Chem. **27** [1965] 105/13). — [2] U. D. Tripathi, J. M. Batwara, R. C. Mehrotra (J. Chem. Soc. A **1967** 991/2). — [3] B. S. Sankhla, R. N. Kapoor (Australian J. Chem. **20** [1967] 383/6). — [4] A. M. El-Aggan, D. C. Bradley, W. Wardlaw (J. Chem. Soc. **1958** 2092/9). — [5] M. A. Saad, H. T. Soliman (J. Chem. U.A.R. **5** [1962] 29/39).

[6] A. Merbach, M.-N. Pitteloud, P. Jaccard (Helv. Chim. Acta **55** [1972] 44/52). — [7] F. R. Hartley, A. W. Wylie (J. Chem. Soc. **1962** 679/81). — [8] M. A. Saad, H. T. Soliman (J. Chem. U.A.R. **5** [1962] 21/8). — [9] L. M. Brown, K. S. Mazdiyasni (Inorg. Chem. **9** [1970] 2783/6). — [10] R. P. N. Sinha (Sci. Cult. [Calcutta] **25** [1960] 494).

[11] S. N. Misra, T. N. Misra, R. N. Kapoor, R. C. Mehrotra (Chem. Ind. [London] **1963** 120). — [12] S. Sankhla, S. N. Misra, R. N. Kapoor (Chem. Ind. [London] **1965** 382/3). — [13] S. N. Misra, T. N. Misra, R. C. Mehrotra (Australian J. Chem. **21** [1968] 797/800). — [14] J. M. Batwara, U. D. Tripathi, R. K. Mehrotra, R. C. Mehrotra (Chem. Ind. [London] **1966** 1379). — [15] A. Mehrotra, R. C. Mehrotra (Indian J. Chem. **10** [1972] 532/5).

[16] U. D. Tripathi, J. M. Batwara, R. C. Mehrotra (J. Chem. Soc. A **1967** 991/2). — [17] K. S. Mazdiyasni, C. T. Lynch, J. S. Smith (Inorg. Chem. **5** [1966] 342/6). — [18] D. C. Bradley, A. K.

Chatterjee, W. Wardlaw (J. Chem. Soc. **1956** 2260/4). — [19] M. Singh, S. N. Misra (J. Indian Chem. Soc. **55** [1978] 643/4). — [20] D. C. Bradley, A. K. Chatterjee, W. Wardlaw (J. Chem. Soc. **1956** 3469/72).

[21] R. A. Andersen, D. H. Templeton, A. Zalkin (Inorg. Chem. **17** [1978] 1962/5). — [22] R. C. Mehrotra, J. M. Batwara (Inorg. Chem. **9** [1970] 2505/10). — [23] L. M. Brown, K. S. Mazdiyasni (Inorg. Chem. **9** [1970] 2783/6). — [24] R. C. Mehrotra, T. N. Misra, S. N. Misra (J. Indian Chem. Soc. **42** [1965] 351/8). — [25] R. C. Mehrotra, M. M. Agrawal (J. Chem. Soc. Chem. Commun. **1968** 469/70).

[26] R. C. Mehrotra, M. M. Agrawal, A. Mehrotra (Syn. Inorg. Metal-Org. Chem. **3** [1973] 181/91). — [27] R. C. Mehrotra, M. M. Agrawal, A. Mehrotra (Syn. Inorg. Metal-Org. Chem. **3** [1973] 407/14). — [28] J. G. Oliver, I. J. Worrall (J. Chem. Soc. A **1970** 845/8). — [29] R. C. Mehrotra, A. Mehrotra (Proc. 10th Rare Earth Res. Conf., Carefree, Ariz., 1973, pp. 354/61). — [30] A. Mehrotra, R. C. Mehrotra (J. Chem. Soc. Chem. Commun. **1972** 189/90).

[31] F. Petrů, B. Hajek, F. Jost (Croat. Chem. Acta **29** [1957] 457/60).

2.6 With 1-Butanol $CH_3CH_2CH_2CH_2OH$ ($=C_4H_9OH$)

2.6.1 Scandium Compounds

Tris and bis 1-butanol adducts of scandium chloride were prepared by the desiccator method described for methanol [1]. Similarly, the tris and bis adducts were also prepared by the second method discussed in that section. The solubility of scandium chloride in 1-butanol is 25.2 mg/100 mg solvent at 25°C [2].

References to 2.6.1:

[1] F. Petrů, B. Hajek, F. Jost (Croat. Chem. Acta **29** [1957] 457/60). — [2] E. M. Kirmse (Z. Chem. [Leipzig] **1** [1961] 332/4).

2.6.2 Lanthanide Compounds

2.6.2.1 Chlorides

Adducts of 1-butanol with the lanthanide chlorides were prepared in the same manner as the methanol adducts, using the anhydrous salts. The tris adducts were isolated with the formulation, $MCl_3 \cdot 3C_4H_9OH$ with M = La, Ce [1, 3] and Sm [2]. Hydrated samarium chloride yielded $SmCl_3 \cdot 2C_4H_9OH \cdot H_2O$ rather than the tris adduct [2]. The tris adduct of $LaCl_3$ was also prepared by dissolving the tris 2-propanol adduct of $LaCl_3$ in benzene (60 g), adding 1-butanol (2.5 g), and after heating 6 h at reflux, distilling the 2-propanol/benzene azeotrope at 72°C and excess solvent at 80°C. The solid was dried as before, and analyses confirmed the tris adduct [1]. A bis adduct, $LaCl_3 \cdot 2C_4H_9OH \cdot H_2O$, was also prepared, using the azeotropic method [1]. The azeotropic method described for the methanol adduct also yielded the tris adduct for cerium(III) chloride [3]. The tris 1-butanol adducts of $PrCl_3$ and $NdCl_3$ were also prepared by dissolving the tris ethanolate adducts in a benzene/butanol mixture (45 g/5.0 g), heating at reflux for 3 h and distilling the ethanol/benzene azeotrope (68°C), followed by excess solvent. After drying at 27°C at 1 Torr, the adducts $MCl_3 \cdot 3C_4H_{10}O$, with M = Pr, Nd, were obtained [13].

The solubilities of three of the lanthanide chloride salts in 1-butanol have been reported [4, 5].

The compositions of the solid phases found from the solubility data are $CeCl_{3.05} \cdot 3.07C_4H_9OH$ [4], $2PrCl_3 \cdot 3C_4H_9OH$ [5], and $2NdCl_3 \cdot 3C_4H_9OH$ [5]. The melting point of $2PrCl_3 \cdot 3C_4H_9OH$ occurs over the range 110 to 113°C, whereas that of the analogous Nd adduct occurs over the range 119 to 124°C [5]. Thermal decomposition of the praseodymium adduct revealed a loss of butanol molecules at 164, 185, and 266°C, whereas a similar thermal curve for the neodymium adduct showed loss of butanol at 158, 184, and 268°C [5].

2.6.2.2 1-Butanolates

The 1-butanolates of praseodymium [6, 9], neodymium [6, 9], samarium [7], gadolinium [8], erbium [8], and ytterbium [8, 10] were prepared by the alcohol interchange reaction starting with the 2-propanolate in the manner already described for the methanolates. The 1-butanolate of cerium(IV), $Ce(OC_4H_9)_4$, was prepared either by passing ammonia through a butanol/benzene mixture containing $(C_5H_6N)_2CeCl_6$, as described on p. 22, or by azeotropic distillation of the cerium(IV) 2-propanolate and a mixture of 1-butanol and benzene, in analogy with the preparation of the 1-propanolates, see p. 21. Removal of the remaining solution after the azeotropic distillation yielded the 1-butanolate [11]. The cerium(IV) derivative is nonvolatile and has a molecular weight of 1820 g/mol (molecular complexity = 4.20) in benzene and 1504 g/mol (molecular complexity = 3.48) in toluene [11]. The molecular weight was determined ebullioscopically. The molecular weights of the Gd and Er derivatives are 1613 and 1623 g/mol, respectively [12].

References to 2.6.2:

[1] S. N. Misra, T. N. Misra, R. C. Mehrotra (J. Inorg. Nucl. Chem. **27** [1965] 105/13). — [2] B. S. Sankhla, R. N. Kapoor (Australian J. Chem. **20** [1967] 383/6). — [3] M. A. Saad, H. T. Soliman (J. Chem. U.A.R. **5** [1962] 29/39). — [4] F. R. Hartley, A. W. Wylie (J. Chem. Soc. **1962** 679/81). — [5] Z. I. Grigorovich (Zh. Neorgan. Khim. **8** [1963] 986/9; Russ. J. Inorg. Chem. **8** [1963] 507/9).

[6] S. N. Misra, T. N. Misra, R. N. Kapoor, R. C. Mehrotra (Chem. Ind. [London] **1963** 120). — [7] S. Sankhla, S. N. Misra, R. N. Kapoor (Chem. Ind. [London] **1965** 382/3). — [8] J. M. Batwara, U. D. Tripathi, R. K. Mehrotra, R. C. Mehrotra (Chem. Ind. [London] **1966** 1379). — [9] S. N. Misra, T. N. Misra, R. C. Misra (Australian J. Chem. **21** [1968] 797/800). — [10] U. D. Tripathi, J. M. Batwara, R. C. Mehrotra (J. Chem. Soc. A **1967** 991/2).

[11] D. C. Bradley, A. K. Chatterjee, W. Wardlaw (J. Chem. Soc. **1956** 2260/4). — [12] R. C. Mehrotra, J. M. Batwara (Inorg. Chem. **9** [1970] 2505/10). — [13] R. C. Mehrotra, T. N. Misra, S. N. Misra (J. Indian Chem. Soc. **42** [1965] 351/8).

2.7 With Isobutyl Alcohol $CH_3CH(CH_3)CH_2OH$ (= 2-Methyl-1-propanol = $i\text{-}C_4H_9OH$) and sec-Butyl Alcohol $CH_3CH_2CH(OH)CH_3$ (= 1-Methyl-1-propanol = $s\text{-}C_4H_9OH$)

Crystals of $YbCl_3 \cdot 3i\text{-}C_4H_9OH$ and $YbCl_3 \cdot 3s\text{-}C_4H_9OH$ were obtained by dissolving the chloride salt in the corresponding alcohol, heating at reflux for 1 h, and after standing overnight, drying in vacuo at 35°C and 0.5 Torr [1]. The azeotropic method described for the methanol adduct also yielded the $CeCl_3 \cdot 3i\text{-}C_4H_9OH$ [2]. The 2-methyl-1-propanolates, $M(i\text{-}OC_4H_9)_3$ of praseodymium [4] and samarium [3] and the 1-methyl-1-propanolates of lanthanum [5], praseodymium [4, 5], neodymium [4, 5], gadolinium [6], erbium [6], and ytterbium [6] have been obtained by alcohol interchange with the 2-propanolate derivative, in analogy with the methanolate. Similarly, $Ce(s\text{-}OC_4H_9)_4$ has been prepared in this manner starting with the cerium(IV) 2-propanolate. The compound is nonvolatile and has a molecular weight of 1810 g/mol (molecular complexity = 4.20) in benzene and 1470 g/mol (molecular complexity = 3.40) in toluene. The molecular weights were determined ebullioscopically [7]. The molecular weights of the Gd and Er 1-methyl-1-propanolates are 1243 and 1353 g/mol, respectively [8].

References to 2.7:

[1] U. D. Tripathi, J. M. Batwara, R. C. Mehrotra (J. Chem. Soc. A **1967** 991/2). — [2] M. A. Saad, H. T. Soliman (J. Chem. U.A.R. **5** [1962] 29/39). — [3] S. Sankhla, S. N. Misra, R. N. Kapoor (Chem. Ind. [London] **1965** 382/3). — [4] S. N. Misra, T. N. Misra, R. C. Mehrotra (Australian J. Chem. **21** [1968] 797/800). — [5] S. N. Misra, T. N. Misra, R. N. Kapoor, R. C. Mehrotra (Chem. Ind. [London] **1963** 120).

[6] J. M. Batwara, U. D. Tripathi, R. K. Mehrotra, R. C. Mehrotra (Chem. Ind. [London] **1966** 1379). — [7] D. C. Bradley, A. K. Chatterjee, W. Wardlaw (J. Chem. Soc. **1956** 3469/72). — [8] R. C. Mehrotra, J. M. Batwara (Inorg. Chem. **9** [1970] 2505/10).

2.8 With tert-Butyl Alcohol $(CH_3)_3CHOH$ (= 2-Methyl-2-propanol = $t\text{-}C_4H_9OH$)

An adduct of tert-butyl alcohol with cerium(III) chloride was prepared in the same manner as for the methanol adduct [1]. The tris adduct was isolated with the formulation, $CeCl_3 \cdot 3\ t\text{-}C_4H_9OH$ [1]. A report of a nonstoichiometric adduct, $YbCl_3 \cdot 1.5\ t\text{-}C_4H_9OH$ has also been made, prepared by dissolving the chloride salt in tert-butyl alcohol. After standing overnight, the resulting white solid was separated and dried in vacuo at 35°C and 0.5 Torr [2].

Salts of tert-butyl alcohol with lanthanum [3, 5], praseodymium [3, 6], neodymium [3], gadolinium [4], erbium [4], and ytterbium [4, 7] were prepared by the alcohol interchange reaction starting with the analogous 2-propanolate. The lanthanum compound is suggested to be polymeric although it sublimes at 280°C at 10^{-4} Torr [5]. Similarly, the cerium(IV) derivative, $Ce(t\text{-}OC_4H_9)_4$, was prepared by the alcohol interchange reaction, using distillation to remove the isopropanol/benzene azeotrope [8]. The cerium derivative sublimes at 140 to 150°C at 0.1 Torr and has a molecular weight of 1062 g/mol (molecular complexity = 2.5) in benzene and 968 g/mol (molecular complexity = 2.2) in toluene, determined ebullioscopically [8]. The derivatives of Gd and Er were prepared by an alcohol interchange reaction where a cyclohexane/isopropylacetate azeotrope is distilled from a mixture of the metal 2-propanolate and tert-butylacetate [9]. The molecular weights of the Gd and Er derivatives are 1130 and 1121 g/mol, respectively [9].

References to 2.8:

[1] S. N. Misra, T. N. Misra, R. C. Mehrotra (J. Inorg. Nucl. Chem. **27** [1965] 105/13). — [2] U. D. Tripathi, J. M. Batwara, R. C. Mehrotra (J. Chem. Soc. A **1967** 991/2). — [3] S. N. Misra, T. N. Misra, R. N. Kapoor, R. C. Mehrotra (Chem. Ind. [London] **1963** 120). — [4] J. M. Batwara, U. D. Tripathi, R. K. Mehrotra, R. C. Mehrotra (Chem. Ind. [London] **1966** 1379). — [5] D. C. Bradley, M. M. Faktor (Chem. Ind. [London] **1958** 1332).

[6] S. N. Misra, T. N. Misra, R. C. Mehrotra (Australian J. Chem. **21** [1968] 797/800). — [7] U. D. Tripathi, J. M. Batwara, R. C. Mehrotra (J. Chem. Soc. A **1967** 991/2). — [8] D. C. Bradley, A. K. Chatterjee, W. Wardlaw (J. Chem. Soc. **1957** 2600/4). — [9] R. C. Mehrotra, J. M. Batwara (Inorg. Chem. **9** [1970] 2505/10).

2.9 With 1-Pentanol $CH_3CH_2CH_2CH_2CH_2OH$ (= $C_5H_{11}OH$)

The compositions of the solid phases from the solution data are reported to be $CeCl_{3.10} \cdot 3.14\ C_5H_{11}OH$ [1], $2\,PrCl_3 \cdot 3\,C_5H_{11}OH$ [2], and $NdCl_3 \cdot C_5H_{11}OH$ [2]. The melting point of $2\,PrCl_3 \cdot 3\,C_5H_{11}OH$ occurs over the range 105 to 106°C while that of $NdCl_3 \cdot C_5H_{11}OH$ occurs over the range 123 to 125°C [2]. Thermal decomposition of the praseodymium adduct reveals a loss of pentanol molecules at 168 and 178°C while a similar curve for the neodymium adduct shows a loss of pentanol at 164, 190, and 261°C [2].

Tris and bis 1-pentanol adducts of scandium chloride were prepared by the desiccator method described for methanol [5]. Similarly, the tetrakis and bis adducts were also prepared by the second method described in that section [6].

1-Pentanolates of praseodymium and neodymium have been prepared by the alcohol interchange reaction starting with the 2-propanolato derivative in analogy with the methanolates [3]. Similarly, the tetrakis(1-pentanolato) cerium(IV) compound, $Ce(OC_5H_{11})_4$, was prepared by azeotropic distillation of the cerium(IV) 2-propanolate and a 1-pentanol/benzene mixture in analogy with the 1-propanolato derivative [4]. The cerium derivative is nonvolatile and has a molecular weight of 2050 g/mol (molecular complexity = 4.20) in benzene and 1660 g/mol (molecular complexity = 3.40) in toluene, determined ebullioscopically [4].

References to 2.9:

[1] F. R. Hartley, A. W. Wylie (J. Chem. Soc. **1962** 679/81). — [2] Z. I. Grigorovich (Zh. Neorgan. Khim. **8** [1963] 986/9; Russ. J. Inorg. Chem. **8** [1963] 507/9). — [3] S. N. Misra, T. N. Misra, R. C. Mehrotra (Australian J. Chem. **21** [1968] 797/800). — [4] D. C. Bradley, A. K. Chatterjee, W. Wardlaw (J. Chem. Soc. **1956** 2260/4).

2.10 With Other Monohydric Alcohols

Several alcohol adducts of scandium chloride were prepared by the desiccator method described for methanol, including the tris and bis adducts of 1-hexanol and benzyl alcohol, the tetrakis and bis adducts of allylalcohol, and the bis adduct of cyclohexanol [2]. Similarly, the second method discussed in that section was used to prepare the bis and tris adducts of 1-hexanol and the tetrakis and tris adducts of 1-heptanol and 1-octanol, as well as the tris adduct of 1-nonanol. The solubilities of scandium chloride in 1-hexanol, 1-heptanol, 1-octanol, and 1-nonanol (21.5, 19.3, 16.9, and 13.5 mg/100 mg solvent, respectively) at 25°C are close to a 1:6 molar ratio of $ScCl_3$ to alcohol in solution. X-ray powder diffraction data are reported for the bis 1-hexanol adduct [8].

The yttrium salts of tert-butyl alcohol, $Y(t\text{-}OC_4H_9)_3$ which sublimed at 242°C and 2 Torr; 3-methyl-2-butanol, $Y(OC_5H_{11})_3$ (m.p. ≈ 225°C, dec, molecular complexity = 1.5); 2-pentanol, $Y(OC_5H_{11})_3$ (m.p. ≈ 225°C, dec, molecular complexity = 1.7), 3-ethyl-3-pentanol, $Y(OC_7H_{15})_3$ (which distilled at 224°C at 760 mm with partial decomposition), 3-hexanol, $Y(OC_6H_{13})_3$ (m.p. ≈ 265°C, dec); and 2-ethyl-1-hexanol, $Y(OC_8H_{17})_3$ (m.p. ≈ 275°C, dec) were prepared by azeotropic distillation of the yttrium 2-propanolate with a mixture of the appropriate alcohol and benzene. The 2-pentanol tris Dy and Yb derivatives were also prepared [6].

The tris (alcoholato) derivatives of gadolinium, erbium, and ytterbium with iso-pentyl alcohol and tert-pentyl alcohol have been prepared using the alcohol interchange reaction starting with the analogous 2-propanolato derivatives [1]. Similarly, the praseodymium and neodymium alcoholato derivatives of 2-pentanol and tert-butyl alcohol have been prepared, and the praseodymium derivative of 2-methyl-2-pentanol has been prepared [2]. Several tetrakis (alcoholato)cerium(IV) compounds have been prepared by azeotropic distillation of the 2-propanolato compound of Ce^{IV} and a mixture of the appropriate alcohol and benzene in analogy with the preparation of the 1-propanolato derivative of Ce^{IV}. This method was used to prepare the derivative of neopentyl alcohol $Ce(OC_5H_{11})_4$ [3]; the 2- and 3-pentanol derivatives, $Ce(OC_5H_{11})_4$ [4]; the 3-methyl-2-butanol derivative, $Ce(OC_5H_{11})$ [4]; the tert-pentyl alcohol derivative, $Ce(OC_5H_{11})_4$ [5]; the 2-methyl-2-pentanol derivative, $Ce(OC_6H_{13})_4$ [5]; the 3-methyl-3-pentanol derivative, $Ce(OC_6H_{13})_4$ [5]; the 2,3-dimethyl-2-butanol derivative, $Ce(OC_6H_{13})_4$ [5]; the 3-ethyl-3-pentanol derivative, $Ce(OC_7H_{15})_4$ [5]; and the 3-methyl-3-hexanol derivative, $Ce(OC_7H_{15})_4$ [5]. The 3-pentanol and 3-methyl-2-pentanol derivatives were also prepared by the reaction of dipyridinium hexachlorocerate(IV) $[(C_5H_6N)_2CeCl_6]$ with the alcohol and ammonia, in analogy to the preparation of the tetrakis(2-propanolato) cerium(IV) compound [4]. Molecular weight and boiling point data for the cerium alcoholates are listed in Table 2/4.

The molecular weight of the tris(tert-pentyl alcoholato) derivative of Gd is reported to be 1252 g/mol [7].

Yttrium and lanthanide tris(hexafluoro-2-propanolato) compounds were prepared by the reactions of the anhydrous chlorides and hexafluoro-2-propanol in the presence of anhydrous ammonia in diethyl ether at 5°C, using 1:3 molar ratios. The crystalline products obtained were purified by recrystallization from hot alcohol or by vacuum distillation at 120 to 140°C and 0.08 Torr. In all cases the diammines were obtained, $M[OCH(CF_3)_2]_3 \cdot 2NH_3$, with M = Y, La, Pr, Sm, Eu, Dy, Ho, Er, Yb, Lu. All reactions were carried out under reduced pressure or under an atmosphere of dry helium to exclude moisture. The melting point ranges of the compounds are listed in Table 2/5. The infrared spectra of the compounds were reported, taken as either Nujol or Fluorolube mulls between CsI plates. Care was taken to exclude moisture during sample preparation (drybox techniques were used). Tentative assignments of the bands observed were made with a metal-N stretching mode assigned to bands at ≈ 520 cm^{-1}. Thermal decomposition curves are presented but no detailed discussion of the mode of decomposition is given. The yttrium derivative undergoes significant decomposition below 100°C, whereas the lanthanide derivatives do not [9].

The ^1H and ^{19}F NMR data for the diamagnetic derivatives, Y, La, and Lu, in $CFCl_3$ are presented. The CH proton resonance appears as a septet at −4.75, −4.60, and −4.90 ppm (δ), respectively, compared to −4.17 ppm for the neat ligand. The ^{19}F spectra consist of doublets at 77.0, 76.0, and 77.0 ppm, respectively, compared to 76.8 ppm for the neat ligand. The coupling constant is 6 Hz. The NH proton resonances were also observed and appear at −1.99, −2.30, and −2.10 ppm (δ), referenced to tetramethylsilane [9].

Table 2/4
Molecular Weight and Boiling Point Data for Miscellaneous Alcoholato Derivatives of Cerium(IV) $[Ce(OR)_4]$ *).

R	R	in benzene M(g/mol)	in benzene molecular complexity	in toluene M(g/mol)	in toluene molecular complexity	b. p./mm	Ref.
2,2-dimethyl-1-propyl- = neopentyl	$(CH_3)_3CCH_2$-	1236	2.53	1200	2.45	260/0.05	[3]
2-pentyl-	$CH_3(C_3H_7)CH$-	1520	3.11	1363	2.79	—	[4]
3-pentyl-	$(C_2H_5)_2CH$-	1420	2.90	1370	2.80	—	[4]
3-methyl-2-butyl-	$CH_3(i\text{-}C_3H_7)CH$-	1464	3.00	1377	2.82	—	[4]
2-methyl-2-butyl-	$(CH_3)_2(C_2H_5)C$-	1150	2.4	1050	22	240/0.1 **)	[5]
3-methyl-3-pentyl-	$(C_2H_5)_2(CH_3)C$-	746.2	1.4	580.4	1.1	140/0.06	[5]
2-methyl-2-pentyl-	$(CH_3)_2(C_3H_7)C$-	761.7	1.4	600	1.1	146/0.05	[5]
2,3-dimethyl-2-butyl-	$(CH_3)_2(i\text{-}C_3H_7)C$-	584	1.1	541.6	1.0	132/0.05	[5]
3-ethyl-3-pentyl-	$(C_2H_5)_3C$-	650.3	1.1	604.7	1.0	154/0.05	[5]
3-methyl-3-hexyl-	$CH_3(C_2H_5)(C_3H_7)C$-	604.5	1.0	—	—	150/0.05	[5]

*) Molecular weights obtained ebullioscopically, boiling points in °C; **) sublimed.

Table 2/5

Melting Points of Rare Earth Hexafluoro-2-propanolato Complexes, $M[OCH(CF_3)_2]_3 \cdot 2NH_3$.

M	m. p. in °C	M	m. p. in °C
Y	68 to 70	Dy	61 to 64
La	63 to 65	Ho	70 to 73
Pr	58 to 60	Er	63 to 64
Sm	68 to 70	Yb	72 to 74
Eu	65 to 67	Lu	78 to 80

References to 2.10:

[1] J. M. Batwara, U. D. Tripathi, R. K. Mehrotra, R. C. Mehrotra (Chem. Ind. [London] **1966** 1379). — [2] F. Petrů, B. Hajek, F. Jost (Croat. Chem. Acta **29** [1957] 457/60). — [3] D. C. Bradley, A. K. Chatterjee, W. Wardlaw (J. Chem. Soc. **1956** 2260/4). — [4] D. C. Bradley, A. K. Chatterjee, W. Wardlaw (J. Chem. Soc. **1956** 3469/72). — [5] D. C. Bradley, A. K. Chatterjee, W. Wardlaw (J. Chem. Soc. **1957** 2600/4).

[6] K. S. Mazdiyasni, C. T. Lynch, J. S. Smith (Inorg. Chem. **5** [1966] 342/6). — [7] R. C. Mehrotra, J. M. Batwara (Inorg. Chem. **9** [1970] 2505/10). — [8] E. M. Kirmse (Z. Chem. [Leipzig] **1** [1961] 332/4). — [9] K. S. Mazdiyasni, B. J. Schaper (J. Less-Common Metals **30** [1973] 105/12).

2.11 With 1,2-Ethanediol (= Ethylene Glycol = $C_2H_6O_2$ = H_2L) and Other Alkanediols

Solvates of ethanediol with lanthanum and cerium(III) acetate have been obtained by simply crystallizing the compounds from solutions of the acetate salts in ethanediol, resulting in $M(CH_3COO)_3 \cdot 2C_2H_6O_2$ [1]. The samarium alkanolates of several diols have been prepared by the alcohol interchange reaction starting with samarium 2-propanolate. The 2-propanolate was dissolved in solutions of the diols in benzene. After heating at reflux for 5 h, the isopropyl alcohol/benzene azeotrope was removed by distillation. The excess solvent was also distilled, and the wet solid products were dried at reduced pressure. The derivatives obtained depend on the mole ratio of 2-propanolate to diol used and are listed in Table 2/6 [2].

The apparent formation constants, for the equilibria $M^{3+} + H_2L \rightleftharpoons MHL^{2+} + H^+$ were determined for yttrium and the lanthanide compounds with 1,2-ethanediol and 1,2-propanediol. The values obtained are listed in Table 2/7 [3]. The cerium(III) solution turns yellow on exposure to air, presumably due to oxidation to cerium(IV) [1].

Table 2/6

Diol Derivatives of Samarium*).

diol	molar ratio	compound
1,2-Ethanediol	1:1	$Sm(OC_3H_7)(O_2C_2H_4)$
1,2-Ethanediol	1:2	$Sm(O_2C_2H_4)(O_2C_2H_5)$
1,2-Ethanediol	1:3	$Sm(O_2C_2H_4)(O_2C_2H_5)(O_2C_2H_6)$
1,3-Propanediol	1:1	$Sm(OC_3H_7)(O_2C_3H_6)$
1,3-Propanediol	1:2	$Sm(O_2C_3H_6)(O_2C_3H_7)$

Table 2/6 (Continued)

diol	molar ratio	compound
1,3-Propanediol	1:3	$Sm(O_2C_3H_6)(O_2C_3H_7)(O_2C_3H_8)$
1,3-Butanediol	1:1	$Sm(OC_3H_7)(O_2C_4H_8)$
1,3-Butanediol	1:2	$Sm(O_2C_4H_8)(O_2C_4H_9)$
2,3-Butanediol	1:1	$Sm(OC_3H_7)(O_2C_4H_8)$
2,3-Butanediol	1:2	$Sm(O_2C_4H_8)(O_2C_4H_9)$
2,3-Butanediol	1:3	$Sm(O_2C_4H_8)(O_2C_4H_9)(O_2C_4H_{10})$
2,2-Dimethyl-1,3-propanediol	1:1	$Sm(OC_3H_7)(O_2C_5H_{10})$
2,2-Dimethyl-1,3-propanediol	1:2	$Sm(O_2C_5H_{10})(O_2C_5H_{11})$
2,2-Dimethyl-1,3-propanediol	1:3	$Sm(O_2C_5H_{10})(O_2C_5H_{11})(O_2C_5H_{12})$
2,3-Dimethyl-2,3-butanediol (= Pinacol)	1:1	$Sm(OC_3H_7)(O_2C_6H_{12})$
2,3-Dimethyl-2,3-butanediol	1:2	$Sm(O_2C_6H_{12})(O_2C_6H_{13})$
2,3-Dimethyl-2,3-butanediol	1:3	$Sm(O_2C_6H_{12})(O_2C_6H_{13})(O_2C_6H_{14})$
1,5-Hexanediol-2,4-pentanediol	1:1	$Sm(OC_3H_7)(O_2C_6H_{12})$
1,5-Hexanediol-2,4-pentanediol	1:2	$Sm(O_2C_6H_{12})(O_2C_6H_{13})$
1,5-Hexanediol-2,4-pentanediol	1:3	$Sm(O_2C_6H_{12})(O_2C_6H_{13})(O_2C_6H_{14})$

*) Compounds formed using (a) 1:1 ratio—contain 2-propanolate (i-OC_3H_7) and the dinegative anion of the diol; (b) 1:2 ratio—contain the dinegative and uninegative anion of the diol; (c) 1:3 ratio—could contain either neutral, uninegative and dinegative anions of the diol as suggested or three uninegative anions of the diol [2].

Table 2/7

Apparent Formation Constants of Yttrium and Lanthanide Compounds with 1,2-Ethanediol and 1,2-Propanediol.

M	pK*) (1,2-ethanediol)	pK*) (1,2-propanediol)	M	pK*) (1,2-ethanediol)	pK*) (1,2-propanediol)
Y	6.95	6.95	Tb	7.20	7.15
La	8.50	8.65	Dy	7.15	7.10
Ce	8.00	8.05	Ho	7.05	6.95
Pr	7.90	7.90	Er	6.95	6.85
Nd	7.80	7.70	Tm	6.80	6.85
Sm	7.50	7.55	Yb	6.70	6.55
Eu	7.30	7.20	Lu	6.50	6.40
Gd	7.25	7.15			

*) $K=[MHL^{2+}][H^+]/[M^{3+}][H_2L]$; I = 0.1 M $NaClO_4$; T = 22 ± 1°C determined by potentiometric titration. Average error, ± 0.05 to 0.08.

References to 2.11:

[1] T. Muniyappan, B. Anjaneyalu (Current Sci. [India] **26** [1957] 319/20). — [2] B. S. Sankhla, R. N. Kapoor (Inorg. Chim. Acta **1** [1967] 182/4). — [3] G. S. Manku, R. C. Chadha (J. Inorg. Nucl. Chem. **34** [1972] 357/9).

2.12 With 1,2,3-Propanetriol (= Glycerol = $C_3H_8O_3$ = H_3L)

Each rare earth alcoholate ion was prepared by mixing a sixfold excess (by volume) of glycerol with a solution of the rare earth nitrate salt dissolved in 96% ethanol. An ammoniacal acetone solution (1 volume 25% NH_3/5 volumes acetone) was added to raise the pH to 8.8 to 9.6. The resulting precipitate was washed with acetone to remove nitrate ion and dried in a vacuum oven at 40°C for 8 h. Compounds with the formulation $M(C_3H_7O_3)_3 \cdot nH_2O$ resulted, where M = La, Pr, Nd, Sm, Eu, Gd, Dy, Ho, Er, Yb, Lu, Y with n = 1 and M = Lu with n = 0 [1]. Loss of the water of crystallization began at 60°C, and the anhydrous compound formed at ≈120°C. Partial decomposition began at 160°C. The infrared spectra were recorded as KBr pellets and showed a sharp band at ≈1390 cm^{-1} attributed to the bending mode of the primary OH groups. The most intense band occurred at 1010 cm^{-1} and was attributed to the ν(CO) stretching vibrations of the secondary alcohol groups. Complete assignments of the spectral bands are listed in Table 2/8. There is a general shift of ≈40 to 80 cm^{-1} towards lower energy in the spectra of the glycerates compared to the spectrum of glycerol [1]. Table 2/9, p. 34, lists the densities measured for these compounds determined pycnometrically in toluene at 25°C.

The apparent formation constants of rare earth compounds with glycerol listed in Table 2/10, p. 34, are for the displacement of one proton of glycerol by a metal ion, $M^{3+} + H_3L \rightarrow MH_2L^{2+} + H^+$ rather than the true association constant. The values are listed for this reaction rather than for the reaction, $M^{3+} + H_2L^- \rightarrow MH_2L^{2+}$, due to the difficulty in determining the true value of Ka for glycerol. In addition the association constant of the neutral ligand with praseodymium has been determined (K = 0.3) at 40°C in D_2O by NME spectroscopy [4].

Electronic absorption spectroscopy of the praseodymium, neodymium, and samarium chlorides with glycerol suggests that a 1:1 complex exists of the type $M(OH)(C_3H_6O_3)$ at pH ≈7 where glycerol exists as a dinegative ion. For erbium, $Er(C_3H_5O_3)$ is suggested to form, where the glycerol forms a trinegative ion [5].

Table 2/8
Infrared Bands*) for the Yttrium and Lanthanide Glycerolates $[M(C_3H_7O_3)_3 \cdot H_2O]$.

M	CH_2 rocking	C-C-C skeletal	δ(CH)	ν(CO) primary alcohol	ν(CO) secondary alcohol	C-C-C skeletal
glycerol	650 med.	820 med.	900 med.	1010 v. s.	1110 s.	1200 med.
Y	710 med.	860 med.	940 med.	1050 s.	1110 v. s.	1260 med.
La	710 med.	860 med.	940 med.	1050 s.	1110 v. s.	1265 med.
Pr	710 med.	855 med.	937 med.	1060 s.	1105 v. s.	1260 med.
Nd	710 med.	860 med.	935 med.	1055 s.	1110 v. s.	1263 med.
Sm	710 med.	860 med.	940 med.	1050 s.	1110 v. s.	1255 med.
Eu	710 med.	862 med.	940 med.	1055 s.	1105 v. s.	1260 med.
Gd	710 med.	860 med.	940 med.	1050 s.	1110 v. s.	1260 med.
Dy	710 med.	862 med.	940 med.	1050 s.	1110 v. s.	1265 med.
Ho	710 med.	865 med.	937 med.	1055 s.	1110 v. s.	1263 med.
Er	710 med.	860 med.	940 med.	1050 s.	1110 v. s.	1260 med.
Yb	710 med.	860 med.	940 med.	1050 s.	1110 v. s.	1260 med.
Lu	710 med.	860 med.	940 med.	1050 s.	1110 v. s.	1260 med.

Table 2/8 (Continued)

M	δ(OH) secondary alcohol	δ(OH) primary alcohol	δ(HOH)	ν(CH)	ν(CH_2) sym.	ν(OH) alcohol
glycerol	1310 med.	1390 med.	1640 med.	2810 med.	2860 med.	3340 v. s. br.
Y	1370 sh.	1390 s.	1660 med.	2880 med.	2940 med.	3360 v. s. br.
La	1370 sh	1390 s.	1660 med.	2880 med.	2940 med.	3360 v. s. br.
Pr	1370 sh.	1390 s.	1660 med.	2885 med.	2945 med.	3360 v. s. br.
Nd	1370 sh.	1390 s.	1660 med.	2890 med.	2950 med.	3360 v. s. br.
Sm	1370 sh.	1390 s.	1660 med.	2880 med.	2940 med.	3355 v. s. br.
Eu	1370 sh.	1390 s.	1660 med.	2880 med.	2940 med.	3360 v. s. br.
Gd	1370 sh.	1390 s.	1660 med.	2890 med.	2950 med.	3360 v. s. br.
Dy	1370 sh	1390 s.	1660 med.	2885 med.	2945 med.	3360 v. s. br.
Ho	1370 sh.	1390 s.	1660 med.	2885 med.	2945 med.	3360 v. s. br.
Er	1370 sh.	1390 s.	1660 med.	2890 med.	2950 med.	3360 v. s. br.
Yb	1370 sh.	1390 s.	1660 med.	2890 med.	2950 med.	3360 v. s. br.
Lu	1370 sh.	1390 s.	1660 med.	2890 med.	2945 med.	3360 v. s. br.

*) In cm^{-1} obtained as KBr pellets.

Table 2/9

Densities of Yttrium and Lanthanide Glycerolates, $M(C_3H_7O_3)_3 \cdot H_2O$.

M	d in g/cm^3	M	d in g/cm^3
Y	2.22	Dy	2.28
La	2.24	Ho	2.28
Pr	2.26	Er	2.27
Nd	2.24	Yb	2.29
Sm	2.25	Lu	2.28
Eu	2.25	Lu	2.29*)
Gd	2.26		

*) Density of anhydrous compound.

Table 2/10

Apparent Formation Constants of Rare Earth Glycerolate Ions.

M	pK* (a)	pK* (b)	M	pK* (a)	pK* (b)
Sc(c)	1.58	—	Gd	7.10	7.17
Y	6.85	6.79	Tb	6.95	6.93
La	8.35	8.28	Dy	6.90	6.95
Ce	7.95	—	Ho	6.80	6.85
Pr	7.75	7.67	Er	6.75	6.73
Nd	7.60	7.62	Tm	6.75	6.50
Sm	7.25	7.25	Yb	6.45	6.42
Eu	7.15	7.15	Lu	6.30	6.50

$K^* = [MH_2L]^{2+}[H^+]/[M^{3+}][H_3L]$.
(a) I = 0.1 M ($NaClO_4$), T = 22 ± 1°C [2].—(b) I = 0.1 M (NaCl), T = 25°C [3].—(c) I = 0.1 M ($NaClO_4$), T = 25°C [6].

References to 2.12:

[1] D. V. Pakhomova, V. N. Kumok, V. V. Serebrennikov (Zh. Neorgan. Khim. **16** [1971] 2991/5; Russ. J. Inorg. Chem. **16** [1971] 1588/90). — [2] G. S. Manku, R. C. Chadha (J. Inorg. Nucl. Chem. **34** [1972] 357/9). — [3] D. V. Pakhomova, V. N. Kumok, V. V. Serebrennikov (Zh. Neorgan. Khim. **15** [1970] 1211/3; Russ. J. Inorg. Chem. **15** [1970] 622/3). — [4] A. P. G. Keiboom, A. Sinnema, J. M. Van der Toorn, H. Van Bekkum (Rec. Trav. Chim. **94** [1975] 53/9). — [5] N. S. Poluektov, N. P. Efryushina (Ukr. Khim. Zh. **35** [1969] 31/7; Soviet Progr. Chem. **35** [1969] 30/5).

[6] L. N. Usherenko, N. V. Kulikova, N. A. Skorik, V. N. Kumok (Zh. Neorgan. Khim. **16** [1971] 3230/2; Russ. J. Inorg. Chem. **16** [1971] 1711/2).

2.13 With Polyols and Sugars

Although no complexes of rare earth ions with sugars and other simple carbohydrates have been isolated, considerable evidence for interaction in solution has been obtained. Primarily nuclear magnetic resonance, spectrophotometric, potentiometric and polarographic data on complex formation are available. Spectrophotometric and polarimetric data are both consistent with complex formation of several of the lanthanides (Pr^{3+}, Nd^{3+}, Ho^{3+}, Er^{3+}) with arabinose, galactose, laevulose [1, 3], glucose, xylose [8], lactose, maltose, saccharose [2, 3], mannitol, inositol [16], and glucitol [6, 16, 17], and various other penta- and hexahydroxy compounds [6, 7, 8]. Job plots of the spectral data yield apparent stability constants for these ligands based on the displacement of 2 to 6 protons by 2 or 3 lanthanide ions [1 to 3, 16, 17]. The exact nature of the complex ions existing in solutions is unclear, because of hydrolysis of the uncomplexed or complexed species.

Nuclear magnetic resonance data are available for a variety of polyalcohols and sugars bound to lanthanum and several of the paramagnetic lanthanide ions. In most instances the data have been obtained in order to simplify the NMR spectrum of the particular ligand and in some cases, by use of the usual lanthanide shift reagent techniques, to estimate the molecular structures of the sugars in solution. The polyols studied by this technique generally have three OH groups capable of binding to the metal ion and are as follows: D-glycero-D-gulo-hepitol [4], L-iditol [4, 6], D-glucitol [4, 6], xylitol [4, 6], galacitol [4, 6], D-mannitol [4, 6], *d*-altritol [4], arabinitol [4, 6], l-threitol [4, 6], allitol [4], ribitol [4, 6], erythritol [4, 6], cis-inositol [5, 10], epi-inositol [10, 12], *d*,l-2-deoxy-2-methyl-epi-inositol [5], C-hydroxymethyl-scyllo-inositol [5], 1,4-anhydro-epi-inositol [5], bornane-2 endo, 3 endo-diol [5], bornane-2 exo, 3 exo-diol [5], D-allose [9], D-gulose [9], D-glycero-D-gulo-heptose [9], D-ribose [9], 1,6-anhydro-β-D-mannopyranoside [10], 1,6-anhydro-β-D-talopyranoside [10], methyl β-D-hamamelopyranoside (= methyl 2-C-hydroxymethyl-β-D-ribopyranoside) [11], 1,6-anhydro-β-D-allopyranoside [12], and 1-O-methyl-D-gulopyranose [13]. Structural evidence obtained by the shift reagent technique using europium and praseodymium are somewhat uncertain due to the possibility of both 1:1 and 1:2 complexes being present and the possibility that both contact and pseudocontact shift mechanisms are operating. The strongest complexes appear to form when there are three hydroxyl groups present in an ax-eq-ax arrangement on three successive carbon atoms of the ring. Structures which have this arrangement appear to exist in higher concentration compared to their tautomers when the metal ion is present [9 to 12]. A more comprehensive analysis of the effect of the praseodymium, neodymium, and europium chloride salts on xylitol is available. Stability constants (K_1 and K_2), and the fully bound isotropic shifts for the 1:1 and 1:2 complexes are shown in [7]. A detailed discussion of the isotropic shifts produced in 1,6-anhydromannose and epi-inositol by several lanthanide ions is also available. By using Ce^{3+}, Pr^{3+}, Nd^{3+}, Sm^{3+}, Eu^{3+}, Tb^{3+}, Dy^{3+}, Ho^{3+}, Er^{3+}, Tm^{3+}, and Yb^{3+}, the isotropic shifts observed have been separated into pseudocontact and contact shifts and the geometry of the complexes in solution have been determined. A comparison between the theoretical and experimental values was made and good agreement reported [14]. The effect and possible use of gadolinium nitrate as a relaxation reagent on the hydrogen atoms of D-allose, D-glucopyranose, maltose, cellobiose, and maltotriose has also been discussed [15]. The relaxation rates of several hydrogen atoms of these sugars in the presence of gadolinium nitrate were measured. Determinations of the stability constants of the Ce(IV) complexes with glucose and xylose were reported at low pH. The hemiacetal oxygen atom was implicated in the coordination [8].

References to 2.13:

[1] N. P. Efryushina, R. S. Laver, N. S. Poluektov (Zh. Neorgan. Khim. **12** [1967] 933/8; Russ. J. Inorg. Chem. **12** [1967] 491/4). — [2] N. P. Efryushina, N. S. Poluektov (Zh. Neorgan. Khim. **12** [1967] 1855/61; Russ. J. Inorg. Chem. **12** [1967] 976/80). — [3] N. S. Poluektov, N. P. Efryushina (Zh. Neorgan. Khim. **15** [1970] 1203/7; Russ. J. Inorg. Chem. **15** [1970] 618/20). — [4] S. J. Angyal, D. Greeves, J. A. Mills (Australian J. Chem. **27** [1974] 1447/56). — [5] S. J. Angyal, D. Greeves, V. A. Pickles (Carbohyd. Res. **35** [1974] 165/73).

[6] A. P. G. Kieboom, T. Spoormaker, A. Sinnema, J. M. Van der Toorn, H. van Bekkum (Rec. Trav. Chim. **94** [1975] 53/9). — [7] G. A. Elgavish, J. Reuben (J. Am. Chem. Soc. **99** [1977] 1765/8). — [8] V. I. Krupenskii, I. I. Korol'lov, T. V. Dolgaya (Izv. Vysshikh Uchebn. Zavedenii Les. Zh. **20** [1977] 102/5 from C.A. **87** [1977] No. 136191). — [9] S. J. Angyal (Australian J. Chem. **25** [1972] 1957/66). — [10] S. J. Angyal (Tetrahedron **30** [1974] 1695/702).

[11] S. J. Angyal (Carbohyd. Res. **26** [1973] 271/3). — [12] S. J. Angyal, D. Greeves, V. A. Pickles (J. Chem. Soc. Chem. Commun. **1974** 589/90). — [13] H. Grasdalen, T. Anthonsen, B. Larsen, O. Smidsrod (Acta. Chem. Scand. B **29** [1975] 17/21). — [14] H. A. Bergen, R. M. Golding (Australian J. Chem. **30** [1977] 2361/9). — [15] L. D. Hall, C. M. Preston (Carbohyd. Res. **41** [1975] 53/61).

[16] N. P. Efryushina, R. S. Lauer, N. S. Poluektov (Ukr. Khim. Zh. **32** [1966] 1038/43; Soviet Progr. Chem. **32** [1966] 781/5). — [17] N. S. Poluektov, N. P. Efryushina (Ukr. Khim. Zh. **36** [1970] 164/9; Soviet Progr. Chem. **36** [1970] 51/7).

3 Salts and Complexes with Aromatic Hydroxy Compounds

Edward R. Birnbaum
Department of Chemistry, New Mexico State University
Las Cruces, New Mexico, USA

3.1 With Phenol C_6H_6O

The stability constants of the scandium, yttrium, and lanthanum complexes with the phenol anion have been determined at 25°C, I = 0.10 M ($NaClO_4$) by a spectral method. For the scandium complex lg K_1 = 7.17 ± 0.07 is reported [1]. Values for the yttrium and lanthanum complexes are lg K_1 = 2.40 and 1.51, respectively [2]. The pKa reported for phenol was 9.78 under these conditions. A considerably different value of lg K_1 for the La^{3+} complex under the same conditions has been reported as 4.17 ± 0.03. The use of this value yields a straight line plot with the equation lg β = 0.48 lg K_a −0.85, where K_a is the acid dissociation constant, for a variety of substituted aromatic alcohols [3]. A similar plot exists for Sc^{3+}, lg β = 0.777 lg K_a −0.42.

Yttrium phenolate was prepared by the reaction of phenol (55 ml) with yttrium metal shavings (2.7 g) in the presence of 7 mg of mercury(II) chloride. The mixture was heated at reflux at 182°C for 24 h, cooled, and filtered. The excess alcohol was distilled off until the product was nearly dry. After recrystallization from hot cyclohexane, the product, $Y(OC_6H_5)_3 \cdot 4\,C_6H_{12}$, decomposed at ≈310°C [4]. The phenolates of the lanthanides were prepared by the alcohol interchange reaction, adding phenol to the appropriate 2-propanolate. The phenolates are insoluble in the benzene/phenol mixture used, allowing the isolation of mixed derivatives corresponding to the molar ratio of 2-propanolate to phenol used. Thus, the compounds $M(i\text{-}OC_3H_7)_n(OC_6H_5)_{3-n}$ were isolated for M = Gd, Er with n = 2, 1 and 0 when a 1:1, 1:2, and 1:3 molar ratio, respectively, was used [5]. Similar methods yielded the Pr and Nd derivatives with n = 2 to 0 [6]. A mixed methanolate-phenolate derivative has also been reported, $Pr(OCH_3)(OC_6H_5)_2$, presumably starting with the tris methanolate rather than the 2-propanolate [6]. All of the compounds reported tend to be brown or black solids. The infrared spectra of several of the compounds are reported as Nujol mulls. The phenoxy group is identified by characteristic ring deformation bands at 1450, 1270, 1162, and 995 cm^{-1} [5]. The phenolate of cerium(IV), $Ce(OC_6H_5)Cl_3$, was prepared by adding phenol (0.94 g) to hexachloroceric acid H_2CeCl_6 in ethanol and heating at reflux until the evolution of HCl stopped. The solution was filtered, and excess ethanol was removed by distillation under reduced pressure (20 to 25 Torr). The brown paste which resulted yielded a brown crystalline compound after repeated crystallizations from petroleum ether (60 to 80°C) [7]. The infrared spectrum of this compound showed the characteristic aromatic ν(C-H) stretching vibration at ≈3050 cm^{-1}, the phenolic ν(C-O) stretching vibration at ≈1400 cm^{-1}, the ν(C-H) bending mode at ≈810 and 759 cm^{-1}, and several C-C framework vibrations. The shift in the phenolic γ(C-O) stretching vibration to higher wave number is typical of coordination of the phenolate oxygen atom to the metal ion. The UV absorption spectrum of this derivative has two peaks at 212 and 270 nm with lg extinction coefficients of 4.21 and 3.6, respectively. Ebullioscopic molecular weight measurements show the derivative to be monomeric in boiling benzene. The molar conductance in methanol is typical of a nonelectrolyte (2 to 4 $\Omega^{-1} \cdot cm^2 \cdot mol^{-1}$) [7].

References to 3.1:

[1] L. N. Usherenko, N. V. Kulikova, N. A. Skorik, V. N. Kumok (Zh. Neorgan. Khim. **16** [1971] 3230/2; Russ. J. Inorg. Chem. **16** [1971] 1711/2). — [2] C. Rostmus, Jr., L. B. Magnusson, C. A. Craig (Inorg. Chem. **5** [1966] 1154/7). — [3] V. N. Kumok (Zh. Neorgan. Khim. **23** [1978] 1792/7; Russ. J. Inorg. Chem. **23** [1978] 985/8). — [4] K. S. Mazdiyasni, C. T. Lynch, J. S. Smith (Inorg. Chem. **5** [1966] 342/6). — [5] R. C. Mehrotra, J. M. Batwara (Inorg. Chem. **9** [1970] 2505/10).

[6] S. N. Misra, T. N. Misra, R. C. Mehrotra (Australian J. Chem. **21** [1968] 797/800). — [7] S. K. Gupta, P. S. Bassi (Current Sci. [India] **46** [1977] 139/40).

3.2 With Nitrophenols (= $C_6H_5NO_3$)

Stability constants of the 2- and 4-nitrophenol complexes with several of the rare earths have been determined by spectral methods. The lg K_1 for La^{3+} in aqueous solution at 25°C at I = 0.10 M, adjusted with $NaClO_4$, was reported as 2.20 for 2-nitrophenol [1]. The value of the pKa of 2-nitrophenol used was 7.06. Using similar conditions, values of lg K_1 for Sc^{3+} and Ce^{3+} were reported as 5.0 ± 0.1 and 2.4 ± 0.1, respectively. The same method and conditions yielded lg K_1 values for the 4-nitrophenol complexes with Sc^{3+} and Ce^{3+} of 5.2 ± 0.1 and 1.1, respectively, using a value of 7.02 for the pKa of 4-nitrophenol [2]. The effect of lanthanum and cerium(III) nitrate on the absorption spectrum of 2-nitrophenol in aqueous ethanolic solutions of different compositions has also been reported [3]. Job's method suggests an 1:3 adduct of La^{3+} with 2-nitrophenol and an 1:1 adduct of Ce^{3+} in 95% ethanol [4].

The instability constants of 2- and 4-nitrophenol with neodymium ethylenediaminetetraacetate have also been reported. The isotropic shifts (both contact and pseudocontact) found were used to determine values of 50 ± 10 mM and 160 ± 40 mM for the instability constants, respectively. The 2-nitrophenol complex appears to be devoid of symmetry, suggesting that the chelating potential of the ortho groups plays an important part in the formation of the complex species [5].

Nitrophenol derivatives of samarium were prepared by the alcohol interchange reaction by adding the particular nitrophenol to samarium 2-propanolate dissolved in benzene in the appropriate molar ratio. After heating at reflux for 2 to 3 h, the binary azeotrope (2-propanol/benzene) was collected, and the compounds were dried under reduced pressure. The isolated adducts were $Sm(i\text{-}OC_3H_7)_nL_{3-n}$ with n = 0 to 2 and L = 2-, 3-, and 4-nitrophenolate anions. The 2-nitrophenol derivatives were red and sparingly soluble in benzene, the 3-nitrophenol derivatives were brown and insoluble in benzene, and the 4-nitrophenol derivatives were yellow and sparingly soluble (1:1 adduct) or insoluble (1:2 and 1:3 adducts) in benzene [6].

References to 3.2:

[1] C. Rostmus, L. B. Magnusson, C. A. Craig (Inorg. Chem. **5** [1966] 1154/7). — [2] V. N. Kumok (Zh. Neorgan. Khim. **23** [1978] 1792/7; Russ. J. Inorg. Chem. **23** [1978] 985/8). — [3] Ch. Bobtelsky, S. Kertes (Bull. Soc. Chim. France **1954** 143/7). — [4] Ch. Bobtelsky, S. Kertes (Bull. Soc. Chim. France **1954** 148/51). — [5] J. Reuben (J. Am. Chem. Soc. **98** [1976] 3726/8).

[6] M. Hasan, S. N. Misra, R. N. Kapoor (U.A.R. J. Chem. **13** [1970] 235/42).

3.3 With 2-Bromophenol (= C_6H_5BrO)

Several 2-bromophenol derivatives of samarium were prepared by the alcohol interchange reaction as described for nitrophenol using samarium 2-propanolate as the starting material. Three adducts were isolated, $Sm(i\text{-}OC_3H_7)_n(C_6H_4BrO)_{3-n}$ with n = 0 to 2. The 1:1 compound is orange and soluble in benzene, the 1:2 compound is red and also soluble, and the 1:3 compound is red but only sparingly soluble in benzene. Orange and red tert-butylalcoholate-bromophenolate compounds were also isolated, having a 2:1 and a 1:2 ligand ratio of tert-butanolate to bromophenolate, respectively. Both compounds were soluble in benzene and were prepared by the addition of tert-butyl alcohol to the reaction mixture described earlier, M. Hasan, S. N. Misra, R. N. Kapoor (U.A.R. J. Chem. **13** [1970] 235/42).

3.4 With 2,4- and 2,6-Dinitrophenols (= $C_6H_4N_2O_5$)

The stability constants (lg K_1) of 2,4-dinitrophenol with scandium [1], lanthanum [2], and cerium [1] have been determined by a spectral method in aqueous solution at 25°C and I = 0.1 M ($NaClO_4$), as 2.4 ± 0.3, 1.01, and 1.0 ± 0.3, respectively. The stability constants (lg K_1) of 2,6-dinitrophenol complexes with scandium and cerium(III)ions under the same conditions are 2.6 ± 0.1 and 1.6 ± 0.2, respectively [3]. The effects of lanthanum and cerium(III) nitrates on the spectrum of 2,4-dinitrophenol in aqueous ethanolic solutions of different compositions have also been studied. An 1:1 complex was reported with a K_1 of 9.0×10^{-2} using Job's method [3].

Compounds of 2,4-dinitrophenol with the rare earth ions were prepared by treating each rare earth oxide with the ligand to yield dark yellow, prismatic crystals with the formulation $M(C_6H_3N_2O_5)_3 \cdot H_2O$ where M = La, Pr, Nd, Sm, Eu, Gd, Tb, Dy, Ho, and Y [4]. The solubilities of the compounds prepared increased with increasing temperature [4]. The solubility product of the lanthanum derivative in $HClO_4$ at 25°C and $\mu = 3$ M ($NaClO_4$) is $(3.55 \pm 0.51) \times 10^{-22}$. The acid dissociation constant of the compound is $(4.23 \pm 0.72) \times 10^{-10}$ [5].

References to 3.4:

[1] V. N. Kumok (Zh. Neorgan. Khim. **23** [1978] 1792/7; Russ. J. Inorg. Chem. **23** [1978] 985/8). — [2] C. Rostmus, L. B. Magnusson, C. A. Craig (Inorg. Chem. **5** [1966] 1154/7). — [3] M. Bobtelsky, S. Kertes (Bull. Soc. Chim. France **1955** 328/34). — [4] N. Duc Van (Tap San Hoa Hoc **13** No. 4 [1975] 22/7 from C.A. **88** [1978] No. 57696). — [5] N. Duc Van (Tap Chi Hoa Hoc **15** No. 2 [1977] 11/5 from C.A. **88** [1978] No. 111313).

3.5 With 2,4,6-Trinitrophenol (= Picric Acid = $C_6H_3N_3O_7$)

The effects of lanthanum and cerium(III) nitrates on the spectrum of picric acid in aqueous ethanolic solutions of different compositions have been studied. Tris complexes were reported with $K_1 = 1.2 \times 10^{-1}$ and 8.8×10^{-2}, respectively, and $K_3 = 6.0 \times 10^{-4}$ and 8.2×10^{-4}, respectively, using Job's method [1]. The luminescence of the picrate salt of europium has been studied. The luminescence is weakly enhanced over the chloride salt at room temperature and increases at liquid air temperatures. No enhanced luminescence is observed at room temperature in an alcohol/ether/isopentane solution, but roughly the same intensity is observed at liquid air temperatures as for the solid compound [2].

References to 3.5:

[1] M. Bobtelski, S. Kertes (Bull. Chim. Soc. France **1955** 328/34). — [2] S. I. Weissman (J. Chem. Phys. **10** [1942] 214/7).

3.6 With Dihydroxybenzenes

3.6.1 With 1,2-Dihydroxybenzene (= Pyrocatechol = $C_6H_6O_2$)

Stability constants for the pyrocatecholate complexes of several of the rare earth ions have been determined by potentiometric pH titration at 25°C and I = 0.1 M (NaCl). The lg β_1 values generally increase with increasing atomic number for the complex $[M(C_6H_4O_2)]^+$, where pyrocatechol is a dinegative ion. The pH range chosen for the studies (≈5 to 7) suppresses metal ion hydrolysis and formation of $[M(C_6H_5O_2)]^{2+}$ [1].

M	Sc	La	Pr	Nd	Sm	Eu	Gd
lg β_1	17.04	9.46	10.31	10.50	11.06	11.17	11.20

M	Dy	Ho	Er	Tm	Yb	Lu
lg β_1	11.34	11.42	11.43	11.56	11.67	11.31

In the case of the scandium complex, allowance was made for hydrolysis in the calculation [2]. A second set of stability constants for the lighter lanthanide ions, scandium, and yttrium at 30°C, I = 0.2 M ($NaClO_4$) has also been reported which differs significantly from the values listed above: Sc (14.20, 14.16), La (7.81, 7.88), Pr (8.60, 8.66), Nd (8.90, 8.85), Sm (9.29, 9.27) [3]. Values for protonation of the dianion used were: lg $\beta_1 = 12.6$ and lg $\beta_2 = 21.94$ [2]; lg $\beta_1 = 11.64$ (11.67) and lg $\beta_2 = 20.71$ (20.79) [3]. Stepwise formation constants and thermodynamic parameters for lanthanide complexes with pyrocatechol are also given in [10]. The stability of the metal complexes follows the expected trend: Ho > Dy > Tb > Gd > Sm > Nd > Pr > Ce.

The addition of ethylenediamine (= en) to aqueous solutions of rare earth ions and pyrocatechol precipitated pyrocatecholate derivatives, $(enH_2)_3[M(C_6H_4O_2)_3]_2$, with M = La, Y, Gd, and Er. These derivatives are less soluble than the corresponding oxalate derivatives [4]. Similar compounds were formed when pyridine, piperidine, or quinoline was used instead of ethylenediamine. However, when ammonia was used, the precipitate was produced only by boiling.

Pyrocatechol derivatives were also investigated using the isomolar method by adding ammonia to a solution of catechol and each rare earth ion to raise the pH to 8.5 [5]. The resulting increase in optical density suggested the presence of a complex having the ratio 1:1.5 (rare earth ion : catecholate) for Y^{3+}, La^{3+}, Pr^{3+}, Nd^{3+}, and Er^{3+}. Potentiometric titrations of these solutions were consistent with the same composition. The sparingly soluble salts were also isolated by centrifugation, and were washed with 50% methanol and then with pure methanol. The freshly separated pale gray precipitates became dark green when dried in air. All the solids had the same composition, $M_2(C_6H_4O_2)_3$ [5].

A cerium(IV) pyrocatechol derivative, $Na_4[Ce(O_2C_6H_4)_4] \cdot 21\,H_2O$, was prepared by techniques typically used to prepare uranium(IV) and thorium(IV) compounds. Sodium hydroxide pellets (0.12 mol) were dissolved in oxygen-free water, and then catechol (0.06 mol) was added. To the alkaline solution, a filtered, oxygen-free solution of cerium(IV) ammonium nitrate was added. Precipitation began immediately [6]. Crystals for X-ray diffraction were grown by cooling the above filtered solution slowly, and a crystal structure determination was done. The deep red cerium(IV) compound has $\bar{4}$ site symmetry in the space group $I\bar{4}$-S_4^2 (No. 82), with 2 molecules to the unit cell. The cell dimensions are a = 14.649(2), c = 9.976(1) Å. The weighted R factor for 3106 independent reflections was 0.053. The cell volume was 2140.8 Å^3, with a density determined by flotation of 1.64(1) g/cm^3 compared to a calculated density of 1.65 g/cm^3. The hafnium(IV) derivative is isomorphous with the cerium(IV) derivative. The coordination polyhedron is an eight-coordinate dodecahedron for the $[Ce(O_2C_6H_4)_4]^{4-}$ anion. The 21 water molecules form a hydrogen-bonded network thoughout the crystal, with each sodium ion coordinated to two pyrocatecholate oxygen atoms and four water oxygen atoms. The Ce-O distances are 2.357(4) and 2.362(4) Å with an internal O-Ce-O angle of 68.3(1)°. The complex anion viewed down the twofold axis is shown in **Fig. 2-2**. For figure of the complex anion viewed along the molecular mirror plane with the A and B sites of the dodecahedron and for shape parameters indicating how far the structure deviates from a perfect dodecahedron and square antiprism, see original paper [7]. The compound is diamagnetic, indicating that the complex is not a semiquinone. The formal potential for the one-electron reduction of the compound was −692 mV vs. SCE, as determined by cyclic voltammetry [7].

Fig. 2-2

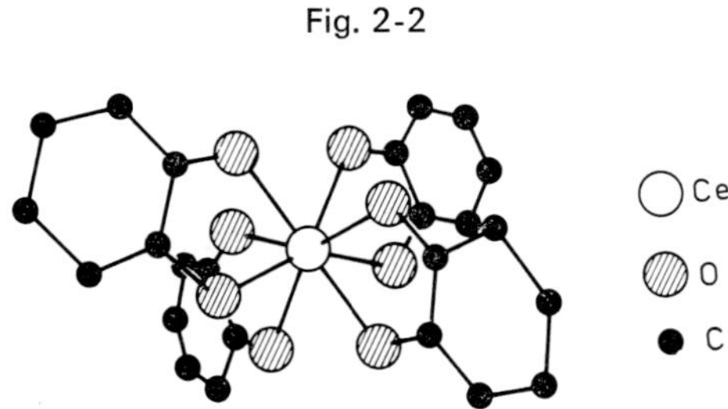

The $[Ce(O_2C_6H_4)_4]^{4-}$ anion viewed down the twofold axis ($\bar{4}$ axis is vertical).

A scandium compound, $[Sc(OH)(C_6H_4O_2)] \cdot 2H_2O$, was reported to form during the titration of a solution of a fourfold excess of pyrocatechol and scandium(III) with NaOH, CsOH, LiOH, or NH_4OH. Potassium hydroxide was reported to give $K_{0.7}H_{0.3}[Sc_3(OH)(C_6H_4O_2)] \cdot 2H_2O$ instead. Compound formation was determined potentiometrically to begin at pH 6.7 to 7.2. The compounds were chemically and thermally unstable [8].

Stability constants of mixed complexes derived from pyrocatechol and N'-(2-hydroxyethyl)-ethylenediamine-N,N,N'-triacetic acid, ethylenediaminetetraacetic acid, or 1,2-cyclohexanediamine-tetraacetic acid are reported in "Rare Earth Elements" D 1, 1980, p. 235, 239, and 241. References for complexes containing pyrocatechol and nitrilotriacetic acid or ethylenediaminetetraacetic acid see [3, 9].

References to 3.6.1:

[1] D. V. Pakhomova, V. N. Kumok, V. V. Serebrennikov (Zh. Neorgan. Khim. **14** [1969] 1434/5; Russ. J. Inorg. Chem. **14** [1969] 752/3). — [2] L. N. Usherenko, N. V. Kulikova, N. A. Skorik, V. N. Kumok (Zh. Neorgan. Khim. **16** [1971] 3230/2; Russ. J. Inorg. Chem. **16** [1971] 1711/2). — [3] S. D. Makhijani, S. P. Sangal (J. Indian Chem. Soc. **55** [1978] 987/8). — [4] G. Beck (Mikrochim. Acta **1954** 337/9). — [5] N. S. Poluektov, K. V. Tserkasevich (Zh. Neorgan. Khim. **12** [1967] 1862/6; Russ. J. Inorg. Chem. **12** [1962] 980/2).

[6] S. R. Sofen, K. Abu-Dari, D. P. Freyberg, K. N. Raymond (J. Am. Chem. Soc. **100** [1978] 7882/7). — [7] S. R. Sofen, S. R. Cooper, K. N. Raymond (Inorg. Chem. **18** [1979] 1611/6). — [8] P. P. Mel'nikov, L. N. Komissarova, E. A. Mel'nik (Izv. Vysshikh Uchebn. Zavedenii Khim. Khim. Tekhnol. **16** [1973] 1002/5 from C.A. **79** [1973] No. 111268). — [9] T. H. Mhaske, K. N. Munshi (J. Indian Chem. Soc. **53** [1976] 11/4). — [10] S. Nagpal, S. N. Dubey, H. L. Kalra, D. M. Puri (Indian J. Chem. A **18** [1979] 270/1).

3.6.2 With 4,5-Dihydroxy-1,3-benzenedisulfonic Acid

(= Pyrocatechol-3,5-disulfonic Acid = $C_6H_6O_8S_2$)
(Disodium salt = Tiron)

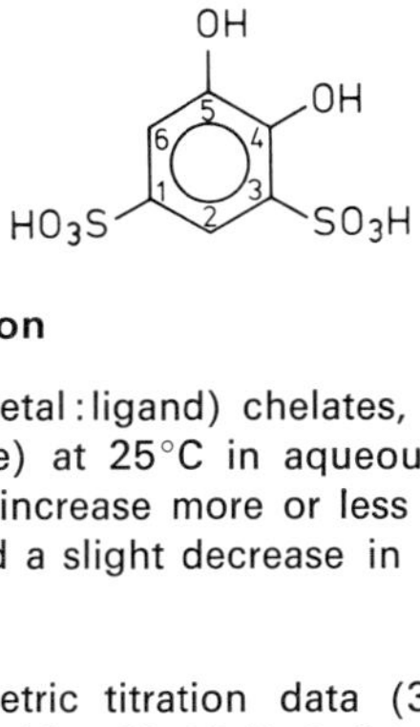

OH
OH
6 5 4
1 2 3
HO_3S SO_3H

3.6.2.1 $M(C_6H_2O_8S_2)_n^{3-4n}$ Ions (n = 1, 2). Formation in Solution

The stability and protonation constants for the 1:1 and 1:2 (metal:ligand) chelates, recorded in Table 3/1, were determined potentiometrically (glass electrode) at 25°C in aqueous media. The values of the stability constants for the 1:1 and 1:2 chelates increase more or less regularly across the rare earth series [1, 3]. The data of Lajunen [3] revealed a slight decrease in lg K_1 for the gadolinium chelate compared with the samarium chelate.

Maiti, Mahadevan, and Sathe [4] interpreted their potentiometric titration data (30°C, I = 0.2 (KNO_3)) in terms of formation of a polynuclear complex of composition $M_3(C_6H_2O_8S_2)_3(OH)_2^{5-}$, for M = La, Pr, Nd, Eu, Gd, Dy, and Er. Tserkasevich and Poluektov [5], concluded on the basis of spectrophotometric studies, that the rare earth chlorides (M = Pr, Nd, Ho, Er) react with Tiron in solution to form two complexes, formulated as $M_2(C_6H_2O_8S_2)_3^{6-}$ and $M(C_6H_3O_8S_2)_2(C_6H_2O_8S_2)_2^{11-}$, at pH 6 to 7 and 12 to 14, respectively. Borkowski and coworkers [6] also proposed the formation of a polynuclear complex of lanthanum with Tiron (2:3 metal ion:Tiron mole ratio) in the pH range 6.5 to 10.8, based on spectrophotometric studies.

Stability constants of mixed complexes derived from pyrocatechol-3,5-disulfonic acid and N'-(2-hydroxyethyl)ethylenediaminetriacetic acid, ethylenediaminetetraacetic acid, or 1,2-cyclohexanediaminetetraacetic acid are reported in "Rare Earth Elements" D 1, 1980, p. 235, 239, and 241.

Table 3/1

Stability and Protonation Constants for 1:1 and 1:2 Rare Earth Chelates with Pyrocatechol-3,5-disulfonic Acid at 25°C.

metal ion	ionic strength ($NaClO_4$)	lg K_1	lg K_1 *)	lg K_2	lg β_2	lg K_{11} **)	lg K_{12} ***)	Ref.
Sc^{3+} ****)	0.1	18.96						[2]
Y^{3+}	0.1	13.72	5.13			3.89		[1]
Y^{3+}	0.5	12.54			21.71			[7]
La^{3+}	0.1	12.87						[1]
Pr^{3+}	0.1	13.47						[1]
Pr^{3+}	0.5	11.8		7.8	19.6		8.4	[3]
Nd^{3+}	0.1	13.69	5.61			4.40		[1]
Nd^{3+}	0.5	11.88		7.75	16.93		8.36	[3]
Sm^{3+}	0.1	13.92	5.72			4.28		[1]
Sm^{3+}	0.5	12.37		8.34	20.71		8.04	[3]
Eu^{3+}	0.5	12.54		8.38	20.92		7.88	[3]
Gd^{3+}	0.1	14.10	5.92			4.30		[1]
Gd^{3+}	0.5	12.50		8.64	21.14		7.87	[3]
Tb^{3+}	0.1	14.14	5.71			4.05		[1]
Tb^{3+}	0.5	12.69		9.00	21.69		7.75	[3]
Dy^{3+}	0.1	14.36	5.59			3.71		[1]
Dy^{3+}	0.5	12.79		9.17	21.96		7.73	[3]
Ho^{3+}	0.1	14.39	5.42			3.51		[1]
Ho^{3+}	0.5	12.88		9.48	22.36		7.64	[3]
Er^{3+}	0.1	14.48	5.45			3.45		[1]
Er^{3+}	0.5	12.89		9.51	22.40		7.61	[3]
Tm^{3+}	0.1	14.36	5.67			3.79		[1]
Tm^{3+}	0.5	12.96		9.54	22.50		7.56	[1]
Yb^{3+}	0.1	14.43	5.65			3.70		[1]
Yb^{3+}	0.5	13.25		9.51	22.76		7.51	[3]
Lu^{3+}	0.5	13.29		9.56	22.85		7.44	[3]

*) $K_1 = [M(C_6H_3O_8S_2)] \cdot [M^{3+}]^{-1} \cdot [(C_6H_3O_8S_2)^{3-}]^{-1}$.

**) $K_{11} = [M(C_6H_3O_8S_2)] \cdot [M(C_6H_2O_8S_2)^{-}]^{-1} \cdot [H^+]^{-1}$.

***) $K_{12} = [M(C_6H_3O_8S_2)(C_6H_2O_8S_2)^{4-}] \cdot [M(C_6H_2O_8S_2)_2^{5-}]^{-1} \cdot [H^+]$.

****) Determined potentiometrically using a mercury electrode.

3.6.2.2 Isolated Compounds

$Na[Nd(C_6H_2O_8S_2)]$ and $Na_6[Nd_2(C_6H_2O_8S_2)_3]$. Neodymium nitrate was added to aqueous solutions of Tiron (variable mole ratios) and the mixture neutralized with sodium hydroxide to pH 6 to 7. Precipitation of the product was effected with ethanol. A compound of composition $Na[Nd(C_6H_2O_8S_2)]$ was obtained using Nd^{3+}:Tiron mole ratios of 1:1 and more, whereas a compound of composition $Na_6[Nd_2(C_6H_2O_8S_2)_3]$ was obtained with Nd^{3+}:Tiron mole ratios of 1:1.5 or 1:3. The mixtures were left to settle, and the precipitates were collected by filtration, washed with ethanol, and dried in air. Both compounds were obtained as pale lilac powders, soluble in water. The compounds were characterized by neodymium and sodium analyses [5].

References to 3.6.2:

[1] L. D. Shtenke, N. A. Skorik, V. N. Kumok (Zh. Neorgan. Khim. **15** [1970] 1214/7; Russ. J. Inorg. Chem. **15** [1970] 623/5). — [2] R. P. Guseva, V. N. Kumok (Zh. Neorgan. Khim. **17** [1972] 3195/8; Russ. J. Inorg. Chem. **17** [1972] 1680/2). — [3] L. H. J. Lajunen (Finn. Chem. Letters **1976** 31/5). — [4] B. Maiti, N. Mahadevan, R. M. Sathe (Indian J. Chem. **9** [1971] 1152/4). — [5] K. V. Tserkasevich, N. S. Poluektov (Zh. Neorgan. Khim. **9** [1964] 128/33; Russ. J. Inorg. Chem. **9** [1964] 69/73).

[6] B. Borkowski, A. Borkowska, K. Banczyk (Poznan. Towarz. Przyjaciol Nauk Wydz. Mat. Przyr. Prace Kom. Pr. Chem. **12** [1971] 3/12 from C.A. **75** [1971] No. 44405). — [7] L. H. J. Lajunen (Finn. Chem. Letters **1976** 63/6).

3.6.3 With 1,2-Dihydroxy-4-methylbenzene (= $C_7H_8O_2$)

Formation constants for 1,2-dihydroxy-4-methylbenzene complexes with the lanthanide ions have been determined by potentiometric pH titration, using Bjerrum's method. Three solvent systems were used: 50% aqueous dimethyl sulfoxide (DMSO), 50% ethanol, and 50% dioxane, see Table 3/2. Values of ΔH were determined from plots of lg K_1 vs. 1/T, and ΔS values were subsequently calculated. A 1:3 complex also apparently forms, but the presence of a large precipitate prevented determination of K_3. The ligand is apparently acting as a monoprotic acid under the conditions of the experiment. The positive ΔH values for the DMSO work suggest that DMSO solvates the lanthanide ions in preference to water, unlike the other two solvent systems, J. L. Bear, M. E. Clark (J. Inorg. Nucl. Chem. **31** [1969] 2811/4).

Table 3/2
Thermodynamic Parameters for 1,2-Dihydroxy-4-methylbenzene Complexes of the Lanthanide Ions*).

M	[50% DMSO]				[50% ethanol]				[50% dioxane]			
	lg K_1	lg K_2	ΔH_1	ΔS_1	lg K_1	lg K_2	ΔH_1	ΔS_1	lg K_1	lg K_2	ΔH_1	ΔS_1
La	3.82	3.50	6.2	38	3.95	2.75	−1.9	12	4.60	3.47	−2.5	13
Nd	4.00	3.70	7.3	43	4.66	3.45	−3.9	8	5.09	4.08	−3.9	10
Gd	4.13	3.82	6.2	40	4.83	3.73	−1.5	17	5.28	4.13	−3.2	14
Tb	4.25	3.97	5.0	37	5.00	3.78	−2.7	14	5.51	4.49	−3.3	14
Dy	4.40	4.08	4.4	35	5.09	3.86	−2.8	14	5.67	4.48	−3.5	14
Ho	4.49	4.23	4.1	35	5.15	4.09	−3.3	13	5.76	4.49	−2.5	18
Er	4.60	4.12	5.5	40	5.27	4.33	−3.4	13	5.77	4.62	−3.8	14
Tm	4.74	4.55	7.1	48	5.36	4.50	−3.5	13	5.80	4.81	−3.1	16
Yb	4.82	4.59	9.4	54	5.49	4.62	−3.9	12	5.86	4.64	−3.4	16

*) The error limits in K_1 are ±10% and in K_2 ±30%. Values of ΔH are in kcal/mol ±0.75%, and values of ΔS are in entropy units ±15%.

3.6.4 With 1,3-Dihydroxybenzene (= Resorcinol = $C_6H_6O_2$)

A derivative of cerium(IV) was prepared by adding resorcinol to a solution of hexachloroceric acid, H_2CeCl_6, prepared from $Ce(OH)_4$ by passing dry HCl gas through the suspension in ethanol. The solution was heated at reflux until HCl gas was no longer evolved, after which the solution was filtered and excess ethanol removed by distillation under reduced pressure (20 to 25 Torr). The brown paste that resulted gave a brown crystalline compound, $CeCl_3(OC_6H_4OH)$, after repeated crystallizations from petroleum ether (60 to 80°C) followed by drying for 48 h in a vacuum oven. The infrared spectrum of the compound shows a band at ≈3620 cm^{-1} due to the presence of a free phenolic OH group and a band at ≈1180 cm^{-1} due to an OH deformation, indicating that only one hydroxo group is ionized and that the molecular structure is monomeric. The ultraviolet absorption spectrum of the derivative contains two peaks at 215 nm (lg $\varepsilon_{max}=4.26$) and 280 nm (lg $\varepsilon_{max}=3.78$). Ebullioscopic molecular weight measurements in boiling benzene also show the compound to be monomeric in solution. The molar conductance of the compound in methanol is very low (2 to 4 $\Omega^{-1}\cdot cm^2\cdot mol^{-1}$), typical of a nonelectrolyte, S. K. Gupta, P. S. Bassi (Current Sci. [India] **46** [1977] 139/40).

3.7 With Trihydroxybenzenes

3.7.1 With 1,2,3-Trihydroxybenzene (= Pyrogallol = $C_6H_6O_3$)

Stability constants (lg β_1) of pyrogallol complexes of several of the rare earths have been determined by potentiometric pH titration at 30°C and I = 0.2 M ($NaClO_4$): Sc (14.79, 14.76), Y (10.68, 10.66), La (9.14, 9.12), Pr (9.99, 9.93), Nd (10.12, 10.05), Sm (10.57, 10.53). Values for the protonation of the dianion used were lg K_1 = 10.98 (10.95) and lg K_2 = 8.82 (8.87). Numbers in parentheses calculated at half $\bar{n}$-values, other numbers calculated by interpolation of various $\bar{n}$-values [1]. Stability constants of mixed ternary complexes containing nitrilotriacetic acid and pyrogallol [1, 2] or ethylenediaminetetraacetic acid (H_4 edta) and pyrogallol [2] have also been reported. For ternary complexes of Pr^{3+}, Nd^{3+}, and Sm^{3+} with H_4 edta and pyrogallol see also [3].

Pyrogallol derivatives of yttrium and the lanthanides were prepared in the same way as the complexes with pyrocatechol, see p. 40. The sparingly soluble salts were isolated by centrifugation, washed with 50% methanol, and then washed with pure methanol. Potassium hydroxide was also used to precipitate the complexes of composition $M_2(C_6H_4O_3)_3$ [4].

A pyrogallol derivative of cerium(IV) was prepared according to the procedure used for the resorcinol compound, see p. 43, by reaction of pyrogallol with hexachloroceric acid in ethanol. The infrared spectrum of the brown crystalline compound, $CeCl_3(OC_6H_3(OH)_2)$, shows a band at ≈3620 cm^{-1} due to the presence of a free phenolic OH group and a band at ≈1180 cm^{-1} due to an OH deformation, indicating that only one hydroxo group is ionized and that the molecular structure is monomeric. The ultraviolet absorption spectrum of the derivative contains two peaks at 215 nm (lg $\varepsilon_{max}=4.29$) and 280 nm (lg $\varepsilon_{max}=3.79$). Ebullioscopic measurements in boiling benzene also show the compound to be monomeric in solution. The molar conductance of the compound in methanol is very low (≈2 to 4 $\Omega^{-1}\cdot cm^2\cdot mol^{-1}$), typical of a nonelectrolyte [5].

References to 3.7.1:

[1] S. D. Makhijani, S. P. Sangal (J. Indian Chem. Soc. **55** [1978] 987/8). — [2] T. H. Mhaske, K. N. Munshi (J. Indian Chem. Soc. **53** [1976] 11/4). — [3] S. D. Makhijani, S. P. Sangal (Kem.-Kemi **6** [1979] 746/7 from C.A. **92** [1980] No. 100312). — [4] N. S. Poluektov, K. V. Tserkasevich (Zh. Neorgan. Khim. **12** [1967] 1862/6; Russ. J. Inorg. Chem. **12** [1967] 980/2). — [5] S. K. Gupta, P. S. Bassi (Current Sci. [India] **46** [1977] 139/40).

3.7.2 With 1,3,5-Trihydroxybenzene (= Phloroglucinol = $C_6H_6O_3$)

Stability constants of phloroglucinol with several of the rare earth ions have been determined using the Bjerrum-Calvin pH titration technique in 0.02 M aqueous $NaClO_4$ solution. The variation of lg K as a function of temperature has been used to determine the thermodynamic parameters for these systems. Table 3/3 lists the values obtained. Only 1:1 complex formation was observed, and the stability constants increase with increasing atomic number for the lanthanides with yttrium having the highest value. The value for the scandium constant is much larger, S. D. Makhijani, S. P. Sangal (Indian J. Chem. A **15** [1977] 565/6).

Table 3/3

Stability Constants and Thermodynamic Parameters for Rare Earth Complexes with Phloroglucinol.

M	lg K_1			$-\Delta G$ in kcal/mol			ΔH in kcal/mol	ΔS in $cal \cdot K^{-1} \cdot mol^{-1}$
	20°C	30°C	40°C	20°C	30°C	40°C		
Sc	15.07	15.27	15.50	20.338	21.311	22.346	5.027	86.924
Y	7.92	8.10	8.30	10.689	11.304	11.966	5.112	54.178
La	6.08	6.25	6.40	8.206	8.723	9.227	7.321	51.259
Ce	6.20	6.37	6.52	8.368	8.890	9.400	7.370	53.663
Pr	6.68	6.85	7.10	9.018	9.546	10.236	5.973	51.218
Nd	6.92	7.12	7.25	9.339	9.937	10.596	7.925	58.950

lg K of the three proton association constants of the ligand anion are 11.54, 9.17, and 8.41 at 20°C; 11.12, 8.94, and 8.17 at 30°C; and 10.75, 8.74, and 7.97 at 40°C.

3.7.3 With 2,3,4-Trihydroxybenzenesulfonic Acid

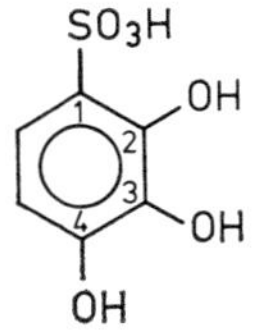

(= Pyrogallol-4-sulfonic Acid = $C_6H_6O_6S$)

The complexation reaction between a rare earth chloride (M = Nd, Ho, Er) and pyrogallol-4-sulfonate (potassium salt) in aqueous media was studied both spectrophotometrically and potentiometrically. The data were interpreted in terms of formation of 1:1 chelates in neutral media and 1:2 (metal:ligand) chelates in alkaline media. The overall formation constants of the 1:2 chelates (determined spectrophotometrically) were 5.05 to 5.6×10^5, 6.4 to 6.6×10^5, and 6.9 to 7.7×10^5 for the neodymium, holmium and erbium complexes, respectively, K. V. Tserkasevich, N. P. Efryushina, N. S. Poluektov (Zh. Neorgan. Khim. **11** [1966] 93/8; Russ. J. Inorg. Chem. **11** [1966] 49/53).

3.8 With 2,3-Dihydroxynaphthalene (= 2,3-Naphthalenediol = $C_{10}H_8O_2$)

Stability constants of 2,3-dihydroxynaphthalene with several of the lanthanides have been determined using the Bjerrum-Calvin pH titration technique in aqueous solution with an ionic strength of 0.1 M adjusted with $NaClO_4$. The variation of lg K as a function of temperature has been used to determine the thermodynamic parameters for these systems. Table 3/4 lists the values obtained. Precipitation prevented the calculation of values for 1:3 complexes. The values of the stability constants increase with increasing atomic number of the lanthanide ion. It should be noted that the ΔH and ΔS values reported agree with the ΔG values only if ΔH is negative [1]. Stability constants and thermodynamic parameters for complexes with Gd^{3+}, Y^{3+}, Tb^{3+}, Dy^{3+}, and Ho^{3+} are given in [4].

Stability constants of 2,3-dihydroxynaphthalene with the lanthanide-ethylenediaminetetra-acetate complexes have also been determined by the potentiometric technique at I = 0.1 M ($NaClO_4$) and 20°C [2].

The luminescence of the 1:1 Tb^{3+} complex as a function of pH is reported [3].

Table 3/4

Stability Contants and Thermodynamic Parameters for Lanthanide Complexes with 2,3-Dihydroxy-naphthalene (ΔG and ΔH in kcal/mol, ΔS in cal · K^{-1} · mol^{-1}).

M	lg K_1		lg K_2		$-\Delta G_1$		$-\Delta G_2$		ΔH_1	ΔH_2	ΔS_1	ΔS_2
	25°C	45°C	25°C	45°C	25°C	45°C	25°C	45°C	35°C		35°C	
La	9.81	9.01	8.97	7.95	13.38	12.11	12.23	11.57	17.40	22.06	13	33
Ce	10.27	9.53	9.17	8.56	14.01	13.87	12.50	12.46	16.10	13.10	7	2
Pr	10.43	9.76	9.39	8.77	14.27	14.20	12.80	12.77	15.31	13.25	3	1
Nd	10.68	9.97	9.60	8.98	14.56	14.51	13.09	13.06	16.30	13.54	2	1
Sm	11.20	10.67	10.07	9.46	15.27	15.24	13.82	13.78	15.72	14.42	1	2

The lg of the two proton association constants of the ligand anion were 12.35 and 8.95 at 25°C and 11.58 and 8.68 at 45°C.

References to 3.8:

[1] S. Arora, H. L. Kalra, S. N. Dubey, D. M. Puri (Acta Ciencia Indica **4** [1978] 16/20). — [2] N. S. Poluektov, L. A. Alakaeva, M. A. Tishchenko (Zh. Neorgan. Khim. **18** [1973] 81/5; Russ. J. Inorg. Chem. **18** [1973] 40/3). — [3] N. S. Poluektov, L. A. Alakaeva, M. A. Tishchenko (Zavodsk. Lab. **37** [1971] 1077/9 from C.A. **75** [1971] No. 156577). — [4] S. Arora, S. N. Dubey, H. L. Kalra, D. M. Puri (Acta Cienc. India **5** [1979] 130/5 from C.A. **92** [1980] No. 117365).

3.9 With 6,7-Dihydroxy-2-naphthalenesulfonic Acid

(= $C_{10}H_8O_5S$)

$M(C_{10}H_5O_5S)_n^{3-3n}$ (n = 1 to 3) and **$M(C_{10}H_6O_5S)_n^{3-2n}$** (n = 1, 2) **Species. Formation in Solution.** Formation constants (Table 3/5) for rare earth chelates with 6,7-dihydroxy-2-naphthalenesulfonic acid were determined potentiometrically (glass electrode) at 25°C in aqueous media, I = 0.5 ($NaClO_4$). The formation of the $M(C_{10}H_6O_5S)^+$ complex is favored in solutions of low ionic strength, as predicted by the Debye-Hückel equation, and containing a large excess of the ligand. The increase in stability of the $M(C_{10}H_5O_5S)$ species is smaller from terbium to lutetium than from praseodymium to gadolinium. The stepwise constants for the bis complex, $M(C_{10}H_5O_5S)_2^{3-}$, as well as the overall formation constants, β_2, increase almost linearly with decreasing ionic radius [1].

Table 3/5

Formation Constants for Rare Earth Chelates Derived from 6,7-Dihydroxy-2-naphthalene-6-sulfonic Acid at 25°C, I = 0.5 ($NaClO_4$).

metal ion	complex						
	$M(C_{10}H_5O_5S)_n^{3-n}$					$M(C_{10}H_6O_5S)_n^{3-2n}$	
	lg K_1	lg K_2	lg β_2	lg K_3	lg β_3	lg β_1	lg β_2
Pr^{3+}	9.19	7.08	16.27				11.85
Nd^{3+}	9.26	7.14	16.40				11.96
Sm^{3+}	9.82	7.33	17.15			3.6	12.68
Eu^{3+}	9.90	7.35	17.25			3.7	12.76
Gd^{3+}	9.92	7.57	17.49			2.4	13.04
Tb^{3+}	10.20	7.90	18.10				13.29

Table 3/5 (Continued)

metal ion	complex						
	$M(C_{10}H_5O_5S)_n^{3-n}$					$M(C_{10}H_6O_5S)_n^{3-2n}$	
	lg K_1	lg K_2	lg β_2	lg K_3	lg β_3	lg β_1	lg β_2
Dy^{3+}	10.25	8.04	18.29				13.41
Ho^{3+}	10.26	8.14	18.40				13.48
Er^{3+}	10.31	8.36	18.67				13.74
Tm^{3+}	10.37	8.55	18.92	4.4	23.3		13.98
Yb^{3+}	10.43	8.65	19.08	4.5	23.6		14.10
Lu^{3+}	10.44	8.84	19.28	4.8	24.1		14.30

Stability constants of mixed complexes derived from 6,7-dihydroxy-2-naphthalenesulfonic acid and N'-(2-hydroxyethyl)ethylenediaminetriacetic acid or ethylenediaminetetraacetic acid are reported in "Rare Earth Elements" D 1, 1980, p. 235 and 238.

Reference to 3.9:

[1] L. H. J. Lajunen (Finn. Chem. Letters **1976** 53/7).

3.10 With 1,8-Dihydroxy-3,6-naphthalenedisulfonic Acid

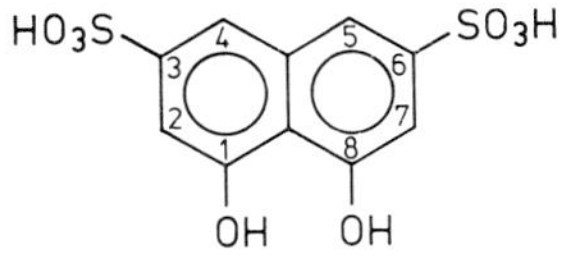

(= Chromotropic Acid = $C_{10}H_8O_8S_2$)

Complexes in Solution

$M(C_{10}H_5O_8S_2)$. Formation constants for the neutral 1:1 rare earth chelates were determined potentiometrically (glass electrode) at 25°C in aqueous media. The values obtained at I = 0.1 and the thermodynamic values (I = 0) are recorded below [1].

metal ion	Sm^{3+}	Eu^{3+}	Gd^{3+}	Tb^{3+}	Dy^{3+}	Ho^{3+}
lg K_1 (I = 0.1)	2.26	2.37	2.41	2.41	2.35	2.20
lg K_1 (I = 0)	3.71	3.87	3.76	3.83	3.74	3.62

$M(C_{10}H_4O_8S_2)^-$ Ion. Spectrophotometric data, obtained for aqueous solutions maintained at constant ionic strength, I = 0.1 ($NaClO_4$), over the pH range 3.8 to 8.05, were interpreted in terms of formation of 1:1 chelates derived from the quadrinegative form of the ligand. Titration data indicated that the protons of the two hydroxo groups were replaced by the rare earth ion on complexation. The negative charge on the complex was confirmed by electromigration studies. The absorption maximum of the complexes in aqueous media was observed at 525 nm, with maximum absorbance obtained at pH 5.07. The stability constants of the complexes are recorded below [2].

metal ion	La^{3+}	Pr^{3+}	Nd^{3+}	Gd^{3+}	Tb^{3+}
K	2.6×10^{-4}	5.2×10^{-4}	1.3×10^{-3}	4.9×10^{-4}	5.5×10^{-4}

Formation constants for mixed complexes derived from 1,2-cyclohexanediaminetetraacetic acid and 1,8-dihydroxy-3,6-naphthalenedisulfonic acid see "Rare Earth Elements" D 1, 1980, p. 241.

References to 3.10:

[1] O. Makitie, L. H. J. Lauri, H. Saarinen (Finn. Chem. Letters **1974** 96/8). — [2] S. Zielinski, W. Puacz (Zh. Analit. Khim. **25** [1970] 2342/6; Russ. J. Anal. Chem. **25** [1970] 2016/20).

4 Complexes with Aldehydes and Monoketones

Edward R. Birnbaum
Department of Chemistry, New Mexico State University
Las Cruces, New Mexico, USA

John H. Forsberg
Department of Chemistry, Monsanto Hall, Saint Louis University
Saint Louis, Missouri, USA

4.1 General Aspects

Several studies have been made on the behavior of the luminescence (referred to as sensitized luminescence) of the lanthanide ions in the presence of fluorescent or phosphorescent aromatic aldehydes and ketones [1 to 6]. In some cases the mechanism of the sensitized luminescence that results is said to be due to a purely collisional mechanism, whereas other reports indicate that complex formation is responsible. In the earlier papers [1 to 3], the mechanism is described as diffusion-controlled, suggesting that complex formation is not taking place. However, at higher concentrations, it appears that complex formation is taking place [4, 5]. Rate constants for inter- and intramolecular energy transfer from triplet states of aromatic ketone triplet states to europium ions are reported in [5].

For the preparation and the properties of complexes with ligands containing both keto groups and N-donor atoms, see "Rare Earth Elements" D 2, for thioketones, see Vol. D 4.

References to 4.1:

[1] E. Matovich, C. K. Suzuki (J. Chem. Phys. **39** [1963] 1442/4). — [2] A. Heller, E. Wasserman (J. Chem. Phys. **42** [1965] 949/55). — [3] W. J. McCarthy, J. D. Winefordner (Anal. Chem. **38** [1966] 848/53). — [4] V. A. Gevorkyan, D. Kh. Grigoryan (Zh. Prikl. Spektrosk. **12** [1970] 876/9; J. Appl. Spectrosc. [USSR] **12** [1970] 652/4). — [5] E. B. Sveshnikova, V. F. Morina, V. L. Ermolaev (Opt. Spektroskopiya **36** [1974] 725/32; Opt. Spectrosc. [USSR] **36** [1974] 421/5).

[6] S. I. Weissman (J. Chem. Phys. **10** [1942] 214/7).

4.2 Complexes with Monoketones and Benzaldehyde

Adducts of several ketones with scandium chloride have been reported [1]. Each of these adducts was prepared under anhydrous conditions by dissolving scandium chloride in an excess of the particular ketone (1:6 molar ratio). Subsequent cooling resulted in the formation of white crystals, which were filtered off, washed with ether, and dried over P_4O_{10} at 10 to 15 Torr. The preparations of the acetone and butanone adducts were maintained below 4°C to prevent liberation of HCl [1]. Table 4/1 lists the adducts obtained, together with the available physical data. All the adducts are very hygroscopic. Infrared spectra, taken as mineral oil mulls, show a considerable lowering of the ν(CO) stretching mode upon adduct formation of ≈40 to 80 cm^{-1}, depending on the ketone. Calorimetric measurements show heats of reaction from ≈9 to 18 kcal/mol [1].

A bis acetone adduct of $ScCl_3$ was prepared by mixing 250 mg of the chloride salt in 25 ml acetone. An oil resulted from evaporation of the excess acetone in a vacuum desiccator over P_4O_{10} and after several days, crystallization began. A tris acetophenone adduct, $ScCl_3 \cdot 3\,C_6H_5COCH_3$, was prepared in a similar fashion by the addition of benzene to an acetophenone solution of the chloride salt until turbidity appeared. Cooling in an ice/salt bath resulted in crystallization. Additional crystals could be obtained by further addition of benzene, followed by cooling. The crystals were washed with benzene and dried in vacuum. A benzophenone adduct, $ScCl_3 \cdot 2\,C_6H_5COC_6H_5$, was prepared by mixing 300 mg of the chloride salt with 3 g of benzophenone in an oil bath at 200 to 220°C. To the warm reaction mixture, 20 to 30 ml of benzene was added, and after standing in a closed vessel for 2 to 3 h, the benzene was removed by filtration. The solid brown reaction product was washed with benzene and dried in vacuum [2].

A tris adduct of benzaldehyde with scandium chloride was also prepared by adding 450 mg of the salt to 20 ml of freshly distilled benzaldehyde and by stirring for 30 min at 5°C. After removal of the undissolved chloride salt, benzene was added to the filtrate until turbidity occurred. Cooling this mixture produced a white, microcrystalline powder having the formulation, $ScCl_3 \cdot 3\,C_6H_5CHO$ [2].

Table 4/1

Physical Data for Ketone Adducts of Scandium Chloride $ScCl_3 \cdot nL$ [1].

No.	ketone	n	ν(CO)*) in cm^{-1}	$\Delta H^{**)}_{mixing}$
1	acetone	3	1210, 1667 (1710)	17.5
2	2-butanone	2	1713, 1658 (1710)	13.3
3	cyclopentanone	3	1745, 1686 (1737)	17.1
4	cyclohexanone	2	1713, 1658 (1710)	10.1
5	acetophenone	3	1684, 1625 (1684)	17.5
6	4'-fluoroacetophenone	2	1693, 1627 (1684)	9.3
7	4'-chloroacetophenone	2	1689, 1628 (1686)	10.8
8	4'-methoxyacetophenone	3	1672, 1590 (1674)	16.4 (35°C)
9	4'-nitroacetophenone	2	1628 (1690)	—
10	4'-methylacetophenone	3	1687, 1626 (1689)	—

*) Numbers in parentheses correspond to ν(CO) for the free ligand; **) kcal/mol at 25°C.

As is the case for alcohols, kinetic investigations of the oxidation of ketones by cerium(IV) have led to a determination of the stability constants and thermodynamic parameters for the cerium(IV)-acetone and methyl ethyl ketone adducts which are reported in Table 4/2 [3].

Table 4/2

Stability Constants for Adducts of Ketones with Cerium(IV)perchlorate.

ketone	K_1	method	t in °C	ΔH in kcal/mol	ΔG cal/mol	ΔS in cal·mol⁻¹·K⁻¹
acetone	4.0	kinetic	25	—	—	—
acetone	3.5	kinetic	30	9.212	826.5	33.7
acetone	3.0	kinetic	35	—	—	—
ethyl methyl ketone (= 2-butanone)	17.25	kinetic	25	—	—	—
ethyl methyl ketone	5.0/4.5	kinetic/spectral	30	40.03	236.7	136.9
ethyl methyl ketone	1.6	kinetic	35	—	—	—

References to 4.2:

[1] N. L. Chikina, Yu. V. Kolodazhnyi, O. A. Osipov (Zh. Obshch. Khim. **45** [1975] 1354/8; J. Gen. Chem. [USSR] **45** [1975] 1324/8). — [2] V. H. Funk, B. Köhler (Z. Anorg. Allgem. Chem. **325** [1963] 67/71). — [3] S. Venkatakrishnan, M. Santappa (Z. Physik. Chem. [Frankfurt] **16** [1958] 73/84).

4.3 Complexes with Hydroxyaldehydes and Hydroxyketones

4.3.1 With Salicylaldehyde (= 2-Hydroxybenzaldehyde = $C_7H_6O_2$)

The stability constants of salicylaldehyde complexes with scandium, yttrium, and lanthanum have been determined by absorption spectrophotometry. For the scandium complex is reported $K \approx 10^4$ at pH = 4.1, 2.6×10^4 at pH = 5.45 and 7.6×10^4 at pH = 6.0 in a 28% ethanol/H_2O mixture containing an acetate buffer (I = 0.025) [1]. Assuming, that only $[M(C_7H_5O_2)]^{2+}$ species are present, lg K_1 = 4.34 is calculated for the yttrium complex and lg K_1 = 3.40 for the lanthanum complex at 25.0°C and I = 0.1 M ($NaClO_4$) [2].

Formation constants of praseodymium, neodymium, and europium complexes, determined by potentiometric pH titrations in an 80% (v/v) aqueous methanol mixture, at 24 ± 1°C, I = 0.1 M (NaCl) are shown below [3]:

M	lg K_1 (± 0.03)	lg K_2 (± 0.05)	lg K_3 (± 0.1)
Pr	5.77	4.38	2.9
Nd	5.82	4.44	3.0
Eu	6.25	4.65	3.1

Each lanthanide salicylaldehyde derivative was prepared by adding an aqueous lanthanide chloride solution (10 ml/0.5 M) dropwise at room temperature over a period of 1/2 h to salicylaldehyde (2 gm) dissolved in 15 ml of 0.1 M aqueous NaOH, 50 ml water, and 15 ml 95% ethanol. The yellow flocculent precipitate which formed was stirred in the reaction mixture for an additional 1/2 h, removed by filtration, washed repeatedly with water, and dried to constant weight in a vacuum desiccator at room temperature. Compounds of composition $M(C_7H_5O_2)_3$ with M = La to Eu were isolated. The compounds of the heavier lanthanide ions are reported to retain significant amounts of sodium. The compounds all decompose over a broad range from 200 to 350°C [4]. The yttrium adduct was prepared in a similar fashion, using alcoholic ammonia to titrate the proton of salicylaldehyde instead of sodium hydroxide. The pH was adjusted to 5.0 with the ammonia solution. The yttrium compound was then washed with ether and dried over $CaCl_2$ (fused) for 48 h, resulting in $Y(C_7H_5O_2)_3$. The compound has the ν(CO) stretching mode at 1649 cm^{-1} in a KBr pellet or a Nujol mull [5]. The infrared spectra of the solid lanthanide compounds in Nujol mulls are nearly identical, with the ν(CO) stretching mode at 1657 ± 4 cm^{-1}. Thermogravimetric curves for the lanthanide compounds are presented [4].

Salicylaldehyde complexes of samarium were prepared by the alcohol interchange reaction by adding salicylaldehyde to samarium 2-propanolate dissolved in benzene in the appropriate molar ratio. After heating at reflux for 2 to 3 h, the binary isopropanol/benzene azeotrope was collected and the compounds dried under reduced pressure. Three yellow adducts were isolated, $Sm(OCH(CH_3)_2)_n(C_7H_5O_2)_{3-n}$ with n = 0 to 2. All three compounds are insoluble in benzene [6].

The fluorescence spectra of several of the rare earth salicylaldehyde complexes have been reported [7], with the europium compound receiving special attention [8, 9]. Energy transfer from the ligand excited state to one of the metal ion excited states is responsible for the occasionally intense luminescence of these compounds. A mixed complex of neodymium with ethylenediaminetetraacetic acid and salicylaldehyde has been reported [10]. The isotropic shifts (both contact and pseudocontact) found were used to determine a value of 6 ± 1 mM for the instability constant. The complex appears to be devoid of axial symmetry, suggesting that the chelating potential of the salicylaldehyde anion plays an important part in the formation of the complex species.

Adducts of salicylaldehyde complexes with 2-hydroxybenzohydrazide (= $C_7H_8N_2O_2$), are described in "Rare Earth Elements" D 2.

References to 4.3.1:

[1] Z. Holzbecher (Collection Czech. Chem. Commun. **28** [1963] 716/27). — [2] C. Postmus, L. B. Magnusson, C. A. Craig (Inorg. Chem. **5** [1966] 1154/7). — [3] A. A. Zholdakov, N. K. Davidenko (Zh. Neorgan. Khim. **13** [1968] 3233/6; Russ. J. Inorg. Chem. **13** [1968] 1662/4). — [4] R. G. Charles (J. Inorg. Nucl. Chem. **26** [1964] 2298/300). — [5] Y. H. Deshpande, V. Ramachandra Rao (Current Sci. [India] **40** [1971] 432/3).

[6] M. Hasan, S. N. Misra, R. N. Kapoor (U.A.R. J. Chem. **13** [1970] 235/42). — [7] A. N. Sevchenko, V. V. Kuznetsova, V. S. Khomenko (Izv. Akad. Nauk SSSR Ser. Fiz. **27** [1963] 710/6; C.A. **59** [1963] 14759). — [8] S. I. Weissman (J. Chem. Phys. **10** [1942] 214/7). — [9] V. V. Kuznetsova, A. N. Sevchenko (Izv. Akad. Nauk SSSR Ser. Fiz. **23** [1959] 2/8; C. A. **1959** 11987). — [10] J. Reuben (J. Am. Chem. Soc. **98** [1976] 3726/8).

4.3.2 With Derivatives of Salicylaldehyde (= HL)

Stability constants of complexes ML with M = La and Y have been determined in aqueous solution by a spectrophotometric method. All values are reported at 25°C and ionic strength I = 0.1 M ($NaClO_4$); estimated error: ± 0.02 lg units [1]:

ligand HL	pKa	lg K_1 (La^{3+})	lg K_1 (Y^{3+})
3-chlorosalicylaldehyde (= $C_7H_5ClO_2$)	6.61	3.16	3.77
4-chlorosalicylaldehyde	7.18	3.38	4.09
5-chlorosalicylaldehyde	7.41	3.20	3.94
6-chlorosalicylaldehyde	8.26	4.08	4.87
3-nitrosalicylaldehyde (= $C_7H_5NO_4$)	5.21	3.03	3.27
5-nitrosalicylaldehyde	5.32	2.73	3.17
3,5-dinitrosalicylaldehyde (= $C_7H_4N_2O_6$)	2.09	1.84	1.75

Tris complexes of the 3- and 5-nitrosalicylaldehyde with europium(III) have been reported, together with their luminescence spectra. The tris(5-nitrosalicylaldehydato)europium compound was isolated both in an anhydrous form and as the 3-hydrate, whereas the 3-nitrosalicylaldehyde compound was prepared only as the 3-hydrate [2].

References to 4.3.2:

[1] C. Postmus, L. B. Magnusson, C. A. Craig (Inorg. Chem. **5** [1966] 1154/7). — [2] V. V. Kuznetsova, I. I. Razvina, V. S. Khomenko, T. M. Kozhan, R. A. Puko (Vestsi Akad. Navuk Belarusk. SSR Ser. Fiz. Mat. Navuk **1972** No. 5, p. 117/20; C.A. **78** [1973] No. 50029).

4.3.3 With 3-Hydroxybenzaldehyde (= $C_7H_6O_2$)

Several 3-hydroxybenzaldehyde complexes of samarium were prepared by the alcohol interchange reaction as discussed for salicylaldehyde using samarium 2-propanolate as the starting material. Yellow compounds of composition $Sm(OCH(CH_3)_2)_n(C_7H_5O_2)_{3-n}$ with n = 0 to 2 were isolated. All three compounds are insoluble in benzene, M. Hasan, S. N. Misra, R. N. Kapoor (U.A.R. J. Chem. **13** [1970] 235/42).

4.3.4 With 2,4- and 3,4-Dihydroxybenzaldehyde (= $C_7H_6O_3$)

Formation constants of lanthanide complexes with 2,4-dihydroxybenzaldehyde were determined by potentiometric pH titration in 50% (v/v) aqueous dioxane at 35°C and I = 0.10 M ($NaClO_4$); lg K_1 and lg K_2 values for the aldehyde are 10.90 and 8.10 [1]:

M	La	Ce	Pr	Nd	Sm	Eu	Gd	Tb	Dy	Ho	Er	Tm	Yb	Lu
lg K_1	4.00	4.20	4.38	4.60	4.80	4.92	4.40	4.48	4.56	4.67	4.73	4.85	5.01	4.86
lg K_2	3.20	3.40	3.69	3.83	3.97	4.05	3.79	3.92	4.02	4.10	4.23	4.30	4.43	4.31

The lanthanide complexes were prepared by extracting the hydrated solid lanthanide chloride salts (free of acid) with 20 ml of alcohol followed by the addition of a threefold molar excess of the ligand dissolved in 30 ml of alcohol [2]. The pH of each solution was adjusted to 5.8 with alcoholic ammonia and the complex precipitated after evaporation on a steam bath. The resulting complexes were washed with acetone and ether and vacuum dried for 48 h, yielding $M(C_7H_5O_3)_3$, with M = Pr, Nd, Sm, and Gd. A brief discussion is given of the infrared spectra, with the ν(C=O) stretching mode assigned to 1650 cm^{-1} and the ν(C-O) stretching mode assigned to 1295 cm^{-1}. The spectra were recorded in Nujol [2].

Stability constants of mixed praseodymium complexes containing nitrilotriacetic acid or N-hydroxyethylethylenediaminetriacetic acid and 3,4-dihydroxybenzaldehyde have been determined in a 50% (v/v) aqueous dioxane solution at I = 0.2 M ($NaClO_4$) by potentiometric pH titration: lg K_1 for 3,4-dihydroxybenzaldehyde binding to praseodymium nitrilotriacetate are 5.43, 5.34, and 5.28 at 25, 35, and 45°C, respectively. The values of lg K_1 for 3,4-dihydroxybenzaldehyde binding to praseodymium N-hydroxyethylethylenediaminetriacetate are 5.26, 5.13, and 5.02 at the same three temperatures, respectively [3].

References to 4.3.4:

[1] G. C. S. Manku, R. C. Chadha, M. S. Sethi (J. Less-Common Metals **23** [1971] 98/102). — [2] S. F. M. Ali, M. P. Gawande, V. Ramachandra Rao (Current Sci. [India] **42** [1973] 817). — [3] T. H. Mhaske, K. N. Munshi (J. Indian Chem. Soc. **53** [1976] 11/4).

4.3.5 With 2-Hydroxy-1-naphthaldehyde (= $C_{11}H_8O_2$)

The stability constants of the cerium(III) chelate were determined by the Calvin-Bjerrum potentiometric technique at 30°C and ionic strength 0.1 M ($NaClO_4$) in 75% (v/v) dioxane/water medium: lg K_1 = 7.13, lg K_2 = 5.11 [2].

Complexes of 2-hydroxy-1-naphthaldehyde with several of the lanthanide ions have been prepared by adding solutions of the lanthanide nitrates (0.005 mol in 25 ml of 80% ethanol) to an ethanolic solution of the ligand (0.015 mol/150 ml) with stirring. Dilute ammonia (1:20) was then added dropwise to each mixture until a flocculent yellow-green precipitate formed. After stirring for an additional 8 h, each product was collected under suction, washed with hot alcohol, dried in air, and later dried in vacuum. Compounds with the formulation $M(C_{11}H_7O_2)_3$ were isolated with M = La, Ce, Pr, Nd, and Sm. The compounds are stable up to 200°C and decompose without melting above that temperature, finally yielding oxides in the range 500 to 600°C. The infrared spectra, obtained as Nujol mulls, show strong bands at 1640 cm^{-1} and 1600 cm^{-1} that are assigned to the ν(CO) and ν(C=C) stretching vibrations. No shift in the carbonyl band is observed upon complexation. The magnetic moments of the paramagnetic praseodymium, neodymium, and samarium compounds, as obtained by the Faraday method, are 3.31, 3.55, and 1.53 μ_B, respectively, [2].

References to 4.3.5:

[1] V. G. Ratolikar (J. Inorg. Nucl. Chem. **41** [1979] 250). — [2] J. N. Patil, D. N. Sen (J. Indian Chem. Soc. **50** [1973] 413/4).

4.3.6 With Monohydroxyderivatives of Acetophenone

Complexes of 2'-hydroxy-5'-methylacetophenone (= $C_9H_{10}O_2$) with lanthanum, praseodymium, and neodymium, $M(C_9H_9O_2)_3$, were prepared by the addition of ethanol/water (1:1) solutions of the ligand to solutions of the nitrate salts in the same solvent. The complexes precipitated immediately upon stirring in a dilute ammonium hydroxide solution. After addition of excess ammonium hydroxide, the mixtures were heated on a water bath for 1 h and filtered. The solid products were washed three times with a hot water/ethanol mixture and dried at 140°C. The infrared spectra of the compounds, obtained as Nujol mulls, show the ν(CO) stretching vibration shifted ≈15 cm^{-1} to 1650, 1647, and 1652 cm^{-1} in the lanthanum, praseodymium, and neodymium compounds, respectively. In addition the ν(MO) stretching mode, ν(COM) (carbonyl) bending mode, and ν(MOC) (O^-) bending mode have been tentatively assigned to 450, 423, and 539 cm^{-1}, respectively, for the lanthanum compound; 450, 423, and 538 cm^{-1}, respectively, for the praseodymium compound, and 452, 423, and 538 cm^{-1}, respectively, for the neodymium compound. The infrared spectrum of the remainder of the carbon skeleton is also assigned. The compounds are insoluble in water, methanol, ethanol, and acetone and partly soluble in methyl methacrylate [1].

Complexes of 3'-chloro-2'-hydroxyacetophenone and 5'-chloro-2'-hydroxyacetophenone (= $C_8H_7ClO_2$) with lanthanide ions have been prepared by adding solutions of the ligands (3.5 mmol in 50% v/v ethanol/water) with vigorous stirring to solutions of the lanthanide nitrates or chlorides (1 mmol) in the same solvent. The addition of dilute ammonium hydroxide caused precipitation. Each precipitate was washed with the hot ethanol/water mixture and dried at 140°C, resulting in compounds with the formulation $M(C_8H_6ClO_2)_3$, with M = La, Pr, and Nd. The carbonyl stretching mode ν(CO) appears at 1658 cm^{-1} in the infrared spectrum of the free ligand in a Laxetol mull. This band is shifted to ≈1639 to 1644 cm^{-1} in the complexes. Several other ligand bands are assigned, but no detailed analysis is made. Thermogravimetric curves of these compounds show rapid decomposition in the region 200 to 400°C, with thermal stability decreasing with increasing atomic number. The magnetic moments of the paramagnetic compounds obtained by the Gouy method are 3.79, 3.70 μ_B for the praseodymium and neodymium 3-chloro derivatives and 3.74 μ_B for the praseodymium and neodymium 5-chloro derivatives [2].

References to 4.3.6:

[1] C. L. Garg, K. V. Narasimham, B. N. Tripathi (J. Inorg. Nucl. Chem. **33** [1971] 387/92). — [2] M. N. Patel, C. B. Patel, R. P. Patel (J. Indian Chem. Soc. **51** [1974] 805/7).

4.3.7 With 2,4-Dihydroxyacetophenone (= $C_8H_8O_2$)

Formation constants of complexes with 2,4-dihydroxyacetophenone were determined by potentiometric pH titrations in 50% (v/v) aqueous dioxane at 35°C and I = 0.10 M ($NaClO_4$); lg K_1 and lg K_2 values are 11.20 and 8.25, K and ΔG in kcal/mol, G. C. S. Manku, R. C. Chadha, M. S. Sehti (J. Less-Common Metals **23** [1971] 98/102).

M	La	Ce	Pr	Nd	Sm	Eu	Gd	Tb	Dy	Ho	Er	Tm	Yb	Lu
lg K_1 . . .	4.18	4.37	4.57	4.72	4.97	5.02	4.62	4.78	4.92	4.98	5.17	5.29	5.42	5.30
lg K_2 . . .	3.27	3.49	3.68	3.82	3.93	4.05	3.82	3.90	3.97	4.07	4.27	4.38	4.59	4.40

4.3.8 With 2,4,6-Trihydroxyacetophenone (= Phloroacetophenone = $C_8H_8O_4$)

The stability constants and ΔG values for the 1:1 complexes of chloroacetophenone with several lanthanide ions have been determined at 25°C by potentiometric titration in a 50% dioxane/water (v/v) mixture. Solutions studied contained 5 ml 0.5 M $HClO_4$, 4.75 ml 1 M $NaClO_4$ and 15.25 ml H_2O, R. C. Agarwal (J. Indian Chem. Soc. **55** [1978] 984/6).

M	La^{3+}	Ce^{3+}	Pr^{3+}	Nd^{3+}	Gd^{3+}
lg K.	3.19	3.39	3.31	3.71	3.69
−ΔG	4.36	4.61	4.51	5.06	5.04

4.3.9 With 2-Hydroxybenzophenone (= $C_{13}H_{10}O_2$)

Complexes of 2-hydroxybenzophenone with several lanthanide ions have been prepared by dissolving stoichiometric quantities of the lanthanide chloride salts and the ligand in absolute ethanol. Anhydrous ammonia gas was bubbled in until no further precipitation occurred, after which distilled water was added with stirring to complete precipitation. The complexes were quite soluble in absolute ethanol. The products were removed by filtration, washed with water, and dried under vacuum at room temperature for 8 h, resulting in compounds with the formulation $M(C_{13}H_9O_2)_3$ with M = La, Pr, Sm, Eu, and Gd. The absorption spectrum of the lanthanum compound is reported at room temperature (25°C). The emission spectra at 77 K of the La, Pr, Sm, Eu, and Gd compounds are reported, G. A. Crosby, R. E. Whan, J. J. Freeman (J. Phys. Chem. **66** [1962] 2493/9).

4.3.10 With Tropolone and Tropolone Derivatives

4.3.10.1 With 2-Hydroxy-2,4,6-cycloheptatrien-1-one (= Tropolone = $C_7H_6O_2$)

4.3.10.1.1 $M(C_7H_5O_2)_n^{3-n}$ Complexes (n = 1 to 4). Formation in Solution

Stepwise formation constants (Table 4/3) of tropolonato rare earth chelates were determined potentiometrically (glass electrode) at 25.0°C in aqueous media, I = 0.1 (KNO_3) [1]. Only the first three formation constants could be evaluated for M = La to Eu, due to the limited solubilities of the tris chelates. The least soluble tris chelates appeared to be those of cations around neodymium. The tetrakis(tropolonato)chelates for M = Gd to Lu, and Y were present in large enough concentrations in solution to permit evaluation of the fourth formation constants [1]. The first stepwise formation constants were also determined spectrophotometrically at 25°C in aqueous media, I = 0.1 ($NaClO_4$), for the yttrium (lg K_1 = 7.4) and lanthanum (lg K_1 = 6.2) chelates [3].

Table 4/3

Formation Constants and Thermodynamic Parameters for Formation of Tropolonato Chelates of the Rare Earths, $M(C_7H_5O_2)_n^{3-n}$, n = 1 to 4.

metal	formation constants*) [1]				thermodynamic parameters**) [2]		
ion	lg K_1	lg K_2	lg K_3	lg K_4	$-\Delta G_1$ in kcal/mol	$-\Delta H_1$ in kcal/mol	$-\Delta S_1$ in cal·mol^{-1}·K^{-1}
Y^{3+}	7.18	6.08	5.01	3.42	9.78	2.56	24.2
La^{3+}	6.19	4.93	4.19		8.44	2.48	20.0
Ce^{3+}	6.56	5.20	4.36		8.94	2.70	21.0
Pr^{3+}	6.61	5.33	4.45		9.00	2.82	20.7
Nd^{3+}	6.77	5.44	4.40		9.23	2.85	21.8
Sm^{3+}	6.91	5.68	4.60		9.41	2.81	22.2
Eu^{3+}	7.10	5.71	4.81		9.68	2.76	23.2
Gd^{3+}	7.04	5.86	4.82	3.30	9.59	2.72	23.1
Tb^{3+}	7.15	6.03	4.82	3.38	9.74	2.74	23.5
Dy^{3+}	7.23	6.15	5.02	3.72	9.86	2.80	23.7
Ho^{3+}	7.41	6.20	5.20	3.77	10.09	2.76	24.6
Er^{3+}	7.54	6.37	5.24	3.96	10.28	2.79	25.2
Tm^{3+}	7.60	6.39	5.40	3.80	10.36	2.94	24.9
Yb^{3+}	7.85	6.50	5.48	3.90	10.69	3.12	25.5
Lu^{3+}	7.69	6.64	5.44	3.96	10.49	2.90	25.5

*) Determined potentiometrically (glass electrode) at 25°C in aqueous media, I = 0.1 (KNO_3).— **) Enthalpy values were determined calorimetrically at 25°C in 0.100 M KNO_3. The enthalpy values were combined with the lg K_1 value to calculate free energies and entropies of formation for the 1:1 chelates.

Campbell and Moeller [1] proposed that the magnitude of the formation constants indicates that the tropolonate ion is bidentate, even in the tetrakis chelates. The formation of relatively stable tetrakis chelates indicates that this ligand can stabilize high coordination numbers in solution. The fact that the average value of lg (K_3/K_4) is about 1.5, whereas the average values of lg (K_1/K_2) and lg (K_2/K_3) are 1.2 and 1.0, respectively, was attributed to destabilization of the 4:1 chelate as a consequence of steric crowding of the ligands about the rare earth ions. A plot of lg K_n versus the reciprocal of the metal ion radius revealed that K_1, K_2 and K_3 increased linearly with decreasing ionic radius, with no irregularities observed for the heavier half of the lanthanide series. A general increase in K_4 with increasing r^{-1} values was also observed, but the reduced precision of the measurements obscured any absolute linearity in the plot. The lg K values for the yttrium chelate are consistent with the size of the yttrium ion [1].

The enthalpies of formation for the 1:1 tropolonato-rare earth chelates (Table 4/3) were measured calorimetrically at 25°C in aqueous media, I = 0.1 (KNO_3). The limited solubility of the tris tropolonato complexes prevented formation of the higher order species in concentrations sufficient to allow calculation of their enthalpies of formation. The enthalpy values were combined with the lg K_1 values reported in [1] to calculate the free energies and entropies of formation of the 1:1 chelates. The ligational enthalpy changes are all exothermic and showed little variation across the rare earth series. The ligational entropy changes are positive and increased regularly across the series. The lack of a discontinuity in the trend of ΔH_1 and ΔS_1 values near the middle of the series was discussed in relation to the hydration spheres of both the free and complexed rare earth ions [2].

4.3.10.1.2 Tris Chelates

Scandium Compound $Sc(C_7H_5O_2)_3$

Muetterties, Wright [4] effected synthesis of the compound according to the following procedure: To a solution of tropolone (0.015 mol) in methanol was added an aqueous solution of scandium nitrate hexahydrate (0.004 mol), giving a yellow solution. On standing, yellow crystals of $H[Sc(C_7H_5O_2)_4]$ slowly separated. After 2 h, the slurry was filtered, and the solid was dissolved in a hot methanol/chloroform mixture. Concentration of this solution effected separation of the tris chelate, which was collected and dried under vacuum at 80°C. The compound thus prepared was soluble in chloroform and melted at 365 to 375°C [4].

Melson and coworkers [5] obtained crystals of a compound suitable for X-ray studies by the following method: An aqueous solution (30 ml) of scandium nitrate hydrate (0.34 g) was added to a solution of tropolone (0.46 g) in methanol (5 ml). The solution was stirred at room temperature for 2 to 3 h and then filtered. The solid was dried at room temperature under vacuum over phosphorus(V) oxide. The dried substance consisted of the pale yellow acid salt, $H[Sc(C_7H_5O_2)_4]$, and the red-brown tris chelate, $Sc(C_7H_5O_2)_3$. Crystals of the latter compound, suitable for the crystal structure study, readily separated from this mixture [5].

Charalambous [6] prepared the tris scandium chelate by heating, under reflux, a mixture of scandium chloride and tropolone (1:3 mol ratio) in ethanol. The crude product was purified by sublimation at 250°C under vacuum (0.1 Torr).

The crystal and molecular structures of the compound obtained by Melson and coworkers [5] were determined by single crystal X-ray diffraction (R = 0.033 for 783 reflections). The compound crystallizes in the trigonal space group $R\bar{3}c$-D^6_{3d} (No. 167) with six molecules in a unit cell of dimensions a = 10.455 and c = 32.595 Å (D(calculated) = 1.318 g/cm³; D(observed) = 1.33 g/cm³). The structure consists of discrete $Sc(C_7H_5O_2)_3$ molecules with the scandium ion located at the 32 special position, (0, 0, $^1/_4$). Thus, the molecule has crystallographically imposed D_3 symmetry. There were no unusually short intermolecular contacts, with the shortest H···H, H···O, H···C, and C···C being 2.48, 2.68, 2.97, and 3.44 Å, respectively. The coordination environment of the six-coordinate scandium ion is depicted in **Fig. 4-1**, p. 56. The coordination polyhedron, defined by the six ligated oxygen atoms, is intermediate between trigonal antiprismatic and trigonal prismatic as determined from a value of 33° for the projected twist angle Φ (Φ was defined by projection of the metal/oxygen bond vectors onto a plane normal to the C_3 axis). The values of Φ for the idealized trigonal antiprismatic and trigonal prismatic polyhedra are 60° and 0°, respectively. The scandium/oxygen bond length is 2.102 Å. The tropolonate group is planar, with a ligand bite of 2.523 Å. The latter value falls within the narrow range observed for compounds containing chelated tropolonate groups. Thus, the ligand behaves as a rigid, inflexible chelating agent, with ligation to the scandium ion causing little change in its structure [5].

Fig. 4-1

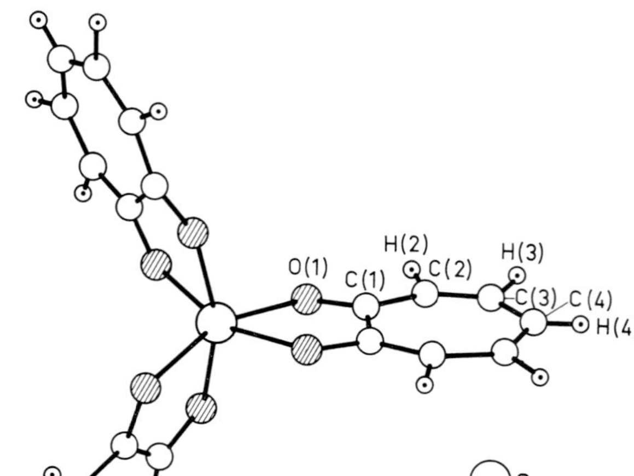

View of tris(tropolonato)scandium(III) down the C_3 axis.

The mass spectrum of the compound (prepared according to the procedure in [6]) was obtained with the spectrometer operating at 80 eV and the insertion probe at ≈250°C. The spectrum revealed a peak corresponding to the molecular ion, $Sc(C_7H_5O_2)_3^+$, which fragments exclusively by loss of a ligand radical to give the even-electron ion, $Sc(C_7H_5O_2)_2^+$. The latter ion presents the most intense peak in the spectrum. Further fragmentation produced a peak corresponding to $Sc(C_7H_5O_2)(OH)^+$. The very low abundance of $Sc(C_7H_5O_2)^+$ and the absence of Sc^+ in the spectrum are in accord with the high stability of the scandium(III) ion [6].

Yttrium Compound $Y(C_7H_5O_2)_3$

A solution of yttrium chloride hexahydrate (0.0066 mol) in 70 ml of methanol and 30 ml of water was added to a solution of tropolone (0.046 mol) in about 50 ml of ethanol. The resulting mixture was heated on a steam bath for 1 h, during which time a solid separated. The yellow solid was collected, and an attempt was made to recrystallize it from a hot mixture of acetonitrile, ethanol, and chloroform. The insoluble fraction from the recrystallization process had an infrared spectrum identical with those of the lighter rare earth chelates (M = La to Ho). It was shown by an X-ray powder pattern to be isomorphous with this group of chelates. The yellow solid, dried under vacuum at 40°C, melted with decomposition at 275 to 277°C.

The pale yellow solution, obtained from the recrystallization attempt, was concentrated, and a yellow solid was obtained, which was dried under vacuum at 80°C (m.p. = 276 to 278°C, with decomposition). The infrared spectrum of this solid was similar to, but not identical with that of the insoluble fraction. An X-ray powder pattern of this material revealed that the sample was poorly crystalline. However, this substance is metastable, and undergoes a transition at ≈180°C in the solid phase to the polymeric lattice characteristic of the insoluble tris tropolonate yttrium chelate described above [4].

Lanthanide Compounds $M(C_7H_5O_2)_3$

The lanthanum complex was obtained by adding a solution of tropolone (0.045 mol) in a mixed solvent of methanol and 50 ml of water to a solution of 0.01 mol of lanthanum chloride heptahydrate in a mixture of 50 ml of water and 25 ml of methanol. Aqueous ammonia was added, and the reaction mixture was heated to 70°C for 45 min, during which time a solid separated. The solid was collected, washed, and dried. A similar procedure was followed to prepare the europium chelate [7]. Several complexes (M = Pr, Nd, Sm, Gd, Tb, Dy, Ho) were prepared by allowing the respective rare earth nitrates (0.01 mol) to react with tropolone (0.032 mol) in warm (50°C) 95% ethanol (250 ml). After about 5 min, solids began to separate. The solutions were heated at 50°C for another 25 min, and then the slurries were filtered. The solids were washed with ethanol and dried under vacuum at 150°C [4]. The erbium, thulium, ytterbium, and lutetium chelates were prepared by vacuum pyrolysis of the ammonium salts of the respective tetrakis tropolonate anions at 200°C (see for preparation of the ammonium salts, p. 60). These compounds were also formed by vacuum pyrolysis of the acid salts, $H[M(C_7H_5O_2)_4]$, at ≈170°C (see preparation of the hydrogen compound, p. 58). The derivatives of the heavier rare earths (M = Er, Tm, Yb, Lu) melted and decomposed at about 390 to 395°C, whereas the other tris chelates did not melt below 400°C [4].

The compounds derived from the larger rare earth ions (M = La to Ho) are difficultly soluble in organic solvents. Strongly basic solvents like dimethyl sulfoxide dissolved these chelates, but only at elevated temperatures and then at a low rate. The intractable character of these chelates, as well as the fact that they were readily obtained in anhydrous form, even though the rare earth metal ions are seldom satisfied with coordination numbers less than seven, prompted the assumption that the tropolonate lattices are not based on discrete six-coordinate chelates. Muetterties, Wright [4] proposed that there is sharing of some of the chelate oxygen atoms between metal atoms (three-coordinate oxygen atoms), forming a polymeric lattice, in which the coordination number of the metal ion exceeds six. X-ray powder patterns revealed that these species (M = La to Ho) are isomorphous.

Size of the metal ion does give rise to some variation in properties within the isostructural group, e. g., the most stable polymeric lattice was formed by the largest ion, lanthanum(III). The lanthanum chelate, when formed from solution, is colloidal and reveals no solubilization phenomena in contact with a water/ethanol/acetonitrile mixture after one week. The lanthanum chelate did not react with sodium tropolonate in solution (see Section 4.3.10.1.4), whereas the remaining members of the isomorphous group formed the sodium salts of the tetrakis chelate anions. The cerium chelate is more tractable than that of lanthanum, and maximum solubility was observed for the holmium chelate among the isomorphous series [4].

The tris chelates of erbium, thulium, ytterbium, and lutetium, formed by vacuum pyrolysis of the acid salts, $H[M(C_7H_5O_2)_4]$, were also isomorphous, as revealed by X-ray powder patterns. However, these chelates have a distinctly different lattice from those of the larger rare earth ions and show greater tractability (they dissolved rapidly in dimethyl sulfoxide to yield rather concentrated solutions). The tris tropolonate of ytterbium was heated to 250°C, with no evidence of a thermal transition to the polymeric lattice characteristic of the lanthanum chelate. Although the authors presented no definitive data bearing on the aggregation in this second type of lattice, a very weak band at 1590 cm^{-1} was interpreted as being characteristic of a polymeric lattice. The lattice energies of the erbium to lutetium group of chelates appeared to be significantly lower than those of the lighter rare earth chelates as evidenced by (1) rapid dissolution in dimethylsulfoxide, (2) facile reaction with ammonium tropolonate to give $NH_4[M(C_7H_5O_2)_4]$ (the tris chelates of europium and gadolinium did not react with ammonium tropolonate), and (3) reaction of the thulium chelate with 1,10-phenanthroline to give a neutral eight-coordinate complex, see p. 58 [4].

There is evidence for two crystalline modifications of the tris chelates of thulium and lutetium. When these compounds were prepared by pyrolysis of the ammonium salts of the tetrakis chelates at 170 to 200°C, the solids appeared to be isomorphous with a soluble form of the tris yttrium chelate, which is described below, and distinct from the crystalline forms described above [4].

Proton NMR data for each of the chelates in $(CD_3)_2SO$ are presented. The tris tropolonates of europium and terbium did not visibly fluoresce, with excitation at 3660 and 2537 Å [4].

The modes of thermal decomposition of the rare earth metal tropolonates were studied by differential thermal analysis and thermogravimetric techniques. The compounds decomposed in two steps, with the first step yielding a basic tropolonate complex and the second a basic carbonate species for the samarium and europium chelates and the sesquioxides for the lanthanum, neodymium, erbium, thulium, and ytterbium tropolonates. The activation energies of decomposition of these compounds ranged from 21.0 to 96.5 kcal/mol [8].

$Er(C_7H_5O_2)_3 \cdot H_2O$

The compound slowly separated in crystalline form from a solution prepared by mixing a methanolic solution (5 ml) of tropolone (0.0082 mol) with a solution of erbium nitrate hexahydrate (0.0022 mol) in 30 ml of water and 10 ml of methanol. The solid was collected and heated in vacuo at 70°C. The hydrate was dried under vacuum for 2 h, yielding the anhydrous tris chelate. The latter compound is isomorphous with the anhydrous chelates of the heavier rare earths (M = Tm, Yb, Lu) obtained by pyrolysis of the hydrogen compounds. When the methanol content of the nitrate solution was reduced to 5 ml, the crystalline phase that separated from solution was $H[Er(C_7H_5O_2)_4]$ [4].

4.3.10.1.3 Mixed Complex Derived from Tropolone and 1,10-Phenanthroline ($= C_{12}H_8N_2$ = phen)

$Tm(C_7H_5O_2)_3 \cdot$ phen. A slurry containing the tris tropolonate of thulium (0.001 mol) and 1,10-phenanthroline (0.0011 mol) in 250 ml of a 50:50 mixture of ethanol and acetonitrile was heated to 70°C for 6 h. The slurry was filtered, and the solid collected was washed with acetonitrile and ethanol and dried under vacuum at 100°C. The product melted with decomposition at 375 to 380°C. Under the same conditions, the tris tropolonate of europium underwent no reaction in a period of 2 days [4].

4.3.10.1.4 Tetrakis Chelates

Rare Earth(III) Compounds

Hydrogen Compound $H[M(C_7H_5O_2)_4]$. Each complex (M = Sc, Y, Tm, Yb, Lu) was prepared by adding an aqueous solution of the rare earth nitrate (0.005 mol) to an ethanolic solution of tropolone (0.02 mol) and then warming the mixture to about 40°C, whereupon a solid separated. The solid was collected, washed with ethanol and water, and dried under vacuum at 70°C. Under similar conditions, the larger rare earth ions (M = La to Ho) yielded polymeric forms of the tris tropolonates [4].

Following their initial communication [9], Melson and coworkers [10] published simultaneously with Davis, Einstein [11] on the crystal and molecular structures of the scandium chelate, as determined independently by the two groups of investigators using single crystal X-ray diffraction techniques (R = 0.027 for 2539 reflections in the structural analysis by Melson and coworkers [10], and R = 0.059 for 2732 reflections in the analysis by Davis, Einstein [11]). The results of the two crystal structures are essentially the same within limits of the combined errors [11]. The complex crystallizes in the triclinic space group $P\bar{1}$-C_i^1 (No. 2) with two molecules in a unit cell of dimensions a = 10.022, b = 11.515, c = 12.004 Å; α = 72.74°, β = 84.58°, and γ = 65.04°, as reported by Melson and coworkers [11]. The unit cell reported by Davis and Einstein [11] (a = 11.624, b = 11.986, c = 10.004 Å, α = 95.33°, β = 116.27°, and γ = 102.32°) is related to that of Melson and coworkers by the transformation 0, 0, −1; −1, 0, −1; 0, 1, 0 [11]. The hydrogen tetrakis(tropolonato)scandium(III) molecules exist as centrosymmetrically related hydrogen-bonded dimers, with an intermolecular O-O bond distance of 2.484 Å. A view of the dimer, excluding ring protons, is depicted in **Fig. 4-2**. Each scandium ion is coordinated by eight oxygen atoms from four bidentate ligands, arranged at the vertices of a dodecahedron (**Fig. 4-3**). The bidentate ligands span equivalent m edges of the polyhedron, and thus, the dodecahedron has approximate D_{2d} symmetry. Dimensions of the coordination polyhedron are given in Fig. 4-3. The hydrogen atoms attached to the 01(4) oxygen atoms in the dimer are hydrogen bonded to the 01(1) oxygen atom across the inversion center [10, 11]. The hydrogen bonds are nearly linear with an O-H distance of 1.00 Å, a hydrogen-bonded O ··· H distance of 1.49 Å, and an O-H ··· O angle of 175.9°. The Sc-O bond distances (Å) are: Sc-01(1) = 2.314, Sc-02(1) = 2.209, Sc-01(2) = 2.180, Sc-02(2) = 2.164, Sc-01(3) = 2.178, Sc-02(3) = 2.183, Sc-01(4) = 2.260, Sc-02(4) = 2.228 Å [10]. The average intraligand O···O separation (chelate bite) is 2.505 Å [10].

Fig. 4-2

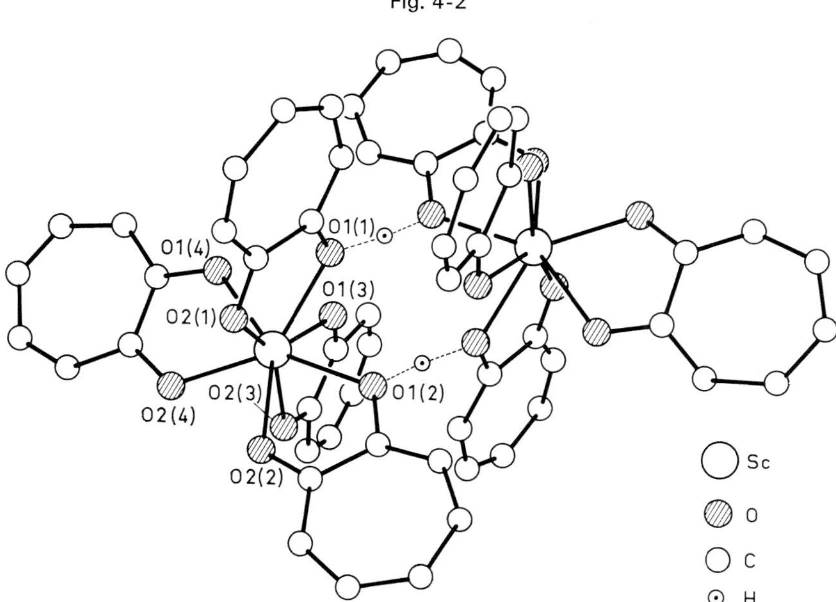

Structure of the hydrogen tetrakis(tropolonato)scandium(III) dimer $[HSc(C_7H_5O_2)_4]_2$.

Fig. 4-3

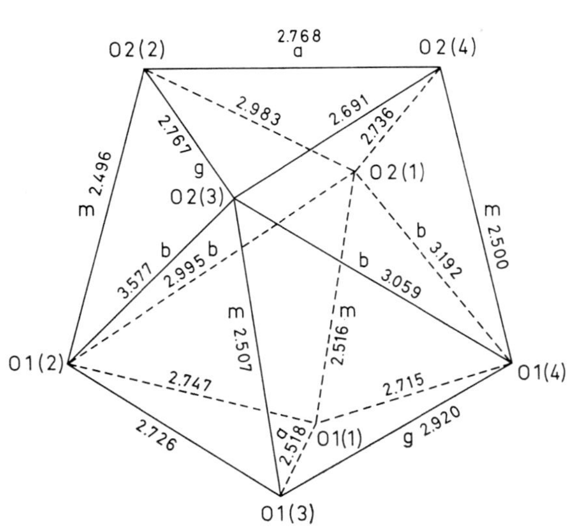

Coordination polyhedron for the scandium(III) ion in $[HSc(C_7H_5O_2)_4]_2$.

The ν(C=O) stretching mode was observed at 1593 cm^{-1} in the infrared spectrum of the scandium chelate (method of sample preparation not described). The ν(O-H) stretching mode, observed at ≈3500 cm^{-1} in the spectrum of the free ligand, was not detected in the spectrum of the scandium chelate [9]. The ESCA spectrum of $HSc(C_7H_5O_2)_4$ obtained in the oxygen (1s) region (KE = 716.4 ±0.2 eV; BE = 532.0±0.2 eV) using MgKα radiation, revealed a symmetrical peak, with a width at half the height of 2.1 eV, indicating that the eight oxygen atoms are equivalent [9].

The compounds are not very soluble in aqueous media. Saturated water/acetonitrile solutions had a pH between 4 and 5. The thulium, ytterbium, and lutetium derivatives are stable up to 150°C under vacuum. Above this temperature, they lose tropolone to yield the anhydrous tris tropolonates. X-ray powder patterns revealed that all complexes are isomorphous [4].

Alkali Metal Salts $M'[M(C_7H_5O_2)_4]$ (M' = Li, Na). Each sodium salt (M = Sc, Y, Ce to Lu except for Pm) was prepared by heating at reflux for 15 h a slurry of the tris tropolonato rare earth chelate (0.001 mol) and sodium tropolonate (0.0011 mol) in a solvent mixture containing 50 ml of acetonitrile, 50 ml of water, and 30 ml of methanol. The slurry was filtered while hot, and the solid was washed with a hot mixture of water and methanol and dried under vacuum at 100 to 150°C. None of the compounds melted below 400°C. Under similar conditions, there was no reaction between the tris tropolonate of lanthanum and sodium tropolonate, within a period of one week at 80°C [4]. The lithium salts (M = La to Lu except for Pm) were prepared by a similar procedure, except that the slurries were heated to 70°C for 2 days and then filtered [4, 12].

Infrared spectra of both series of compounds were obtained in the region 1650 to 300 cm^{-1} on samples dispersed as Nujol and hexachlorobutadiene mulls [12, 13]. Data obtained for the lithium salts of the tetrakis chelates were tabulated in [12]. The ν(C-C) stretching modes were assigned to bands at 1597 to 1600 cm^{-1} and 1432 to 1444 cm^{-1}, whereas the ν(C-O) stretching modes were assigned to bands at 1515 (1510 sh) to 1524 (1510 sh) cm^{-1}, coupled with ν(C-C), and 1357 to 1371 cm^{-1}. The spectra of the series of isostructural chelates exhibited identical band patterns in the region of the ν(M-O) stretching modes, where four peaks were observed. The wave numbers of one significantly metal sensitive ν(M-O) stretching mode in the spectra of the lithium salts are presented below. The variation of ν_3 was discussed in relation to the 4f-orbital population and was considered to provide evidence for a small, but finite, crystal-field stabilization in all the complexes except those of the $4f^0$, $4f^7$, and $4f^{14}$ ions. Plots of ν_3(M-O) versus the 4f-orbital population for the sodium and lithium salts are depicted in [12]. The relationship between lg β_4 (M = Gd to Lu) and ν_3(M-O) was also discussed for both series of salts [12].

M	La	Ce	Pr	Nd	Sm	Eu	Gd
ν_3 (M-O) in cm^{-1} .	490.5	492	494.5	496	499	502	500

M	Tb	Dy	Ho	Er	Tm	Yb	Lu
ν_3 (M-O) in cm^{-1} .	504	505.5	506.5	508.5	510.5	512.5	512.5

Ammonium Salts $NH_4[M(C_7H_5O_2)_4]$. The europium, thulium, and lutetium complexes were each prepared by adding an aqueous solution (100 ml) of the appropriate rare earth nitrate to a warm mixture of tropolone (0.04 mol) and 10 ml of concentrated aqueous ammonia in 50 ml of ethanol and 100 ml of water. Solids separated immediately on mixing the two solutions. Each slurry was heated to 70°C for a period of 2 h. The solid was collected, washed with water and ethanol, and then dried under vacuum at 70°C. Lanthanum nitrate under these conditions yielded only the tris tropolonato chelate [4].

The tris tropolonates of several rare earths were heated to 70°C in contact with a solution of ammonium tropolonate in a water/methanol mixture. In a period of 2 h, the tris tropolonates of ytterbium, holmium, and dysprosium reacted to yield the ammonium salt of the respective tetrakis chelate anions. Under the same conditions, tris tropolonates of europium, gadolinium, and terbium were unreactive.

The crystal and molecular structures of the holmium compound were determined by single-crystal X-ray diffraction (R = 0.075 for 3228 independent reflections). The salt crystallizes in the monoclinic space group C2/c-C_{2h}^6 (No. 15) with sixteen formula weights in a unit cell of dimensions a = 22.498, b = 18.176, c = 29.073 Å, and β = 114.40°; density D (calculated) = 1.638 g/cm³; D (observed) = 1.635 g/cm³. There are two crystallographically independent $Ho(C_7H_5O_2)_4^-$ complex anions in the asymmetric unit of the ammonium salt. In each anion, the holmium ion is eight-coordinate, with the eight oxygen atoms of the four bidentate tropolonate groups situated, to a fair approximation, at the vertices of an idealized D_{2d} dodecahedron. Each ligand spans m edges of the dodecahedron, thus, giving m m m m stereoisomers. The eight independent Ho-O bonds present in each complex averaged to 2.35 and 2.36 Å, respectively. No systematic differences were observed between dodecahedral A- and B-type metal-oxygen bond lengths [14]. X-ray powder patterns revealed that each of the ammonium salts was isomorphous [4].

All the ammonium salts are stable under vacuum to about 150°C. Above this temperature, ammonium tropolonate volatilizes to leave a residue of the tris tropolonate chelate [4].

Cerium(IV) Compound

$Ce(C_7H_5O_2)_4$. A solution of ammonium cerium(IV) sulfate dihydrate (0.01 mol) in a minimum of ≈5% sulfuric acid was added to a solution of 0.04 mol of tropolone in a mixture of 60 ml of methanol, 80 ml of water, and 1 ml of concentrated sulfuric acid. A copious brick-red precipitate appeared immediately. The reaction mixture was stirred for 15 min and then filtered. The solid was recrystallized from dichloromethane and dried under vacuum at 40°C. The product was a black crystalline substance that did not melt below 400°C [4].

References to 4.3.10.1:

[1] D. L. Campbell, T. Moeller (J. Inorg. Nucl. Chem. **31** [1969] 1077/82). — [2] D. L. Campbell, T. Moeller (J. Inorg. Nucl. Chem. **32** [1970] 945/51). — [3] M. Hirai, Y. Oka (Bull. Chem. Soc. Japan **43** [1970] 778/82). — [4] E. L. Muetterties, C. M. Wright (J. Am. Chem. Soc. **87** [1965] 4706/17). — [5] T. J. Anderson, M. A. Neuman, G. A. Melson (Inorg. Chem. **13** [1974] 158/63).

[6] J. Charalambous (Inorg. Chim. Acta **18** [1976] 241/6). — [7] E. L. Muetterties, C. M. Wright (J. Am. Chem. Soc. **86** [1964] 5132/7). — [8] M. D. Taylor, R. Panayappan (J. Therm. Anal. **7** [1975] 385/96 from C.A. **83** [1975] No. 52661). — [9] D. J. Olszanski, T. J. Anderson, M. A. Neuman, G. A. Melson (Inorg. Nucl. Chem. Letters **10** [1974] 137/41). — [10] T. J. Anderson, M. A. Neuman, G. A. Melson (Inorg. Chem. **13** [1974] 1884/90).

[11] A. R. Davies, F. W. B. Einstein (Inorg. Chem. **13** [1974] 1880/4). — [12] L. G. Hulett, D. A. Thornton (J. Mol. Struct. **13** [1972] 115/27). — [13] L. G. Hulett, D. A. Thornton (Chimia [Aarau] **26** [1972] 72/4). — [14] V. W. Day (Diss. Abstr. Intern. B **30** [1970] 4079/80).

4.3.10.2 With 4-Methyltropolone (= $C_8H_8O_2$)

$M(C_8H_7O_2)^{2+}$ Ion. Formation in Solution. The formation constants for the 1:1 4-methyltropolonato rare earth chelates were determined spectrophotometrically in aqueous media, I = 0.5 ($NaClO_4$) (temperature not specified). The values of lg K_1 generally increase with decreasing metal ion radius, and the values of lg (K_n/K_{n+1}) lie in the range 1.3 to 1.5, B. P. Gupta, Y. Dutt, R. P. Singh (Indian J. Chem. **5** [1967] 214/5).

metal ion	Y^{3+}	La^{3+}	Ce^{3+}	Pr^{3+}	Nd^{3+}	Sm^{3+}
lg K_1	7.38	6.12	6.42	6.71	6.78	7.07
metal ion	Gd^{3+}	Dy^{3+}	Ho^{3+}	Er^{3+}	Yb^{3+}	Lu^{3+}
lg K_1	7.10	7.36	7.42	7.57	7.74	8.01

4.3.10.3 With 3-Carboxy-4-methyltropolone (= $C_9H_8O_4$)

$M(C_9H_6O_4)_n^{3-2n}$ Ions. Formation in Solution. Formation constants for rare earth complexes derived from 3-carboxy-4-methyltropolone are recorded in Table 4/4. The values of lg K_n generally increase with decreasing metal ion radius, except for a decrease at gadolinium [1]. It was also observed that the wavelength of maximum absorbance in the electronic spectrum of the complex decreases (range of 388 to 382 nm) with decreasing metal ion radius [2].

Table 4/4
Formation Constants for 3-Carboxy-4-methyltropolonato Rare Earth Chelates, $M(C_9H_6O_4)_n^{3-2n}$ at 25°C, I = 0.2 ($NaClO_4$).

metal ion	method*)	lg K_1	lg K_2	lg β_2	lg K_3	lg β_3	Ref.
Y^{3+}	gl	8.26	6.62	14.88	3.96	18.84	[1]
Y^{3+}	sp	8.47					[2]
La^{3+}	gl	7.20	5.56	12.76	3.40	16.16	[1]
La^{3+}	sp	7.07					[2]
Ce^{3+}	gl	7.42	5.72	13.14	3.34	16.48	[1]
Ce^{3+}	sp	7.20					[2]
Pr^{3+}	gl	7.74	5.96	13.70	3.56	17.26	[1]
Pr^{3+}	sp	7.45					[2]
Nd^{3+}	gl	7.76	6.04	13.80	3.70	17.50	[1]
Nd^{3+}	sp	7.69					[2]
Sm^{3+}	gl	7.98	6.42	14.40	3.78	18.18	[1]
Sm^{3+}	sp	7.99					[2]
Gd^{3+}	gl	8.02	6.38	14.40	3.82	18.22	[1]
Gd^{3+}	sp	7.91					[2]
Dy^{3+}	gl	8.28	6.78	15.06	4.00	19.06	[1]
Ho^{3+}	gl	8.24	6.76	15.00	3.98	18.98	[1]
Ho^{3+}	sp	8.58					[2]
Er^{3+}	gl	8.35	6.85	15.20	4.18	19.38	[1]
Er^{3+}	sp	8.65					[2]
Yb^{3+}	gl	8.60	7.00	15.60	4.42	20.02	[1]
Yb^{3+}	sp	8.86					[2]
Lu^{3+}	gl	8.64	7.06	15.70	4.60	20.30	[1]
Lu^{3+}	sp	8.85					[2]

*) Abbreviation: gl = potentiometric measurement with a glass electrode, sp = spectrophotometric determination.

References to 4.3.10.3:

[1] B. P. Gupta, Y. Dutt, R. P. Singh (J. Indian Chem. Soc. **43** [1966] 610/2). — [2] B. P. Gupta, Y. Dutt, R. P. Singh (J. Inorg. Nucl. Chem. **29** [1967] 1806/10).

4.3.10.4 With 3-Isopropyltropolone (α-Isopropyltropolone = $C_{10}H_{12}O_2$)

$Lu(C_{10}H_{11}O_2)_3$. The lutetium chelate was prepared according to the procedure described in Section 4.3.10.6 for preparation of the analogous γ-isopropyltropolonate lutetium chelate. For this isomer, the molecular weight in benzene (ebullioscopic) was 1813, in chloroform and dichloromethane (vapor phase osmometry) the values were 1806 and 1750, respectively, and in benzene (cryoscopic) the value was 2100 (dimer = 1328, trimer = 1992), E. L. Muetterties, H. Roesky, C. M. Wright (J. Am. Chem. Soc. **88** [1966] 4856/61).

4.3.10.5 With 4-Isopropyltropolone (= β-Isopropyltropolone = $C_{10}H_{12}O_2$)

$M(C_{10}H_{11}O_2)_n^{3-n}$ Complexes (n = 1 to 4). **Formation in Solution.** The stepwise formation constants (Table 4/5) for the 4-isopropyltropolonato rare earth chelates were determined potentiometrically (glass electrode) in 80% (v/v) methanol/water, I = 0.1 (NaCl) [1, 2]. A plot of lg K_n versus atomic number revealed a general increase in stability across the rare earth series, with two maxima at europium and erbium and a minimum at gadolinium [1]. The formation constants of the 1:1 chelates in aqueous media, recorded below, were determined spectrophotometrically, I = 0.5 ($NaClO_4$) (temperature not specified). The value of lg K_1 showed a more or less general increase with decreasing metal ion radius [3].

metal ion	Y^{3+}	La^{3+}	Ce^{3+}	Pr^{3+}	Nd^{3+}	Sm^{3+}
lg K_1	7.28	6.29	6.53	6.74	6.70	7.04
metal ion	Gd^{3+}	Dy^{3+}	Ho^{3+}	Er^{3+}	Yb^{3+}	Lu^{3+}
lg K_1	7.15	7.33	7.40	7.49	7.62	7.95

Table 4/5
Formation Constants of 4-Isopropyltropolonato Rare Earth Chelates, $M(C_{10}H_{11}O_2)_n^{3-n}$ (n = 1 to 4).*)

metal ion	t in °C	lg K_1	lg K_2	lg K_3	lg K_4	Ref.
Y^{3+}	24	9.0	7.5	6.2	4.8	[1]
La^{3+}	24	7.88	6.35	5.0	3.4	[1]
Pr^{3+}	24	8.3	6.85	5.5	4.1	[1]
Pr^{3+}	23	8.31	6.88	5.48	4.20	[2]
Nd^{3+}	24	8.4	6.99	5.7	4.2	[1]
Sm^{3+}	24	8.7	7.29	6.0	4.6	[1]
Eu^{3+}	24	8.9	7.45	6.1	4.7	[1]
Gd^{3+}	24	8.8	7.33	6.1	4.6	[1]
Tb^{3+}	24	9.1	7.6	6.3	4.9	[1]
Dy^{3+}	24	9.2	7.7	6.4	5.0	[1]
Ho^{3+}	24	9.2	7.7	6.5	5.2	[1]
Er^{3+}	24	9.4	7.9	6.7	5.4	[1]

*) Determined potentiometrically (glass electrode) at 24°C in 80% (v/v) methanol/water, I = 0.1 (NaCl) [1].

References to 4.3.10.5:

[1] N. K. Davidenko, A. A. Zholdakov (Zh. Neorgan. Khim. **13** [1968] 2955/9; Russ. J. Inorg Chem. **13** [1968] 1522/4). — [2] K. B. Yatsimirskii, N. K. Davidenko, L. N. Lugina (Dokl. Akad. Nauk SSSR **170** [1966] 864/7; Dokl. Chem. Proc. Acad. Sci. USSR **170** [1966] 954/7). — [3] B. P. Gupta, Y. Dutt, R. P. Singh (Indian J. Chem. **5** [1967] 214/5).

4.3.10.6 With 5-Isopropyltropolone (= γ-Isopropyltropolone = $C_{10}H_{12}O_2$)

$M(C_{10}H_{11}O_2)_3$. The lanthanum chelate was prepared by adding 5-isopropyltropolone (0.015 mol) to a solution of lanthanum chloride heptahydrate (0.005 mol) in a mixed solvent of 100 ml of water and 70 ml of methanol. The clear solution was slowly evaporated on a steam bath, yielding a pale yellow precipitate. The precipitate was collected and then slurried in a hot mixture of methanol, water, and acetonitrile. The slurry was warmed on a steam bath for 0.5 h and then filtered while hot. The solid was dried under vacuum at 60°C. The europium and gadolinium compounds were similarly prepared from their nitrate salts (0.0022 mol). In each case, a yellow solid began to separate several minutes after mixing the reactants. After 24 h, the solid was isolated by filtration and purified according to the procedure described above. All of the above compounds had melting points above 400°C, and were insoluble in common organic solvents.

The terbium compound was prepared by adding a solution of terbium nitrate hexahydrate (0.0011 mol) in 50 ml of water to a methanolic solution (20 ml) of 5-isopropyltropolone (0.0037 mol). During the addition, the solution turned yellow, and then a yellow solid separated. The slurry was stirred for 6 h, and during this time, 50 ml of water was added to effect complete precipitation. The solid was collected, washed with water, and dried at 25°C under vacuum (1 μ). The compound was soluble in methanol and chloroform, but insoluble in acetone.

The anhydrous holmium chelate was obtained by adding a solution of 5-isopropyltropolone (0.0073 mol) in 20 ml of methanol to an aqueous solution (40 ml) of the nitrate salt (0.0022 mol) and warming the mixture on a steam bath. Acetonitrile (50 ml) and methanol (30 ml) were then added, followed by the addition of aqueous ammonia to bring the reaction mixture to pH 6. Addition of the ammonia solution produced a somewhat gummy precipitate. The reaction slurry was further heated on a steam bath for 1 h and then filtered. The solid was dried under vacuum at 80°C. The infrared spectrum of the solid was identical with those of the lanthanum and europium chelates.

Erbium nitrate hexahydrate was treated with 5-isopropyltropolone according to the procedure described for holmium nitrate. In this case, the crude product was added to a hot mixture of 30% cyclohexane and 70% chloroform, and the slurry was filtered. The insoluble material, after drying under vacuum at 80°C, had an infrared spectrum identical with those of the tris chelates of lanthanum, europium, and holmium. The compound melted with decomposition at 380 to 386°C. The cyclohexane/chloroform filtrate was concentrated to yield a crystalline solid, which melted with decomposition at 380 to 389°C.

The same procedure was followed with lutetium nitrate hexahydrate as with erbium nitrate. In this case, the crude product obtained was completely soluble in chloroform. It was recrystallized from a hot mixture of chloroform/methanol, and dried under vacuum at 80°C for 40 h. The compound melted, with decomposition, at 375 to 385°C. The soluble fraction of the erbium chelate and the lutetium compound exhibited identical infrared spectra.

The insoluble nature of the γ-isopropyltropolonate chelates of the larger rare earth ions in common organic solvents supports the view that these metal ions are seven- or eight-coordinate in a polymeric lattice formed through bridging tropolonate oxygen atoms. The degree of association apparently drops significantly in going from the large rare earth cations to the smaller ions such as lutetium(III). Molecular weight studies (vapor pressure osmometry) of the soluble erbium and lutetium chelates in nonpolar solvents indicated an association that approximates a trimer, e. g., for the erbium complex in chloroform, molecular weight = 1686 (dimer = 1312, trimer = 1968), and for the lutetium complex in chloroform, molecular weight = 1521 (dimer = 1328, trimer = 1992), E. L. Muetterties, H. Roesky, C. M. Wright (J. Am. Chem. Soc. **88** [1966] 4856/61).

$M(C_{10}H_{11}O_2)_3 \cdot nH_2O$ (n = 1, 2). The hydrated holmium (n = 1) and erbium (n = 2) chelates were prepared by adding a solution of the respective rare earth nitrate hexahydrates (0.0011 mol) in 25 ml of water to a methanolic solution (30 ml) of 5-isopropyltropolone (0.003 mol). Triethylamine was added to the reaction mixtures until the pH was about 6. The mixture was stirred for 3 h, and during this time, 60 ml of water was added to effect precipitation of the product. The collected solids were dried under vacuum (1 μ) at 25°C. The infrared spectra revealed strong bands associated with water. It was suggested that the compounds represent seven- or eight-coordinate, polymeric structures. Heating the erbium chelate under vacuum at 80°C produced the anhydrous compound, which was isomorphous with the lanthanum (through terbium) chelate as judged by comparison of X-ray powder patterns, E. L. Muetterties, H. Roesky, C. M. Wright (J. Am. Chem. Soc. **88** [1966] 4856/61).

5 Complexes with Diketones and Polyketones

John H. Forsberg
Department of Chemistry, Monsanto Hall, Saint Louis University
Saint Louis, Missouri, USA

Preliminary remarks

Complexes with diketones derived from ferrocene are described in "Eisen-Organische Verbindungen" A 3. Complexes with diketones containing the oxo groups in N-heterocyclic ring systems, e. g., derivatives of pyrazolone, are described in "Rare Earth Elements" D 2.

5.1 Complexes with 1,3-Diketones $RC(O)CH_2C(O)R$ (= β-Diketones)

General Literature:

A. B. P. Sinha, Fluorescence of Laser Action in Rare Earth Chelates, in: C. N. R. Rao, J. R. Ferraro, Spectroscopy in Inorganic Chemistry, Vol. II, Academic Press, New York 1971, pp. 255/88.

R. E. Sievers, Nuclear Magnetic Shift Reagents, Academic Press, New York 1973.

W. DeW. Horrocks, Lanthanide Shift Reagents and Other Analytical Applications in: G. N. LaMar, W. DeW. Horrocks, R. H. Holm, NMR of Paramagnetic Molecules, Principles and Applications, Academic Press, New York 1973, pp. 479/519.

B. C. Mayo, Lanthanide Shift Reagents in NMR Spectroscopy, Chem. Soc. Rev. **2** [1973] 49/74.

R. von Ammon, R. D. Fischer, Shift Reagents in NMR Spectroscopy, Angew. Chem. Intern. Ed. Engl. **10** [1972] 675/92.

A. F. Cockerill, G. L. O. Davies, R. C. Harden, D. M. Rackham, Lanthanide Shift Reagents for NMR Spectroscopy, Chem. Rev. **73** [1973] 553/88.

J. Reuben, Paramagnetic Lanthanide Shift Reagents in NMR Spectroscopy, Progr. Nucl. Magn. Reson. Spectrosc. **9** [1975] 1/70.

J. Reuben, G. A. Elgavish, Shift Reagents and NMR of Paramagnetic Lanthanide Complexes, in: K. A. Gschneidner, Jr., L. Eyring, Handbook on the Physics and Chemistry of Rare Earths, Vol. 4, Pt. II, Amsterdam-New York-Oxford 1979, pp. 483/514.

5.1.1 General Aspects. Luminescence Studies

Weissman [1] first reported that irradiation of certain rare earth chelates, including diketonates, with ultraviolet light produced narrow line emissions in the visible region characteristic of the metal ion. It was established that the metal ion fluorescence was due to an intramolecular transfer of energy from electronic states associated with the ligand to 4f energy states of the metal ion, rather than direct excitation of the rare earth ion [1 to 3]. The discovery of laser action [4 to 6] in rare earth diketonates prompted extensive investigations of fluorescence among these compounds including studies of the mode of intramolecular energy transfer and metal ion excitation [7 to 54, 77 to 80, 144], measurement of quantum efficiencies and lifetimes (radiative and nonradiative decay, rise times) [14, 19, 22, 27, 34, 35, 37, 38, 42 to 44, 48 to 73, 75, 76, 82, 83], and studies of quenching mechanisms [13, 14, 17, 34, 48, 49, 57, 60, 65 to 67, 69 to 73, 83 to 85]. Valuable information relating to the symmetry of the complex, coordination number, nature of bonding, and dissociation of the chelates in solution has also been obtained from fluorescence measurements and is presented together with the characterization of individual compounds. Reviews by Sinha [100, 140] of fluorescence and laser action in rare earth chelates indicate that fluorescence has been observed for diketonato chelates with various stoichiometries, e. g., tris chelates, with or without adduct molecules, tetrakis chelates, or mixed ligand complexes containing a diketonate ion.

The first systematic studies and interpretation of the mechanism of fluorescence (**Fig. 5-1**, p. 66) in these compounds were reported by Crosby and coworkers [3, 7 to 12]. In the initial step, absorption of energy by the ground state singlet of the ligand (S_0) results in an excited state singlet (S_1). Energy may be lost from this state either by a nonradiative intrasystem cross-over to give an excited state ligand triplet or by a radiative emission (ligand fluorescence) back to the ground state. The latter process quenches any subsequent metal ion fluorescence. If the ligand triplet is formed, it may undergo either a nonradiative transfer of energy to the metal ion or energy loss by phosphorescence. The latter process again quenches metal ion fluorescence. Energy transfer to the metal ion produces an excited state (referred to hereafter also as the resonance level) which may lose energy either by a radiative (ionic fluorescence) or a nonradiative (metal ion quenching) process.

Fig. 5-1

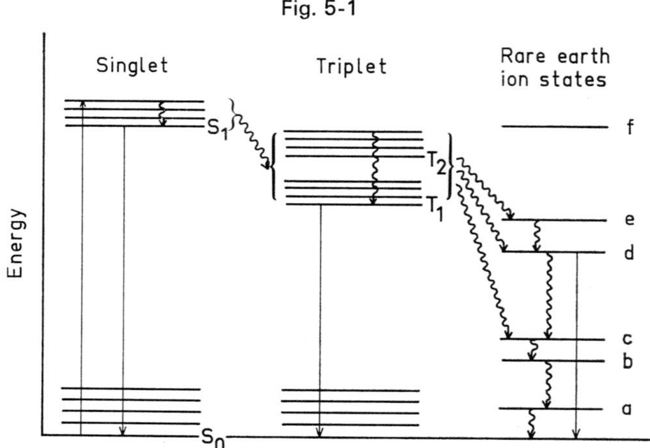

Schematic energy level diagram for a rare earth chelate possessing low-lying 4f electronic states (radiative and nonradiative transitions are indicated by straight arrows and wavy arrows, respectively).

Whan, Crosby [12], in a manner consistent with this interpretation, have classified complexes of the rare earth ions into three categories in terms of their fluorescence characteristics. The first category includes the derivatives of the lanthanum(III), gadolinium(III), and lutetium(III) ions, which exhibit no metal ion fluorescence, but strong ligand fluorescence and phosphorescence. Both the lanthanum(III) ($4f^0$) and the lutetium(III) ($4f^{14}$) ions have single energy terms and cannot undergo intra-4f transitions. With the gadolinium(III) ($4f^7$) ion, the lowest lying excited term appears at a higher energy level than the triplet of any diketonate investigated, thus precluding ligand to metal ion energy transfer and subsequent metal ion fluorescence. The second category includes complexes of the samarium(III), europium(III), terbium(III), and dysprosium(III) ions, which exhibit strong ion fluorescence and weak ligand fluorescence and phosphorescence. Each of these ions has an excited energy term lying close to and slightly below the ligand triplet, allowing for efficient intramolecular energy transfer. The third category includes complexes of praseodymium(III), neodymium(III), holmium(III), erbium(III), thulium(III), and ytterbium(III), which exhibit weak ion fluorescence and weak ligand fluorescence and phosphorescence. The weak ligand luminescence indicates efficient intramolecular energy transfer. However, each of the ions has only relatively small energy differences between the various terms, thus increasing the probability of nonradiative deactivation of the excited state as a consequence of the need to dissipate a smaller amount of energy in each nonradiative decay step.

Crosby, Whan, Alire [10] showed that the intramolecular energy transfer efficiency is a sensitive function of the relative positions of the resonance energy levels and the metastable triplet states of the complexes. The requirement for metal ion sensitization is that the lowest triplet state energy level of the complex must be nearly equal to or must lie above the resonance level of the rare earth ion. For example, the complex of Dy^{III} with dibenzoylmethane ($T_1 = 20520\ cm^{-1}$) exhibited only ligand fluorescence and phosphorescence (resonance level of dysprosium(III) $= 20958\ cm^{-1}$), whereas those with benzoylacetone ($T_1 = 21480\ cm^{-1}$) and tribenzoylmethane exhibited emissions characteristic of the metal ion. These authors proposed that intramolecular energy transfer occurs through vibrational coupling between the 4f excited states of the metal ion and lowest ligand triplet [10].

In accord with the Whan and Crosby mechanism, Sato, Wada [22] observed that the fluorescence yields of europium(III) diketonates decrease with increasing energy differences between the triplet state of the diketonate and the resonance level of the europium(III) ion. Time-resolved spectroscopic measurements showed that the excitation energy is transferred from the ligand triplet state to the lower and nearest resonance level of the europium ion [22]. The decay time for the energy transfer from the lowest triplet is generally in the range of 10^{-8} to 10^{-9} s [19, 21, 25].

Matsuda, Makishima, Shionoya [18] observed that in mixed solutions of the europium(III) and lanthanum(III) complexes with dibenzoylmethane, the intensity and decay time of the ligand phosphorescence of the lanthanum chelate decreased with increasing mole fraction of the europium chelate at 77 K. Nearly in proportion to this, the intensity of emission of the europium chelate is enhanced, verifying the occurrence of energy transfer from the lanthanum chelate to the europium chelate. This observation was interpreted as evidence for intramolecular energy transfer via the triplet state. These investigators proposed that intramolecular energy transfer is dominated by a resonance mechanism owing to the exchange interaction of the π electron system of the ligand triplet state with the 4f electron system of the coordinated rare earth ion [19]. Brittain, Richardson [72] observed similar intermolecular energy transfer between the terbium(III) and europium(III) chelates derived from 2,2,6,6-tetramethyl-3,5-heptanedione in thirteen different organic solvents at room temperature.

The efficiency of energy transfer may also depend on the nature of the metal-ligand bond. Filipescu, Sager, Serafin [47] observed that meta substitution of electron donating methoxy groups in the phenyl rings of the complex $Eu(dbm)_3$ (with dibenzoylmethane = Hdbm) resulted in enhanced ion fluorescence intensity, whereas either meta- or para-substitution of electron withdrawing groups resulted in decreased intensity. These variations were attributed to changes in the donor ability of the diketonate oxygen atoms toward the europium(III) ion [47]. Schimitschek, Nehrich, Trias [68], on the other hand, reported that ortho-, meta-, or para-substitution of fluorine, chlorine, or bromine atoms in the phenyl rings of the tetrakis(benzoyltrifluoroacetonato) chelate of europium(III) (ten different chelates were studied) showed little effect on the quantum yield measured for each of the chelates. Shionoya and coworkers [17] observed that the efficiency of intramolecular energy transfer in several europium(III) diketonates increases with the Eu-O stretching frequency.

Volshin and coworkers [50 to 54] studied the effect of pressure on the luminescence spectra of tris and tetrakis(benzoylacetonato)chelates of europium(III). They showed that the quantum yield increased 2.4 times when the pressure was increased to 23 kbar and then decreased to zero when the pressure reached 55 to 60 kbar [51]. The results were explained by assuming that the energy of the triplet level is reduced by approximately 1500 cm^{-1} upon increasing the pressure to 55 kbar [50]. The mean lifetime of luminescence decreased with increasing pressure (5.6×10^{-4} s at 0 kbar and 3.3×10^{-4} s at 17 kbar). These investigators concluded that the shorter quenching time constant at the higher pressure resulted from a decrease in energy between the ligand triplet and emitting level of the metal ion, with consequent enhancement of transition probability [53].

According to the mechanism proposed by Whan, Crosby [12], ligand sensitized fluorescence of the rare earth ion occurs only when the ligand triplet lies above the emitting level of the rare earth ion. Kleinerman [32, 48] notes, however, that the absence of sensitized emission in the case where the ligand triplet lies below the resonance level of the rare earth ion could be due to quenching of the emitting level of the rare earth by the lower lying triplet of the ligand. He proposed that energy transfer occurs via the ligand-excited singlet state, with rate constants of about 10^{11} s^{-1} and higher. Transfer via the lowest triplet would predominate when the rate of energy transfer from the singlet excited state is smaller than the rate of intersystem crossing to the triplet level. Tanaka and coworkers [21] rejected the mechanism proposed by Kleinerman on the basis of their observation that the time for intramolecular energy transfer is only about 10^{-8} s. Bhaumik, El-Sayed [30] reported additional experimental evidence indicating that direct excitation of the rare earth ion through the ligand singlet is not predominant. They selected a number of rare earth diketonates and organic quenchers such that the lowest singlet and triplet states of the quencher bracketed those of the chelate. For example, cis-piperylene quenches the metal ion fluorescence of both the hexafluoroacetylacetonato and thenoyltrifluoroacetonato chelates of europium(III) (stoichiometry of the chelates not reported) in EPA (ether:isopentane:ethanol = 8:3:5) at 25°C (the triplet state of cis-piperylene lies below the triplet levels (T_1) of both ligands). On the other hand, naphthalene quenches only the hexafluoroacetylacetonato europium(III) fluorescence, since the triplet state of naphthalene lies below the triplet state of the hexafluoroacetylacetonato chelate, but above the triplet levels of the thenoyltrifluoroacetonato chelate. Furthermore, it was shown that cis-piperylene and naphthalene do not effectively quench the ligand singlet (S_1) state. As the concentration of organic quencher increases, the intensity of rare earth chelate emission decreases, but levels off at approximately 10% of its initial value. It was proposed that the unquenched ion emission represents an upper limit to the contribution of direct metal excitation via the singlet state. The quenching experiments do not rule out the possibility of a third mechanism, namely, that the ligand singlet (S_1) transfers energy to a rare earth ion level higher than the ligand triplet (T_1) level, followed by a second intramolecular energy transfer back to the ligand triplet, and finally a third intramolecular energy transfer to the resonance level of the rare earth ion ($S_1 \rightsquigarrow$ rare earth $\rightsquigarrow T_1 \rightsquigarrow$ rare earth $\rightarrow$ emission) [30].

Measurement of rise and decay times has provided a further understanding of the fluorescence mechanism. Nardi, Yatsiv [73] observed that the rise time of the $^5D_0 \rightarrow ^7F_J$ emission of crystalline $Eu(dbm)_3$ (Hdbm = dibenzoylmethane) is of about the same order of magnitude as the decay time of the $^5D_1 \rightarrow ^7F_J$ emission (3 μs at room temperature and 5 μs at liquid air temperature), which led them to conclude that the energy from the triplet state reaches the $^5D_0$ state via nonradiative deactivation of the $^5D_1$ level in europium chelates. The $^5D_0$ state was found to have a radiative lifetime of approximately 500 μs. Bhaumik, Nugent [49] measured the rise and decay times of the $^5D_0 \rightarrow ^7F_2$ (613 nm) and decay time of the $^5D_1 \rightarrow ^7F_1$ (535 nm) transitions of several europium(III) diketonates (Table 5/1) using the technique of stroboscopic time-resolved spectroscopy. Their results confirm those of Nardi, Yatsiv [73]. Noting that the relaxation time for the $^5D_1$ state in the diketonate chelates is shorter than the corresponding relaxation time for europium(III) systems in which there are no vibrational modes near 1760 cm^{-1} (the difference in wave numbers between the $^5D_0$ and $^5D_1$ states), they proposed that the ligand C=O and C=C stretching vibrations make important contributions to the nonradiative relaxation of the $^5D_1$ state [49]. Other investigators have also concluded that if the ligand triplet level is above the $^5D_1$ state of the europium(III) ion, then the $^5D_0$ state is populated primarily by nonradiative energy transfer from the $^5D_1$ state, with a nonradiative decay time of the order of a few microseconds [21, 42, 60, 73].

Table 5/1

Fluorescence Rise and Decay Times of the 613 nm ($^5D_0 \rightarrow ^7F_2$), Lines and Decay Times of the 535 nm ($^5D_1 \rightarrow ^7F_1$), Lines of Some Europium(III) Chelates [49].

chelate*)	medium**)	613 nm decay time (±40 μs)		613 nm rise time (±0.3 μs)		535 nm decay time (±0.3 μs)	
		300 K	77 K	300 K	77 K	300 K	77 K
$C_5H_{12}N[Eu(bzac)_4]$	crystal	550	570	1.0	2.0	1.0	2.0
	EPA	140	420	≦0.4	1.8	—	2.0
	EtOH/MeOH	140	430	≦0.6	2.0	—	1.8
$(C_2H_5)_2NH[Eu(bzac)_4]$	crystal	523	548	1.2	1.6	—	1.7
$Eu(bzac)_3 \cdot 2H_2O$	crystal	296	364	≦0.2	1.4	—	1.5
	EPA	150	440	—	1.6	—	—
	EtOH/MeOH	118	427	≦0.6	1.8	—	—
$C_5H_{12}N[Eu(dbm)_4]$	crystal	414	488	1.0	1.8	—	1.9
	EPA	97	442	—	1.8	—	—
	EtOH/MeOH	55	348	0.5	2.0	—	—
$Eu(acac)_3 \cdot 3H_2O$	crystal	—	320	—	1.1	—	—
$Eu(ttac)_3 \cdot 2H_2O$	crystal	329	376	0.6	1.0	0.5	1.5
	EPA	335	375	1.0	1.9	—	—
	EtOH/MeOH	327	360	1.0	1.8	—	2.0
$NH_4[Eu(hfacac)_4]$	crystal	819	825	1.4	1.5	1.2	1.5
	EPA	377	443	1.8	3.0	1.8	3.0
	EtOH/MeOH	433	435	2.0	2.7	1.1	—

*) Abbreviations of diketones see p. 73, $C_5H_{12}N$ = piperidinium ion.— **) EPA = 3:3:1 (v/v) diethyl ether/isopentane/ethyl alcohol.— EtOH/MeOH = 3:1 (v/v) ethanol/methanol.

Sensitization of the rare earth chelate via intermolecular energy transfer from the excited state of an organic species has also been described. Bhaumik, El-Sayed [36] observed intermolecular energy transfer from the lowest triplet state of benzophenone to the ligand in tris(hexafluoroacetylacetonato) europium(III) dihydrate, which in turn, intramolecularly transferred its energy to the europium(III) ion with subsequent metal ion emission. The intermolecular energy transfer was shown to be diffusion controlled. Wildes, White [142] described sensitized chemiluminescence of mono(1,10-phenanthroline) and tris(benzoyltrifluoroacetonato)europium(III) by energy transfer from the excited species generated by dissociation of dioxetane. Hemingway, Park, Bard [143] reported the observation of intermolecular energy transfer from an excited charge-transfer complex (exciplex) directly to tris chelates of europium(III) derived from anions of dibenzoylmethane or dinaphthoylmethane, with subsequent emission of the europium(III) ion. The exciplex was produced by the electron transfer reaction of electrogenerated radical ions, e.g., reaction of tris(p-tolylamine) radical cation and either benzophenone or dibenzoylmethane radical anion.

Tanaka and coworkers [21] observed a rise time of 10^{-8} s (the time required for intramolecular energy transfer) for the $^5D_0$ state of europium(III) in the tetrakis(benzylidenetrifluoroacetylacetonato) chelate (spectrum obtained for the piperidinium salt in EPA (ether:isopentane:ethanol = 5:5:2) at 77 K), with a radiative lifetime of 408 μs. In this case, the lowest ligand triplet is situated between the $^5D_1$ and $^5D_0$ levels, precluding efficient energy transfer to the $^5D_1$ level. They also measured a rise time of 10^{-8} s for the $^5D_4$ level in several terbium diketonates in EPA at 77 K, with decay time of the order of about 500 μs [21].

The data in Tables 5/2 [34], 5/3 [64], and 5/4 [69] indicate that quantum yields are markedly dependent on the specific rare earth ion, the detailed ligand and solvent environment, and temperature. In accord with the mechanism proposed by Whan, Crosby [12], the intensity of metal ion fluorescence depends upon the amount of energy available in the triplet state, the efficiency of energy transfer to the metal ion, and the probability that ionic emission will occur rather than nonradiative deactivation. Dawson, Kropp, Windsor [34] measured fluorescence yields of solutions of several europium(III) and terbium(III) chelates upon excitation of the ligand and direct excitation of the rare earth ion (Table 5/2). Quantum yields obtained upon ligand excitation are, in most instances, substantially lower than values obtained upon direct excitation of the emitting level of the metal ion, indicating that only a small fraction ($P_{L,I}$) of energy absorbed by the ligands succeeds in reaching the emitting level of the rare earth ions. Quantum yields obtained upon excitation of upper ion levels were lower than yields obtained upon direct excitation of the emitting level, indicating that energy is lost within the ion manifold by nonradiative pathways from the upper levels directly to the ground state. Further loss of energy occurs between the emitting ion level and the ground state, although in certain europium chelates the fluorescence efficiency, following direct excitation of the emitting level, approached unity.

Bhaumik [64] studied the temperature dependence of the fluorescence quantum yield and decay time (Table 5/3) of various rare earth chelates in an attempt to understand the quenching mechanisms in these compounds. Since the decay time of the europium(III) chelates tended to attain the same value near 77 K, he concluded that the rate of quenching at the rare earth ion site approached a constant value at these temperatures. However, the quantum yields varied for each compound at 77 K, indicating that quenching also occurred at the ligand site to a different degree in each of the compounds. The latter proposal was also substantiated by the fact that the quantum yield of the system decreased more rapidly than the decay time with increasing temperature. Substitution of fluorine atoms for hydrogen atoms in the ligand diminished the quenching at both the emitting rare earth ion site and ligand site, as evidenced by greater decay times and quantum yields in the fluorinated compounds. The dominant part of the quenching in both the ligand and the ion sites was attributed to coupling of the electronic states to the environment through molecular vibrations. The electronic energy levels of either the ligand or metal ion are expected to be less susceptible to nonradiative quenching in the fluorinated chelates, since the energy of vibration of the C-F bond is small compared to the C-H bond [64]. The results of Filipescu et al. [69] (Table 5/4) confirmed that the highest quantum yields are obtained for rare earth chelates derived from fluorinated diketones.

Table 5/2
Fluorescence Lifetimes and Yields of Europium(III) and Terbium(III) Chelates with Excitation of the Ligand and Individual Metal Ion Levels [34].

chelate*)	solvent	excitation of ligand		quantum yield upon direct excitation of the metal ion			$P_{L,I}$**)
		fluorescence lifetime, τ (μs) at 23°C	quantum yield, Q_L	$Q(^5D_0)$	$Q(^5D_1)$	$Q(^5D_4)$	
$Eu(acac)_3 \cdot H_2O$	methanol	160	<0.002	0.18	0.10	—	0.01
$C_5H_{12}N[Eu(bzac)_4]$	methanol	170	0.004	0.28	0.09	—	0.014
$Eu(dbm)_3$	methanol	66	0.005	0.12	0.015	—	0.04
$C_5H_{12}N[Eu(ttac)_4]$	ethanol	370	0.13	0.84	0.41	—	0.15
$C_5H_{12}N[Eu(ttac)_4]$	acetone	550	0.43	0.82	0.62	—	0.52
$TbCl_3$	water	380	—			0.046	
$Tb(acac)_3 \cdot H_2O$	ethanol	820	0.19			0.33	0.6
$C_5H_{12}N[Tb(bzac)_4]$	methanol	***)	<0.001			<0.005	
$C_5H_{12}N[Tb(ttac)_4]$	methanol	***)	<0.001			<0.005	

*) Abbreviations of diketones see p. 73, $C_5H_{12}N^+$ = piperidinium ion. — **) $P_{L,I}$ represents efficiency of energy transfer from the ligand to the ion; $P_{L,I} = Q_L/Q^5D_0$ for europium(III); $P_{L,I} = Q_L/Q^5D_4$ for terbium(III). — ***) Fluorescence at 23°C too weak for measurement.

Table 5/3
Fluorescence Quantum Efficiency and Decay Times of Rare Earth Diketonates [64].

chelate*)	solvent**)	quantum efficiency in %			decay time in μs		
		300 K	100 K	77 K	300 K	100 K	77 K
$Tb(acac)_3 \cdot 2H_2O$	EPA	14	—	43	600	—	825
	EM	20	43	—	830	1000	—
$Eu(bzac)_3 \cdot H_2O$	EPA	<1	—	20	150	—	440
	EM	<1	30	—	118	427	—
$C_5H_{12}N[Eu(bzac)_4]$	EPA	<1	—	21	140	—	420
	EM	≈2	30	—	150	430	—
$Eu(dbm)_3$	EPA	≈1	—	32	150	—	396
$Eu(ttac)_3 \cdot H_2O$	EPA	22	—	40	335	—	375
$Eu(hfacac)_3 \cdot 2H_2O$	EPA	35	—	39	430	—	470

*) Abbreviations of diketones see p. 73, $C_5H_{12}N^+$ = piperidinium ion. — **) EPA = 5:5:2 (v/v) ether/isopentane/ethanol, EM = 3:1 (v/v) ethanol/methanol.

The decreased quantum efficiency observed at higher temperatures has also been accounted for in terms of quenching of the emitting level via thermal excitation to the lowest triplet state of the ligand [34, 57, 65] and quenching due to higher rates of collisions between complex species and solvent molecules [74]. Jacobs, Weber, Pearson [71] measured the fluorescence decay time for tris(2,2,6,6-tetramethyl-3,5-heptanedionato)terbium(III) (emission from the $^5D_4$ state of Tb^{3+}) in the vapor phase as a function of temperature (230 to 300°C) and pressure (1.6 to 5.8 Torr) and demonstrated that intermolecular collisional deactivation is unimportant for this chelate in the gaseous state, and that quenching of the terbium(III) emitting level occurs through nonradiative transfer of energy to the lowest triplet level of the ligand. Dawson, Kropp, Windsor [34] similarly accounted for the fact that the fluorescence intensity of a terbium(III) chelate is generally more dependent on temperature than that of the corresponding europium(III) chelate, e. g., fluorescence of piperidinium tetrakis(thenoyltrifluoroacetonato)europate(III) at 23°C is relatively efficient (Table 5/2), whereas no fluorescence was observed for the corresponding chelate of terbium(III). The fluorescence of the europium chelate at 23°C was considered possible because the energy difference between the ion emitting level and the triplet state is larger and therefore the thermal population of the triplet state is smaller than for the terbium chelate (the emitting state ($^5D_4$) of terbium(III) lies 3200 cm^{-1} above the emitting state ($^5D_0$) of europium(III), and in general, more closely matches the energy of the ligand triplet). Thermal population of the triplet state from the emitting level at 23°C is approximately a million times larger in a terbium chelate than in the corresponding europium chelate, and accordingly, the rate of triplet quenching of the $^5D_4$ level of terbium(III) is expected to be orders of magnitude larger than that of the $^5D_0$ state of europium(III). Mikula, Salomon [65] reported that increased ion emission was observed at low temperatures (−196°C) with tris and tetrakis chelates of terbium(III) with dibenzoylmethane in EPA (ether/isopentane/ethanol = 5:5:2), but not with the corresponding complexes of europium(III) upon deuteration of the phenyl groups of the ligand. This difference was also explained in terms of quenching of the ion-emitting level through thermal population of the energy level immediately above the emitting level. In the europium chelates this level is the ionic $^5D_1$ level, which is unaffected by deuteration of the ligand (the ligand triplet is 3400 cm^{-1} above the $^5D_0$ state). In the case of the terbium chelates, the level that becomes thermally populated is the lowest ligand triplet, which is raised by 50 cm^{-1} upon deuteration of the ligand (the undeuterated ligand triplet lies 250 to 350 cm^{-1} above the $^5D_4$ state). Thermal quenching through the triplet is thereby lessened and as a result the intensity of emission obtained from the deuterated terbium chelates is greater than that of the undeuterated chelate. Shepherd and coworkers [37, 101] also accounted for the temperature dependency of the radiative lifetime of the metal ion emitting states in crystalline tert-butylammonium tetrakis(hexafluoroacetylacetonato)europate(III) and tris(2,2,6,6-tetramethyl-3,5-heptanedionato)terbium(III) in terms of back donation of energy from the emitting level to the ligand triplet.

Table 5/4
Quantum Efficiency of Octacoordinate Rare Earth Diketonato Chelates and Adducts of Tris Diketonato Chelates in Organic Solvents at 25°C [69].

chelate*)	DMF	CH_3CN	C_6H_6	C_3H_7CN	CH_3CH_2OH
$C_5H_{12}N[Eu(ttac)_4]$	0.615	0.615	0.221	0.633	0.151
$Eu(ttac)_3 \cdot bpy$	0.503	0.163			
$Eu(ttac)_3 \cdot phen$	0.365	0.123			
$C_5H_{12}N[Eu(bzac)_4]$	0.063	0.038			
$Eu(bzac)_3 \cdot bpy$	0.074	insol.			
$Eu(bzac)_3 \cdot phen$	0.075	0.092			
$C_5H_{12}N[Eu(dbm)_4]$	0.074	0.067			
$Eu(dbm)_3 \cdot bpy$	0.058	insol.			
$Eu(dbm)_3 \cdot phen$	0.040	0.049			
$C_5H_{12}N[Eu(da)_4]$	0.035	insol.			
$C_5H_{12}N[Eu(bztac)_4]$	0.473	0.578	0.295	0.624	0.230
$C_5H_{12}N[Eu(ntac)_4]$	0.580	insol.	0.585	0.561	0.354

*) Abbreviations of diketones see p. 73, $C_5H_{12}N^+$ = piperidinium ion, bpy = 2,2'-bipyridine, phen = 1,10-phenanthroline.

With reference to solvent effects, fluorescence intensity is affected by either collisions with or coordination to solvent molecules. In the first instance, energy is dissipated as translational or vibrational energy. In the second case, it may be dissipated by vibrational coupling. In general, fluorescence intensity increases with the polarity of the solvent [74], with the exception that alcoholic solutions generally exhibit lower quantum yields [69]. Filipescu, McAvoy [74] accounted for the increase in fluorescence intensity with an increase in the dipole moment of the solvent in terms of two effects, namely (1) an increase in energy transfer from the ligand to the ion and/or an increase in the intra-4f radiative transition probability as a result of the new crystal field introduced by the complexing solvent molecules, which has a higher intensity for the more polar solvents, and (2) a cloud formation around the ion by the polar solvent molecules, which decreases collisional quenching efficiency. Filipescu et al. [69] attributed the low quantum efficiencies of the chelates in alcoholic solutions (Table 5/3) to the presence of the high energy vibrations of the O-H bonds of the solvent molecules, allowing for effective nonradiative decay. Halverson, Brinen, Leto [66, 67] applied the concept of a protective sheath to overcome the quenching by solvent molecules and increase fluorescence intensity through the use of alkyl substituted 1,3-diketones in addition to synergistic agents such as trioctylphosphine. On the other hand, Dawson, Kropp, Windsor [34], on the basis of values measured for rate constants for radiative and nonradiative deactivation of the emitting levels of both chelates and simple salts of europium(III) and terbium(III), concluded that enhanced fluorescence yields obtained with chelates, as compared to the simple salt, were due primarily to the enhancement of the radiative transition by the chelate environment rather than a protective influence by the ligands against quenching.

Coordination of Lewis bases to tris diketonato rare earth(III) chelates generally results in enhancement of metal ion fluorescence quantum yields [86 to 97]. Kleinerman, Hovey, Hoffman [86] noted that the Lewis base can increase the fluorescence yield by (a) decreasing the extent of radiationless decay of the emissive level of the ion and (b) increasing the efficiency of energy transfer to the rare earth ion by minimizing nonradiative deactivation of the diketonate ligand. The mechanism of enhancement of emission quantum yield of a methanolic solution of $Eu(dbm)_3$ (Hdbm = dibenzoylmethane) by triethylamine was investigated by Fukuzawa and coworkers [87, 88] over the temperature range 77 to 289 K. The nature of the temperature dependence of the emission intensity was similar for both the tris chelate and the tris chelate-adduct under direct excitation of the metal level, but differed under excitation of the ligand. It was suggested that the amine sensitization is mainly due to a suppression of nonradiative process in the ligand triplet state. Brittain, Richardson [97] observed that the emission intensity of the $^5D_0 \rightarrow {}^7F_2$ europium(III) transition in tris(6, 6, 7, 7, 8, 8, 8-heptafluoro-2,2-dimethyl-3,5-octanedionato)europium(III), is sufficiently sensitive to interactions between the complex and substrate molecules that stepwise addition (titration) of a substrate to a solution of this complex in carbon tetrachloride leads to an emission titration curve the shape and limiting values of which are diagnostic of the adduct composition.

Charles, Riedel [98] compared the fluorescence properties (intensity, band shape, number of emission lines) of fifteen tetrakis chelates of composition $Q^+[Eu(bztac)_4]^-$ (Q^+ is an organic cation and bztac = benzoyltrifluoroacetonate ion) with those of the corresponding tris chelate, $Eu(bztac)_3 \cdot 2H_2O$. In each case, the fluorescence intensity of the tetrakis compound was greater than that of the tris chelate, both in the solid phase and in solution. However, the fluorescence properties of the tetrakis chelates in the solid phase are dependent on the nature of the cation. The differences were attributed to crystal packing requirements of the cation and consequent changes in the local symmetry of the europium(III) ion created by the eight coordinated oxygen atoms. In acetonitrile, nearly all compounds exhibited the same high quantum efficiencies (ca. 70 to 75%), except for the pyridinium, quinolinium, and isoquinolinium salts, which gave lower fluorescence intensities. The lower efficiencies of solutions prepared from the last three compounds were attributed to partial dissociation of the tetrakis chelates to the tris chelates, which are substantially less fluorescent.

Shepherd [99] also observed changes in the fluorescence properties of several tetrakis(benzoyltrifluoroacetonato)europate(III) salts derived from organic cations (M^+) in benzene, which he attributed to the ability of the cation and anionic complex to form an ion pair. In contrast to the behavior in the nonpolar solvent, benzene, spectra of these salts are not cation dependent in the more polar solvent, acetonitrile. The difference in behavior was attributed to complete ionic dissociation of the type, $M[Eu(bztac)_4] \rightarrow Eu(bztac)_4^- + M^+$, in the more polar solvent. Ross, Blanc, Pressley [55] reported that deuteration of the cation in piperidinium tetrakis(benzoyltrifluoroacetonato)europate(III) had little effect on either the lifetime or quantum yield of the europium(III) luminescence.

Lempicki, Samelson [5] first reported laser action of a rare earth chelate. They achieved stimulated emission of the 6131 Å line ($^5D_0 \rightarrow {}^7F_2$) of europium(III) in a cold (−150°C) alcoholic (3:1 ethanol: methanol) solution of a benzoylacetonate (latter species were shown to be the tetrakis chelate [141]) by pumping with a xenon flash lamp. Threshold energy is between 1740 and 1920 J. Other investigators noted laser action in several europium(III) diketonates including benzoylacetonates [4, 33, 102 to 115, 141], complexes with dibenzoylmethane [6, 105, 115 to 118], benzoyltrifluoroacetonate [16, 55, 68, 98, 105, 119 to 123, 134, 135], thenoyltrifluoroacetonate [103, 121, 123 to 126, 135], trifluoroacetylacetonate [114, 123, 127, 135], and monohalogenobenzoylacetonates [68, 128, 129]. The majority of these studies involved solutions of the tetrakis chelates. Laser action has also been noted for the tris chelates of terbium(III) derived from thenoyltrifluoroacetonate [130, 131] and trifluoroacetylacetonate [132], and the tetrakis chelate of neodymium(III) derived from thenoyltrifluoroacetonate [133]. The laser characteristics of the rare earth metal chelates have been reviewed in detail [100, 136 to 139].

Abbreviations of diketones:

Hacac	= acetylacetone (= 2,4-pentanedione)
Hhfacac	= hexafluoroacetylacetone (= 1,1,1,5,5,5-hexafluoro-2,4-pentanedione)
Httac	= thenoyltrifluoroacetone (= 4,4,4-trifluoro-1-(2-thienyl)-1,3-butanedione)
Hbzac	= benzoylacetone (= 1-phenyl-1,3-butanedione)
Hbztac	= benzoyltrifluoroacetone (= 4,4,4-trifluoro-1-phenyl-1,3-butanedione)
Hntac	= 2-naphthoyltrifluoroacetone (= 4,4,4-trifluoro-1-(2-naphthyl)-1,3-butanedione)
Hdbm	= dibenzoylmethane (= 1,3-diphenyl-1,3-propanedione)
Hda	= dianisoylmethane (= 1,3-dianisyl-1,3-propanedione)

References to 5.1.1:

[1] S. I. Weissman (J. Chem. Phys. **10** [1942] 214/7). — [2] A. N. Sevchenko, A. K. Trofimov (Zh. Experim. Teor. Fiz. **21** [1951] 220/9 from C.A. **1952** 2403). — [3] G. A. Crosby, M. Kasha (Spectrochim. Acta **10** [1958] 377/82). — [4] E. J. Schimitschek (Appl. Phys. Letters **3** [1963] 117/8). — [5] A. Lempicki, H. Samelson (Phys. Letters **4** [1963] 133/5).

[6] A. Lempicki, H. Samelson (Appl. Phys. Letters **2** [1963] 159/61). — [7] G. A. Crosby, R. E. Whan (Naturwissenschaften **47** [1960] 276/7). — [8] G. A. Crosby, R. E. Whan (J. Chem. Phys. **32** [1960] 614/5). — [9] G. A. Crosby, R. E. Whan, J. J. Freeman (J. Phys. Chem. **66** [1962] 2493/9). — [10] G. A. Crosby, R. E. Whan, R. M. Alire (J. Chem. Phys. **34** [1961] 743/8).

[11] G. A. Crosby, R. E. Whan (J. Chem. Phys. **36** [1962] 863/5). — [12] R. E. Whan, G. A. Crosby (J. Mol. Spectrosc. **8** [1962] 315/27). — [13] J. J. Freeman (SC-DC-3564 [1963] 1/219 from C.A. **61** [1964] 11436). — [14] J. J. Freeman, G. A. Crosby (J. Phys. Chem. **67** [1963] 2717/23). — [15] N. McAvoy, N. Filipescu, M. R. Kagan, F. A. Serafin (J. Phys. Chem. Solids **25** [1964] 461/8).

[16] C. Brecher, A. Lempicki, H. Samelson (U.S. 3454901 [1969] from C.A. **71** [1969] No. 65946). — [17] S. Shionoya, Y. Matsuda, M. Morita, S. Makishima (Proc. Intern. Conf. Lumin., Budapest 1966 [1968], Vol. 2, p. 1709/13 from C.A. **70** [1969] No. 52102). — [18] Y. Matsuda, S. Makishima, S. Shionoya (Bull. Chem. Soc. Japan **41** [1968] 1513/8). — [19] G. Yamaguchi, S. Okumura, C. Yamanaka (Technol. Rept. Osaka Univ. **19** [1969] 853/75 from C.A. **71** [1969] No. 65725). — [20] Y. Matsuda, S. Makishima, S. Shionoya (Bull. Chem. Soc. Japan **42** [1969] 356/62).

[21] M. Tanaka, G. Yamaguchi, J. Shiokawa, C. Yamanaka (Bull. Chem. Soc. Japan **43** [1970] 549/50). — [22] S. Sato, M. Wada (Bull. Chem. Soc. Japan **43** [1970] 1955/62). — [23] S. J. Lyle, J. E. Newbery, A. D. Witts (J. Chem. Soc. Dalton Trans. **1972** 1726/9). — [24] J. S. Curran, T. M. Shepherd (J. Chem. Soc. Faraday Trans. II **69** [1973] 126/31). — [25] B. D. Joshi, A. G. Page, B. M. Patel (Z. Physik. Chem. [Leipzig] **255** [1974] 103/8).

[26] C. R. S. Dean, T. M. Shepherd (Chem. Phys. Letters **32** [1975] 480/2). — [27] J. Pantoflicek, L. Parma (Czech. J. Phys. **18** [1968] 1610/21). — [28] V. L. Ermolaev, E. A. Saenko, G. A. Domrachev, Yu. K. Khudenskii, V. G. Aleshin (Opt. Spektroskopiya **22** [1967] 854/5; Opt. Spectrosc. [USSR]

22 [1967] 466/7). — [29] K. Kreher (Wiss. Z. Karl Marx Univ. Leipzig Math. Naturw. Reihe **15** [1966] 319/25 from C.A. **66** [1967] No. 42075). — [30] M. L. Bhaumik, M. A. El-Sayed (J. Chem. Phys. **42** [1965] 787/8).

[31] M. L. Bhaumik, M. A. El-Sayed (Appl. Opt. Suppl. No. 2 [1965] 214/5). — [32] M. Kleinerman (AD-447468 [1964] 1/19 from C.A. **63** [1965] 17356). — [33] H. Samelson, A. Lempicki, C. Brecher (J. Chem. Phys. **40** [1964] 2553/8). — [34] W. R. Dawson, J. L. Kropp, M. W. Windsor (J. Chem. Phys. **45** [1966] 2410/8). — [35] M. E. Movsesyan, V. A. Gevorkyan, D. Kh. Grigoryan (Zh. Prikl. Spektrosk. **9** [1968] 260/3 from C.A. **70** [1969] No. 52808).

[36] M. L. Bhaumik, M. A. El-Sayed (J. Phys. Chem. **69** [1965] 275/80). — [37] R. S. C. Dean, M. T. Shepherd (J. Chem. Soc. Faraday Trans. II **71** [1975] 146/55). — [38] R. S. C. Dean, M. T. Shepherd (Proc. 11th Rare Earth Res. Conf., Traverse City, Mich., 1974, p. 854/5). — [39] N. Filipescu, S. Bjorklund, N. McAvoy, C. R. Hurt (Can. J. Chem. **45** [1967] 1714/6). — [40] W. E. Ohnesorge (Fluoresc. Phosphorescence Anal. **1966** 151/67).

[41] W. F. Sager, N. Filipescu, F. A. Serafin (J. Phys. Chem. **69** [1965] 1092/100). — [42] W. M. Watson, R. P. Zerger, J. T. Yardley, G. D. Stucky (Inorg. Chem. **14** [1975] 2675/80). — [43] P. Yuster, S. I. Weissman (J. Chem. Phys. **17** [1949] 1182/8). — [44] H. Samelson, A. Lempicki, V. A. Brophy, C. Brecher (J. Chem. Phys. **40** [1964] 2547/53). — [45] A. N. Sevchenko, V. V. Kuznetsova, R. A. Puko, V. S. Komenko, T. I. Razvina, T. M. Kozhan (Izv. Akad. Nauk SSSR Ser. Fiz. **36** [1972] 1013/7 from C.A. **77** [1972] No. 81694).

[46] G. D. R. Napier, J. D. Neilson, T. M. Shepherd (Chem. Phys. Letters **31** [1975] 328/30). — [47] N. Filipescu, W. F. Sager, F. A. Serafin (J. Phys. Chem. **68** [1964] 3324/46). — [48] M. Kleinerman (J. Chem. Phys. **51** [1969] 2370/81). — [49] M. L. Bhaumik, L. J. Nugent (J. Chem. Phys. **43** [1965] 1680/8). — [50] V. A. Voloshin, A. I. Savutskii, A. I. Kas'yanov (Opt. Spektroskopiya **40** [1976] 607/8; Opt. Spectrosc. [USSR] **40** [1976] 346).

[51] V. A. Voloshin, A. I. Savutskii (Ukr. Fiz. Zh. **20** [1975] 1498/504 from C.A. **83** [1976] No. 210887). — [52] V. A. Voloshin, M. G. Krimnus, A. I. Savutskii (Tezisy Dokl. 1st Vses. Soveshch. Fiz. Tekhn. Vys. Davlenii, Donetsk 1973, p. 141 from C.A. **83** [1975] No. 50279). — [53] V. A. Voloshin, A. A. Galkin, A. I. Savutskii (Dokl. Akad. Nauk SSSR **189** [1969] 511/2 from C.A. **72** [1970] No. 84617). — [54] V. A. Voloshin, A. I. Savutskii (High Temp.-High Pressures **8** [1976] 607/9 from C.A. **88** [1978] No. 143745). — [55] D. L. Ross, J. Blanc, R. J. Pressley (Appl. Phys. Letters **8** [1966] 101/2).

[56] H. Samelson, A. Lempicki (J. Chem. Phys. **39** [1963] 110/2). — [57] M. Metlay (J. Electrochem. Soc. **111** [1964] 1253/5). — [58] C. W. Kuhlman, S. I. Weissman (J. Chem. Phys. **40** [1964] 1776/7). — [59] M. L. Bhaumik, L. J. Nugent, L. Ferder (J. Chem. Phys. **41** [1964] 1158/9). — [60] M. L. Bhaumik (J. Chem. Phys. **41** [1964] 574/5).

[61] M. L. Bhaumik, H. Lyons, P. C. Fletcher (J. Chem. Phys. **38** [1963] 568/9). — [62] F. F. Rieke, R. Allison (J. Chem. Phys. **37** [1962] 3011/2). — [63] H. Winston, O. J. Marsh, C. K. Suzuki, C. L. Telk (J. Chem. Phys. **39** [1963] 267/71). — [64] M. L. Bhaumik (J. Chem. Phys. **40** [1964] 3711/5). — [65] J. J. Mikula, R. E. Salomon (J. Chem. Phys. **48** [1968] 1077/85).

[66] F. Halverson, J. S. Brinen, J. R. Leto (J. Chem. Phys. **41** [1964] 157/63). — [67] F. Halverson, J. S. Brinen, J. R. Leto (J. Chem. Phys. **41** [1964] 2752/60). — [68] E. J. Schimitschek, R. B. Nehrich, J. A. Trias (J. Chim. Phys. **64** [1967] 173/82). — [69] N. Filipescu, G. W. Mushrush, C. R. Hurt, N. McAvoy (Nature **211** [1966] 960/1). — [70] A. J. Twarowski, D. S. Kliger (Chem. Phys. Letters **41** [1976] 329/32).

[71] R. R. Jacobs, M. J. Weber, R. K. Pearson (Chem. Phys. Letters **34** [1975] 80/3). — [72] H. G. Brittain, F. S. Richardson (J. Chem. Soc. Faraday Trans. II **73** [1977] 545/50). — [73] E. Nardi, S. Yatsiv (J. Chem. Phys. **37** [1962] 2333/5). — [74] N. Filipescu, N. McAvoy (J. Inorg. Nucl. Chem. **28** [1966] 253/9). — [75] M. Syczewski (Roczniki Chem. **44** [1970] 721/34 from C.A. **73** [1970] No. 103957).

[76] L. I. Kononenko, S. V. Bel'tyukova, S. B. Meshkova, V. N. Drobyazko, N. S. Poluektov (Dopovidi Akad. Nauk. Ukr.RSR B **1975** 816/9 from C.A. **84** [1976] No. 10577). — [77] G. Perichet, B. Pouyet (J. Chim. Phys. **74** [1977] 373/5 from C.A. **87** [1977] No. 13791). — [78] T. A. Shakhverdov (Izv. Akad. Nauk SSSR Ser. Fiz. **36** [1972] 1018/23 from C.A. **77** [1972] No. 81724). — [79] V. J. Rao, A. P. B. Sinha (Indian J. Chem. **9** [1971] 152/6). — [80] V. L. Ermolaev, E. B. Sveshnikova (Opt. Spektroskopiya **24** [1968] 293/6; Opt. Spectrosc. [USSR] **24** [1968] 153/4).

[81] V. L. Ermolaev, E. B. Sveshnikova, E. A. Saenko (Opt. Spektroskopiya **22** [1967] 165/7; Opt. Spectrosc. [USSR] **22** [1967] 86/7). — [82] L. I. Kononenko, E. V. Melent'eva (Spektrosk. At. Mol. **1969** 304/5 from C.A. **74** [1971] No. 17708). — [83] E. Butter, K. Kreher (Proc. Intern. Conf. Lumin., Budapest 1966 [1968], p. 1719/22 from C.A. **70** [1969] No. 52791). — [84] V. L. Ermolaev, T. A. Shakhverdov (Opt. Spektroskopiya **30** [1971] 648/54; Opt. Spectrosc. [USSR] **30** [1971] 351/4). — [85] L. I. Kononenko, N. S. Poluektov, R. A. Vitkun (Zh. Prikl. Spektrosk. **17** [1972] 256/60 from C.A. **78** [1973] No. 36016).

[86] M. Kleinerman, R. J. Hovey, D. O. Hoffman (J. Chem. Phys. **41** [1964] 4009/10). — [87] T. Fukuzawa, N. Ebara (Bull. Chem. Soc. Japan **45** [1972] 1324/9). — [88] T. Fukuzawa, N. Ebara, M. Katayama, H. Koizumi (Bull. Chem. Soc. Japan **48** [1975] 3460/3). — [89] E. L. Fink (Appl. Opt. **7** [1968] 29/31). — [90] E. V. Melent'eva, L. I. Kononenko, E. G. Koltunova, N. S. Poluektov (Zh. Prikl. Spektrosk. **5** [1966] 328/34 from C.A. **66** [1967] No. 60503).

[91] E. V. Melent'eva, L. I. Kononenko, N. S. Poluektov (Spektrosk. Tr. Sib. 4th Soveshch., Tomsk 1965 [1969], p. 180/2 from C.A. **74** [1971] No. 17719). — [92] K. Kreher, E. Butter (Z. Physik. Chem. [Leipzig] **243** [1970] 161/5). — [93] E. Butter, K. Kreher (Z. Naturforsch. **20a** [1965] 408/12). — [94] M. Kleinerman (U.S. 3398099 [1968] from C.A. **69** [1968] No. 87190). — [95] L. I. Kononenko, V. N. Drobyazko, N. S. Poluektov (Ural'sk 7th Konf. Spektrosk. Tezisy Dokl., Sverdlovsk 1971, p. 8/10 from C.A. **81** [1974] No. 129274).

[96] N. S. Poluektov, I. I. Zheltvai, G. I. Gerasimenko, M. A. Tishchenko, A. A. Kucher (Zh. Prikl. Spektrosk. **24** [1976] 276/80 from C.A. **84** [1976] No. 171648). — [97] H. G. Brittain, F. S. Richardson (J. Chem. Soc. Dalton Trans. **1976** 2253/7). — [98] R. C. Charles, E. P. Riedel (J. Inorg. Nucl. Chem. **28** [1966] 3005/18). — [99] T. M. Shepherd (Nature **212** [1966] 745). — [100] A. P. B. Sinha (in: C. N. R. Rao, J. R. Ferraro, Spectroscopy in Inorganic Chemistry, Vol. II, Academic Press, New York 1971, p. 255/88).

[101] T. D. Brown, T. M. Shepherd (J. Chem. Soc. Dalton Trans. **1972** 1616/9). — [102] A. Lempicki, H. Samelson, C. Brecher (J. Chem. Phys. **41** [1964] 1214/24). — [103] C. Brecher, A. Lempicki, H. Samelson (Nature **202** [1964] 580/1). — [104] H. Samelson, V. A. Brophy, C. Brecher, A. Lempicki (J. Chem. Phys. **41** [1964] 3998/4000). — [105] H. Samelson, A. Lempicki (J. Chim. Phys. **64** [1967] 165/72).

[106] Y. Meyer, R. Astier, J. Simon (Compt. Rend. **259** [1964] 4604/7). — [107] Y. Meyer, H. Poncet, M. Verron (Compt. Rend. **259** [1964] 103/6). — [108] Y. H. Meyer (J. Phys. [Paris] **26** [1965] 558/60). — [109] R. Seitz (U.S. 3477037 [1969] from C.A. **72** [1970] No. 17200). — [110] M. Verron, Y. Meyer (Fr. 1453474 [1966] from C.A. **67** [1967] No. 16686).

[111] O. de Witte, Y. Meyer (J. Chim. Phys. **64** [1967] 186/90). — [112] P. Crozet, Y. H. Meyer (Nature **213** [1967] 1115/6). — [113] A. V. Aristov, Yu. S. Maslyukov (Zh. Prikl. Spektrosk. **8** [1968] 717/22 from C.A. **69** [1968] No. 101396). — [114] R. B. Nehrich, E. J. Schimitschek, J. A. Trias (Phys. Letters **12** [1964] 198/9). — [115] C. Brecher, H. Samelson, A. Lempicki, V. A. Brophy (AD-447468 [1964] 1/19 from C.A. **63** [1965] 17356).

[116] E. J. Schimitschek, R. B. Nehrich (J. Appl. Phys. **35** [1964] 2786/7). — [117] C. Brecher, A. Lempicki, H. Samelson (U.S. 3405372 [1968] from C.A. **69** [1968] No. 112077). — [118] A. Lempicki, K. H. Weise (U.S. 3319183 [1967] from C.A. **67** [1967] No. 103874). — [119] E. P. Riedel, R. G. Charles (J. Appl. Phys. **36** [1965] 3954/5). — [120] H. Samelson, A. Lempicki, C. Brecher, V. Brophy (Appl. Phys. Letters **5** [1964] 173/4).

[121] M. L. Bhaumik, M. A. El-Sayed (U.S. 3562173 [1971] from C.A. **74** [1971] No. 105591). — [122] E. J. Schimitschek, J. A. Trias, R. B. Nehrich (J. Appl. Phys. **36** [1965] 867/8). — [123] E. J. Schimitschek, R. B. Nehrich, J. A. Trias (J. Chem. Phys. **42** [1965] 788/90). — [124] D. E. Altman, M. Geller (U.S. 3470493 [1969] from C.A. **71** [1969] No. 118282). — [125] G. E. Malashkevich, V. V. Kuznetsova (Zh. Prikl. Spektrosk. **22** [1975] 230/3 from C.A. **83** [1975] No. 18602).

[126] G. E. Malashkevich, V. V. Kuznetsova, V. S. Khomenko (Zh. Prikl. Spektrosk. **22** [1975] 1006/8 from C.A. **83** [1975] No. 155352). — [127] R. B. Nehrich, E. J. Schimitschek, J. A. Trias (U.S. 3388071 [1968] from C.A. **69** [1968] No. 31989). — [128] E. J. Schimitschek, R. B. Nehrich, J. A. Trias (Appl. Phys. Letters **9** [1966] 103/4). — [129] E. J. Schimitschek, R. B. Nehrich, J. A. Trias (U.S. 3450641 [1969] from C.A. **71** [1969] No. 44376). — [130] E. H. Hufman (Phys. Letters **7** [1963] 237/9).

[131] E. H. Huffman (Nature **200** [1963] 158/9). — [132] S. Bjorklund, G. Kellermeyer, C. R. Hurt, N. McAvoy, N. Filipescu (Appl. Phys. Letters **10** [1967] 160/2). — [133] B. Whittaker (Nature **228** [1970] 157/9). — [134] D. L. Ross (U.S. 3451009 [1969] from C.A. **71** [1969] No. 55453). — [135] E. J. Schimitschek, R. B. Nehrich, J. A. Trias (U.S. 3360478 [1967] from C.A. **68** [1968] No. 55320).

[136] A. Lempicki, H. Samelson, C. Brecher (Appl. Opt. Suppl. No. 2 [1965] 205/13). — [137] K. K. Rohatgi (J. Sci. Ind. Res. [India] **24** [1965] 456/61). — [138] R. B. Green (Am. Lab. **3** [1971] 47/50, 52, 54, 56/7). — [139] J. P. Fackler (Progr. Inorg. Chem. **7** [1966] 361/425). — [140] S. P. Sinha (Chim. Chronika [Athens, Greece] **29** [1964] 223/9 from C.A. **62** [1965] 140).

[141] M. L. Bhaumik, P. C. Fletcher, L. J. Nugent, S. M. Lee, S. Higa, C. L. Telk, M. Weinberg (J. Phys. Chem. **68** [1964] 1490/3). — [142] P. D. Wildes, E. H. White (J. Am. Chem. Soc. **93** [1971] 6286/8). — [143] R. E. Hemingway, Su-Moon Park, A. J. Bard (J. Am. Chem. Soc. **97** [1975] 200/1). — [144] F. Halverson, J. S. Brinen, J. R. Leto (J. Chem. Phys. **40** [1964] 2790/2).

5.1.2 Complexes with Acetylacetone $CH_3C(O)CH_2C(O)CH_3$ (= 2,4-Pentanedione = $C_5H_8O_2$ = Hacac)

5.1.2.1 $M(acac)_n^{3-n}$ (n = 1 to 3). Formation and Properties in Solution

Formation constants for mono, bis, and tris(acetylacetonato)rare earth(III) chelates in aqueous solution are recorded in Table 5/5. Grenthe, Fernelius [1] observed that the lg K_1 values increased linearly as a function of ionic potential for the lighter rare earth ions (M = La to Eu), whereas lg K_1 is nearly constant for the heavier rare earth ions (M = Gd to Lu). The difference in stability between consecutive elements in the first half of the series is about 0.15 lg K units, whereas the total variation from terbium(III) to lutetium(III) is only 0.2 lg K units. The lg K_1 value for the yttrium(III) chelate falls in the neodymium(III) to europium(III) region [1]. The effect of ionic strength on stability of the chelates in solution is revealed by comparing the data obtained at 30°C, of Grenthe, Fernelius [1], I = 0.1 ($NaClO_4$), with that of Izatt and coworkers [2, 6], I = 0, or the data obtained at 25°C, of Quagliano and coworkers [3], I = 2.0 ($NaClO_4$), with that of Dadgar, Choppin [4], I = 0.1 ($NaClO_4$). Dadgar, Choppin [4] observed that the formation constants determined in solutions with 0.1 ($NaClO_4$) ionic strength are about twice as large as those determined in solutions with 2.0 ($NaClO_4$) ionic strength.

Table 5/5
Formation Constants for Acetylacetonato Rare Earth Chelates.

metal ion	method*)	temp. in °C	medium	lg K_1	lg K_2	lg K_3	Ref.
Sc^{3+}	gl	30	0	8.0	7.2		[2]
Sc^{3+}	dist	25	0.1 ($NaClO_4$)	8.3			[9]
Y^{3+}	gl	30	0.1 ($NaClO_4$)	5.87	4.98	3.25	[1]
Y^{3+}	gl	30	0	6.4	4.7	2.8	[2]
Y^{3+}	gl	25	2.0 ($NaClO_4$)	5.57	4.59		[3]
Y^{3+}	gl	25.0	0.1 ($NaClO_4$)	5.90			[4]
Y^{3+}	gl	30	0.1 ($NaClO_4$)**)	7.70	5.92	4.91	[5]
La^{3+}	gl	30	0.1 ($NaClO_4$)	4.96	3.45	2.5	[1]
La^{3+}	gl	30	0	5.1	3.8	3.0	[2]
La^{3+}	gl	25	2.0 ($NaClO_4$)	4.71	3.45		[3]
La^{3+}	gl	25.0	0.1 ($NaClO_4$)	4.92			[4]
La^{3+}	gl	30	0.1 ($NaClO_4$)**)	6.12	4.55	3.83	[5]
Ce^{3+}	gl	30	0.1 ($NaClO_4$)	5.09	3.3	2.9	[1]
Ce^{3+}	gl	25.0	0.1 ($NaClO_4$)	5.20			[4]
Ce^{3+}	gl	30	0	5.28	3.98		[6]
Ce^{3+}	gl	20		5.30	3.97	3.38	[7]

Table 5/5 (Continued)

metal ion	method*)	temp. in °C	medium	lg K_1	lg K_2	lg K_3	Ref.
Pr^{3+}	gl	30	0.1 ($NaClO_4$)	5.27	3.93	3.2	[1]
Pr^{3+}	gl	25	2.0 ($NaClO_4$)	5.01	3.83		[3]
Pr^{3+}	gl	25.0	0.1 ($NaClO_4$)	5.43			[4]
Pr^{3+}	gl	30	0.1 ($NaClO_4$)**)	6.85	5.59	4.61	[5]
Pr^{3+}	gl	30	0	5.43	4.13	2.96	[6]
Nd^{3+}	gl	30	0.1 ($NaClO_4$)	5.30	4.10	3.20	[1]
Nd^{3+}	gl	30	0	5.6	4.3	3.2	[2]
Nd^{3+}	gl	25.0	0.1 ($NaClO_4$)	5.41			[4]
Nd^{3+}	gl	30	0.1 ($NaClO_4$)**)	6.84	5.62	4.64	[5]
Sm^{3+}	gl	30	0.1 ($NaClO_4$)	5.59	4.46	2.90	[1]
Sm^{3+}	gl	30	0	5.9	4.5	3.2	[2]
Sm^{3+}	gl	25	2.0 ($NaClO_4$)	5.20	5.32		[3]
Sm^{3+}	gl	25.0	0.1 ($NaClO_4$)	5.74			[4]
Eu^{3+}	gl	30	0.1 ($NaClO_4$)	5.59	4.46	2.90	[1]
Eu^{3+}	gl	30	0	6.0	4.5	3.5	[2]
Eu^{3+}	gl	25	2.0 ($NaClO_4$)	5.41	4.30	4.30	[3]
Eu^{3+}	gl	25.0	0.1 ($NaClO_4$)	5.99			[4]
Gd^{3+}	gl	30	0.1 ($NaClO_4$)	5.90	4.48	3.41	[1]
Gd^{3+}	gl	25	2.0 ($NaClO_4$)	5.42	4.39		[3]
Gd^{3+}	gl	25.0	0.1 ($NaClO_4$)	5.90			[4]
Tb^{3+}	gl	30	0.1 ($NaClO_4$)	6.02	4.61	3.41	[1]
Tb^{3+}	gl	25.0	0.1 ($NaClO_4$)	6.02			[4]
Dy^{3+}	gl	30	0.1 ($NaClO_4$)	6.03	4.67	3.34	[1]
Dy^{3+}	gl	25	2.0 ($NaClO_4$)	5.74	4.48	4.48	[3]
Dy^{3+}	gl	25.0	0.1 ($NaClO_4$)	6.08			[4]
Ho^{3+}	gl	30	0.1 ($NaClO_4$)	6.05	4.68	3.40	[1]
Ho^{3+}	gl	25.0	0.1 ($NaClO_4$)	6.07			[4]
Ho^{3+}	ion ex	30	0.1	5.65	4.76		[8]
Er^{3+}	gl	30	0.1 ($NaClO_4$)	5.99	4.68	3.38	[1]
Er^{3+}	gl	25	2.0 ($NaClO_4$)	5.70	4.38		[3]
Er^{3+}	gl	25.0	0.1 ($NaClO_4$)	6.16			[4]
Tm^{3+}	gl	30	0.1 ($NaClO_4$)	6.09	4.76	3.48	[1]
Tm^{3+}	gl	25	2.0 ($NaClO_4$)	6.03	4.72		[3]
Tm^{3+}	gl	25.0	0.1 ($NaClO_4$)	6.18			[4]
Yb^{3+}	gl	30	0.1 ($NaClO_4$)	6.18	4.86	3.60	[1]
Yb^{3+}	gl	25.0	0.1 ($NaClO_4$)	6.18			[4]
Yb^{3+}	ion ex	30	0.1	5.7	4.45		[8]
Lu^{3+}	gl	30	0.1 ($NaClO_4$)	6.23	4.77	3.63	[1]
Lu^{3+}	gl	25	2.0 ($NaClO_4$)	5.98	4.78		[3]
Lu^{3+}	gl	25.0	0.1 ($NaClO_4$)	6.14			[4]

*) Abbreviations: dist = distribution between two phases, gl = glass electrode, ion ex = ion exchange. —**) Determined in 3:1 (v/v) acetone/water.

Formation constants have also been determined in aqueous organic solvent mixtures. The data in Table 5/6 reveal that the stabilities of the mono, bis, and tris chelates increase with increasing concentration of methanol in aqueous methanol mixtures [11,12]. Romanenko, Kostromina [13] observed that lg K_1 values increased linearly with the mole fraction of dioxane in aqueous dioxane mixtures.

Stepwise and overall formation constants of La^{3+}, Pr^{3+}, Nd^{3+}, and Sm^{3+} have been determined potentiometrically by Mehrotra et. al [40] in dioxane/water mixture (75% v/v) at 30°C and ionic strength I = 0.1 (M) $NaClO_4$.

Table 5/6

Formation Constants for Mono-, Bis-, and Tris(acetylacetonato)Rare Earth(III) Chelates in Methanol/Water Mixtures*).

metal ion	lg K_1			lg K_2	lg K_3	Ref.
	5% (v/v) CH_3OH-H_2O	50% (v/v) CH_3OH-H_2O	80% (v/v) CH_3OH-H_2O	80% (v/v) CH_3OH-H_2O	80% (v/v) CH_3OH-H_2O	
Y^{3+}			7.7	5.78	4.0	[11]
La^{3+}			6.54	4.90	3.1	[11]
La^{3+}	4.98	5.88	6.95			[12]
Ce^{3+}	5.29	6.26	7.38			[12]
Pr^{3+}			7.12	5.36	3.7	[11]
Pr^{3+}	5.43	6.47	7.58			[12]
Nd^{3+}			7.23	5.47	3.8	[11]
Nd^{3+}	5.49	6.50	7.72			[12]
Sm^{3+}			7.5	5.64	3.9	[11]
Sm^{3+}	5.81	6.89	8.12			[12]
Eu^{3+}			7.7	5.69	3.9	[11]
Eu^{3+}	5.95	7.03	8.29			[12]
Gd^{3+}			7.6	5.66	3.9	[11]
Gd^{3+}	6.00	6.98	8.23			[12]
Tb^{3+}			7.8	5.78	4.0	[11]
Tb^{3+}	6.13	7.20	8.39			[12]
Dy^{3+}			7.8	5.82	4.0	[11]
Dy^{3+}	6.19	7.23	8.38			[12]
Ho^{3+}			7.8	5.82	4.1	[11]
Ho^{3+}	6.15	7.15	8.38			[12]
Er^{3+}			7.8	5.83	4.1	[11]
Er^{3+}	6.21	7.25	8.41			[12]
Yb^{3+}	6.32	7.40	8.62			[12]
Lu^{3+}	6.25	7.34	8.59			[12]

*) Values from [11] determined potentiometrically (glass electrode) at 24°C; I = 0.1 (NaCl). Values from [12] determined by potentiometric titration (glass and silver-silver chloride electrodes) of a 5×10^{-3} M rare earth(III) chloride solution with a 5×10^{-3} M acetylacetone solution in water (no supporting electrolyte).

The ligational enthalpy and entropy changes for the chelates are recorded in Table 5/7 [4, 10]. The chelates are entropy stabilized, with no large variations in either the enthalpy or entropy terms across the rare earth series. For a thermochemical study of formation reactions for acetylacetonates in an aqueous solution, see also [37, 38].

The proton NMR spectrum of the diamagnetic complex, $Lu(acac)_3 \cdot 2H_2O$, has been recorded in several solvents. The spectral profile is temperature dependent when the spectrum is obtained in nonpolar solvents such as benzene and toluene. The multiplicity of methyl proton resonances

Table 5/7

Free Energy, Enthalpy, and Entropy Changes Corresponding to Formation of the (Acetylacetonato)Rare Earth(III) Chelates [4, 10].

metal ion	$-\Delta G$ in $kcal \cdot mol^{-1}$			$-\Delta H$ in $kcal \cdot mol^{-1}$			ΔS in $cal \cdot mol^{-1} \cdot K^{-1}$			Ref.
	LnL^{2+}	LnL_2^+	LnL_3	LnL^{2+}	LnL_2^+	LnL_3	LnL^{2+}	LnL_2^+	LnL_3	
Y^{3+}	8.05 ± 0.04			0.86 ± 0.05			24.1 ± 0.30			[4]
La^{3+}	6.72 ± 0.04			0.07 ± 0.05			22.4 ± 0.30			[4]
La^{3+}	6.56 ± 0.14	11.13 ± 0.20	14.47 ± 0.20	0.31 ± 0.16	0.65 ± 0.35	1.4 ± 2.0	21.0 ± 0.7	35.1 ± 1.4	44 ± 7	[10]
Ce^{3+}	7.09 ± 0.04			0.02 ± 0.05			23.8 ± 0.30			[4]
Pr^{3+}	7.41 ± 0.04			0.23 ± 0.05			24.1 ± 0.30			[4]
Nd^{3+}	7.38 ± 0.04			0.30 ± 0.05			23.8 ± 0.30			[4]
Nd^{3+}	7.03 ± 0.14	12.48 ± 0.20	16.78 ± 0.20	0.60 ± 0.16	1.30 ± 0.35	3.7 ± 2.0	21.5 ± 0.7	37.5 ± 1.4	44 ± 7	[10]
Sm^{3+}	7.83 ± 0.04			0.79 ± 0.05			23.6 ± 0.30			[4]
Sm^{3+}	7.42 ± 0.14	13.37 ± 0.20	17.26 ± 0.20	0.78 ± 0.16	2.75 ± 0.35	4.8 ± 2.0	22.3 ± 0.7	35.6 ± 1.4	42 ± 7	[10]
Eu^{3+}	8.17 ± 0.04			0.85 ± 0.05			24.6 ± 0.30			[4]
Gd^{3+}	8.05 ± 0.04			0.99 ± 0.05			23.6 ± 0.30			[4]
Gd^{3+}	7.84 ± 0.14	13.82 ± 0.20	18.40 ± 0.20	1.04 ± 0.16	3.20 ± 0.35	8.0 ± 2.0	22.8 ± 0.7	35.6 ± 1.4	42 ± 7	[10]
Tb^{3+}	8.22 ± 0.04			1.10 ± 0.05			23.9 ± 0.30			[4]
Dy^{3+}	8.29 ± 0.04			1.08 ± 0.05			24.2 ± 0.30			[4]
Dy^{3+}	8.02 ± 0.14	14.26 ± 0.20	18.74 ± 0.20	1.25 ± 0.16	3.55 ± 0.35	7.5 ± 2.0	22.7 ± 0.7	35.9 ± 1.4	38 ± 7	[10]
Ho^{3+}	8.28 ± 0.04			1.05 ± 0.05			24.3 ± 0.30			[4]
Ho^{3+}	8.05 ± 0.14	14.30 ± 0.20	18.87 ± 0.20	1.13 ± 0.16	3.38 ± 0.35	7.4 ± 2.0	23.2 ± 0.7	36.6 ± 1.4	38 ± 7	[10]
Er^{3+}	8.40 ± 0.04			1.15 ± 0.05			24.3 ± 0.30			[4]
Tm^{3+}	8.43 ± 0.04			1.04 ± 0.05			24.8 ± 0.30			[4]
Yb^{3+}	8.43 ± 0.04			1.01 ± 0.05			24.9 ± 0.30			[4]
Yb^{3+}	8.23 ± 0.14	14.72 ± 0.20	19.56 ± 0.20	1.01 ± 0.16	3.11 ± 0.35	5.5 ± 2.0	24.2 ± 0.7	38.9 ± 1.4	47 ± 7	[10]
Lu^{3+}	8.38 ± 0.04			0.88 ± 0.05			25.2 ± 0.30			[4]

Values of [4] for ΔG calculated from lg K_1 values obtained potentiometrically at 25°C in solutions maintained at constant ionic strength; I = 0.1 ($NaClO_4$), for ΔH measured calorimetrically at 25°C in solutions maintained at constant ionic strength; I = 0.1 ($NaClO_4$). Values of [10] for ΔH measured calorimetrically at 25°C in solutions maintained at constant ionic strength; I = 0.2 ($NaNO_3$).

observed in such solvents was attributed to the existence, in solution, of a dimeric form of the complex, with slow exchange (NMR time scale) between the nonequivalent methyl groups inherent to the dimeric structure. A molecular-weight measurement (vapor pressure osmometry) confirmed the existence of the dimer in benzene. The temperature dependence of the spectrum for the complex in acetone is also consistent with the proposal of a monomer⇌dimer equilibrium, with $\Delta H^\circ = -28 \pm 1.5$ kJ · mol^{-1} and $\Delta S^\circ = -74.5 \pm 4.5$ J · mol^{-1} · K^{-1}. The observation of a single resonance for both the methyl and 3-H proton resonances in the strongly coordinating solvents, dimethylsulfoxide and pyridine, indicated that the complex exists as the monomer in these solvents [14]. The methyl proton resonances in the spectra of the paramagnetic tris chelates of europium(III) and terbium(III) were observed at −421 and −2829 Hz (with respect to TMS at 60 MHz), respectively, while the 3-H proton resonances were observed, respectively, at −360 and −8647 Hz. The isotropic shifts were considered to be primarily of dipolar origin [15].

Freed, Weissman, Fortress [16] interpreted the absorption spectrum of the tris(acetylacetonato)europium(III) in terms of a monomer⇌dimer equilibrium in solution (benzene and carbon tetrachloride). Moeller, Ulrich [17] observed that the intensity and resolution of the absorption bands for tris chelates were dependent on the solvent used, contending that the nature of the bands is determined by the ligand field imposed on the 4f-electrons. The effects of various organic solvents on the absorption spectra of the tris chelates of neodymium(III) [18 to 20], samarium(III) [20], gadolinium(III), dysprosium(III), erbium(III), and ytterbium(III) [21] have also been reported. For spectra and structure of rare earth acetylacetonates and other β-diketonates with M = Pr, Nd, Eu, Er in water/methanol solutions, see also [39].

The oscillator strengths of absorption bands for the tris chelates of praseodymium(III) [22] and neodymium(III) [23] (samples dispersed in methanol, dimethylformamide, 3:1 (v/v) ethanol/methanol, 1:1:3 (v/v) methanol/ethanol/dimethylformamide), and erbium(III) [24] (in methanol and dimethylformamide) were measured and evaluated in terms of the Judd-Ofelt theory—see B. R. Judd (Phys. Rev. **127** [1962] 750/61) and G. S. Ofelt (J. Chem. Phys. **37** [1962] 511/20). Oscillator strengths were also determined for the hypersensitive transitions in the tris chelates of neodymium(III) ($^4I_{9/2} \rightarrow {}^2G_{7/2}$, $^4G_{5/2}$), holmium(III) ($^5I_8 \rightarrow {}^5G_8$), and erbium(III) ($^4I_{15/2} \rightarrow {}^2H_{11/2}$) in solutions of acetone-water mixtures and were found to increase with the increasing dielectric constant of the solvent [24, 27, 28]. Mehta, Tandon, Swami [23], and Isobe, Misumi [24] concluded that the intensities of the hypersensitive transitions in the spectra of the tris chelates were best estimated by the magnitude of the τ_2 parameter, among the three parameters (τ_2, τ_4, τ_6) of the Judd-Ofelt theory. Mason and coworkers [29] attributed the oscillator strengths of the hypersensitive transitions to coulombic correlation between transient induced electric dipoles in the ligands and the **f**-electron quadrupole moments of the metal ion [29].

The values of the Slater-Condon (F_2, F_4, F_6) and Racah (E^1, E^2, E^3) interelectronic repulsion, and Lande spin-orbit interaction parameters have been computed for the tris chelates of neodymium(III) [25] and praseodymium(III) [26]. Crystal field calculations were carried out for the tris ytterbium(III) chelate, and good agreement between the calculated energy levels and those observed in the absorption spectrum was reported [30].

Brecher, Samelson, Lempicki [31] recorded the emission spectra of $Eu(acac)_3$ and $Eu(acac)_4^-$ at −180°C in alcoholic solutions (3:1 (v/v) ethanol/water) and in a 4:1 volume mixture of the alcoholic solvent with dimethylformamide. The spectrum of the tetrakis chelate in both solvents revealed lines corresponding to the tris chelate as well as the tetrakis species, indicating partial dissociation of the tetrakis complex in solution. The degree of dissociation (0.24 and 0.87, respectively, in the alcoholic and alcohol/dimethylformamide solutions) was determined from the intensity of the $^5D_0 \rightarrow {}^7F_0$ transition due to the tris complex. In the alcoholic solution, the number of lines observed in the spectrum of the tetrakis chelate was accounted for in terms of a distorted dodecahedral arrangement of oxygen atoms surrounding the central ion (emission lines were tabulated for the $^5D_0 \rightarrow {}^7F_2$, $^5D_0 \rightarrow {}^7F_1$, and $^5D_0 \rightarrow {}^7F_0$ transitions of both the tris and tetrakis chelates). The spectrum of the tetrakis chelate in the alcohol-dimethylformamide mixture was accounted for in terms of a nine-coordinate central ion, derived from eight oxygen atoms of the diketone located at the vertices of an Archimedean antiprism with a dimethylformamide molecule coordinating axially to produce a species of C_{4v} symmetry. Structural assignments were not proposed in the case of the tris chelates, inasmuch as a multiplicity of lines was observed in the spectrum, indicating the existence of a number of different, low-symmetry species in solution [31].

Phosphorescence quenching of $Tb(acac)_3 \cdot 3H_2O$ by several $M(acac)_3 \cdot 3H_2O$ species (M = Pr, Nd, Sm, Eu, Dy, Ho, Er) in n-butanol at 293 K was attributed to intermolecular energy transfer between the terbium(III) chelate and other lanthanide(III) chelates (refer to section 5.1.1 for discussion of emission spectra of rare earth diketonates and intra- and intermolecular energy transfer mechanisms). Lifetime measurements indicated that the transfer occurs from the $^5D_4$ level of terbium(III) to excited levels of the other rare earth ions, with bimolecular rate constants within the range 0.5 to 4.9×10^5 $dm^3 \cdot mol \cdot s^{-1}$ [32]. The efficiency of intermolecular energy transfer was shown to be markedly solvent dependent. Neilson, Shepherd [33] proposed that the terbium(III) to lanthanide(III) energy transfer occurs through formation of mixed dimers, $MTb(acac)_6$, where the Tb-M distance is approximately 0.4 nm. The solvent dependent behavior was related to the relative concentrations of monomeric and dimeric species in the various solvents, as determined by molecular weight measurements (vapor pressure osmometry) at 310 K for 0.01 $mol \cdot dm^{-3}$ solutions. The molecular weight measurements revealed that in benzene, the $M(acac)_3 \cdot 3H_2O$ (M = Pr, Nd, Sm, Eu, Gd, Tb, Ho, Er) species exist predominantly in the dimeric form, whereas in acetone and acetonitrile, the complexes (M = Sm, Eu, Tb) gave $\bar{n}$ values of 1.47 and 1.67, respectively ($\bar{n}$ is the average number of monomers per solute molecule). They proposed that the rate controlling step in the intermolecular transfer of energy is formation of the mixed dimer, which occurs with a rate on the order 10^5 $dm^3 \cdot mol^{-1} \cdot s^{-1}$. In strongly coordinating solvents such as pyridine, only monomeric species exist, and no significant energy transfer occurs [33]. Dimer formation was also proposed for the anhydrous yttrium and erbium chelates in chloroform [36].

The polarogram of the tris europium(III) chelate in a 0.1 M solution of tetraethylammonium perchlorate in dimethylformamide revealed a single wave with a half-wave potential of −1.55 V vs. SCE. The electrode reaction was diffusion controlled. Addition of up to 5% (v/v) water had no effect on the half-wave potential or the diffusion current. However, addition of more water caused the diffusion current to decrease. The addition of cations (Li^+ und Mg^{2+}) to the solutions shifted the reduction potentials to more positive potentials [34]. The polarogram of the europium(III) chelate in dimethylacetamide was similar, showing a single reduction wave ($E_{1/2}$ = −1.52 V vs. SCE) corresponding to an irreversible, diffusion controlled, one-electron reduction [35].

References to 5.1.2.1:

[1] I. Grenthe, W. C. Fernelius (J. Am. Chem. Soc. **82** [1960] 6258/60). — [2] R. M. Izatt, W. C. Fernelius, C. G. Haas, B. P. Block (J. Phys. Chem. **59** [1955] 170/4). — [3] H. Yoneda, G. R. Choppin, J. L. Bear, J. V. Quagliano (Inorg. Chem. **3** [1964] 1642/4). — [4] A. Dadgar, G. R. Choppin (J. Coord. Chem. **1** [1971/72] 179/82). — [5] N. K. Dutt, P. Bandyopadhyay (J. Inorg. Nucl. Chem. **26** [1964] 729/36).

[6] R. M. Izatt, C. G. Haas, B. P. Block, W. C. Fernelius (J. Phys. Chem. **58** [1964] 1133/6). — [7] R. M. Izatt, W. C. Fernelius, B. P. Block (J. Phys. Chem. **59** [1955] 235/7). — [8] J. Prasilova (J. Inorg. Nucl. Chem. **26** [1964] 661/2). — [9] T. Sekine, A. Koizumi, M. Sakairi (Bull. Chem. Soc. Japan **39** [1966] 2681/4). — [10] B. E. Kozer, G. A. Prik (Zh. Fiz. Khim. **50** [1976] 626/9; Russ. J. Phys. Chem. **50** [1976] 367/9).

[11] A. A. Zholdakov, N. K. Davidenko (Zh. Neorgan. Khim. **13** [1968] 3223/6; Russ. J. Inorg. Chem. **13** [1968] 1662/5). — [12] N. A. Kostromina, E. D. Romanenko (Zh. Neorgan. Khim. **16** [1971] 1267/71; Russ. J. Inorg. Chem. **16** [1971] 671/3). — [13] E. D. Romanenko, N. A. Kostromina (Zh. Neorgan. Khim. **13** [1968] 1840/7; Russ. J. Inorg. Chem. **13** [1968] 958/62). — [14] J. A. Kemlo, J. D. Neilson, T. M. Shepherd (Inorg. Chem. **16** [1977] 1111/4). — [15] D. R. Eaton (J. Am. Chem. Soc. **87** [1965] 3097/102).

[16] S. Freed, S. I. Weissman, F. E. Fortress (J. Am. Chem. Soc. **63** [1941] 1079). — [17] T. Moeller, W. F. Ulrich (J. Inorg. Nucl. Chem. **2** [1956] 164/75). — [18] N. K. Davidenko, A. A. Zholdakov (Zh. Neorgan. Khim. **14** [1969] 83/9; Russ. J. Inorg. Chem. **14** [1969] 44/7). — [19] N. K. Davidenko, A. A. Zholdakov (Zh. Neorgan. Khim. **14** [1969] 408/11; Russ. J. Inorg. Chem. **14** [1969] 210/2). — [20] M. Radoitchitch (Ann. Chim. [Paris] [11] **13** [1940] 5/10).

[21] D. Purushotham, V. Ramachandra Rao, Bh. S. V. Raghava Rao (Anal. Chim. Acta **33** [1965] 182/97). — [22] S. P. Tandon, P. C. Mehta (J. Chem. Phys. **52** [1970] 4313/5). — [23] P. C. Mehta, S. P. Swami (J. Chem. Phys. **53** [1970] 414/7). — [24] T. Isobe, S. Misumi (Bull. Chem. Soc. Japan **47** [1974] 281/4). — [25] S. P. Tandon, P. C. Mehta (J. Chem. Phys. **52** [1970] 4896/902).

[26] S. P. Tandon, P. C. Mehta (J. Chem. Phys. **52** [1970] 5417/20). — [27] N. S. Poluektov, L. I. Kononenko, S. V. Bel'tyukova, V. N. Drobyazko, E. V. Melent'eva (Zh. Fiz. Khim. **49** [1975] 1379/81; Russ. J. Phys. Chem. **49** [1975] 814/6). — [28] N. S. Poluektov, S. V. Bel'tyukova (Dopovidi Akad. Nauk Ukr.RSR B **36** [1974] 451/4; C.A. **81** [1974] No. 70648). — [29] S. F. Mason, R. D. Peacock, B. Stewart (Mol. Phys. **30** [1975] 1829/41). — [30] W. G. Perkins, G. A. Crosby (J. Chem. Phys. **42** [1965] 407/14).

[31] C. Brecher, H. Samelson, A. Lempicki (J. Chem. Phys. **42** [1965] 1081/96). — [32] G. D. R. Napier, J. D. Neilson, T. M. Shepherd (J. Chem. Soc. Faraday Trans. II **71** [1975] 1487/96). — [33] J. D. Neilson, T. M. Shepherd (J. Chem. Soc. Faraday Trans. II **72** [1976] 557/64). — [34] S. Misumi, M. Aihara, Y. Nonaka (Bull. Chem. Soc. Japan **43** [1970] 774/8). — [35] M. Aihara, H. Hasuo, S. Misumi (Mem. Fac. Sci. Kyushu Univ. C **9** [1975] 269/76; C.A. **83** [1975] No. 199447).

[36] J. K. Przystal, W. G. Bos, I. B. Liss (J. Inorg. Nucl. Chem. **33** [1971] 679/89). — [37] G. A. Prik, B. E. Kozer (Str. Svoistva Primen. Beta Diketonatov Metal. Mater. 3rd Vses. Semin., Moscow 1977 [1978], pp. 99/101 from C.A. **91** [1979] No. 79726). — [38] G. A. Prik, B. E. Kozer, T. A. Tselyapina (Izv. Vysshikh Uchebn. Zavedenii Khim. Khim. Tekhnol. **22** [1979] 559/63 from C.A. **91** [1979] No. 79718). — [39] N. K. Davidenko, A. G. Goryushko, K. B. Yatsimirskii (Str. Svoistva Primen. Beta Diketonatov Metal. Mater. 3rd Vses. Semin., Moscow 1977 [1978], pp. 58/75 from C.A. **91** [1979] No. 150411). — [40] R. C. Mehrotra, B. P. Bachlas, B. P. Gupta (Indian J. Chem. **18** A [1979] 370/2).

5.1.2.2 Tris(acetylacetonato) Compounds

5.1.2.2.1 M(acac)$_3$

Tris(acetylacetonato)scandium(III) was prepared by heating at reflux an aqueous solution containing scandium(III) nitrate and a moderate excess of acetylacetone and ammonia. The product was extracted into benzene, and the benzene layer was separated and concentrated, depositing the product in the form of colorless plates. The product was purified by dissolving it in benzene and effecting precipitation by the addition of petroleum ether. When dried at ambient temperatures, the compound melted over the range 177 to 187°C; however, after subliming the product under vacuum (8 to 10 Torr), the compound melted sharply at 187 to 187.5°C [1]. The scandium compound was also prepared by the reaction of scandium(III) chloride with ammonium acetylacetone (1:4 mole ratio) in 60% (v/v) ethanol/water [2].

The values for the heat of fusion and heat and entropy of sublimation, as evaluated by Melia, Merrifield [3, 4] at 25°C and 1 atm pressure, are 28.8 kJ/mol, 23.8 kcal/mol, and 40.0 cal · mol^{-1} · K^{-1}. The calculated absolute entropy of gaseous Sc(acac)$_3$ at 25°C is 164.7 cal · mol^{-1} · K^{-1} [3, 4]. The value of the heat of sublimation for Sc(acac)$_3$ reported by Komissarova and coworkers (5] is 13.9 kcal/mol.

The crystal and molecular structures of the compound were determined by single crystal X-ray diffraction techniques. The compound crystallizes in the space group Pbca-D_{15}^{2h} with cell parameters a = 15.38, b = 13.73, c = 16.72 Å from [6]; a = 15.45, b = 13.81, c = 16.847 Å from [17]. The structure consists of discrete Sc(acac)$_3$ molecules containing six coordinate scandium(III) ions. The coordination polyhedron was described as a slightly distorted octahedron, with only slight deviation from D_3 symmetry. The intrachelate oxygen-oxygen distance (ligand bite) is 2.715 Å and the intrachelate separation ranges from 2.94 to 3.09 Å. The Sc-O interatomic bond distance is 2.070 Å [6].

Richardson, Wagner, Sands [7] attempted syntheses of the anhydrous rare earth acetylacetonates (M = La, Nd, Ho) by drying the hydrated chelates under vacuum over magnesium perchlorate for several days. However, the resulting products were contaminated with an impurity, which was identified as the hydrolysis product, M(acac)$_2$OH. The compounds obtained are hygroscopic and were handled only under a dry atmosphere. X-ray diffraction studies showed that these compounds are amorphous.

Lis, Bos [8] effected synthesis of the anhydrous chelates (M = La, Nd, Sm, Eu, Gd, Tb) by carefully heating the corresponding hydrated chelates under vacuum. Dehydration was carried out at 60 to 80°C and 10^{-5} to 10^{-6} Torr for periods of 5 to 8 days. The compounds were obtained in the form of amorphous solids.

Bos and coworkers [9, 10] prepared pure crystalline compounds (M = Gd, Tb, Dy, Ho, Er, Y) by allowing the rare earth(III) hydrides (prepared by treating the rare earth metals with hydrogen at 250°C for 2 h) to react directly with acetylacetone. The resulting reaction mixtures were kept at 75°C for 1 to 2 days. Occasionally, when the reaction of the hydrate with acetylacetone was progressing very slowly, a small amount of glacial acetic acid was introduced into the reaction mixture. The crystalline nature of these compounds was demonstrated by X-ray powder patterns, which revealed that the compounds are isomorphous. Attempts to obtain anhydrous chelates of the lighter rare earths (M = La, Nd, Eu) by the reaction of their hydrides with acetylacetone were unsuccessful [9].

The anhydrous europium(III) chelate was obtained in crystalline form following two procedures. In the first, europium metal was allowed to react directly with an excess of acetylacetone. The reaction mixture was stirred and heated at 70 to 80°C under a hydrogen atmosphere of 250 to 300 Torr. The reaction was completed in 1 to 4 days. After filtering the reaction mixture under anhydrous conditions, crystallization of the product was effected by evaporating most of the excess acetylacetone from the filtrate under vacuum and cooling the remaining solution. The resulting precipitate was collected and stored under vacuum in a sealed tube. In the second method, the crystalline europium chelate, as well as the crystalline forms of the neodymium and gadolinium chelates, were obtained by recrystallizing the amorphous form of the corresponding anhydrous chelate under anhydrous conditions from acetylacetone. X-ray powder patterns showed that these crystalline compounds are isomorphous [8].

The direct reaction of acetylacetone with a solution of anhydrous rare earth(III) 2-propanolate (M = Pr [11], Nd [11], Sm [12]) in benzene yielded a fine, crystalline product. Although precipitation appeared to be complete at room temperature, the reaction mixture was heated at reflux for 4 h at 115°C to ensure completion of the reaction. The isopropyl alcohol formed during the reaction was removed from the mixture by distillation as the azeotrope with benzene. Excess solvent was removed by distillation at atmospheric pressure, and the product obtained was dried under vacuum (0.05 Torr) at 28°C.

The synthesis of the anhydrous tris chelates of erbium(III), dysprosium(III), and holmium(III) was effected using the metal evaporation technique. The rare earth metal was evaporated by heating under vacuum (10^{-5} Torr), then co-condensed with acetylacetone at −196°C. Recrystallization of the products from toluene-ether mixtures yielded pure compounds [13]. The anhydrous tris chelate of samarium(III) was prepared by the direct reaction of the metal with acetylacetone [14].

Characterization of the compounds prepared according to the various procedures described by Bos and coworkers [8 to 10] revealed the existence of two distinct forms of each complex, one which is crystalline and the other which is amorphous. The crystalline compounds melt sharply at relatively low temperatures [M, m.p. (°C): Y, 102; Eu, 94; Gd, 100; Tb, 101.5; Dy, 103; Ho, 104; Er, 103], in contrast to the amorphous forms which decomposed at higher temperatures and over a broader range when heated (decomposition range of the amorphous form of the europium chelate is 160 to 170°C) [8, 9]. The crystalline chelates could be melted and resolidified without significant changes in their properties, provided they were protected from moist air. Upon exposure to moist air, the melting points of the crystalline products increased to approximately 130°C and exhibited melting ranges of 6 to 8°C [9, 10]. Significant mass loss was also observed, and changes in the infrared spectra and diffraction patterns occurred upon exposure of the samples to moist air. An examination of the infrared spectra of the crystalline compounds after exposure revealed that peaks associated with the acetylacetonate ion were still present, but a large peak due to hydroxide ion had appeared. The hydrolysis reaction $M(acac)_3 + H_2O \rightleftharpoons M(acac)_2(OH) + C_5H_8O_2$ was proposed to account for the changes [9]. In contrast to the behavior of the crystalline chelates, the anhydrous amorphous products gained weight when exposed to the atmosphere. In the case of the europium(III) chelate, the weight change corresponded to formation of the monohydrate [8]. Infrared analysis provided further evidence that hydration had taken place. Although the infrared spectra of the various forms of the tris chelates have many features in common, enough differences exist to allow identification of each form. Absorption bands for the anhydrous crystalline, anhydrous amorphous, trihydrated, and hydrolyzed forms of the europium(III) tris chelate are recorded and discussed in [8]. The spectra of the various forms of the other rare earth metals are similar to those obtained for europium [8].

Visible spectra were recorded for chloroform solutions of the crystalline neodymium [8], erbium [9], and holmium [9] chelates and compared with the spectra of the amorphous, hydrated, and hydrolyzed forms of the chelates. Of particular interest are the bands corresponding to the hypersensitive transitions (at 521 and 578 nm for M = Nd; at 520 to 522 nm for M = Er; at 450 to 460 nm for M = Ho), since these bands show differences in intensity and fine structure with changes in the environment of the ion. The positions and intensities of the hypersensitive absorption bands are different in the spectra of the various forms of the chelates, suggesting structural differences in these compounds [8, 9]. The fluorescence spectrum of the crystalline terbium compound (solid samples or solutions in ethanol and chloroform) was characterized by sharp lines, but the complexity of the spectrum indicated that the crystallographic environment of the terbium ion sites was of very low symmetry [9]. The emission spectrum of the europium complex (prepared according to [7]) revealed the number of lines ($^5D_0 \rightarrow {}^7F_0$: 579.5 nm; $^5D_0 \rightarrow {}^7F_1$: 588.8, 592.5, 596.8 nm; $^5D_0 \rightarrow {}^7F_2$: 612.0, 613.8, 617.2 nm) consistent with the assignment of a distorted octahedral coordination environment about a six coordinate metal ion [18].

On the basis of cryoscopic measurements, the apparent molecular weight of the crystalline erbium chelate in chloroform is 728 g/mol (the monomer and dimer have molecular weights of 464.6 and 929.2 g/mol, respectively). The intermediate value observed is equally well explained by the occurrence of a monomer-dimer equilibrium in solution, or a predominantly dimeric species in solution, with partial dissociation of the ligand from the dimer accounting for the apparent low molecular weight [9]. The PMR spectra (60 MHz) of the crystalline yttrium chelate in deuteriochloroform exhibited two absorption peaks at 118 Hz (methyl protons) and 325 Hz (olefinic protons) with respect to TMS. The peak at 118 Hz revealed a shoulder at approximately 121 Hz, which was attributed to free acetylacetone produced by dissociation of the chelate (dimer) in solution [9]. Based on the fact that the molecular structures of tris(acetylacetonato)diaqualanthanum(III) [15] and tris(acetylacetonato)diaquayttrium(III) monohydrate [16] are distorted square antiprisms containing eight coordinate rare earth ions, the molecular structure proposed for each of the crystalline dimeric chelates consists of two square antiprisms sharing a square face, with four oxygen atoms of two acetylacetonato ions acting as the bridging atoms. Each rare earth ion is eight coordinate in the proposed dimer [10]. The amorphous compounds, with their broad, higher melting ranges, are likely to be polymeric, possibly with varying degrees of polymerization [9].

References to 5.1.2.2.1:

[1] G. T. Morgan, H. W. Moss (J. Chem. Soc. **1914** 189/201). — [2] M. Z. Gurevich, B. D. Stepin, L. N. Komissarova, N. E. Lebedeva, T. M. Sas (Zh. Neorgan. Khim. **16** [1971] 93/8; Russ. J. Inorg. Chem. **16** [1971] 48/51). — [3] T. P. Melia, R. Merrifield (J. Inorg. Nucl. Chem. **32** [1970] 1489/93). — [4] T. P. Melia, R. Merrifield (J. Inorg. Nucl. Chem. **32** [1970] 2573/9). — [5] L. N. Komissarova, M. Z. Gurevich, T. S. Sas, B. D. Stepin (Zh. Neorgan. Khim. **23** [1978] 3145/7; Russ. J. Inorg. Chem. **23** [1978] 1745/6).

[6] T. J. Anderson, M. A. Neuman, G. A. Melson (Inorg. Chem. **12** [1973] 927/30). — [7] M. F. Richardson, W. F. Wagner, D. E. Sands (Inorg. Chem. **7** [1968] 2495/500). — [8] I. B. Liss, W. G. Bos (J. Inorg. Nucl. Chem. **39** [1977] 443/7). — [9] J. K. Przystal, W. G. Bos, I. B. Liss (J. Inorg. Nucl. Chem. **33** [1971] 679/89). — [10] J. M. Koehler, W. G. Bos (Inorg. Nucl. Chem. Letters **3** [1967] 545/8).

[11] R. C. Mehrotra, T. N. Misra, S. N. Misra (Indian J. Chem. **3** [1965] 525/7). — [12] B. S. Sankhla, R. N. Kapoor (Australian J. Chem. **20** [1967] 685/8). — [13] J. R. Blackborow, C. R. Eady, E. A. Koerner v. Gustorf, A. Scrivanti, O. Wolfbeis (J. Organometal. Chem. **108** [1976] C32/C34). — [14] L. R. Crisler (U.S. 3919274 [1975] from C.A. **84** [1976] No. 90310). — [15] T. Phillips, D. E. Sands, W. F. Wagner (Inorg. Chem. **7** [1968] 2295/9).

[16] J. A. Cunningham, D. E. Sands, W. F. Wagner (Inorg. Chem. **6** [1967] 499/503). — [17] V. I. Ivanov, O. M. Petrukhin (Kristallografiya **15** [1970] 576/7; Soviet Phys.-Cryst. **15** [1970] 492). — [18] T. I. Razvina, V. S. Khomenko, V. V. Kuznetsova, R. A. Puko (Zh. Prikl. Spektrosk. **19** [1973] 866/71; J. Appl. Spectrosc. [USSR] **18/9** [1973] 1461/5).

5.1.2.2.2 M(acac)$_3$ · n H_2O (n = 1 to 3)

Preparation and Stability

Stites, McCarty, Quill [1] obtained the tris chelates (M = Y, La, Ce, Pr, Nd, Sm, Eu) by adding, slowly and with stirring, a solution of ammonium acetylacetonate to an aqueous solution of the rare earth(III) chloride. The pH of each reaction mixture was maintained at a value just below that required for precipitation of the corresponding rare earth hydroxide by the addition of aqueous ammonia or hydrochloric acid, as required. The mixture was stirred for 12 h, and the resulting crystalline precipitates were air dried for 24 h, and then placed in a desiccator over magnesium perchlorate for four days. Although the degree of hydration was not given, Pope, Steinbach, Wagner [2] later showed that the air-dried compounds obtained by the method of Stites and coworkers were the trihydrates and those dried over magnesium perchlorate were the monohydrates. The melting points (°C) of the compounds prepared by Stites and coworkers [1] were: Y, 131 to 132; La, 142 to 143; Ce, 146 to 147; Pr, 143 to 146; Nd, 144 to 145; Sm, 144 to 145; Eu, 144 to 145. Pope and coworkers [2] followed a procedure similar to that described above to effect synthesis of a series of mono and trihydrated chelates (n = 3 for M = Y, Pr, Nd, Sm, Gd, Dy, Ho, Er, Yb; n = 1 for M = Y, Gd, Dy, Ho, Er, Yb). The trihydrates were obtained when the products, which had been collected by filtration, washed with water, and air-dried, were recrystallized from 60% (v/v) ethanol/water (several ml of acetylacetone were added to the ethanol/water solution to prevent hydrolysis). The resulting crystals were collected by filtration, washed with successive portions of 50% (v/v) ethanol/water, then air-dried for several hours. The monohydrates were obtained following two procedures: (1) by dehydrating the trihydrate in a desiccator over magnesium perchlorate, and (2) by recrystallizing the trihydrate from 95% ethanol and air-drying the resulting solid for 6 to 8 h. The average melting points (°C) of the monohydrated compounds were: Y, 130; Pr, 139; Nd, 140; Sm, 138; Gd, 139; Dy, 136; Ho, 133; Er, 129; Yb, 125 [36]. Lyle and Witts [3] obtained the trihydrated europium chelate by saturating a methanolic solution containing hydrated europium(III) nitrate (0.05 mmol) and acetylacetone (3 mmol) with ammonia. The resulting solution was evaporated under reduced pressure to near dryness. The residual yellow oil and crystals were washed with several portions of water. The white crystals remaining were dried in air [3]. Compounds of composition M(acac)$_3$ · 2 H_2O (M = La, Pr, Nd, Sm) were prepared by dissolving the trihydrates (prepared according to the procedure of [1]) in 95% ethanol at 50°C, then immediately evaporating the solutions, partially, by aspiration, in order to cool the solutions quickly. After a large portion of the product had precipitated, aspiration was ceased and the solution was filtered. The precipitate was collected in the form of small plates [4].

Thermogravimetric analyses indicated that the trihydrates readily lose two moles of water when heated at 80 to 100°C, yielding the monohydrates. The monohydrate was observed to be stable only over a fairly narrow temperature range (10 to 20°C), and on further heating in air, weight losses corresponding to decomposition, rather than simple dehydration, were observed [5]. Gas chromatographic analyses of the decomposition products (sample heated in a stream of helium) indicated that decomposition involved a hydrolysis reaction, with formation of the hydroxo species, M(acac)$_2$(OH), and liberation of acetylacetone [6]. Karasev et al. [7] observed that thermal decomposition of Eu(acac)$_3$ · H_2O at 120°C yielded Eu_2(acac)$_5$OH, together with formation of acetylacetone, acetic acid, acetone, Eu(acac)$_2$(CH_3COO) · H_2O, and Eu(acac)(CH_3COO)$_2$ · H_2O (acetylacetone liberated in the presence of water at high temperature undergoes partial decomposition to acetone and acetic acid). At 140°C the dehydrated product, Eu_2(acac)$_5$OH, is transformed further, yielding Eu(acac)$_3$ and Eu(acac)$_2$OH.

Crystal and Molecular Structures

Crystallographic data for several M(acac)$_3$ · n H_2O compounds (n = 1 to 4) are recorded in Table 5/8.

The compound of empirical composition Y(acac)$_3$ · 3 H_2O crystallizes in a monoclinic unit cell with symmetry $P2_1/n$ (standard setting $P2_1/c$-C^5_{2h}, No. 14) and dimensions a = 11.24, b = 22.20, c = 8.44 Å, and β = 100.5°. The crystal is composed of [Y(acac)$_3$(H_2O)$_2$] molecules linked by a series of hydrogen bonds (2.90 Å) through the third water molecule associated with each complex. The yttrium ion is eight coordinate, with oxygen atoms from three bidentate acetylacetonate groups and two water molecules O(1) and O(9) as donors (see **Fig. 5-2**). The coordination polyhedron (**Fig. 5-3**) was described as a distorted square antiprism, with the distortion attributed to the combined effects of coordination of two chemically distinct ligands and the constraints due to bidentate ligation of the acetylacetonate groups. The average Y-O (acac) bond distance (range 2.302 to 2.456 Å) is only about 0.04 Å shorter than the average Y-O (H_2O) bond distance (2.409 Å) [8].

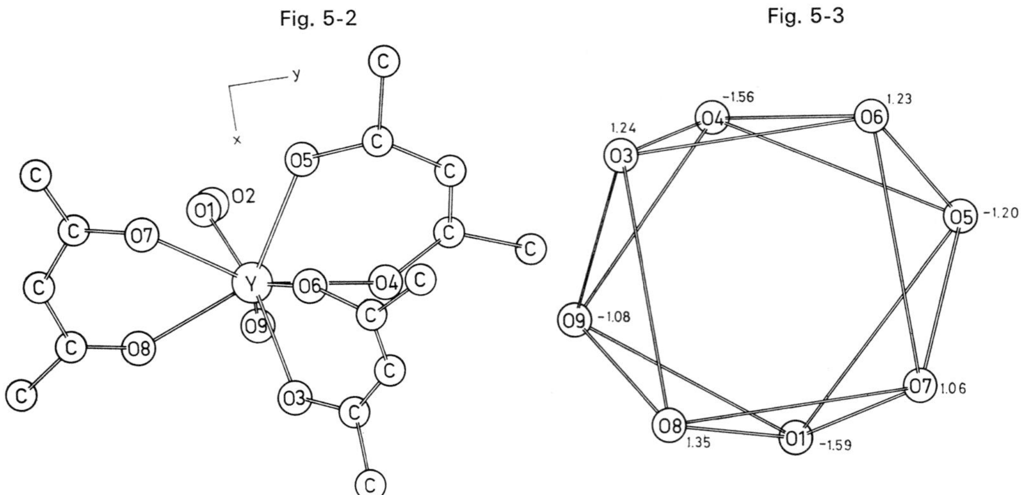

Molecular structure of $Y(acac)_3 \cdot 3H_2O$ projected onto [001].

The coordination polyhedron formed by the eight oxygen atoms bonded to the yttrium atom in $Y(acac)_3 \cdot 3H_2O$. The distance of each atom from the central plane is given.

Aslanov and coworkers [9] determined the crystal (space group and unit cell dimensions recorded in Table 5/8) and molecular structure of $Eu(acac)_3 \cdot 3H_2O$ (R=0.149 for 1780 independent reflections) and noted that the structure is similar to that of the analogous yttrium(III) chelate. The metal chelate molecules are linked into polymeric chains by a system of hydrogen bonds, and the chains in turn are joined into sheets by a series of hydrogen bonds contributed by the noncoordinated water molecule associated with each complex. The coordination polyhedron surrounding the eight-coordinate metal ion was described as a tetragonal antiprism, with one square face of the antiprism formed from two pairs of oxygen atoms from acetylacetonate groups, whereas the other face is formed from the oxygen atoms of the third acetylacetonate group and two water molecules. The interatomic Eu-O bond distances [Eu-O(acac) = 2.36 to 2.49 Å; Eu-O(H_2O) = 2.55, 2.58 Å] are noticeably longer than the corresponding values reported for $[Y(acac)_3(H_2O)_2] \cdot H_2O$ in [9]. Furthermore, all of the Eu-O(acac) bond distances are less than the two Eu-O(H_2O) bond distances, which is a regularity that was not observed in the yttrium(III) chelate [9]. Aslanov and coworkers [10] observed that the structure of the neodymium complex, $[Nd(acac)_3)H_2O)_2] \cdot H_2O$, was similar to that found for $[Eu(acac)_3(H_2O)_2] \cdot H_2O$, with interatomic Nd-O bond distances of 2.38 to 2.48 Å (R = 0.11 for 2515 reflections).

The $Ho(acac)_3 \cdot 4H_2O$ compound crystallizes in the monoclinic space group $P2_1/n$ (standard setting $P2_1/c$-C_{2h}^5, No. 14) with unit cell dimensions a = 10.65, b = 20.0, c = 11.04 Å, β = 90°45′ (R = 0.137 for 2300 independent reflections). The crystal structure consists of $[Ho(acac)_3(H_2O)_2]$ molecules linked to each other through a series of hydrogen bonds formed between the two inner sphere and two outer sphere water molecules. The holmium ion is eight-coordinate with oxygen atoms from three bidentate acetylacetonate groups and two water molecules as donors. The coordination polyhedron was described as a slightly distorted dodecahedron with the three chelate rings spanning three m-type edges of the polyhedron. The Ho-O(acac) bond distances are in the range 2.23 to 2.39 Å and the H-O(H_2O) bond distances are 2.38 and 2.50 Å. The difference between the two metal-oxygen (water) bond distances suggests that one water molecule is pushed slightly out of the inner coordination sphere [11].

Table 5/8

Description of Unit Cell, Space Group, and Unit Cell Dimensions for Tris(acetylacetonato) Rare Earth(III) Hydrates, $M(acac)_3 \cdot nH_2O$ (n = 1 to 4).

metal	n	unit cell	space group	unit cell dimensions and angles a (Å)	b (Å)	c (Å)	α	β	γ	Ref.
Y	3	monoclinic	$P2_1/n\text{-}C_{2h}^5$ (No. 14)	11.24	22.20	8.44		100.5°		[8]
Y	3	monoclinic	$P2_1\text{-}C_2^2$ (No. 4) or $P2_1/m\text{-}C_{2h}^2$ (No. 11)	8.52	21.8	11.17	90°	102°	90°	[15]
La	2	triclinic	$P\bar{1}\text{-}C_i^1$ (No. 2)	9.044	11.456	10.724	95.57°	114.66°	100.28°	[12]
Pr	2	triclinic	$P\bar{1}\text{-}C_i^1$ (No. 2)	16.51	10.68	11.38	74.53°	85.61°	99.93°	[14]
Nd	3	triclinic	$P1\text{-}C_1^1$ (No. 1) or $P\bar{1}\text{-}C_i^1$ (No. 2)	9.04	10.80	11.43	105°	101°	63°	[15]
Nd	2	triclinic	$P\bar{1}\text{-}C_i^1$ (No. 2)	9.16	11.39	10.91	105°45′	63°27′	101°37′	[13]
Eu	3	monoclinic	$P2_1/n\text{-}C_{2h}^5$ (No. 14)	8.60	22.6	11.17		98°		[9]
Tb	3	monoclinic	$P2_1\text{-}C_2^2$ (No. 4) or $P2_1/m\text{-}C_{2h}^2$ (No. 11)	10.65	20.0	11.04	90°	90°	90°	[15]
Ho	4	monoclinic	$P2_1/n\text{-}C_{2h}^5$ (No. 14)	10.65	20.0	11.04		90°45′		[11]
Er*)	2		$P2_1\text{-}C_2^2$ (No. 4) or $P2_1/m\text{-}C_{2h}^2$ (No. 11)	10.65	20.2	11.08	90°	90°	90°	[15]
Yb	1	triclinic	$P\bar{1}\text{-}C_i^1$ (No. 2)	13.01	18.33	8.32	100.40°	102.12°	105.19°	[16]
Yb	1**)	triclinic	$\bar{1}$	8.728	8.435	15.623	95.95°	107.72°	96.18°	[17]

*) Composition = $Er(acac)_3 \cdot 2H_2O \cdot C_2H_6O$; C_2H_6O = ethanol. — **) Composition = $Yb(acac)_3 \cdot H_2O \cdot 0.5\,C_6H_6$; C_6H_6 = benzene.

Tris(acetylacetonato)diaqualanthanum(III) crystallizes in space group $P\bar{1}$-C_i^1 (No. 2) with unit cell dimensions a = 9.044, b = 11.456, c = 10.724 Å, α = 95.57°, β = 114.66°, and γ = 100.28° (diffraction data were collected for two crystals with R = 0.091 for 1559 reflections for the first crystal and R = 0.059 for 1394 reflections for the second crystal). The crystal consists of discrete $[La(acac)_3(H_2O)_2]$ molecules linked by hydrogen bonds in zigzag chains which are nearly parallel to the [101] direction. No close contact exists between chains, with the shortest interchain distances equal to 3.39 Å. The lanthanum(III) ion is eight-coordinate (see **Fig. 5-4**), with oxygen atoms from three bidentate acetylacetonate groups and two water molecules O(7) and O(8) as donors. The coordination polyhedron was described as a distorted square antiprism, with the distortion attributed to the presence of chemically different ligands bonded to the same metal atom. The authors noted that the structure could also be described as a distorted rhombic dodecahedron. The average La-O(acac) bond distance is 2.473 Å (range 2.427 to 2.496 Å), and the two La-O(H_2O) distances are 2.562 and 2.584 Å [12]. Single-crystal X-ray analysis (R = 0.107) of $[Nd(acac)_3(H_2O)_2]$ revealed that the compound had the same general structural features as the corresponding lanthanum compound (description of unit cell given in Table 5/8, p. 87). The geometry of the coordination sphere was discussed in terms of a distorted tetragonal antiprism, a two-centered trigonal prism, and a distorted dodecahedron. The average M-O(acac) and M-O(H_2O) interatomic bond distances are 2.44 and 2.54 Å, respectively [13].

Fig. 5-4

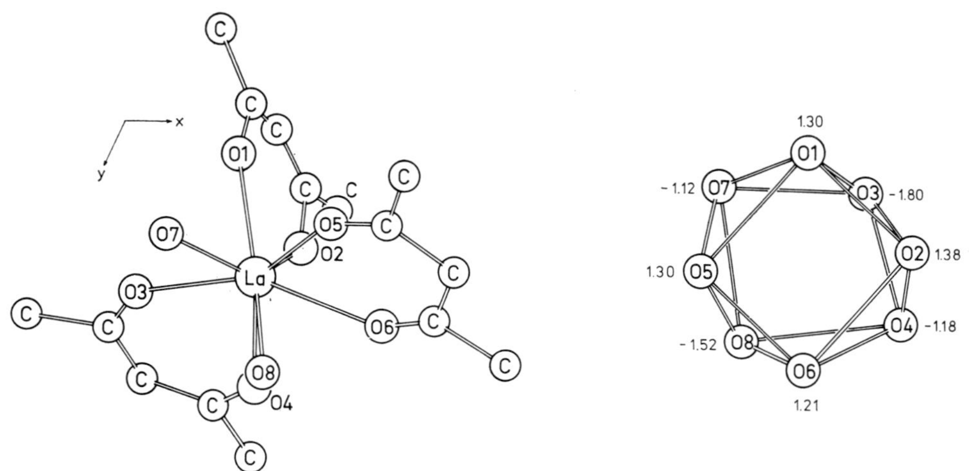

Molecular structure of $[La(acac)_3(H_2O)_2]$ projected along [010] and the coordination polyhedron formed by the eight oxygen atoms bonded to lanthanum. The distance of each atom from the central plane is given.

The $[Pr(acac)_3(H_2O)_2]$ compound crystallizes in space group $P\bar{1}$-C_i^1 (No. 2) with unit cell dimensions a = 16.51, b = 10.68, and c = 11.38 Å and angles α = 74.53°, β = 85.61°, and γ = 99.93° (four molecules per unit cell; structure refined to R = 0.28). The slightly distorted square antiprismatic coordination sphere is formed by eight oxygen atoms, with an average Pr-O bond distance of 2.69 Å. The atoms of the acetylacetonate ring are nearly coplanar [14].

Tris(acetylacetonato)aquaytterbium(III) crystallizes in a triclinic unit cell of symmetry $P\bar{1}$-C_i^1 (No. 2) and dimensions a = 13.01, b = 18.83, c = 8.32 Å, α = 100.40°, β = 102.12°, and δ = 105.19° (R = 0.076 for 2992 independent reflections). Two crystallographically distinct and independent $[Yb(acac)_3H_2O]$ species were observed and are depicted in **Fig. 5-5**. A hydrogen bond (2.73 Å) between the water molecule designated OII_1 and the carbonyl oxygen atom OI_6 links the two independent molecules in pairs. Contact distances between these two molecules are 3.03, 3.15, and 3.84 Å for OI_1-OII_4, OI_1-OII_1, and OII_3-CI_{11}, respectively, whereas the shortest contact distance between pairs of molecules is 3.98 Å. Each ytterbium ion is seven-coordinate, with oxygen atoms from three bidentate acetylacetonate groups and one water molecule as donors. The average

Yb-O(acac) interatomic bond distance [range is 2.173 to 2.326 Å for Yb(1) and 2.147 to 2.248 Å for Yb(2)] is 0.10 Å shorter than the average Yb-O(H_2O) bond distance (2.336 Å). The coordination polyhedra formed about the independent ytterbium ions were both described as monocapped trigonal prisms (**Fig. 5-6**), with a carbonyl oxygen atom capping one of the lateral faces. The authors noted that the coordination polyhedron about Yb_I or Yb_{II} can be derived from the square antiprismatic structure of $Y(acac)_3 \cdot 3H_2O$ (see Fig. 5-2) by removing the coordinated water molecule O(1) and shifting the carbonyl oxygen atom O(4) to this vertex [16]. The structure of tris(acetylacetonato)aquaytterbium(III) hemibenzene, $Yb(acac)_3 \cdot H_2O \cdot 0.5\,C_6H_6$, is similar to that described above for tris(acetylacetonato)aquaytterbium(III) with the benzene molecule occupying a center of symmetry site in the crystal lattice. The sum of the molecular volume of benzene (147.6 $Å^3$) and the volume (931.8 $Å^3$) occupied by the two $[Yb(acac)_3(H_2O)]$ molecules in the unit cell of the unsolvated structure is 1079.4 $Å^3$, as compared with a unit cell volume of 1077.9 $Å^3$ determined for the benzene solvate, indicating that the efficiency of packing of the molecules is about the same for the solvated and unsolvated structures [17].

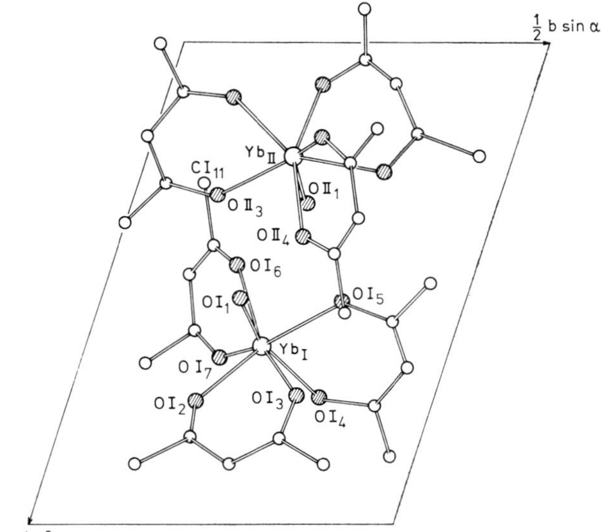

Fig. 5-5

The molecular structure of $[Yb(acac)_3(H_2O)]$ showing the two crystallographically distinct and independent molecules projected along [001]. Molecules shown are linked by a hydrogen bond.

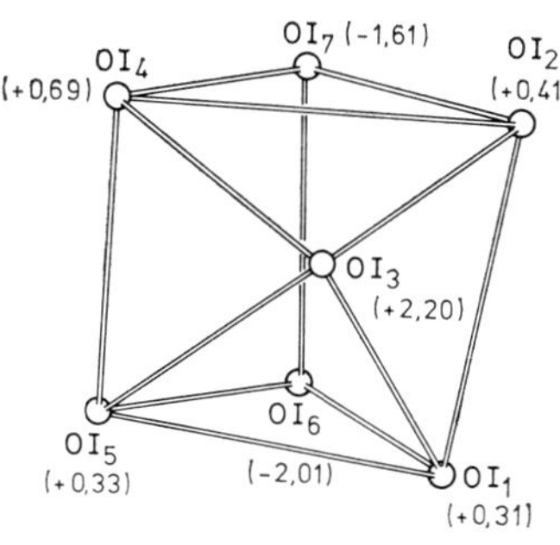

Fig. 5-6

Coordination polyhedron formed by the seven oxygen atoms bonded to Yb_I. The distance of each atom from the central plane is given. The view is along a vector 30° from Yb_I-O I_6 in the plane of Yb_I-O I_6-O I_7. The coordination polyhedron about Yb_{II} is similar to that of Yb_I.

Spectral Studies

The general aspects of the emission spectra of rare earth diketonates are presented in section 5.1.1. Fluorescence spectral studies of the tris acetylacetonato complexes include the mechanism of metal ion excitation via ligand to metal intramolecular energy transfer [18 to 27], measurement of luminescence efficiency and quantum yields [28 to 34], and determination of metal ion site symmetries [36 to 38]. The emission spectrum of solid $[Eu(acac)_3 \cdot H_2O]$ (compound prepared according to [1]) revealed the number of lines ($^5D_0 \rightarrow {}^7F_0$: 579.5 nm; $^5D_0 \rightarrow {}^7F_1$: 588.8, 593.6,

595.3 nm; $^5D_0 \rightarrow {}^7F_2$: 613.4, 616.5, 618.9, 621.4 nm) consistent with the assignment of a trigonal prismatic coordination environment for the seven-coordinate metal ion [37]. The spectra of $[Eu(acac)_3 \cdot 2H_2O]$ (compound prepared according to [4]) and $[Eu(acac)_3 \cdot 2H_2O] \cdot H_2O$ (compound prepared according to [1]) were similar (for the dihydrate, $^5D_0 \rightarrow {}^7F_0$: 579.5 nm; $^5D_0 \rightarrow {}^7F_1$: 591.1, 592.2, 593.6 nm; $^5D_0 \rightarrow {}^7F_2$: 612.7, 615.1, 616.8, 618.1, 622.1 nm; for the trihydrate, $^5D_0 \rightarrow {}^7F_0$: 579.5 nm; $^5D_0 \rightarrow {}^7F_1$: 589.9, 591.7, 595.8 nm; $^5D_0 \rightarrow {}^7F_2$: 612.1, 614.1, 617.1, 618.3, 624.3 nm) and interpreted in terms of ligation of eight oxygen atoms arranged at the vertices of a distorted antiprism [37]. Similar results for the emission spectra of $Eu(acac)_3 \cdot H_2O$ and $Eu(acac)_3 \cdot 3H_2O$ are reported in [36]. The emission spectrum of solid $Eu(acac)_3 \cdot 2H_2O$ reported by Brecher, Samelson, Lempicki [35] (compound prepared by combining a stoichiometric amount of europium chloride with sodium acetylacetonate in water or 50% ethanol, washing the resulting precipitate with several portions of water and allowing it to air dry before drying under vacuum at room temperature) revealed a multiplicity of lines, which are more than allowed for a single chemical species ($^5D_0 \rightarrow {}^7F_0$: 579.5, 579.8 nm; $^5D_0 \rightarrow {}^7F_1$: 588.1, 589.6, 591.7, 592.2, 595.5, 596.7 nm; $^5D_0 \rightarrow {}^7F_2$: 611.1, 611.6, 612.1, 613.0, 613.9, 616.2, 616.8, 617.9, 625.0 nm). Frequencies of electronic-vibrational transitions for the emission spectrum of the europium chelate are tabulated in [39].

Irradiation of the $M(acac)_3 \cdot nH_2O$ (M = La, Gd, Lu; value of n not specified) species in an ethanol/methanol glass at 77 K produced a single blue phosphorescence band (M, band maximum, phosphorescence lifetime): La, 24.4 kK, 70.66 ms; Gd, 124.6 kK, 1.03 ms; Lu, 24.8 kK, 66.88 ms. The structure and position of the band, and lifetime of the transition indicated that the band was due to a triplet-singlet (π-π^*) phosphorescence originating from a ligand localized state [40].

The values of the Slater-Condon ($F_2 = 303.5\ cm^{-1}$, $F_4 = 41.90\ cm^{-1}$, $F_6 = 4.586\ cm^{-1}$), Racah interelectronic repulsion ($E^1 = 4366\ cm^{-1}$, $E^2 = 23.32\ cm^{-1}$, $E^3 = 450.52\ cm^{-1}$), and Lande ($695.9\ cm^{-1}$) parameters were calculated from the diffuse reflectance spectrum of the praseodymium tris chelate. The nephelauxetic ratio, β, has the value 0.9423 [41]. The diffuse reflectance spectrum is also reported for the neodymium tris chelate [42, 43], with $\beta = 0.9839$ [42].

Moeller, Gulyas, Marshall [44] observed that adsorption of the yttrium(III) or gadolinium(III) tris chelate (degree of hydration not specified) on columns of D-lactose hydrate or D-lactose, followed by elution with benzene/petroleum ether, effected partial resolution of the chelates into optically active fractions. In the absence of moisture, solutions in benzene/petroleum ether were chemically stable and racemized only very slowly. The racemization process was observed to be first order. Although resolution of the yttrium chelate was also effected in chloroform, it was difficult to obtain reproducible results, and racemization was always rapid.

The Mössbauer spectrum of $Eu(acac)_3 \cdot 3H_2O$ consisted of a single broad absorption band (line width = $3.0\ mm \cdot s^{-1}$) devoid of any resolved fine structure (isomer shift at 173.0 K = $-0.37\ mm \cdot s^{-1}$ relative to the ^{151}Eu source in an Sm_2O_3 lattice). The isomer shifts and line widths do not vary significantly over the temperature range studied (79.3 to 173.0 K). Some 6 s electron participation in the bonding of the complexes was proposed, since the isomer shift for the complex was positive relative to the isomer shift ($-0.78\ mm \cdot s^{-1}$) of the predominantly ionic $EuF_3 \cdot 0.5H_2O$ species [62]. The isomer shift of $Tm(acac)_3 \cdot 3H_2O$ was approximately zero, within the experimental error of measurement, $\pm 0.2\ mm \cdot s^{-1}$ [45].

Assignments of vibrational modes in the infrared and Raman spectra (1600 to 200 cm^{-1} region) of the hydrated (n = 2, 3) tris chelates of europium(III) are recorded in Table 5/9, p. 91 [46 to 48]. Normal coordinate analyses have been carried out for the in-plane vibrations of the acetylacetonate chelate rings, using a simplified model of single ring analysis (C_{2v} symmetry) [46, 47]. The calculated wave numbers of the in-plane vibrational modes are recorded in Table 5/9. Although normal coordinate analysis based on a single chelate ring model provided, to a good approximation, an interpretation of the infrared spectra [46, 47], some bands observed in the infrared and Raman spectra of the hydrated europium tris chelate were better interpreted by including all three chelate rings (D_3 symmetry) in the analysis of the spectrum [48]. Normal coordinate analysis of the in-plane chelate ring vibrational modes has been carried out also for the tris chelates of lanthanum [49], praseodymium [46, 49], neodymium [46, 49], samarium [49], gadolinium, dysprosium, and erbium [46]. Assignments for the bands observed in the spectrum of the erbium complex are tabulated in [50]. Additional infrared data and assignments of vibrational modes are reported in references [51 to 61].

Table 5/9

Assignments of Vibrational Modes in the Infrared and Raman Spectra of the Tris(acetylacetonato)europium(III) Chelates. Infrared (IR) and Raman Bands in cm^{-1}.

Misumi, Iwasaki [46] IR[a) obs	IR[a) calc	assignment	Liang, Schimitschek, Stephens, Trias [47] IR[b) obs	IR[b) calc	assignment	Liang, Schimitschek, Trias [48] IR[c)	Raman[d)	assignment
1601 vs	1596	C=C str (ν_8)	1600 vs	1624	C=C str (ν_9)	1600 vs	1600 m	C=O str (ν_2) (1600 m, 1545 vs)
1553 sh	1514	C=O str (ν_1)	—	—			1545 vs	
1510 sh	1486	C=O str + CH bend (ν_9)	1515 vs	1562	C=O str (ν_2)	1515 vs	1515 m	C=C str (ν_9)
—	—		1450	1500	CH bend + C=O str (ν_{10})	1450 sh	1460 m	CH bend + C=O str (ν_{10})
1385 vs	—	CH_3 deg def	1390 vs (1450, 1390, 1353 vs)		CH_3 antisym bend	1390 vs		CH_3 asym bend
1354 sh	—	CH_3 sym def	1353		CH_3 sym bend	1353 sh	1370 vs	CH_3 sym bend
1255 s	1255	C=C str (ν_2) + C-CH_3 str	1258 s	1223	C=C str + C-CH_3 str (ν_3)	1258 s	1260 s	C=C str + CCH_3 str (ν_3)
1184 vw	1236	CH bend (ν_{10})	1190 w	1231	CH bend + C=O str (ν_{11})	1190 w	1190 m	CH bend + C=O str (ν_{11})
1011 m	—	CH_3 rock	1014 s	—	CH_3 rock	1014 s	1020 w	CH_3 rock
—	—		—	—		—	945 w	complicated mode involving C=O str (ν_4) (945 w, 917 s)
914 m	917	CCH_3 str + C=O str (ν_3)	917 s	920	complicated mode involving C=O str (ν_4)	917 s	—	
845 vw	885	CCH_3 str (ν_{11})	900 sh	892	CCH_3 str (ν_{12})	900 sh	890 vw	CCH_3 str (ν_{12})
—	—		—	—		—	860 vw	
—	—		787 w		—	777 w	790 vw	
761	—	CH out-of-plane bend	764, 756 m		CH out-of-plane bend	760 m, 750 w	750 vw	CH out-of-plane bend
650 w	649	ring def + MO str (ν_4)	650 m	631	ring def + CCH_3 str (ν_5)	650 m	660 m	ring def + CCH_3 str (ν_5)

Table 5/9 (Continued)

Misumi, Iwasaki [46]			Liang, Schimitschek, Stephens, Trias [47]			Liang, Schimitschek, Trias [48]		
IR[a)]		assignment	IR[b)]		assignment	IR[c)]	Raman[d)]	assignment
obs	calc		obs	calc				
615 vw	610	CCH_3 bend + MO str (ν_{12})	—					
—			560 vw		—	560 w	570 vw	CCH_3 bend (ν_{13}) (570 vw, 535 sh)
527 w	—	C-C(C)(O) out-of-plane bend	530 m	578	CCH_3 bend (ν_{13})	530 m	535 sh	
424 sh	422	M-O str (ν_5)	410 m (410, 395)	412	complicated mode involving M-O str (ν_6)	410 m	410 s	complicated mode involving M-O str (ν_6)
394 vw	347	ring deform (ν_{13})	395	376	ring deform (ν_{14})	395 sh		ring def (ν_{14})
373 w	—	e)	—	—		—	—	
317 br	291	M-O str (ν_{14})	305 vw	—	—	—		
—	—		265 vw	252	C-CH_3 bend (ν_7)	265		CCH_3 bend (ν_7)
236 sh	260	CCH_3 bend (ν_6)	225 w	241	M-O str (ν_{15})	225		M-O str (ν_{15})
203 br	203	ring def (ν_7)	200 w	208	ring def + M-O str (ν_8)	200		ring def + M-O str (ν_8)

a) Samples dispersed in Nujol mulls and KBr discs for spectra obtained in 4000 to 400 cm^{-1} region; spectra measured by the polyethylene paraffin method in the 400 to 100 cm^{-1} region.— b) Samples dispersed in CsI discs; in the region 500 to 200 cm^{-1}, samples cast from acetonitrile solution on polyethylene sheets were also used.— c) Samples dispersed in KBr or CsI pellets.— d) Powdered crystalline sample.— e) Out-of-plane displacement of central metal ion perpendicular to the plane.

The assignments of absorption bands which are recorded in Table 5/9 are not in total agreement, particularly in the regions of the ν(C=O), ν(C=C), and ν(M-O) stretching vibrations. Liang, Schimitschek, Trias [48] noted that the error in the calculated wave numbers for the C=C and C=O stretching modes, relative to the observed values, is still within the range of a few percent if the assignment of these two modes is interchanged. An attempt to distinguish the two modes by Raman polarization measurements was not successful [48]. Mathur, Surana, Tandon [50] assigned bands at 1605 (vs) and 1460 (sh) cm^{-1} to the ν_{as}(C=O) and ν_s(C=O) + δ(C-H) modes, respectively, in the spectrum of $Er(acac)_3 \cdot 2H_2O$ (KBr discs), and bands at 1515 (vs) and 1265 (m) cm^{-1} to the ν_{as}(C=C) and ν_s(C=C) + ν($C-CH_3$) modes, respectively.

Liang and coworkers [47] assigned the metal-oxygen stretching vibrations to bands near 410, 225, and 200 cm^{-1}, but noted that the two bands at 410 and 200 cm^{-1} are coupled modes. Misumi, Iwasaki [46] assigned the pure metal-oxygen stretching modes to bands at 420 to 432 cm^{-1} (ν_5) and 304 to 322 cm^{-1} (ν_{14}), in the spectra of several hydrated tris chelates (M = Pr, Nd, Eu, Gd, Dy, Er), whereas the coupled vibrations involving metal-oxygen stretching and vibrational modes other than C=O stretching modes were assigned to bands at 650 to 651 cm^{-1} (ν_4) and 611 to 623 cm^{-1} (ν_{12}). The bands at 304 to 322 cm^{-1} were considered too broad to be analyzed precisely. Based on these assignments, the authors noted that the metal-oxygen stretching mode (ν_5) shifted to higher wave numbers with decreasing metal ion radius [M, ν_5 (cm^{-1}): Pr, 420; Nd, 420; Eu, 424; Gd, 428; Dy, 428; Er, 432]. Accordingly, the force constant of the metal-oxygen stretching mode increases with decreasing M-O bons length [M, k(mdyn · $Å^{-1}$): Pr, 2.10; Nd, 2.12; Eu, 2.18; Gd, 2.20; Dy, 2.24; Er, 2.29] [46]. Mathur and coworkers [50] assigned one ν(M-O) stretching mode to a band at 398 cm^{-1} in the spectrum of $Er(acac)_3 \cdot 2H_2O$, and concluded that the other ν(M-O) stretching modes appear in the region 250 to 100 cm^{-1}. Mehta, Tandon [51] recorded wave number assignments of metal-oxygen stretching modes, and the corresponding force constants obtained by normal coordinate analysis, for the chelates of lanthanum, praseodymium, neodymium, and samarium.

References to 5.1.2.2.2:

[1] J. G. Stites, C. N. McCarty, L. L. Quill (J. Am. Chem. Soc. **70** [1948] 3142). — [2] G. W. Pope, J. F. Steinbach, W. F. Wagner (J. Inorg. Nucl. Chem. **20** [1961] 304/13). — [3] S. J. Lyle, A. D. Witts (Inorg. Chim. Acta **5** [1971] 481/4). — [4] M. F. Richardson, W. F. Wagner, D. E. Sands (Inorg. Chem. **7** [1968] 2495/500). — [5] N. K. Khalmurzaev, I. A. Murav'eva, L. I. Martynenko, V. I. Spitsyn, A. S. Berlyand, N. G. Dzyubenko, E. D. Stukel'man (Zh. Neorgan. Khim. **20** [1975] 1752/8; Russ. J. Inorg. Chem. **20** [1975] 980/3).

[6] N. K. Khalmurzaev, I. A. Murav'eva, L. I. Martynenko, V. I. Spitsyn (Zh. Neorgan. Khim. **21** [1976] 1635/7; Russ. J. Inorg. Chem. **21** [1976] 894/5). — [7] V. E. Karasev, N. I. Steblevskaya, E. T. Karaseva (Zh. Neorgan. Khim. **22** [1977] 3234/8; Russ. J. Inorg. Chem. **22** [1977] 1762/4). — [8] J. A. Cunningham, D. E. Sands, W. F. Wagner (Inorg. Chem. **6** [1967] 499/503). — [9] A. L. Il'inskii, L. A. Aslanov, V. I. Ivanov, A. D. Khalilov, O. M. Petrukhin (Zh. Strukt. Khim. **10** [1969] 285/9; Russ. J. Struct. Chem. **10** [1969] 263/6). — [10] L. A. Aslanov, M. O. Dekaprilevich, M. A. Porai-Koshits, V. I. Ivanov (Zh. Strukt. Khim. **8** [1967] 1106; Russ. J. Struct. Chem. **8** [1967] 982).

[11] L. A. Aslanov, E. F. Korytnyi, M. A. Porai-Koshits (Zh. Strukt. Khim. **12** [1971] 661/6; Russ. J. Struct. Chem. **12** [1971] 600/4). — [12] T. Phillips, D. E. Sands, W. F. Wagner (Inorg. Chem. **7** [1968] 2295/9). — [13] L. A. Aslanov, M. A. Porai-Koshits, M. O. Dekaprilevich (Zh. Strukt. Khim. **12** [1971] 470/3; Russ. J. Struct. Chem. **12** [1971] 431/4). — [14] Y. H. Lin, C. S. Mo, Y. Hsing (Hua Hsueh Hsueh Pao **34** [1976] 171/8 from C.A. **88** [1978] No. 43942). — [15] A. I. Byrke, L. A. Aslanov, E. F. Korytnyi (Vestn. Mosk. Univ. Ser. II **21** No. 6 [1966] 117/9; Moscow Univ. Chem. Bull. **21** [1966] 533/4).

[16] J. A. Cunningham, D. E. Sands, W. F. Wagner, M. F. Richardson (Inorg. Chem. **8** [1969] 22/8). — [17] E. D. Watkins, J. A. Cunningham, T. Phillips, D. E. Sands, W. F. Wagner (Inorg. Chem. **8** [1969] 29/33). — [18] G. D. R. Napier, J. D. Neilson, T. M. Shepherd (Chem. Phys. Letters **31** [1975] 328/30). — [19] V. L. Ermolaev, E. A. Saenko, G. A. Domrachev, Yu. K. Khudenskii, V. G. Aleshin (Opt. Spektroskopiya **22** [1967] 854/5; Opt. Spectrosc. [USSR] **22** [1967] 466/7). — [20] W. F. Sager, N. Filipescu, F. A. Serafin (J. Phys. Chem. **69** [1965] 1092/100).

[21] M. Kleinerman (J. Chem. Phys. **51** [1969] 2370/81). — [22] M. Yoshihisa, S. Makishima, S. Shionoya (Bull. Chem. Soc. Japan **42** [1969] 356/62). — [23] V. L. Ermolaev, V. G. Aleshin, E. A. Saenko (Dokl. Akad. Nauk SSSR **165** [1965] 1048/51; Soviet Phys. Dokl. **10** [1965/66] 1186/8). — [24] B. D. Joshi, A. G. Page, B. M. Patel (Z. Physik. Chem. [Leipzig] **255** [1974] 103/8). — [25] K. Kreher, E. Butter (Z. Physik. Chem. [Leipzig] **243** [1970] 152/60).

[26] A. N. Sevchenko, A. G. Morachevskii (Izv. Akad. Nauk SSSR Ser. Fiz. **15** [1951] 628 from C.A. **1952** 8974). — [27] N. Filipescu, W. F. Sager, F. A. Serafin (J. Phys. Chem. **68** [1964] 3324/46). — [28] J. J. Freeman, G. A. Crosby (J. Phys. Chem. **67** [1963] 2717/23). — [29] C. Gorller-Walrand, E. Graauwmans, S. de Jaegere (Proc. 10th Rare Earth Res. Conf., Carefree, Ariz., 1973, p. 839/45 from C.A. **85** [1976] No. 27065). — [30] W. M. Watson, R. P. Zerger, J. T. Yardley, G. D. Stucky (Inorg. Chem. **14** [1975] 2675/80).

[31] F. Halverson, J. S. Brinen, J. R. Leto (J. Chem. Phys. **41** [1964] 157/63). — [32] V. A. Mode, D. H. Sisson (Spectrosc. Letters **7** [1974] 9/14). — [33] M. E. Movsesyan, V. A. Gevorkyan, F. P. Safaryan, P. G. Mezhlumyan (Izv. Akad. Nauk Arm. SSR Ser. Fiz. Mat. Nauk **18** [1965] 101/5 from C.A. **64** [1965] 7546). — [34] M. L. Bhaumik, I. R. Tannenbaum (J. Phys. Chem. **67** [1963] 2500/1). — [35] C. Brecher, H. Samelson, A. Lempicki (J. Chem. Phys. **42** [1965] 1081/96).

[36] E. T. Karaseva, V. E. Karasev, N. I. Steblevskaya (Koord. Khim. **2** [1976] 248/52; Soviet J. Coord. Chem. **2** [1976] 188/91). — [37] T. I. Razvina, V. S. Khomenko, V. V. Kuznetsova, R. A. Puko (Zh. Prikl. Spektrosk. **19** [1973] 866/71; J. Appl. Spectrosc. [USSR] **18/19** [1973] 1461/5). — [38] T. I. Razvina (Mater. Resp. 2nd Konf. Molodykh Uch. Fiz., Minsk 1972 [1972/73], Vol. 2, p. 40/1 from C.A. **82** [1975] No. 91835). — [39] K. I. Gur'ev, N. I. Davydova, I. A. Zhigunova, V. F. Zolin, M. A. Kovner, V. A. Kudryashova, V. I. Tsaryuk (Opt. Spektroskopia **28** [1970] 921/5; Opt. Spectrosc. [USSR] **28** [1970] 499/501). — [40] G. A. Crosby, R. J. Watts, S. J. Westlake (J. Chem. Phys. **55** [1971] 4663/4).

[41] S. P. Tandon, R. C. Govil (Z. Naturforsch. **26a** [1971] 1357/9). — [42] S. P. Tandon, R. C. Govil (Spectrosc. Letters **4** [1971] 73/7). — [43] T. I. Razvina, V. S. Khomenko, V. V. Kuznetsova (Zh. Prikl. Spektroskopii **23** No. 1 [1975] 60/3; J. Appl. Spectrosc. [USSR] **22/23** [1975] 916/9). — [44] T. Moeller, E. Gulyas, R. H. Marshall (J. Inorg. Nucl. Chem. **9** [1958] 82/5). — [45] C. I. Wynter, C. H. Cheek, M. D. Taylor, J. J. Spijkerman (Nature **218** [1968] 1047).

[46] S. Misumi, N. Iwasaki (Bull. Chem. Soc. Japan **40** [1967] 550/4). — [47] C. Y. Liang, E. J. Schimitschek, D. H. Stephens, J. A. Trias (J. Chem. Phys. **46** [1967] 1588/93). — [48] C. Y. Liang, E. J. Schimitschek, J. A. Trias (J. Inorg. Nucl. Chem. **32** [1970] 811/31). — [49] P. C. Mehta, S. S. L. Surana, S. P. Tandon (Can. J. Spectrosc. **18** [1973] 55/60). — [50] R. C. Mathur, S. S. L. Surana, S. P. Tandon (Z. Naturforsch. **30b** [1975] 207/9).

[51] P. C. Mehta, S. P. Tandon (Z. Naturforsch. **26a** [1971] 759/62). — [52] G. A. Domrachev, V. P. Ippolitova (Tr. Khim. Khim. Tekhnol. **1966** 227/40 from C.A. **68** [1968] No. 73665). — [53] R. West, R. Riley (J. Inorg. Nucl. Chem. **5** [1958] 295/303). — [54] L. A. Gribov, Yu. A. Zolotov, M. P. Noskova (Zh. Strukt. Khim. **9** [1968] 448/57; Russ. J. Struct. Chem. **9** [1968] 378/86). — [55] C. Duval, R. Freymann, J. Lecomte (Bull. Soc. Chim. France **1952** 106/13).

[56] J. P. Dismukes, L. H. Jones, J. C. Bailar (J. Phys. Chem. **65** [1961] 792/5). — [57] K. E. Lawson (Spectrochim. Acta **17** [1961] 248/58). — [58] T. I. Razvina, V. S. Khomenko, V. V. Kuznetsova, R. A. Puko, N. N. Mit'kina (Vestsi Akad. Navuk Belarusk. SSR. Ser. Fiz. Mat. Navuk Nr. 2 [1973] 118/21 from C.A. **80** [1974] No. 126428). — [59] G. A. Domrachev, V. P. Ippolitova (Zh. Neorgan. Khim. **12** [1967] 459/62; Russ. J. Inorg. Chem. **12** [1967] 235/7). — [60] A. I. Byrke, N. N. Magdesieva, L. I. Martynenko, V. I. Spitsyn (Zh. Neorgan. Khim. **12** [1967] 666/71; Russ. J. Inorg. Chem. **12** [1967] 348/51).

[61] P. C. Mehta, S. S. L. Surana, S. P. Tandon (Indian J. Pure Appl. Phys. **7** [1969] 767/8). — [62] S. L. Lyle, A. D. Witts (J. Chem. Soc. Dalton Trans. **1975** 185/8).

5.1.2.3 Adducts of Tris(acetylacetonato) Compounds

5.1.2.3.1 Solvates M(acac)$_3$ · n H_2O · m S
(with S = Alcohols, Acetone, Dioxane, Benzene, Pyridine, and Pyridine Derivatives)

Most of the solvates (Table 5/10), except those derived from dioxane, were obtained by recrystallizing the corresponding tris chelate trihydrate from the respective solvent [1]. (The tris chelate trihydrates were prepared according to the method of Stites, McCarty, Quill [2] and purified by recrystallization from 60% ethanol, as described by Pope, Steinbach, Wagner [3].) No special precautions were taken to exclude water from the various solvents. Partial evaporation of the solvent was sometimes necessary to effect crystallization of the product, particularly in the case of the methanol solvates. The dioxane solvates (No. 7) were obtained by recrystallizing the corresponding chelate from hot dioxane to which a little acetylacetone had been added. The hot solutions were diluted with water to 70 to 80% dioxane and then allowed to cool slowly [1].

Table 5/10
Solvates of Tris(acetylacetonato) Rare Earth(III) Chelates, M(acac)$_3$ · n H_2O · m S [1].

No.	M	solvent (= S)	n	m	formula of complex
1	Gd, Y	methanol, (= CH_3OH)	2	1	$M(C_5H_7O_2)_3 \cdot 2H_2O \cdot CH_3OH$
2	Yb	methanol, (= CH_3OH)	0	2	$Yb(C_5H_7O_2)_3 \cdot 2CH_3OH$
3	Y, Yb	ethanol, (= C_2H_5OH)	3	1	$M(C_5H_7O_2)_3 \cdot 3H_2O \cdot C_2H_5OH$
4	Yb	1-propanol, (= C_3H_7OH)	1	0.5	$Yb(C_5H_7O_2)_3 \cdot H_2O \cdot 0.5C_3H_7OH$
5	Yb	1-butanol, (= C_4H_9OH)	1	0.5	$Yb(C_5H_7O_2)_3 \cdot H_2O \cdot 0.5C_4H_9OH$
6	Yb	acetone, (= C_3H_6O)	1	0.5	$Yb(C_5H_7O_2)_3 \cdot H_2O \cdot 0.5C_3H_6O$
7	Nd, Dy, Ho	dioxane, (= $C_4H_8O_2$)	2	0.5	$M(C_5H_7O_2)_3 \cdot 2H_2O \cdot 0.5C_4H_8O_2$
8	Yb	benzene, (= C_6H_6)	1	0.5	$Yb(C_5H_7O_2)_3 \cdot H_2O \cdot 0.5C_6H_6$
9	Pr, Nd, Sm, Eu, Gd Dy, Ho, Er, Yb, Y	pyridine, (= C_5H_5N)	2	1	$M(C_5H_7O_2)_3 \cdot 2H_2O \cdot C_5H_5N$
10	Nd, Gd, Er	2-methylpyridine, (= C_6H_7N)	2	1	$M(C_5H_7O_2)_3 \cdot 2H_2O \cdot C_6H_7N$
11	Dy	3-methylpyridine, (= C_6H_7N)	2	1	$Dy(C_5H_7O_2)_3 \cdot 2H_2O \cdot C_6H_7N$
12	Nd, Dy	2,4-dimethylpyridine, (= C_7H_9N)	2	1	$M(C_5H_7O_2)_3 \cdot 2H_2O \cdot C_7H_9N$
13	Dy	2,6-dimethylpyridine, (= C_7H_9N)	1	2	$Dy(C_5H_7O_2)_3 \cdot H_2O \cdot 2C_7H_9N$

Infrared spectra (Nujol or Fluorolube mulls) of the solvates generally resemble a superposition of the spectrum of the solvating molecule and that of the corresponding rare earth acetylacetonate hydrate. The authors proposed that only the water is coordinated when both water and an organic donor are present and that the organic donor molecules are hydrogen bonded or held in the crystal by lattice forces [1]. The emission spectra of $[Eu(acac)_3(H_2O)_2] \cdot C_5H_5N$ ($^5D_0 \rightarrow {}^7F_0$: 17245 cm^{-1}; $^5D_0 \rightarrow {}^7F_1$: 16962, 16881, 16805 cm^{-1}; $^5D_0 \rightarrow {}^7F_2$: 16360, 16290, 16247, 16160, 16026 cm^{-1}) and $[Eu(acac)_3(H_2O)_2] \cdot H_2O$, see p. 90, were similar, indicating that the outer sphere water molecule of $[Eu(acac)_3(H_2O)_2] \cdot H_2O$ is replaced by pyridine on recrystallization [14].

X-ray powder diffraction data revealed that the ethanol solvates of the yttrium and ytterbium compounds (No. 3) were isomorphous, as were the pyridine adducts of praseodymium through ytterbium (No. 9).

Crystals of the methanol adducts turned opaque rapidly when removed from solution. Infrared spectra, together with chemical analyses, revealed that the adducts decomposed to the trihydrated chelates. Crystals of the ethanol adducts were converted to the monohydrates when allowed to stand in air for several hours. The acetylacetonates of the rare earths below gadolinium do not appear to form ethanol adducts, although they do form methanol solvates. However, the latter compounds could not be isolated in pure form due to coprecipitation of the monohydrates. The pyridine and methyl pyridine adducts all had a faint odor characteristic of the solvating pyridine, and decomposed on standing in air. The compounds with acetone, benzene, *n*-propanol, and *n*-butanol were also unstable and slowly turned opaque in air. The acetone adduct cannot be kept even in a closed container for more than a few days [1].

5.1.2.3.2 Complexes with Bidentate Organic Bases

With Ethylenediamine ($=C_2H_8N_2=$ en), **1,2-Propanediamine** ($=C_3H_{10}N_2=$ pn) **and 1-Amino-2-propanol** ($=C_3H_9NO$)

M(acac)$_3$ · L with L = en, pn, or C_3H_9NO and M = La, Nd were prepared by dissolving M(acac)$_3$ · nH_2O with n = 2 for M = La, n = 1 for M = Nd in 25 ml of warm (60 to 70°C) ethyl acetate containing the respective amine or aminoalcohol. In each case, the insoluble residue was removed by filtration and the filtrate cooled to about 0°C. The crystalline product thus obtained was collected by filtration and dried at room temperature under reduced pressure. The visible absorption spectra of each neodymium compound in chloroform and pyridine and the NMR spectra of the ethylenediamine and propanediamine adducts of the lanthanum chelate in deuteriochloroform were recorded and are discussed in [4]. The infrared spectra of the adducts in the region 2800 to 4000 cm^{-1} were also depicted and discussed in [4]. The compounds M(acac)$_3$ · en and M(acac)$_3$ · pn showed irreversible pyrolytic DTA endotherms and subsequent exotherms in the range 130 to 160°C. The M(acac)$_3$ · C_3H_9NO adducts also decomposed on heating at 123 to 124°C, accompanied by weight loss corresponding, in each case, to the elimination of aminopropanol [4].

With 2,2-Bipyridine (= bpy) **and 4,4'-Bipyridine** ($=C_{10}H_8N_2$)

M(acac)$_3$ · bpy. Precipitation of the samarium, europium, and terbium compounds was effected by adding hexamethylenetetraamine to an acidic solution containing acetylacetone, 2,2'-bipyridine, and a rare earth salt until the reaction mixture was brought to pH 6. Melting ranges (°C) of the compounds are: Sm, 188 to 190; Eu, 200 to 202; Tb, 208 to 210 [5]. The neodymium and erbium adducts were obtained by recrystallizing the respective tris chelate hydrates from an alcoholic solution containing bipyridine [6].

The neodymium complex was also prepared by dissolving the hydrated tris chelate (n = 1) in 25 ml of warm (60 to 70°C) ethyl acetate containing 2,2'-bipyridine. The insoluble residue was removed by filtration and the filtrate cooled to about 0°C, yielding a crystalline product. The compound was collected by filtration and dried at room temperature under reduced pressure [4]. The visible absorption spectrum of the neodymium complex in chloroform and pyridine is discussed in [4].

Nd(acac)$_3$ · $C_{10}H_8N_2$. The adduct was prepared by dissolving the hydrated tris chelate (n = 1) in 25 ml of warm (60 to 70°C) ethyl acetate containing 4,4'-bipyridine. The insoluble residue was removed by filtration and the filtrate cooled to about 0°C. The crystalline product thus obtained was collected by filtration and dried at room temperature under reduced pressure. The visible absorption spectrum of the adduct in chloroform and pyridine was discussed [4].

With 2,7-Dimethyl-1,8-naphthyridine H$_3$C N N CH$_3$ ($=C_{10}H_{10}N_2$)

M(acac)$_3$ · $C_{10}H_{10}N_2$. Each compound (M = Pr to Yb, except for Pm) was prepared according to the following procedure. Acetylacetone (6 mmol) was added to a solution of a rare earth(III) chloride (1 mmol) in ethanol, and the resulting mixture was saturated with ammonia and stirred for 30 min. Evaporation of the solvent under vacuum gave an oily residue which was washed with several portions of pentane until a solid formed. After the powder had been air dried, it was dissolved in 125 ml of a hot ethyl acetate-methanol (4:1) mixture, then 2 mmol of 2,7-dimethyl-1,8-

naphthyridine was added. Each adduct was produced in crystalline form by cooling the reaction mixture at 0°C overnight. The products were collected by filtration, washed with anhydrous ethyl ether, and dried under vacuum for 48 h. The colors of the compounds are (M, color): Pr, green; Nd, violet; Sm, Eu, off-white; Gd, Tb, Dy, Tm, Yb, white; Ho, pale orange, Er, pink.

Infrared spectra (4000 to 200 cm^{-1}, KBr pellets) revealed bands characteristic of acetylacetonate and 2,7-dimethyl-1,8-naphthyridine groups coordinated in a bidentate manner, suggesting a coordination number of eight for the central ion. Molar conductivities for ≈0.6 to 0.7 millimolar solutions of the adducts in nitromethane at 25°C (Λ_M = 6.6 to 8.7 $\Omega^{-1} \cdot cm^2 \cdot mol^{-1}$) are indicative of non-electrolyte behavior [7].

With 1,10-Phenanthroline, (= $C_{12}H_8N_2$ = phen)

and 4,7-Distyrylphenanthroline (= $C_{28}H_{20}N_2$)

M(acac)$_3$ · phen. The europium(III) and terbium(III) compounds were prepared by adding, dropwise and with stirring, a solution of the rare earth salt (2 mmol) in 95% ethanol to a hot solution of 6 mmol acetylacetone, 2 mmol 1,10-phenanthroline, and 6 mmol sodium hydroxide in the same solvent. Each mixture was allowed to cool, and the crystalline product was collected and recrystallized from 95% ethanol. Both compounds melt in the range 250 to 255°C [8].

Precipitation of several adducts (M = Y, La, Pr, Nd, Sm, Eu, Gd, Tb, Er) was effected by adding hexamethylenetetraamine to aqueous solutions containing a rare earth salt, acetylacetone, and 1,10-phenanthroline until the reaction mixture was brought to pH 6. Melting ranges (°C) of the compounds were: Y, 230 to 232; La, 219 to 220; Pr, 221 to 222; Nd, 230 to 232; Sm, 223 to 225; Eu, 226 to 228; Gd, 238 to 240; Tb, 225 to 227; Er, 228 to 230 [5]. Precipitation of the adducts was effected by adding ammonia to an aqueous solution of the reactants [9].

The crystal and molecular structures of the europium compound, prepared by Melby and coworkers [8], were determined by single-crystal X-ray analysis (R = 0.0334 for 1123 reflections) [10]. The compound crystallizes in the monoclinic space group $P2_1/c$-C_{2h}^5 (No. 14) with four molecules in a unit cell of dimensions a = 9.670, b = 21.339, c = 16.497 Å, and β = 116°11′. (The value of both the observed and calculated densities was 1.37 g/cm³.) The monomeric molecules contain an eight-coordinate europium ion, with six oxygen atoms from three bidentate acetylacetonate groups and two nitrogen atoms from the bidentate 1,10-phenanthroline molecule as donors (**Fig. 5-7a**). The

Fig. 5-7

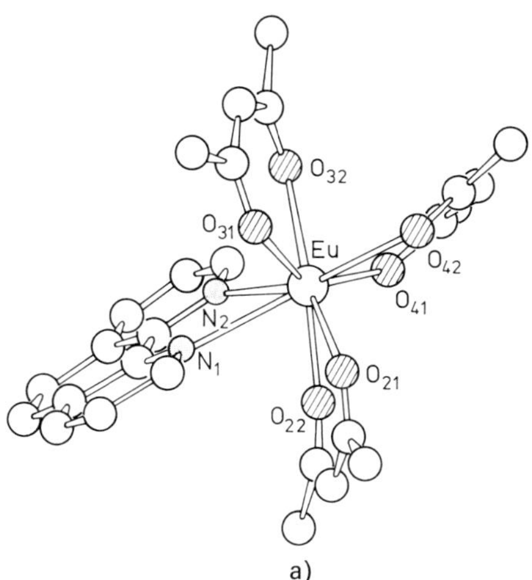

a)
Molecular structure of Eu(acac)$_3$ · phen.

b)
Coordination polyhedron of Eu(acac)$_3$ · phen.

coordination polyhedron (Fig. 5-7b) was described as a slightly distorted square antiprism, with the ligands spanning the s edges of the polyhedron. The symmetry around the europium ion is approximately C_s. The six Eu-O bond lengths average 2.397 Å (range 2.356 to 2.409 Å) and the two Eu-N bond lengths average 2.643 Å (2.645 and 2.641 Å). Because of the unequal Eu-N and Eu-O bond lengths, the upper face of the antiprism is translated parallel relative to the bottom face, with the two oxygen atoms of the upper face lying closer to the rotation axis of the antiprism. The plane of the upper face is 1.44 Å from the europium ion as compared to 1.23 Å for the lower face. The acetylacetonate ligand that shares the square face with the phenanthroline ligand has a folding angle (measure of nonplanarity of chelate ring using the O-O folding line) of 5°, whereas the other two acetylacetonate ligands have angles of 11° and 20°. The phenanthroline group has a folding angle of 10°. All ligands are folded away from the ligand that shares the same square face [10].

Schmidt, Butter [11] found that the europium compound obtained from acetone solutions crystallized in space group $P2_1/n\text{-}C_{2h}^5$ (No. 14) with unit cell dimensions a = 14.6, b = 21.1, c = 9.59 Å, and β = 99.2° (density = 1.43 g/cm³).

Melby and coworkers [8] noted that the fluorescence intensity of the europium compound was markedly brighter than that of the parent hydrate, both in the solid state and in solution, whereas the intensities of the terbium adduct and hydrate were comparable. Similar observations were reported by Butter, Kreher [12].

Eu(acac)$_3$ · $C_{28}H_{20}N_2$. In 200 ml of boiling toluene were dissolved 1.2 mmol of Eu(acac)$_3$ · 3H_2O and 1.2 mmol of 4,7-distyrylphenanthroline. About 50 ml of solvent was distilled from the reaction mixture, and the hot solution was filtered and allowed to stand overnight, yielding the product in the form of yellow needles. Melting range of the compound was 255 to 260°C [8].

5.1.2.3.3 Complexes with 4-Amino-3-penten-2-one $CH_3C(O)CH{=}C(NH_2)CH_3$ (= C_5H_9NO)

M(acac)$_3$ · C_5H_9NO. The adducts (M = Yb, Lu) were obtained as the major reaction products by Richardson, Wagner, Sands [1], when attempting to prepare the tris acetylacetonato chelates according to the procedure described by Stites, McCarty, Quill [2]. A slight excess of aqueous ammonia was used to prepare the ammonium acetylacetonate solution in the case of ytterbium; however, no excess was required for preparation of the lutetium adduct. The products were purified by recrystallization from benzene, acetone, and other organic solvents, including acetylacetone. They are quite stable in air and in a variety of organic solvents, but are decomposed in 60% aqueous ethanol with formation of tris chelate trihydrate [1].

The crystal structure of the ytterbium compound has been determined from three-dimensional X-ray diffraction data. The structure was refined by full-matrix least-squares methods to R = 0.073 for 1301 independent reflections. The compound crystallizes in the monoclinic space group $P2_1/c\text{-}C_{2h}^5$ (No. 14) with four molecules in a unit cell of dimensions a = 18.070, b = 8.538, c = 15.938 Å, and β = 99.13°. The observed and calculated densities are both 1.56 g/cm³. The ytterbium atom is seven-coordinate, being bonded to six acetylacetonate oxygen atoms and the oxygen atom from the 4-amino-3-penten-2-one molecule. The coordination polyhedron is a capped trigonal prism, with an acetylacetonate oxygen as the capping atom. The NH_2 group in 4-amino-3-penten-2-one is hydrogen bonded to the acetylacetonate oxygen atoms in a glide-related molecule, thus linking the molecules in chains parallel to (100). The metal-acetylacetonate rings are folded about the O-O lines, at angles ranging from 10.1° to 19.6°. The Yb-O (4-amino-3-penten-2-one) distance of 2.24(2) Å is practically the same as the average of the six Yb-O (acetylacetonate) distances, 2.23(2) Å [13].

The shift of ν(C=O) stretching mode from 1700 cm^{-1} in the infrared spectrum of 4-amino-3-penten-2-one to 1643 cm^{-1} in the spectra (Nujol or Fluorolube mulls) of the adducts indicates that ligation occurs through the keto group. The ν(N-H) stretching modes were observed at 3355 and 3205 cm^{-1}. The NMR spectrum for the lutetium adduct in $CDCl_3$ revealed peaks which were consistent with the formulation M(acac)$_3$ · C_5H_9NO [1].

5.1.2.3.4 Complex with Diantipyrylmethane

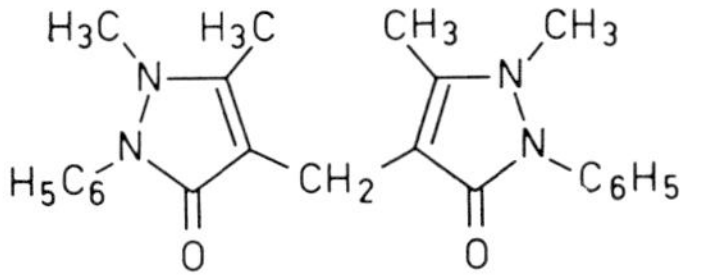

$(= C_{23}H_{24}N_4O_2)$

$Nd(acac)_3 \cdot C_{23}H_{24}N_4O_2$. The compound forms in aqueous acetone from $NdCl_3$, acetylacetone, and diantipyrylmethane. It was characterized by its IR spectrum. The Nd atom is 8-coordinate [15].

5.1.2.3.5 Complexes with Tributyl Phosphate $(C_4H_9O)_3PO$ $(= C_{12}H_{27}O_4P)$, Trioctylphosphine Oxide $(C_8H_{17})_3PO$ $(= C_{24}H_{51}OP)$, and Triphenylphosphine Oxide $(C_6H_5)_3PO$ $(= C_{18}H_{15}OP)$

$M(acac)_3 \cdot L$. Each series of adducts (M = Nd, Er) was prepared by allowing the corresponding hydrated tris chelate to react directly with the neutral donor (1:1 mol ratio). The triphenylphosphine adducts (M = Nd, Er) were also obtained by recrystallizing the hydrated tris chelate from an alcoholic solution containing the ligand. The shift of the $\nu(P{=}O)$ stretching mode (55 cm^{-1} for L = tributylphosphate; 15 cm^{-1} for L = triphenylphosphine oxide, trioctylphosphine oxide) to lower wave numbers in the infrared spectra of the adducts relative to the free ligand values suggests direct ligation of the neutral ligand to the central ion through the phosphoryl or phosphate oxygen atom [6].

References to 5.1.2.3:

[1] M. F. Richardson, W. F. Wagner, D. E. Sands (J. Inorg. Nucl. Chem. **31** [1969] 1417/26). — [2] J. G. Stites, C. N. McCarty, L. L. Quill (J. Am. Chem. Soc. **70** [1948] 3142). — [3] G. W. Pope, J. F. Steinbach, W. F. Wagner (J. Inorg. Nucl. Chem. **20** [1961] 304/13). — [4] I. Yoshida, H. Kobayashi, K. Ueno (Bull. Chem. Soc. Japan **46** [1973] 2140/4). — [5] L. I. Kononenko, E. V. Melent'eva, R. A. Vitkun, N. S. Poluektov (Ukr. Khim. Zh. **31** [1965] 1031/5 from C.A. **64** [1966] No. 9219).

[6] I. A. Murav'eva, L. I. Martynenko, V. I. Spitsyn (Zh. Neorgan. Khim. **22** [1977] 3009/12; Russ. J. Inorg. Chem. **22** [1977] 1636/8). — [7] M. NgSee, H. W. Latz, D. G. Hendricker (J. Inorg. Nucl. Chem. **39** [1977] 71/4). — [8] L. R. Melby, N. J. Rose, E. Abramson, J. C. Caris (J. Am. Chem. Soc. **86** [1964] 5117/25). — [9] E. Butter (Wiss. Z. Karl Marx Univ. Leipzig Math. Natur. Reihe **21** [1972] 3/16 from C.A. **77** [1972] No. 83093). — [10] W. H. Watson, R. J. Williams, N. R. Stemple (J. Inorg. Nucl. Chem. **34** [1972] 501/8).

[11] W. Schmidt, E. Butter (Z. Chem. [Leipzig] **8** [1968] 117). — [12] E. Butter, K. Kreher (Z. Naturforsch. **20a** [1965] 408/12). — [13] M. F. Richardson, P. W. R. Corfield, D. E. Sands, R. E. Sievers (Inorg. Chem. **9** [1970] 1632/8). — [14] E. T. Karaseva, V. E. Karasev, N. I. Steblevskaya (Koord. Khim. **2** [1976] 248/52; Soviet J. Coord. Chem. **2** [1976] 188/91). — [15] M. A. Tishchenko, G. I. Gerasimenko, Yu. N. Anisimov, N. S. Poluektov (Dokl. Akad. Nauk SSSR **250** [1980] 122/5; Dokl. Chem. Proc. Acad. Sci. USSR **250** [1980] 16/9).

5.1.2.4 Tetrakis(acetylacetonato) Compounds of Type $M'[M(acac)_4]$

Alkali Metal Salts $M[Eu(acac)_4]$ (M = Na, K). The sodium compound was prepared by dropwise addition of a solution (20 ml) of sodium acetylacetonate in ethanol (prepared by reacting 3.05 mg-atoms of sodium with 3.05 mmol of acetylacetone in ethanol) to a boiling mixture of 2.87 mmol of $Eu(acac)_3 \cdot H_2O$ in 75 ml of absolute ethanol and 15 ml of benzene. An additional 30 ml of ethanol and 20 ml of benzene were added, and then 50 ml of liquid was removed by distillation. The resulting solid was collected, air dried, and then extracted continuously (Soxhlet apparatus) with ethanol for

2.5 h. The potassium salt, K[Eu(acac)$_4$], was prepared by adding an aqueous solution of europium(III) chloride (2 mmol) to a boiling solution of acetylacetone in a mixture of 50 ml of acetone and 10 ml of 1 N potassium hydroxide. The solution was filtered while hot, and the filtrate was allowed to cool to room temperature. The solid was collected by filtration and washed with acetone-water (1:1). The melting points of the sodium and potassium salts are 345 and 325°C, respectively [1].

The luminescence spectrum of the Eu(acac)$_4^-$ ion was interpreted in terms of D_2 symmetry at the europium(III) site [2]. The splitting of the $^5D_0 \rightarrow {}^7F_1$ transition into three components at 590.9, 593.2, 594.2 nm and of the $^5D_0 \rightarrow {}^7F_2$ transition into five components at 611.8, 614.2, 615.8, 618.7, 620.7 nm indicates the presence of a second-order symmetry axis [8]. The electronic structure of the Eu(acac)$_4^-$ ion has been investigated by extended Wolfsberg-Helmholtz calculations [3, 4].

Tetraethylammonium Salt $(C_2H_5)_4N[Ce(acac)_4]$. The cerium(III) complex was prepared under an atmosphere of pure nitrogen using deoxygenated solvents. A solution of cerium(III) chloride heptahydrate (2 mmol) and 2,2-dimethoxypropane (10 cm^3) in absolute ethanol (15 cm^3) was added to a mixture of acetylacetone (8 mmol), sodium hydroxide (8 mmol), and tetraethylammonium chloride (2 mmol) in absolute ethanol (25 cm^3). The solvent was removed under vacuum, the residue extracted with ethyl acetate (30 cm^3), and the suspended solid removed by filtration. The extract was concentrated under vacuum to approximately 10 cm^3. Slow addition of n-hexane effected separation of a crystalline product, which was collected, washed with n-hexane, and dried under vacuum. The absorption spectrum of the compound in acetonitrile revealed an absorption maximum at 24.9×10^3 cm^{-1} ($\varepsilon = 144$ dm$^3 \cdot$ mol$^{-1} \cdot$ cm^{-1}), with a shoulder at 27.5×10^3 cm^{-1} [5].

5.1.2.5 Cerium(IV) Chelate

Ce(acac)$_4$. The complex was prepared by allowing air oxidation of tris(acetylacetonato)-cerium(III) in hot benzene. After heating for 1.5 h, the dark red-brown solution (250 ml) of the cerium(IV) chelate was boiled down to a volume of 40 ml, diluted to 300 ml with hot hexane, and finally cooled in a dry-ice acetone bath to effect crystallization. The fine, needle-shaped, red-black crystals were quickly collected by filtration and washed with two 25 ml portions of hexane. Two recrystallizations from benzene-hexane were required for purification of the product [6]. The compound was also obtained by the direct reaction of an aqueous cerium(IV) ammonium sulfate solution with acetylacetone (1:4 mole ratio), using a concentrated solution of aqueous ammonia to neutralize the ligand, effecting precipitation of the product. The precipitate was collected by filtration, washed with water, and dried under vacuum. The product was recrystallized twice from benzene-petroleum ether or toluene-petroleum ether solvent mixtures and dried under vacuum [7].

The crystal and molecular structures of $Ce(C_5H_7O_2)_4$ have been the subject of several investigations [9 to 14]. The compound exists in either of two crystalline modifications, designated the α- and β-forms, both of which are in the monoclinic system. Generally the α-form was crystallized from nonpolar solvents such as n-hexane and carbon tetrachloride, preferably by cooling the solutions, while the β-form was crystallized from polar or aromatic solvents, preferably from hot solutions [9].

Titze [10] reported that the α-form crystallizes in space group $P2_1/c$-C_{2h}^5 (No. 14) with four molecules in a unit cell of dimensions a = 11.692, b = 12.6482, c = 16.936 Å, β = 112.339° (R = 0.106 for 2619 independent reflections). The unit cell dimensions obtained by Matković and Grdenić [12, 13] (a = 11.70, b = 12.64, c = 16.93, β = 112°15') agree with those reported by Titze [10]. The cerium(IV) ion is eight-coordinate, with oxygen atoms from four bidentate acetylacetonate groups as donors. The mean Ce-O interatomic distance is 2.33 Å [10]. The coordination polyhedron has been described as predominantly square antiprismatic (idealized D_{4d} symmetry) [10, 13], predominantly dodecahedral (idealized D_{2d} symmetry), [9] and bicapped trigonal prismatic (C_{2v} symmetry) [14]. Allard [9] reported that the ligand configuration corresponds to the dodecahedral stereoisomer III d-*g g g g* [notation of J. L. Hoard, J. V. Silverton (Inorg. Chem. **2** [1963] 235/43)] depicted in **Fig. 5-8**, with a significant distortion towards the antiprismatic isomer, II a-*s s s s*. The ligand wrapping pattern in the bicapped trigonal prismatic structure described by Steffen and Fay [14] lies along the deformation pathway between the *s s s s* square antiprismatic stereoisomer and the *g g g g* dodecahedral stereoisomer.

Fig. 5-8

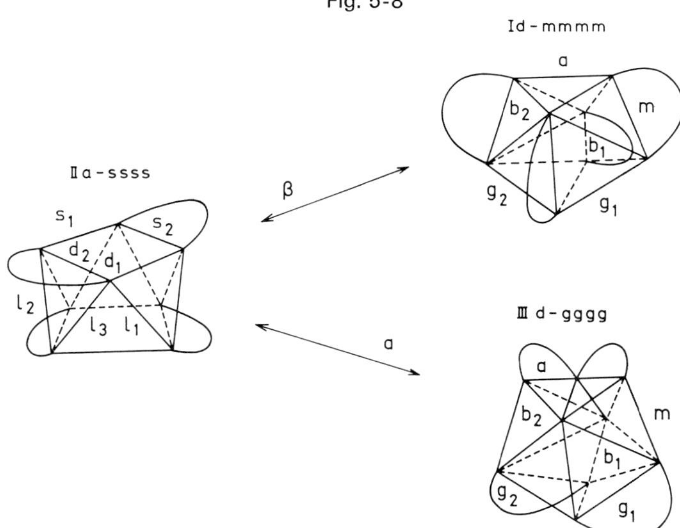

Stereoisomers for α- and β-forms of Ce(acac)$_4$. Hybrids between square antiprism D_{4d}-$\bar{8}2$ m (left) and dodecahedron D_{2d}-$\bar{4}2$ m (right).

Titze [11] reported that the β-form of the complex crystallizes in space group C2/c-C_{2h}^6 (No. 15), with four molecules in a unit cell of dimensions a = 22.006, b = 8.378, c = 14.370 Å, β = 115.78° (R = 0.0093 for 892 independent reflections). The cerium(IV) ion is eight-coordinate with a Ce-O interatomic bond distance of 2.32 Å. The coordination polyhedron was described as primarily square antiprismatic [9, 11, 14] with the ligand configuration II a-*ssss* [9], but with a slight distortion toward the dodecahedral I d-*mmmm* isomer [9, 14]. Allard [9] contends that the main difference between the α- and β-forms, as defined by the ligand parameters, is more pronounced folding of the ligand in the β-form.

Infrared spectral data (samples dispersed in Nujol and hexachlorobutadiene mulls) are tabulated and discussed in [7]. The ν Ce-O stretching mode was assigned to a band at 2500 cm^{-1} [7]. The electronic absorption spectrum of the red-brown compound in acetonitrile revealed bands at 36400 cm^{-1} (31000) and 46900 cm^{-1} (11000) (absorption coefficient, ε in dm^3 · mol^{-1} · cm^{-1}, given in parentheses) [5]. For the polarographic and voltammetric behavior of Ce(acac)$_4$ in acetonitrile, see [15].

References to 5.1.2.4 and 5.1.2.5:

[1] L. R. Melby, N. J. Rose, E. Abramson, J. C. Caris (J. Am. Chem. Soc. **86** [1964] 5117/25). — [2] J. Narusis, A. O. Litinskii, R. Rakauskas, J. Batarunas (Lietuvos Fiz. Rinkinys **9** [1969] 729/36 from C.A. **72** [1970] No. 126981). — [3] O. A. Litinskii, R. Rakauskas, J. Batarunas, J. Glembockis (Intern. J. Quantum Chem. **3** [1969] 373/6). — [4] A. O. Litinskii, J. Batarunas, R. Rakauskas, I. I. Glembotskii (Zh. Strukt. Khim. **10** [1969] 1058/62; Russ. J. Struct. Chem. **10** [1969] 938/41). — [5] M. Ciampolini, F. Maui, N. Nardi (J. Chem. Soc. Dalton Trans. **1977** 1325/8).

[6] T. J. Pinnavaia, R. C. Fay (Inorg. Syn. **12** [1970] 77/80). — [7] T. Yoshimura, C. Miyake, S. Imoto (Bull. Chem. Soc. Japan **46** [1973] 2096/101). — [8] T. I. Razvina, V. S. Khomenko, V. V. Kuznetsova, R. A. Puko (Zh. Prikl. Spektrosk. **19** [1973] 866/71; J. Appl. Spectrosc. [USSR] **19** [1973] 1461/5). — [9] B. Allard (J. Inorg. Nucl. Chem. **38** [1976] 2109/14). — [10] H. Titze (Acta Chem. Scand. A **28** [1974] 1079/88).

[11] H. Titze (Acta Chem. Scand. **23** [1969] 399/408). — [12] D. Grdenić, B. Matković (Acta Cryst. **12** [1959] 817/8). — [13] B. Matković (Acta Cryst. **16** [1963] 456/61). — [14] W. L. Steffen, R. C. Fay (Inorg. Chem. **17** [1978] 779/82). — [15] G. Gritzner, H. Murauer, V. Gutmann (J. Electroanal. Chem. Interfacial Electrochem. **101** [1979] 177/83).

5.1.2.6 Other Acetylacetonato Complexes

Complexes derived from acetylacetone and 5,10,15,20-tetraarylporphyrines or diamine-N-polycarboxylic acid see "Rare Earth Elements" D 1, 1980, p. 94, 234 and 236.

Mixed Chelates with Propionylacetone (= $C_6H_{10}O_2$), **Benzoylacetone** (= $C_{10}H_{10}O_2$), **and Dibenzoylmethane** (= $C_{15}H_{12}O_2$)

Na[M(acac)$_3$($C_6H_9O_2$)], Na[M(acac)$_3$($C_{10}H_9O_2$)], and **Na[M(acac)$_3$($C_{15}H_{11}O_2$)]** with M = La, Pr, Nd, Sm were prepared by slowly adding an alcoholic solution containing 1 mmol of the β-diketone and 1 mmol of sodium hydroxide to a solution of the hydrated tris acetylacetonato chelate in ethanol. The resulting crystalline precipitates were collected by filtration, washed several times with alcohol, and finally recrystallized by dissolving in a minimum volume of dimethylformamide and effecting reprecipitation with nitrobenzene or petroleum ether (40 to 60°C). The compounds which were dried in air are only sparingly soluble in polar organic solvents. The molar conductance values (75 to 80 $\Omega^{-1} \cdot cm^2 \cdot mol^{-1}$) of the compounds in dimethylformamide (10^{-3} M) at 30°C are in accord with the behavior of 1:1 electrolytes. The compounds were further characterized by their electronic absorption spectra in methanol and n-hexane and infrared spectra (Nujol mulls) [1].

With N,N'-Bis(salicylidene)ethylenediamine (OH)$C_6H_4CH{=}NCH_2CH_2N{=}CHC_6H_4$(OH) (= $C_{16}H_{16}N_2O_2$)

M(acac)$_2$($C_{16}H_{14}N_2O_2$)$_{0.5}$. Complexes with M = Y, La, Pr, Nd were prepared by adding a boiling solution of bis(salicylidene)ethylenediamine (1.2 mmol) in 40 ml of ethanol to a boiling ethanolic solution of acetylacetone, followed by passage of ammonia gas through the solution in a slow but steady stream for a period of about 10 min. The reaction vessel was stoppered tightly and left undisturbed for 4 to 5 h. Shining crystals of the mixed chelates began to deposit after about an hour. Each compound was collected by filtration and dried under vacuum over sulfuric acid at room temperature. The colors and melting points (°C) of the compounds were: Y, pale yellow, 226 to 228; La, cream, 228 to 230; Pr, bright green, 224 to 226; Nd, pink, 235 to 236; Sm, cream, 225 to 228. The compounds were insoluble in common organic solvents. The compounds were characterized further by elemental analyses and their infrared spectra (KBr disc). A dinuclear structure was proposed with the bis(salicylidene)ethylenediamine as the bridging ligand [2].

References to 5.1.2.6:

[1] N. K. Dutt, S. Sur (J. Inorg. Nucl. Chem. **33** [1971] 115/9). — [2] N. K. Dutt, K. Nag (J. Inorg. Nucl. Chem. **30** [1968] 2779/83).

5.1.3 Complexes with 1,1,1-Trifluoro-2,4-pentanedione $CF_3C(O)CH_2C(O)CH_3$ (= Trifluoroacetylacetone = $C_5H_5F_3O_2$)

5.1.3.1 Tris Chelates M($C_5H_4F_3O_2$)$_3$ · n H_2O (n = 0, 1, 2)

Each compound (Table 5/11) was prepared by adding an aqueous solution of ammonium trifluoroacetylacetonate to an aqueous solution of the rare earth chloride. The resulting precipitate was collected by filtration, washed with water, air dried, and finally recrystallized from methanol. The colors and melting points of the compounds are recorded in Table 5/11. All the chelates are soluble in methanol. Upon heating, the solids slowly softened and finally melted completely. In each case, the latter temperature was taken as the melting point. The sublimation of the chelates at 1 Torr pressure using deoxygenated nitrogen as the carrier gas is reported in [1]. Thermogravimetric analyses indicated that dehydration of the complexes by heating is accompanied by decomposition [2].

The dihydrated europium chelate was prepared by adding an ethanolic solution (25 ml) of sodium trifluoroacetylacetonate (3 mmol), dropwise, over a period of 2 h, to a well-stirred solution (25 ml) of europium chloride (1 mmol) in water. Needle-like crystals of the compound formed while the reaction mixture was allowed to stand overnight [3]. The dihydrated erbium chelate was prepared by mixing erbium chloride hexahydrate, trifluoroacetylacetone, and ammonia in ethanol. After partial evaporation of the solvent, the crystalline precipitate was collected, washed with ethanol and air dried [4].

Table 5/11
Physical Properties of the Tris(trifluoroacetylacetonato) Rare Earth Chelates, $M(C_5H_4F_3O_2)_3 \cdot n\,H_2O$ [1].

M	n	color	m. p. in °C	M	n	color	m. p. in °C
Sc	3	white	124 to 126	Gd	2	white	145 to 147
Y	0	white	146 to 148	Tb	2	white*)	154 to 156
La	2	white	142 to 144	Dy	2	white	152 to 154
Ce	2	dark brown	127 to 129	Ho	2	cream white	154 to 156
Pr	2	light green	136 to 138	Er	2	pink	148 to 150
Nd	2	lavender	147 to 149	Tm	2	white	156 to 158
Sm	2	cream white	147 to 149	Yb	2	white	140 to 142
Eu	2	white	144 to 146	Lu	2	white	188 to 190

*) Yellow fluorescence.

Infrared data for $Er(C_5H_4F_3O_2)_3 \cdot 2H_2O$ (KBr disc) are tabulated in [4]. The $\nu_{as}(C{=}O)$ and $\nu_{as}(C{=}C)$ modes were assigned to the very strong bands at 1615 and 1530 cm^{-1}, $\nu_s(C{=}O)$, coupled with δ C-CH, to a medium intensity band at 1440 cm^{-1}, $\nu_s(C{=}C)$ (coupled with ν C-CH_3) to a medium intensity band at 1240 cm^{-1}, and $\nu_s(C{=}C)$ (coupled with ν C-CF_3) to a strong band at 1180 cm^{-1}. The ν Er-O stretching mode was assigned to a weak band at 390 cm^{-1} [4]. The mass spectrum of $Eu(C_5H_4F_3O_2)_3 \cdot 3H_2O$ and the volatility of the complex were studied by V. S. Khomenko et al. [13].

5.1.3.2 Lewis Base Adducts of the Tris Chelates

With Tributyl Phosphate $(C_4H_9O)_3PO$ $= C_{12}H_{27}O_4P)$

$M(C_5H_4F_3O_2)_3 \cdot 2(C_4H_9O)_3PO$. Each adduct (M = Nd, Eu, Ho, Tm) was prepared by a solvent extraction procedure. The pH of the aqueous phase containing the rare earth chloride was adjusted to 5.27 with a buffer solution of sodium acetate and acetic acid, and then extracted with a cyclohexane solution containing trifluoroacetylacetone and tributyl phosphate (trifluoroacetone : tributyl phosphate : rare earth mol ratio = 3:2:1). Each complex was isolated as an oil from the extractant solution by evaporation of the organic solvent. Thermograms of the compounds are discussed in [2]. The europium compound was also isolated as an amber oil by extracting an aqueous solution of the nitrate salt with an ether solution containing the chelating agent and tributyl phosphate (m. p. = 10°C) [5].

With Dihexyl Sulfoxide $(C_6H_{13})_2SO$ **and Trioctylphosphine Oxide** $(C_8H_{17})_3PO$

$Eu(C_5H_4F_3O_2)_3 \cdot 2(C_6H_{13})_2SO$ and **$Eu(C_5H_4F_3O_2)_3 \cdot 2(C_8H_{17})_3PO$.** Each compound was prepared by extracting an aqueous solution of europium nitrate with an ether solution containing trifluoroacetylacetone and the neutral oxygen donor (mol ratio = 1:3:2). The compounds were isolated as amber oils from the ether phase. The melting points of the dihexyl sulfoxide and trioctylphosphine oxide adducts were 15 and 0°C, respectively. The luminescence spectra of the adducts in 1:1 (v/v) methylcyclohexane-3-methylpentane were recorded in the region 6100 to 6225 Å ($^5D_0 \rightarrow {}^7F_2$ transition). The data revealed that the luminescence intensity of the adduct was enhanced relative to the tris chelate, an observation attributed to an "insulating sheath effect" which reduces radiationless energy losses from the metal chelate [5].

5.1.3.2 Tetrakis Chelates of the Type $M'[M(C_5H_4F_3O_2)_4]$

Alkali Metal Salts (M' = Na, K, Cs)

The erbium chelates (M' = Na, K) were prepared by neutralizing a mixture of trifluoroacetylacetone and erbium nitrate in 1:1 ethanol-water (v/v) with the alkali metal carbonate. The bulk of the ethanol was removed and the residue extracted with chloroform. The extractant solution was shaken with an excess of the alkali metal carbonate and allowed to stand overnight. The mixture was filtered, the filtrate evaporated to dryness, and the residue sublimed under vacuum [6].

The gadolinium and erbium chelates (M' = Na, K, Cs) were prepared by adding aqueous solutions of the rare earth salt ($GdCl_3$ or $Tb(NO_3)_3$), dropwise, to rapidly stirred solutions containing the ligand and an equivalent amount of the respective alkali metal hydroxides (ligand : metal ion mol ratio = 4.5 : 1). The solutions were warmed to redissolve any precipitated product and allowed to cool. The sodium and potassium salts charred at 250°C before melting (m.p. > 250°C), whereas the two caesium salts both melted at 214 to 215°C. Both compounds were characterized by elemental analyses and the terbium compound by emission spectral data [7].

Ammonium Salts $NH_4[M(C_5H_4F_3O_2)_4]$

Each chelate (M = Gd, Tb) was prepared by adding an aqueous solution of the rare earth salt ($GdCl_3$ or $Tb(NO_3)_3$), dropwise, to a 95% ethanolic solution containing the ligand and an equivalent amount of ammonia (ligand : metal ion mol ratio = 4.5 : 1). The products crystallized from the solutions. The gadolinium chelate melted at 170 to 171°C, and the terbium compound melted at 174 to 175°C [7].

Tetraethylammonium Salt $(C_2H_5)_4N[Ce(C_5H_4F_3O_2)_4]$

The tetrakis cerium(III) chelate was prepared under an atmosphere of pure nitrogen using deoxygenated solvents according to the following procedure. A solution of cerium(III) chloride heptahydrate (2 mmol) and 2,2-dimethoxypropane (10 cm^3) in absolute ethanol (15 cm^3) was added to a mixture of trifluoroacetylacetone (8 mmol), sodium hydroxide (8 mmol), and tetraethylammonium chloride (2 mmol) in absolute ethanol (25 cm^3). The solvent was removed under reduced pressure, the residue extracted with ethyl acetate (30 cm^3), and the suspended solid removed by filtration. The solution was concentrated under vacuum to approximately 10 cm^3. Precipitation of the product was effected by slowly adding n-hexane. The product was collected, washed with n-hexane, and dried under vacuum. The yellow colored compound was characterized by elemental analyses and its electronic absorption spectrum, measured in acetonitrile (23900 [sh], 25500 [136], 28200 [sh], 35300 cm^{-1} [49500]; absorption coefficient [ε in $dm^3 \cdot mol^{-1} \cdot cm^{-1}$] given in bracket) [8].

Piperidinium Salts $C_5H_{10}NH_2[M(C_5H_4F_3O_2)_4]$

The complexes with M = Y or La were prepared by adding an aqueous solution of the respective rare earth chloride (8 mmol) to a boiling mixture of trifluoroacetylacetone (32 mmol) and piperidine (32 mmol) in 95% ethanol. Each reaction mixture was heated at reflux for 1 h and then filtered, and the filtrate was concentrated by allowing evaporation of the solvent until the first sign of turbidity. The solution was allowed to stand at 3°C for 1 day, and the resulting crystals were collected by filtration. The compounds were recrystallized from dichloromethane-hexane (m. p. = 106 to 108°C for M = Y; m. p. = 122 to 124°C for M = La) [9]. The gadolinium (m. p. = 90 to 92°C) and terbium (m. p. = 100 to 101°C) chelates were prepared according to a similar procedure [7].

Proton NMR studies of the lanthanum and yttrium complexes in deuteriochloroform [9], and of mixtures of the tetrakis(trifluoroacetylacetonato) and tetrakis(hexafluoroacetylacetonato)yttrium chelates in the same solvent [10], have been reported and are discussed on p. 113.

The europium compound was prepared by adding an ethanolic solution (10 ml) of the chloride salt (2 mmol), dropwise, to a solution of trifluoroacetylacetone (8 mmol) and piperidine (8 mmol) in ethanol. The solvent was removed, and the viscous, brown, oily residue extracted with acetone. Removal of the acetone yielded a brown oil, which was treated with n-pentane to obtain the solid product [3].

Isoquinolinium Salt $C_9H_7NH[Eu(C_5H_4F_3O_2)_4]$

To a solution of 8 mmol of trifluoroacetylacetone and 2 mmol of europium chloride in 70 ml of hot ethanol was added 8 mmol of isoquinoline. The solution was allowed to stand and cool to room temperature. The crystallized product was collected, washed with water, and dried under vacuum. The compound was recrystallized from cyclohexane (m. p. 111 to 114°C) and characterized by elemental analyses and its emission spectrum [11].

Tetraphenylarsonium Salts $(C_6H_5)_4As[M(C_5H_4F_3O_2)_4]$

Each compound (M = Gd, Tb) was prepared by adding aqueous solutions of the rare earth salt ($GdCl_3$ or $Tb(NO_3)_3$) and a tetraphenylarsonium salt to a 95% ethanolic solution containing trifluoroacetylacetone and an equivalent amount of sodium hydroxide. The volume of the resulting solution was reduced by boiling until crystallization had begun, then sufficient ethanol was added to redissolve the product, and the mixture was allowed to cool. The compounds were recrystallized from aqueous ethanol solutions. The gadolinium chelate melted at 158 to 159°C, and the terbium compound melted at 162 to 163°C [7].

5.1.3.3 Cerium(IV) Compound $Ce(C_5H_4O_2F_3)_4$

The complex was prepared by adding an ethanolic solution of the ligand to an aqueous solution of ceric ammonium nitrate. The chelate thus formed was extracted from the crude precipitate with benzene or toluene. The product crystallized from the extractant solution and was recrystallized twice from benzene and dried under vacuum [12].

Infrared spectral data (samples dispersed in Nujol and hexachlorobutadiene mulls) are tabulated and discussed. The ν(Ce-O) stretching mode was assigned to a band at 240 cm^{-1} [12]. The electronic absorption spectrum of the red-brown compound in acetonitrile revealed bands at 21300 (sh), 26500 (4400), 35500 (29100) (absorption coefficients, ε in $dm^3 \cdot mol^{-1} \cdot cm^{-1}$, are given in parentheses) [8].

References to 5.1.3:

[1] E. W. Berg, J. J. C. Acosta (Anal. Chim. Acta **40** [1968] 101/13). — [2] J. W. Mitchell, C. V. Banks (Anal. Chim. Acta **57** [1971] 415/24). — [3] S. J. Lyle, A. D. Witts (Inorg. Chim. Acta **5** [1971] 481/4). — [4] R. C. Mathur, S. S. L. Surana, S. P. Tandon (Z. Naturforsch. **30 b** [1975] 207/9). — [5] F. Halverson, J. S. Brinen, J. R. Leto (J. Chem. Phys. **41** [1964] 157/63).

[6] R. Belcher, J. R. Majer, R. Perry, W. I. Stephen (J. Inorg. Nucl. Chem. **31** [1969] 471/8). — [7] T. D. Brown, T. M. Shepherd (J. Chem. Soc. Dalton Trans. **1973** 336/41). — [8] M. Ciampolini, F. Mani, N. Nardi (J. Chem. Soc. Dalton Trans. **1977** 1325/8). — [9] R. C. Fay, N. Serpone (J. Am. Chem. Soc. **90** [1968] 5701/6). — [10] N. Serpone, R. Ishayek (Inorg. Chem. **10** [1971] 2650/6).

[11] H. Bauer, J. Blanc, D. L. Ross (J. Am. Chem. Soc. **86** [1964] 5125/31). — [12] T. Yoshimura, C. Miyake, S. Imoto (Bull. Chem. Soc. Japan **46** [1973] 2096/101). — [13] V. S. Khomenko, T. A. Rasshinina, V. M. Suboch (Vestsi Akad. Navuk Belarusk. SSR Ser. Khim. Navuk **1979** No. 3, p. 36/40 from C.A. **91** [1979] No. 100799).

5.1.4 Complexes with 1,1,1,5,5,5-Hexafluoro-2,4-pentanedione $CF_3C(O)CH_2C(O)CF_3$ (= Hexafluoroacetylacetone = $C_5H_2F_6O_2$)

5.1.4.1 Tris Chelates $M(C_5HF_6O_2)_3 \cdot nH_2O$ (n = 0 to 3)

Halverson, Brinen, Leto [1] prepared the $M(C_5HF_6O_2)_3 \cdot nH_2O$ (n = 1 for M = La, n = 2 for M = Eu) complexes by extracting an aqueous solution (pH 4 to 6) of the rare earth nitrate with a diethyl ether solution of ammonium hexafluoroacetylacetonate (prepared by neutralizing hexafluoroacetylacetone with a stoichiometric amount of concentrated aqueous ammonia). After drying with sodium sulfate, the ether phase was allowed to evaporate, leaving a viscous oil from which the product crystallized at a low temperature. The compounds were recrystallized from a 9:1 (v/v) water/methanol mixture. The europium compound thus obtained melted at 125°C. The compositions of the compounds were established by elemental analyses (Richardson and coworkers [2] obtained the lanthanum chelate in the form of the trihydrate according to this procedure and were unable to reproduce the monohydrate formulation—see discussion below).

The anhydrous lanthanum and europium chelates were reportedly prepared from the dihydrates by titration with cold 3:1 (v/v) petroleum ether/methanol, followed by filtration under nitrogen and drying under vacuum (0.02 Torr) over phosphorus(V) oxide for 60 h. Although final purification of the powdery solids by vacuum sublimation was feasible, passing a dilute hexane solution down a neutral alumina chromatography column was recommended. The anhydrous europium chelate thus obtained melted at 196°C. The anhydrous chelates readily absorbed moisture, reforming the hydrated chelates upon exposure to the atmosphere [1]. Steinberg, Mashall, Glasner [3], on the other hand, reported that the hydrated tris chelates were not dehydrated, even after drying for several months under high vacuum over phosphorus(V) oxide, Richardson, Wagner, Sands [2] noted that the hydrated chelates were not dehydrated under vacuum over magnesium perchlorate, but were partially hydrolyzed (see discussion below).

Richardson and coworkers [2] prepared the series of hydrated tris chelates ($n = 3$ for M = La, Pr, Nd; $n = 2$ for M = Y, Nd to Yb except for Pm, Tm) according to a procedure similar to that of Halverson and coworkers [1] described above. Following evaporation of the ether phase, the resulting mixture of oil and crystals was extracted several times with 100-ml portions of boiling hexane. The extractant solutions were combined and allowed to cool, yielding crystalline products. The crystals were collected by filtration and air dried for a few hours. The compositions of the compounds were established by elemental analyses and Karl Fischer titrations. None of the dihydrated chelates lost water to give the monohydrate or anhydrous chelate when stored over magnesium perchlorate, either at atmospheric pressure or under vacuum, although the compounds were partially hydrolyzed under vacuum. The trihydrates readily converted to the corresponding dihydrates over magnesium perchlorate, but these chelates picked up moisture on exposure to the atmosphere to yield the original trihydrated chelates. The hydrated chelates were sublimed under vacuum at 100°C unchanged [2]. Thermogravimetric analyses verified the volatile nature of the chelates, with half-temperatures ranging from 165°C for the lutetium chelate to 200°C for the praseodymium compound (half-temperature is defined as the temperature of 50% weight loss on the TGA curve), but also revealed extensive decomposition of the chelates on heating [4]. The compounds are extremely soluble in oxygen-containing organic solvents such as alcohols, ethers and ketones; insoluble in water, hexane, cold chloroform and carbon tetrachloride; and moderately soluble in benzene. X-ray powder patterns revealed that the trihydrated chelates are isomorphous, as are the dihydrated compounds [2].

Berg, Acosta [5] also prepared a series of tris chelates (M = Sc, Y, La to Lu, except for Pm) according to the procedure of Halverson and coworkers [1], but the degree of hydration of their compounds, with the exception of the scandium and yttrium chelates, which were reportedly anhydrous, was not determined. The chelates of lanthanum, cerium, praseodymium, holmium, erbium, and lutetium were purified by sublimation under vacuum (0.05 Torr) at 180°C, demonstrating the volatile nature of the chelates. The anhydrous chelates were prepared by drying the hydrated compounds under vacuum (0.02 Torr) over phosphorus(V) oxide for 72 h. The colors and melting points (°C) of the compounds thus obtained are recorded in Table 5/12. The complexes demonstrated high thermal stability and also high volatility, especially under vacuum. Sublimation recrystallization temperature zones were determined at 1 atmosphere pressure using deoxygenated nitrogen as the carrier gas [5].

Lyle, Witts [6] effected synthesis of $Eu(C_5HF_6O_2)_3 \cdot 2H_2O$ by adding an ethanolic solution (25 ml) of sodium hexafluoroacetylacetonate (3 mmol), dropwise, over a period of about 2 h, to a stirred solution of europium chloride (1 mmol) in 25 ml of water. Sodium hexafluoroacetylacetonate was prepared by treating hexafluoroacetylacetone with a stoichiometric amount of sodium hydroxide in ethanol. An oil separated during the addition of sodium hexafluoroacetylacetonate, which failed to yield a crystalline product overnight. The oil was extracted with ether, the extractant solution dried, and the ether removed by distillation. The residual oil was extracted twice with 10 ml portions of boiling n-hexane. The crystalline tris chelate was formed on allowing the extractant solution to cool [6]. The Mössbauer spectrum of the compound thus prepared, taken with a ^{151}Eu source in an Sm_2O_3 lattice, was obtained over the temperature range 79 to 220 K. The average isomer shift, $-0.40\ mm \cdot s^{-1}$ (range $= -0.38$ to $-0.43\ mm \cdot s^{-1}$) was positive relative to the predominantly ionic compound, $EuF_3 \cdot 0.5\,H_2O$, indicating some covalency in the Eu-O bond [7].

Table 5/12
Physical Properties of the Tris(hexafluoroacetylacetonato) Rare Earth Chelates Prepared by Berg, Acosta [5].

M	n	color	m.p. in °C	M	n	color	m.p. in °C
Sc	0	white	116 to 117	Gd	*)	white	170 to 173
Y	0	white	142 to 144	Tb	*)	white**)	170 to 172
La	*)	white	143 to 146	Dy	*)	cream-white	185 to 188
Ce	*)	yellow	125 to 126	Ho	*)	cream-white	214 to 215
Pr	*)	light green	148 to 151	Er	*)	pink	194 to 198
Nd	*)	lavender	141 to 142	Tm	*)	white	194 to 196
Sm	*)	cream-white	144 to 145	Yb	*)	white	177 to 178
Eu	*)	light yellow	176 to 177	Lu	*)	white	222 to 223

*) Degree of hydration not determined.—**) Yellow fluorescence.

Bhaumik [8] effected synthesis of $Eu(C_5HF_6O_2)_3 \cdot 2\,H_2O$ by mixing hexafluoroacetylacetone, neutralized with an equivalent amount of aqueous ammonia, with an aqueous solution of europium chloride. The precipitated product was washed with water and recrystallized from a mixture of ether and hexane (m.p. = 181 to 182°C). On heating the compound at 120°C under vacuum, Bhaumik observed that the sublimate revealed a europium fluorescence as strong as $Eu(C_5HF_6O_2)_3 \cdot 2\,H_2O$, suggesting that the substance passed through the vapor phase without decomposition [8].

Infrared spectral data are tabulated and discussed in [2] and [9]. No significant changes in the spectra of the dihydrates were observed from neodymium to ytterbium, except that the ν(O-H) stretching mode near 3500 cm^{-1} shifted regularly to higher wave numbers across the series (3470 cm^{-1} for M = Nd to 3550 cm^{-1} for M = Yb). The peaks at 3695 and 3610 cm^{-1} in the spectrum of $Nd(C_5HF_6O_2)_3 \cdot 3\,H_2O$, due to the third weakly bonded water molecule, disappeared when the chelate was converted to the dihydrate. The broad ν(OH) stretching band at 3550 cm^{-1} in the spectrum of the trihydrated neodymium chelate shifted to 3470 cm^{-1} on conversion to the dihydrate [2]. The ν(C=O), ν(C=C), and ν(C=O) (coupled with a CH bend) stretching modes were assigned to bands at 1657 (s), 1618 (sh), and 1568 cm^{-1}, respectively, in the spectrum (KBr disc) of the hydrated neodymium chelate [9].

The exchange of rare earth ions in chelate solutions was observed by emission spectroscopy. The tris chelate of europium fluoresces strongly at 613 nm in alcohol. On adding an alcoholic solution of terbium nitrate to the europium chelate solution, the strong red fluorescence due to europium was readily replaced by a green terbium fluorescence, characteristic of a hexafluoroacetylacetonato terbium chelate. The decrease in europium fluorescence was directly proportional to the amount of terbium nitrate added. At a terbium concentration of nearly 10^3 times that of europium, the red fluorescence was completely replaced by the green terbium fluorescence. The fluorescence decay time and the activation curve for the green emission followed closely those of a hexafluoroacetylacetonato terbium chelate [10].

5.1.4.2 Lewis Base Adducts of the Tris Chelates

With Acetone $(CH_3)_2CO$ $(=C_3H_6O)$

$M(C_5HF_6O_2)_3 \cdot (CH_3)_2CO$. Compounds of composition $M(C_5HF_6O_2)_3 \cdot (CH_3)_2CO$ (M = Sc, Y, La, Pr, Nd, Eu, Gd) were obtained by treating suspensions of the anhydrous rare earth chlorides in dry acetone with anhydrous hexafluoroacetylacetone. With samples dispersed in sodium chloride discs, a free carbonyl stretching band in the 1700 cm^{-1} region was not observed in the infrared spectra of the chelates. However, the infrared spectra of carbon tetrachloride solutions revealed a band at 1692 cm^{-1}, in addition to the chelate ν(C=O) stretching mode in the 1640 cm^{-1} region, suggesting that dissociation of the adduct occurs in solution. Differential thermal analysis and thermogravimetric analysis showed that the scandium chelate sublimed at about 100°C, both in pure argon and in air. The gadolinium chelate sublimed at about 150°C in argon but decomposed in air. Vapor phase chromatograms (injection port at 150°C, column at 60°) of the chelates in hexafluoroacetylacetone revealed two peaks, the first presumably due to solvent and the second due to the chelate. No third peak, which might be attributed to acetone from the adduct, was observed [3].

With Dimethylformamide $(=C_3H_7NO$, = DMF)

$M(C_5HF_6O_2)_3 \cdot n\,DMF$ (n = 1, 2) and **$M(C_5HF_6O_2)_3 \cdot H_2O \cdot DMF$.** Compounds of compositions $M(C_5HF_6O_2)_3 \cdot DMF$ (M = Y, Pr, Sm, Gd, Dy, Er, Lu), $M(C_5HF_6O_2)_3 \cdot 2DMF$ (M = Y, La, Pr, Nd, Sm, Eu, Gd, Tb, Er, Yb, Lu), and $M(C_5HF_6O_2)_3 \cdot H_2O \cdot DMF$ (M = Y, Pr, Sm, Dy, Er) were prepared according to the following two procedures. The first method involved adding a stoichiometric amount (1:1 or 2:1) of dimethylformamide to the solid chelate and heating gently. Droplets of water appeared in the resulting oil. The substance was heated at ≈60°C under vacuum to yield the anhydrous 2:1 and 1:1 adducts, whereas the hydrated adducts were obtained by exposing the anhydrous compounds to the atmosphere or by allowing the initial preparation to cool and dry in air. The second method was similar to the first, except that benzene was used as a solvent. Evaporation of the benzene yielded the adducts [4].

The ν(C=O) stretching mode of the dimethylformamide molecule is shifted to lower wave numbers (≦1650 cm^{-1}) in the infrared spectra of the adducts (Nujol and Fluorolube mulls) relative to the free ligand value (1685 cm^{-1}), indicating that ligation occurs through the carbonyl group (the ν(C=O) stretching mode of DMF is masked by strong carbonyl absorptions of the chelate in the region 1640 to 1650 cm^{-1}). Eight-coordination in the bis adducts and seven-coordination in the mono adducts were supported by molecular weight measurements. The NMR spectra of the dimethylformamide adducts in carbon tetrachloride revealed a downfield shift of the methyl proton resonance, relative to those of the free amide molecule, characteristic of coordination [4].

Thermogravimetric analyses revealed that the compounds are thermally stable, subliming completely without decomposition. The temperatures corresponding to 50% weight loss (defined as the half temperature) on the TGA curve indicated that the heavier rare earth chelate adducts are more volatile. For the $M(C_5HF_6O_2)_3 \cdot 2DMF$ species, the half-temperature range was 175°C for the lutetium adduct to 214°C for that of lanthanum; for the $M(C_5HF_6O_2)_3 \cdot DMF$ species the range was 164°C for the lutetium complex to 199°C for that of praseodymium; and for the $M(C_5HF_6O_2)_3 \cdot DMF \cdot H_2O$ species the range was 169°C for the erbium compound to 198°C for the praseodymium complex [4].

With 2,2-Dimethoxypropane $(CH_3)_2C(OCH_3)_2$ $(=C_5H_{12}O_2)$

$M(C_5HF_6O_2)_3 \cdot C_5H_{12}O_2$. Each adduct (M = Y, Pm, Sm, Gd, Dy, Ho, Er, Yb, Lu) was obtained by dissolving the hydrated chelate, $M(C_5HF_6O_2)_3 \cdot 2H_2O$, in ≈5 ml of dimethoxypropane, heating the reaction mixture at reflux for a few seconds, and removing the excess solvent under vacuum. Crystals were sometimes obtained, but most frequently the product was an oil. Insufficient boiling of the dimethoxypropane solutions yielded acetone-methanol solvates $M(C_5HF_6O_2)_3 \cdot CH_3OH \cdot (CH_3)_2CO$ instead of dimethoxypropane adducts [4].

Infrared (samples dispersed in Nujol and Fluorolube mulls) and NMR (chloroform solutions) spectral evidence suggested that the dimethoxypropane molecule is chelated to give a four-membered ring and an eight-coordinate metal ion. The compounds are highly volatile but showed slight decomposition on sublimation. The half temperatures (defined as the temperature corresponding to 50% weight loss on the TGA curve) ranged from 132°C for the lutetium adduct to 184°C for that of praseodymium [4].

With Tributyl Phosphate $(C_4H_9O)_3PO$ $(=C_{12}H_{27}O_4P)$

$M(C_5HF_6O_2)_3 \cdot 2(C_4H_9O)_3PO$. Each adduct (M = Nd, Eu, Tm) was prepared by extracting an aqueous solution of the rare earth chloride (buffered at pH 5.27 with sodium acetate-acetic acid) with a cyclohexane solution containing hexafluoroacetylacetone and tributyl phosphate (mole ratio of the reactants = 1:3:2). The extractant solution was separated, and the complexes were isolated as oils by evaporating the organic solvent. Thermograms of the adducts revealed that the compounds are thermally stable and volatile, subliming in the range 200 to 250°C [11]. The overall formation constants, β_2 $(\beta_2 = [M(C_5HF_6O_2)_3 \cdot 2(C_4H_9O)_3PO] \cdot [M(C_5HF_6O_2)_3]^{-1} \cdot [(C_4H_9O)_3PO]^{-2})$, of the adducts in cyclohexane were determined at 25°C by a solvent extraction procedure (lg $\beta_2 = 10.50$ for M = Nd, lg $\beta_2 = 10.84$ for M = Eu; lg $\beta_2 = 10.76$ for M = Tm) [12].

With Tris(2,2,2-trifluoroethyl) Phosphate $(CF_3CH_2O)_3PO$ $(=C_6H_6F_9O_4P)$ and **Tris(2,2,3,3,3-pentafluoropropyl) Phosphate** $(CF_3CF_2CH_2O)_3PO$ $(=C_9H_6F_{15}O_4P)$

$M(C_5HF_6O_2)_3 \cdot 2(CF_3CH_2O)_3PO$ and **$M(C_5HF_6O_2)_2 \cdot 2(CF_3CF_2CH_2O)_3PO$.** Each compound in both series of adducts (M = Pr, Nd, Eu, Ho, Er, Tm) was prepared according to the procedure described above for preparation of the analogous tributyl phosphate adducts. Thermograms of the adducts revealed that the compounds are thermally stable and volatile, subliming in the range ≈150 to 180°C [11].

With Trioctylphosphine Oxide $(C_8H_{17})_3PO$ $(=C_{24}H_{51}OP)$ **and Dihexyl Sulfoxide** $(C_6H_{13})_2SO$ $(=C_{12}H_{26}OS)$

$Eu(C_5HF_6O_2)_3 \cdot 2(C_8H_{17})_3PO$ and **$Eu(C_5HF_6O_2)_3 \cdot 2(C_6H_{13})_2SO$.** Each compound was obtained as an amber oil by extracting an aqueous solution of europium nitrate with an ether solution containing the chelating agent (as the free ligand or as the ammonium salt) and neutral oxygen donor (mole ratio of the reactants = 1:3:2). The trioctylphosphine oxide and dihexyl sulfoxide adducts isolated from the ether phase melted at −10 and 0°C, respectively. The luminescence spectra of the adducts in solution were recorded in the region 6100 to 6225 Å ($^5D_0 \rightarrow {}^7F_2$ transition). The data revealed that the luminescence intensity of the adduct was enhanced relative to the tris chelate, an observation attributed to an "insulating sheath effect" which reduces radiationless energy losses from the metal chelate [13].

5.1.4.3 Tetrakis Chelates of the Type $M'[M(C_5HF_6O_2)_4]$

Alkali Metal Salts (M' = K, Rb, Cs)

Lippard [14] first effected synthesis of an alkali metal salt of a tetrakis(hexafluoroacetylacetonato) rare earth chelate, $Cs[Y(C_5HF_6O_2)_4]$, by treating an aqueous-ethanolic solution of yttrium chloride with caesium hexafluoroacetylacetonate (1:4 mole ratio). The europium compound, $Cs[Eu(C_5HF_6O_2)_4]$, was prepared according to the same procedure [15]. The yttrium salt sublimed in both air and under vacuum, without decomposition, at temperatures ranging from 180 to 230°C. The mass spectrum of the yttrium chelate revealed a very very weak peak corresponding to the molecular ion, $Cs[Y(C_5HF_6O_2)_4]^+$, but a strong peak corresponding to $Cs[Y(C_5HF_6O_2)_3]^+$, the latter species suggesting the existence of a strong ion pair [14].

Compounds of composition $M'[Sc(C_5HF_6O_2)_4]$ (M' = K, Rb, Cs) were prepared by adding an aqueous alcoholic solution of the respective alkali metal salt of hexafluoroacetylacetone to an aqueous solution of scandium chloride. The precipitate that formed was collected by filtration, washed with water and ethyl ether, and dried in air [16]. The corresponding yttrium chelates (M' = K, Rb, Cs) were prepared by adding an alcoholic solution of yttrium chloride to an aqueous-alcoholic solution of potassium, rubidium, or caesium hexafluoroacetylacetonate. The resulting solution was evaporated to one third of its initial volume and then cooled to room temperature. The resulting precipitate was collected by filtration, washed with water and ethyl ether, and dried in air at 80°C [17]. Each compound crystallized in the form of long needles and was sparingly soluble in water, but readily soluble in ethanol. The melting points of the scandium compounds increased slightly with the radius of the alkali metal ion (227°C for M' = K, 231°C for M' = Rb, 240°C for M' = Cs) [16], whereas the melting points of the yttrium compounds are almost identical (233°C for M' = Rb, Cs, and 231°C for M' = K) [17]. Thermogravimetric analyses revealed that the scandium compounds sublimed in air at 145 to 250°C and that the yttrium compounds sublimed at 150 to 230°C [16, 17]. The molar conductance values ($\Lambda_M = 108$ to $109\ \Omega^{-1} \cdot cm^2 \cdot mol^{-1}$ at 25°C) of solutions of the yttrium salts in methanol are indicative of 1:1 electrolyte behavior [17].

The $Cs[M(C_5HF_6O_2)_4]$ (M = La, Eu) compounds were prepared in greater than 80% yield by cation exchange procedures. The rare earth, as $La(NO_3)_3$ or $EuCl_3$, was fixed on Dowex 50 ion exchange resin in either the caesium or hydrogen form (in the latter case the excess acid form resin was converted to the caesium form before elution of the complex), and eluted with 0.4 M caesium hexafluoroacetylacetonate in 50% ethanol. The melting points of the chelates were 186 to 187°C (M = La) and 212 to 214°C (M = Eu) [18].

The crystal and molecular structures of $Cs[Y(C_5HF_6O_2)_4]$ were determined by single-crystal X-ray analysis (R = 0.065 for 1650 independent reflections) [19, 20]. The compound crystallizes in the orthorhombic space group Pbcn-D_{2h}^{14} (No. 60) with four molecules in a unit cell of dimensions a = 8.679, b = 21.518, c = 17.553 Å ($\rho_{obs} = 2.10\ g \cdot cm^{-3}$, $\rho_{calc} = 2.12\ g \cdot cm^{-3}$). The crystal structure consists of discrete monomeric $Y(C_5HF_6O_2)_4^-$ anions and Cs^+ cations, closely packed, in infinite chains parallel to the crystallographic a-axis (Cs-F interatomic distance = 3.2 to 3.7 Å). Each caesium ion is surrounded by eight fluorine atoms from two neighboring $Y(C_5HF_6O_2)_4^-$ complexes. Each yttrium ion is coordinated to essentially eight equivalent oxygen atoms (average Y-O bond distance is 2.323 Å) located at the vertices of a dodecahedron (**Fig. 5-9a** and 5-**9b**). Each ligand spans adjacent vertices between the two bisphenoids (g-edges) constituting the dodecahedron (Fig. 5-9b), resulting in overall D_2 symmetry. The essentially planar ligands exhibited folding along the oxygen-oxygen line, with the dihedral angles between the ligand plane and the plane defined by the O-Y-O group being 7.6° in each case [20].

Fig. 5-9

a)

The molecular structure of $Cs[Y(C_5HF_6O_2)_4]$ projected on the ab-plane. The arrows indicate the crystallographic twofold axis which passes through the yttrium atom.

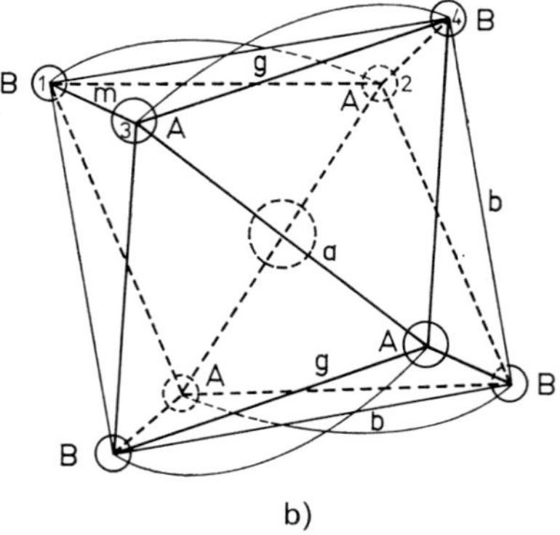

b)

View of coordination dodecahedron down the twofold axis through yttrium (from b direction).

Single-crystal X-ray analysis of $Cs[Eu(C_5HF_6O_2)_4]$ (R = 0.085 for 1893 reflections) revealed that the compound is isomorphous with the yttrium compound. The orthorhombic crystals are in space group Pbcn with unit cell dimensions a = 8.660, b = 21.75, and c = 17.45 Å. The europium compound exhibited the same qualitative features of symmetry and packing as the yttrium compound described above, namely, that the chelate ligands are essentially planar but folded along the O···O line by about 8° with respect to the O-Eu-O plane, and that the type of span by the ligands and the dodecahedron of oxygen atoms results in approximately D_2 symmetry for the complex anion. The slightly longer M-O bond distance observed in the europium compound, 2.38 Å, compared with the yttrium compound, 2.32 Å, was accounted for in terms of the difference in the yttrium(III) and europium(III) radii [15].

Ammonium Salts $NH_4[M(C_5HF_6O_2)_4] \cdot nH_2O$ (n = 0, 1)

The ammonium salts (M = Y, La to Lu except for Ce, Pm) were prepared according to three procedures. In the first, hexafluoroacetylacetone (0.00125 mol) was added to a solution of the tris chelate hydrate (0.00125 mol) in a minimum amount of benzene (10 ml), and ammonia was bubbled into the solution until precipitation ceased. The second method involved dissolving 1 g of the tris chelate hydrate in 1 ml of acetone and adding 1 ml of an acetone solution that contained the stoichiometric amount of ammonium hexafluoroacetylacetonate. The acetone was allowed to evaporate and the residue was recrystallized from benzene. These methods yielded anhydrous compounds for all the rare earths. The third method involved the reaction of an aqueous solution of the rare earth chloride with an aqueous solution of ammonium hexafluoroacetylacetonate (1:4 mole ratio). The precipitated chelate was collected by filtration. The third method yielded the monohydrated salts for M = La, Pr, and Nd, whereas the anhydrous salts were obtained for the remaining members of the series. The monohydrated neodymium tetrakis chelate is not stable in air and loses water over a period of several hours. The lanthanum and praseodymium hydrates appeared stable when exposed to the atmosphere. All of the hydrates lose water over magnesium perchlorate, but the lanthanum and praseodymium compounds pick up water again when exposed to the atmosphere [2].

X-Ray powder patterns revealed four crystalline modifications among the ammonium tetrakis salts. The first series extends from lanthanum through neodymium (Nd = α-form), the second modification consists of a β-form of the neodymium chelate, the third includes compounds of samarium through holmium, and the fourth includes compounds of erbium through lutetium and yttrium. The aqueous method of preparation yielded the β-form of the neodymium chelate, whereas the α-form was prepared by bubbling ammonia into a benzene solution containing the tris chelate hydrate and hexafluoroacetylacetone. Recrystallization of the α-form from benzene yielded the anhydrous β-form [2].

Infrared spectral data are tabulated and discussed in [2]. Thermogravimetric analysis of the M = Sm, Gd, Er, Lu, and Y chelates showed that the compounds are thermally stable and volatile, subliming without decomposition, with half temperatures ranging from 185°C for the lutetium complex to 199°C for the samarium compound (the half temperature is defined as the temperature corresponding to 50% weight loss on the TGA curve) [4].

Alkylammonium Salts

Trimethylammonium Salts, $(CH_3)_3NH[M(C_5HF_6O_2)_4]$, with M = Gd, Tb, were prepared by heating at reflux a 95% ethanolic solution containing the rare earth salt ($GdCl_3$ or $Tb(NO_3)_3$), hexafluoroacetylacetone, and trimethylamine (mole ratio = 1:4:8). Water was then added to effect precipitation of the product. The gadolinium chelate melted at 134 to 135°C and the terbium chelate at 136 to 137°C. Both compounds were characterized by elemental analyses and the terbium compound by its emission spectrum [21].

Tetramethylammonium Salts $(CH_3)_4N[M(C_5HF_6O_2)_4]$. The europium compound was prepared by mixing together in 50 ml of 95% ethanol 20 ml of 1 N sodium hydroxide, 3.2 ml of hexafluoroacetylacetone, an aqueous solution (20 ml) containing 2.2 g of europium nitrate hexahydrate, and an aqueous solution (10 ml) containing 10 mmol of tetramethylammonium chloride. The solution was concentrated by boiling on a steam bath until crystallization had begun. Then just sufficient ethanol was added to dissolve the solid at the boiling point. The solution was filtered and the filtrate allowed to stand overnight at room temperature, yielding an analytically pure, crystalline sample. The compound melted at 190°C [22]. The gadolinium and terbium chelates were prepared, using a similar procedure, from the chloride and nitrate salts, respectively. The gadolinium chelate melted at 235 to 236°C and the terbium compound at 244 to 245°C [21]. All compounds were characterized by elemental analyses and emission spectral data [8, 21, 22].

Triethylammonium Salts $(C_2H_5)_3NH[M(C_5HF_6O_2)_4]$. Each compound (M = La, Pr, Nd, Sm, Eu, Gd, Tb, Dy, Ho, Er, Yb) was prepared by combining 3.6 ml of hexafluoroacetylacetone, 20 mmol of triethylamine, and an aqueous solution (25 ml) of a rare earth nitrate (5 mmol) in 50 ml of 95% ethanol. The solution was boiled down to half its original volume, and 100 ml of water was added to effect precipitation of a sticky solid. Vigorous scratching converted the substance to a granular product, which was collected by filtration, washed with water, and air dried. The compounds were recrystallized from chloroform. Each compound melted at 130°C. The complexes were characterized further by elemental analyses and emission spectroscopy [22].

Tetraethylammonium Salts $(C_2H_5)_4N[M(C_5HF_6O_2)_4]$. The cerium(III) complex was prepared under an atmosphere of pure nitrogen, using deoxygenated solvents, according to the following procedure. A solution of cerium chloride heptahydrate (2 mmol) and 2,2-dimethoxypropane (10 ml) in absolute ethanol (15 ml) was added to a mixture of hexafluoroacetylacetone (8 mmol), sodium hydroxide (8 mmol), and tetraethylammonium chloride (2 mmol) in absolute ethanol (25 ml). The solvent was removed under reduced pressure, the residue extracted with ethyl acetate (30 ml), and the suspended solid removed by filtration. The solution was concentrated under vacuum to approximately 10 ml. Then n-hexane was slowly added to effect crystallization of the product. The yellow-colored crystals were collected, washed with n-hexane, and dried under vacuum. The absorption spectrum of a solution of the compound in acetonitrile revealed bands at 22200 (64), 24300 (95), 26900 (sh), and 33000 cm^{-1} (55000) (absorption coefficients, ε in $dm^3 \cdot mol^{-1} \cdot cm^{-1}$, are given in parentheses) [23].

The europium compound was prepared according to the procedure described for preparation of the analogous tetramethylammonium salt, see p. 111 (m.p. = 153°C). The compound was characterized by elemental analysis and emission spectroscopy [22].

Tert-Butylammonium Salts $(CH_3)_3CNH_3[M(C_5HF_6O_2)_4]$. Each compound (M = Gd, Tb) was prepared according to the procedure described for preparation of the analogous trimethylammonium salts, see p. 111. The gadolinium chelate melted at 95 to 96°C and the terbium compound at 93 to 94°C. The compounds were characterized by elemental analyses and their emission spectra [21].

Pyridinium Salts $C_5H_5NH[M(C_5HF_6O_2)_4] \cdot nH_2O$ (n = 0, 1)

Richardson, Wagner, Sands [2] effected synthesis of a series of chelates (n = 1 for M = La, Pr; n = 0 for M = Y, Sm to Lu) according to the following procedures. One method is essentially the same as described for preparation of the analogous ammonium salts, see p. 111. The resulting compounds were recrystallized from benzene or from a benzene-hexane mixture. The second method involved mixing hot solutions of the tris chelate hydrate (1 g in 5 to 6 ml of benzene) with a stoichiometric amount of pyridinium hexafluoroacetylacetonate in 5 to 6 ml of benzene. Enough benzene was added to dissolve any precipitate which appeared. The solution was filtered and the filtrate allowed to cool, whereupon the salt crystallized. The lanthanum and praseodymium compounds were obtained in the form of plates and the remaining compounds in the form of large needles [2].

The europium salt exhibited bright pink fluorescence under ultraviolet light, and the terbium salt fluoresced greenish yellow. X-ray powder patterns revealed three isomorphous series of pyridinium salts: the first for the hydrated chelates of lanthanum, praseodymium, and neodymium; the second for the anhydrous compounds of samarium through terbium; and the third for the anhydrous chelates of holmium through ytterbium. Infrared spectral data for the compounds are tabulated and discussed in [2]. Thermogravimetric analyses showed that the compounds are thermally stable, subliming without decomposition, and volatile, with half temperatures ranging from 184°C for the lutetium salt to 210°C for the lanthanum compound (the half temperature was defined as the temperature corresponding to 50% weight loss on the TGA curve) [4].

Melby and coworkers [22] prepared the anhydrous chelates of neodymium, europium, and terbium according to the procedure described for preparation of the analogous triethylammonium salts, see p. 112. The neodymium and terbium chelates melted at 185°C, whereas the europium chelate melted at 175°C. The complexes were characterized by elemental analyses and emission spectroscopy.

Salts of Pyridinium Derivatives

3-Methylpyridinium and 4-Methylpyridinium Salts $LH[M(C_5HF_6O_2)_4]\cdot L$. Each complex with L = 4-methylpyridine (= C_6H_7N) and M = Eu, Gd, Tb or 3-methylpyridine with M = Eu was prepared by adding a solution of the rare earth chloride (0.001 mol) in 10 ml of 1:1 (v/v) aqueous ethanol, dropwise, to a solution of hexafluoroacetyl acetone (0.0045 mol) and the pyridine derivative (0.01 mol) in 20 ml of ethanol. The crystalline products were collected, washed with 1:1 ethanol/water, and dried under vacuum. The complexes were sublimed under vacuum with only partial loss of the neutral amine, suggesting that the additional amine molecule was not simply occluded in the crystal structure [21, 24]. The gadolinium and terbium chelates derived from 4-methylpyridine melted at 97 to 98°C [21]. Glidewell and Sheperd [24] proposed that these compounds contain a hydrogen bonded $(L_2H)^+$ unit.

2,6-Dimethylpyridinium Salts, $C_7H_9NH[M(C_5HF_6O_2)_4]$ with M = Eu, Tb were prepared according to the procedure described for preparation of the analogous triethylammonium salts, see p. 112. Each compound melted at 130°C. The compounds were characterized by elemental analyses and emission spectra [22].

The 2,4,6-Trimethylpyridinium Salt, $C_8H_{11}NH[Eu(C_5HF_6O_2)_4]$ was prepared according to the procedure described for preparation of the analogous triethylammonium salts, see p. 112 (m. p. = 130°C) [22].

Piperidinium Salts $C_5H_{10}NH_2[M(C_5HF_6O_2)_4]$

The yttrium chelate was prepared by adding an aqueous solution (30 ml) of yttrium chloride (8 mmol) to a boiling solution of hexafluoroacetylacetone (32 mmol) and piperidine (32 mmol) in 50 ml of 95% ethanol. The mixture was stirred while heating at reflux for 0.5 h, then the solvent was evaporated (ca. 30 ml) by heating over a hot plate for an additional 0.5 h. The evaporation of the solvent resulted in formation of an oil, which did not crystallize after standing at 0°C for 1 day. The oily lower layer was separated and dissolved in approximately 20 ml of dichloromethane and 5 ml of hexane. The solution was concentrated by passing a gentle stream of nitrogen over the surface until the first appearance of crystals. The product was allowed to crystallize at 0°C for 2 days. The crystals were collected by filtration, washed twice with 20 ml portions of hexane, and dried under vacuum at 70°C for 8 h. The compound thus obtained melted at 130 to 132°C [25].

The europium compound was prepared by adding an ethanolic solution (10 ml) of the chloride salt (2 mmol), dropwise, to a mixture of hexafluoroacetylacetone (8 mmol) and piperidine (8 mmol) in ethanol. The ethanol was removed, yielding a brown, viscous oil, interspersed with crystals of piperidinium chloride. The residue was extracted with acetone, the solvent was allowed to evaporate from the extractant solution, and the resulting oil was treated with n-hexane to effect separation of a white solid [6]. The Mössbauer spectrum of the europium chelate, taken with a ^{151}Eu source in an Sm_2O_3 lattice, was obtained over the temperature range 79 to 220 K. The average isomer shift, $-0.56\ mm\cdot s^{-1}$ (range -0.55 to $-0.58\ mm\cdot s^{-1}$) was positive relative to the predominantly ionic compound, $EuF_3\cdot 0.5H_2O$, indicating some covalency in the europium-oxygen bond [7].

The gadolinium and terbium chelates were prepared by heating, at reflux, a 95% ethanolic solution containing the rare earth salt ($GdCl_3$ or $Tb(NO_3)_3$), hexafluoroacetylacetone, and piperidine (1:4:8 mol ratio). Water was added to the resulting mixture to effect precipitation of the products. The gadolinium chelate melted at 128 to 129°C and the terbium compound at 110 to 111°C. The compounds were characterized by elemental analyses and their emission spectra [21].

The ligand exchange equilibrium between $C_5H_{10}NH_2[Y(C_5HF_6O_3)_4]$ and the trifluoro acetylacetonato compound $C_5H_{10}NH_2[Y(C_5H_4F_3O_2)_4]$, (see p. 104, for preparation of the trifluoroacetylacetonato complex, $C_5H_{10}NH_2[Y(C_5H_4F_3O_2)_4]$) was studied in deuteriochloroform by proton NMR (the system was characterized quantitatively in the methylene proton region). In the temperature range −58 to −16°C, the equilibrium quotients for the formation of mixed complexes, $C_5H_{10}NH_2[Y(C_5H_4F_3O_2)_n(C_5HF_6O_2)_{4-n}]$ (n = 1 to 3), are 3 to 5 times larger than statistical, assuming a total random distribution of ligands. Deviations from statistical behavior were attributed to enthalpy changes, with the entropy changes being zero or near zero within experimental error. Relative rates of ligands exchanging between two complexes were discussed in terms of the coalescence behavior of the methylene proton resonances, from which it appeared that the hexafluoroacetylacetonate ligands exchange faster than the trifluoroacetylacetonate ligands [25].

Piperazinium Salt $C_4H_{10}N_2H[Eu(C_5HF_6O_2)_4]$

The complex was prepared according to the procedure described for preparation of the analogous triethylammonium salts. The compound melts at 210°C. It was characterized by elemental analysis and its emission spectrum [22].

N-Methylquinolinium Salt $C_9H_7NCH_3[Eu(C_5HF_6O_2)_4]$ and N-Methylphenazinium Salt $C_{12}H_8N_2CH_3[Eu(C_5HF_6O_2)_4]$

The compounds were prepared according to the procedure described in the section for preparation of the analogous tetramethylammonium salts. The complexes melt at 100 and 120°C, respectively [22].

Tetraphenylarsonium Salt $(C_6H_5)_4As[M(C_5HF_6O_2)_4]$

Each compound (M = Gd, Tb) was prepared by adding an aqueous solution of the rare earth salt ($GdCl_3$ or $Tb(NO_3)_3$) and a tetraphenylarsonium salt to a 95% ethanolic solution containing the ligand and an equivalent amount of sodium hydroxide (metal ion : ligand: $(C_6H_5)_4As^+$ mol ratio = 1:4:2.5). The volume of the resulting mixture was reduced by boiling until crystallization had begun, then sufficient ethanol was added to redissolve the product that had formed, and the solution was allowed to cool. The precipitates were collected, dried, and recrystallized from chloroform. The gadolinium chelate melts at 158 to 159°C and the terbium compound at 136 to 137°C. The compounds were characterized by elemental analyses and their emission spectra [21].

The NMR spectra of mixtures of $(C_6H_5)_4As[Y(C_5HF_6O_2)_4]$ (method of chelate preparation not presented) and $(C_6H_5)_4As[Y(C_5H_4F_3O_2)_4]$ ($C_5H_5F_3O_2$=trifluoroacetylacetone) in deuteriochloroform at −40°C revealed the number of peaks (8) in the methylene-proton region expected for the five species, $M(C_5HF_6O_2)_n(C_5H_4F_3O_2)_{4-n}$ (n = 0 to 4). The ligand distribution did not appear to be random, with mixed complexes forming preferentially in the exchange reactions. As the temperature was raised to +40°C, the resonances corresponding to individual mixed complexes coalesced, producing only one peak for each ligand. It appeared that the coalescence of the proton resonances for the $Y(C_5HF_6O_2)_3(C_5H_4F_3O_2)^-$ and $Y(C_5HF_6O_2)_2(C_5H_4F_3O_2)_2^-$ species in the methylene region was equally fast for both hexafluoroacetylacetone and trifluoroacetylacetone, indicating that the two ligands are exchanging at the same rate [26].

5.1.4.4 Tetrakis Chelate of Cerium(IV) $Ce(C_5HF_6O_2)_4$

The compound was prepared by extracting an aqueous ceric ammonium nitrate solution with a solution of hexafluoroacetylacetone in benzene. The crystals resulting in the organic layer were collected and recrystallized twice from benzene and dried under vacuum. Infrared spectral data (Nujol and hexachlorobutadiene mulls) are tabulated and discussed in [27]. The ν(Ce-O) stretching mode was assigned to a band at 218 cm^{-1} [27].

5.1.4.5 Pentakis Chelates

Pyridinium Salts of composition $(C_5H_5NH)_2M(C_5HF_6O_2)_5$, with M = La, Pr, Nd, were obtained by allowing pyridinium hexafluoroacetylacetone to react with the respective tris chelate hydrate in hot benzene (mole ratio = 2:1). The compounds were characterized by elemental analyses, X-ray powder patterns, and infrared spectral data [2].

5.1.4.6 Complexes Derived from Hexafluoroacetylacetonate and Trifluoroacetate Ions, $M(C_5HF_6O_2)_2(CF_3CO_2)\cdot 2H_2O$

Each mixed ligand complex (M = Y, La, Pr, Nd, Sm, Gd, Dy, Ho, Er, Yb) was obtained as the major by-product from the preparation of $M(C_5HF_6O_2)_3\cdot nH_2O$ according to the procedure of Richardson and coworkers, as described in section 5.1.4.1 [2]. After extracting the aqueous rare earth chloride solution with the ether solution of ammonium hexafluoroacetylacetonate, separating and allowing the ether to evaporate, the residue was extracted several times with small portions of hot benzene in order to remove $M(C_5HF_6O_2)_3\cdot nH_2O$ and $NH_4[M(C_5HF_6O_2)_4]$. The residue was then recrystallized from benzene, yielding the mixed ligand complex. The yield of the mixed ligand complex was increased by adding ammonium trifluoroacetate to the ether solution of ammonium hexafluoroacetylacetonate. The reaction of hexafluoroacetylacetone with water to give trifluoroacetic acid and trifluoroacetone is essentially the reverse of the condensation reaction used to prepare hexafluoroacetylacetone.

The compounds are extremely soluble in organic solvents containing an oxygen atom, slightly soluble in benzene, and insoluble in hexane and in water. X-ray powder patterns revealed that the praseodymium, neodymium, and samarium mixed chelates formed one series of isomorphous compounds, whereas the remaining compounds formed a second isomorphous series. Infrared spectral data were tabulated and discussed. The spectra of compounds in one isomorphous series were quite different from those in the second series, particularly in the ν(O-H) stretching region (3200 to 3700 cm^{-1}), in the ν(C=O) stretching region (1600 to 1700 cm^{-1}), and in the $\nu(C\text{-}CF_3)$ stretching region (795 to 820 cm^{-1}) [2].

References to 5.1.4:

[1] F. Halverson, J. S. Brinen, J. R. Leto (J. Chem. Phys. **40** [1964] 2790/2). — [2] M. F. Richardson, W. F. Wagner, D. E. Sands (J. Inorg. Nucl. Chem. **30** [1968] 1275/89). — [3] M. Steinberg, J. Mashall, A. Glasner (Proc. 8th Intern. Conf. Coord. Chem., Vienna 1964, p. 202/3). — [4] M. F. Richardson, R. E. Sievers (Inorg. Chem. **10** [1971] 498/504). — [5] E. W. Berg, J. J. Chiang Acosta (Anal. Chim. Acta **40** [1968] 101/13).

[6] S. J. Lyle, A. D. Witts (Inorg. Chim. Acta **5** [1971] 481/4). — [7] S. J. Lyle, A. D. Witts (J. Chem. Soc. Dalton Trans. **1975** 185/8). — [8] M. L. Bhaumik (J. Inorg. Nucl. Chem. **27** [1965] 261). — [9] M. L. Morris, R. W. Moshier, R. E. Sievers (Inorg. Chem. **2** [1963] 411/2). — [10] M. L. Bhaumik (J. Inorg. Nucl. Chem. **27** [1965] 243/4).

[11] J. W. Mitchell, C. V. Banks (Anal. Chim. Acta **57** [1971] 415/24). — [12] J. W. Mitchell, C. V. Banks (Talanta **19** [1972] 1157/69). — [13] F. Halverson, J. S. Brinen, J. R. Leto (J. Chem. Phys. **41** [1964] 157/63). — [14] S. J. Lippard (J. Am. Chem. Soc. **88** [1966] 4300/1). — [15] J. H. Burns, M. D. Danford (Inorg. Chem. **8** [1969] 1780/4).

[16] M. Z. Gurevich, B. D. Stepin, V. V. Zelentsov (Zh. Neorgan. Khim. **15** [1970] 890/2; Russ. J. Inorg. Chem. **15** [1970] 454/5). — [17] M. Z. Gurevich, B. D. Stepin, V. V. Zelentsov (Zh. Neorgan. Khim. **15** [1970] 1996/8; Russ. J. Inorg. Chem. **15** [1970] 1028/9). — [18] C. E. Higgins (J. Inorg. Nucl. Chem. **35** [1973] 1941/4). — [19] S. J. Lippard, F. A. Cotton, P. Legzdins (J. Am. Chem. Soc. **88** [1966] 5930/1). — [20] M. J. Bennett, F. A. Cotton, P. Legzdins, S. J. Lippard (Inorg. Chem. **7** [1968] 1770/6).

[21] T. D. Brown, T. M. Sheperd (J. Chem. Soc. Dalton Trans. **1973** 336/41). — [22] L. R. Melby, N. J. Rose, E. Abramson, J. C. Caris (J. Am. Chem. Soc. **86** [1964] 5117/25). — [23] M. Ciampolini, F. Mani, N. Nardi (J. Chem. Soc. Dalton Trans. **1977** 1325/8). — [24] C. Glidewell, T. M. Sheperd (J. Inorg. Nucl. Chem. **37** [1975] 348/50). — [25] N. Serpone, R. Ishayek (Inorg. Chem. **10** [1971] 2650/6).

[26] F. A. Cotton, P. Legzdins, S. J. Lippard (J. Chem. Phys. **45** [1966] 3461/2). — [27] T. Yoshimura, C. Miyake, S. Imoto (Bull. Chem. Soc. Japan **46** [1973] 2096/101).

5.1.5 With 3-(2,4,6-Trimethylphenyl)-2,4-pentanedione $CH_3C(O)-CH(C_6H_2(CH_3)_3)-C(O)CH_3$ (= 3-Mesitylacetylacetone = $C_{14}H_{18}O_2$ = HL)

The complexes ML_3 with M = Sc, Y, and La are prepared by dissolving the metal oxides in concentrated HCl acid and by adding a methanol solution of the ligand. Then concentrated aqueous ammonia is slowly added until the complexes precipitate. They are filtered off, washed with methanol, and dried. The yields of pale yellow crystals of the trihydrates are 76% for the scandium complex and 85% for the yttrium complex, and the yield of the colorless crystals of the trihydrated lanthanum complex is 78%. The scandium complex can be dried at 100°C/0.01 Torr for 4 h to give the hemihydrate, which melts at 239°C. Under the same conditions for 6 h, the yttrium and lanthanum complexes are dehydrated to the monohydrates, which melt with decomposition above 160°C (Y) or above 185°C (La). None of the complexes can be sublimed. The molecular ions are observed in the mass spectra. The ^1H NMR chemical shift differences between the ortho and para methyl signals are very small. In the IR spectra, the ν_{CO} fall between 1578 and 1590 cm^{-1}.

An impure CeL_4 was prepared as follows: An acidified aqueous solution of cerium(IV) nitrate hexahydrate is poured into a methanol solution of the ligand. The solution is brought to pH 6 with aqueous ammonia to precipitate a brownish-yellow solid. This is boiled with additional ligand in benzene in the air for 3 h. Excess ligand is sublimed off, and the complex recrystallized from methanol/hexane to give a 67% yield of the dark red crystals. The crystals do not melt below 350°C and cannot be sublimed. There is an IR absorption at 1572 cm^{-1}.

Reference to 5.1.5:
G. Brill, H. Musso (Liebigs Ann. Chem. **6** [1979] 803/10).

5.1.6 Complexes with 2,4-Hexanedione $CH_3C(O)CH_2C(O)CH_2CH_3$ (= Propionylacetone = $C_6H_{10}O_2$)

5.1.6.1 $M(C_6H_9O_2)_n^{3-n}$ Complexes (n = 1 to 3). Formation in Solution

Formation constants for the propionylacetonato chelates were determined potentiometrically (glass electrode) at 30°C in 3:1 (v/v) acetone-water, I = 0.1 ($NaClO_4$) [1].

Metal Ion	Y^{3+}	La^{3+}	Pr^{3+}	Nd^{3+}
lg K_1	7.14	6.22	6.72	6.94
lg K_2	5.80	4.87	5.84	6.24
lg K_3	5.62	4.16	5.38	5.01
lg β_3	18.56	15.25	17.94	18.19

5.1.6.2 Isolated Compounds

5.1.6.2.1 $M(C_6H_9O_2)_3 \cdot nH_2O$ (n = 0 to 2)

The lanthanum complex (n = 2) was obtained by treating an aqueous solution of lanthanum chloride with a solution of ammonium propionylacetonate according to the procedure described by Stites, McCarty, Quill [2] for preparation of the tris(acetylacetonato) rare earth chelates (see Section 5.1.2.2.2, p. 85) [1]. The addition of an equivalent amount, or more, of ammonium propionylacetonate solution, to solutions of the chloride salts of praseodymium, neodymium, or yttrium yielded copious precipitates of the hydroxo bis(propionylacetonato) compounds (see below). In each case, the precipitate was filtered, washed with water, and dried over concentrated sulfuric acid. The dried mass was extracted with acetone, propionylacetone was added to the extractant solution (slightly more than one equivalent), and the solvent was allowed to evaporate from the mixture at a temperature of 50 to 60°C. The resulting residue was then extracted with acetone and the extractant solution allowed to evaporate, yielding the tris chelate (n = 0 for M = Y, n = 1 for M = Pr, Nd). The complexes were collected by filtration and washed with cold acetone. They are soluble in organic solvents [1]. The tris chelates were similarly prepared from the rare earth nitrate salts [7].

5.1.6.2.2 $M(C_6H_9O_2)_2OH \cdot H_2O$

Each compound (M = Y, Pr, Nd) was prepared by adding an aqueous solution of ammonium propionylacetonate (excess of two equivalents), with stirring, to a neutral solution of the rare earth chloride. Complete precipitation of the product was effected by adding dilute ammonia to the mixture. The crystalline precipitate was collected by filtration, washed with cold water, recrystallized from alcohol, and air dried [3].

5.1.6.3 Complexes Derived from 2,4-Hexanedione and Other β-Diketones

(acetylacetone = Hacac, = $C_5H_8O_2$, benzoylacetone = $C_{10}H_{10}O_2$ and dibenzoylmethane = $C_{15}H_{12}O_2$ = HL)

5.1.6.3.1 Tris Chelates

Each compound (Table 5/13) was prepared by adding a solution of the appropriate diketone in acetone to a solution containing an equimolar amount of the $M(C_6H_9O_2)_2OH \cdot H_2O$ complex in the same solvent. The resulting mixture was heated at reflux for 5 to 6 h. The solution was then cooled and allowed to evaporate very slowly. When the solution had been reduced to one fifth of its original volume, crystals began to appear. After further concentration, each product was collected by filtration and washed carefully with acetone. The compounds were purified by repeated recrystallizations from acetone. The colors and melting points of the compounds are recorded in Table 5/13. The compounds were characterized further by elemental analyses and their infrared and electronic absorption spectra. They were checked for purity by chromatography through activated alumina [4].

Table 5/13

Physical Properties of Tris Chelates Derived from 2,4-Hexanedione and Acetylacetone (I), Benzoylacetone (II) and Dibenzoylmethane (III).

	I $M(C_6H_9O_2)_2(acac)$		II $M(C_6H_9O_2)_2(C_{10}H_9O_2)$		III $M(C_6H_9O_2)_2(C_{15}H_{11}O_2)$	
M	color	m.p. in °C	color	m.p. in °C	color	m.p. in °C
Y	white	120 to 122	pale yellow	89 to 90	light yellow brown	123 to 127
La	white	100 to 102	light yellow	89 to 92	light brown	68 to 72
Pr	green	101 to 104	greenish yellow	92 to 95	greenish brown	82 to 84
Nd	pink	103 to 105	pale brown	94 to 96	pink brown	79 to 81
Sm	light cream	129 to 131	pale yellow	95 to 97	yellow brown	81 to 85

5.1.6.3.2 Tetrakis Chelates - Sodium Salts

$Na[M(C_6H_9O_2)_nL_{4-n}]$ complexes of composition $Na[M(C_6H_9O_2)_2(acac)(C_{10}H_9O_2)]$, $Na[M(C_6H_9O_2)_2(acac)(C_{15}H_{11}O_2)]$ and $Na[M(C_6H_9O_2)_2(C_{10}H_9O_2)(C_{15}H_{11}O_2)]$ with M = La, Pr, Nd, Sm were prepared by adding an alcoholic solution containing 1 mmol of a sodium diketonate salt, slowly, with stirring, to an alcoholic solution of the appropriate tris(propionylacetonato)chelate (section 5.1.6.2.1) or mixed ligand chelate (section 5.1.6.3.1). Slow evaporation of the reaction mixture yielded the product in crystalline form. The complexes were collected and washed several times with drops of ice-cold ethanol. The compounds were further purified by dissolving in minimum volumes of dimethylformamide and effecting precipitation by adding nitrobenzene. The compounds were air dried and analyzed [5].

Infrared spectral data (Nujol mulls) for $Na[Pr(C_6H_9O_2)_3(acac)]$, $Na[Pr(C_6H_9O_2)_2(acac)_2]$ and $Na[Pr(C_6H_9O_2)_2(acac)(C_{10}H_9O_2)$ are presented in [5]. The molar conductances of each compound in dimethylformamide (between 72 and 75 $\Omega^{-1} \cdot cm^2 \cdot mol^{-1}$) corresponded to 1:1 electrolyte behavior. Thermogravimetric analyses of the complexes failed to give any definite mode of decomposition [5].

5.1.6.4 Complexes Derived from 2,4-Hexanedione and N-Heterocyclic Compounds

The mixed ligand chelates derived from 2,4-hexanedione and 2-pyridinecarboxylic acid ($= C_6H_5NO_2$), 2-quinolinecarboxylic acid ($= C_{10}H_7NO_2$), or 2,2'-bipyridine ($= C_{10}H_8N_2$) were prepared by mixing solutions of $M(C_6H_9O_2)_2(OH) \cdot H_2O$ (see p. 117) in acetone with solutions of the respective ligands (1:1 mole ratio) in the same solvent. In each case, the mixture was heated at reflux for several hours (in the case of the complex derived from bipyridine, 22 to 24 h of heating at reflux was required). The solution was filtered and the filtrate cooled and allowed to evaporate slowly. The crystalline product was collected by filtration and washed carefully with acetone. The compounds were purified by repeated recrystallizations from acetone. Each complex with 8-hydroxyquinoline ($= C_9H_7NO$) was obtained by simply mixing the two solutions in acetone, collecting the precipitated product and washing it with acetone. The colors and melting points of all the compounds are recorded in Table 5/14. The compounds were characterized further by elemental analyses and their infrared and electronic absorption spectra. The compounds were checked for purity by chromatography through activated alumina [6].

Table 5/14

Physical Properties of Mixed Rare Earth Chelates Derived from 2,4-Hexanedione and N-Heterocyclic Compounds.

	with 8-hydroxy-quinoline $M(C_6H_9O_2)_2$-(C_9H_6NO)		with 2-pyridine-carboxylic acid $M(C_6H_9O_2)_2$-$(C_6H_4NO_2)$		with 2-quinoline-carboxylic acid $M(C_6H_9O_2)_2$-$(C_{10}H_6NO_2)$		with 2,2'-bipyridine $M(C_6H_9O_2)_2(OH)$-$(C_{10}H_8N_2)$	
M	color	m.p. in °C	color	m.p. in °C	color	m.p. in °C	color	m.p. in °C
Y	yellow	210	white	178 to 188	white	161	white	178 to 181
La	yellow	200*)	white	175 to 180	white	181 to 185	white	172 to 174
Pr	greenish yellow	210*)	greenish yellow	198 to 200	greenish yellow	185 to 189	green	178 to 182
Nd	brownish yellow	210*)	light pink	200 to 205	light pink	188	light pink	183 to 185
Sm	yellow	205*)	cream	203 to 205	cream	185 to 190	cream	202 to 203

*) decomposed

5.1.6.5 Complexes Derived from 2,4-Hexanedione and Bis(salicylidene)ethylenediamine $HOC_6H_4CH{=}NCH_2CH_2N{=}CHC_6H_4OH$ ($= C_{16}H_{16}N_2O_2$)

$M(C_6H_9O_2)_2(C_{16}H_{14}N_2O_2)_{0.5}$. Each mixed ligand complex with M = Y, La, Pr, Nd, Sm was prepared by adding a boiling solution of bis(salicylidene)ethylenediamine (1.2 mmol) in 40 ml of ethanol to a boiling ethanolic solution of acetylacetone, followed by passage of ammonia gas through the solution in a slow but steady stream for a period of about 10 min. The reaction vessel was stoppered tightly and left undisturbed for 4 to 5 h. Shining crystals of the mixed chelates began to deposit after about an hour. The compound was collected by filtration and dried under vacuum over sulfuric acid at room temperature. The colors and melting points (°C) of the products were: Y, pale yellow, 212 to 215; La, pale yellow, 200 to 204; Pr, pale green, 202 to 204; Nd, rosy pink, 208 to 209; Sm, cream, 206 to 210. The complexes were insoluble in common organic solvents. The compounds were further characterized by elemental analyses and their infrared spectra (KBr discs). A dinuclear structure was proposed with the bis(salicylidene)ethylenediamine as the bridging ligand [8].

References to 5.1.6:

[1] N. K. Dutt, P. Bandyopadhyay (J. Inorg. Nucl. Chem. **26** [1964] 729/36). — [2] J. G. Stites, C. N. McCarty, L. L. Quill (J. Am. Chem. Soc. **70** [1948] 3142). — [3] N. K. Dutt, P. Bandyopadhyay (J. Inorg. Nucl. Chem. **26** [1964] 1610/2). — [4] N. K. Dutt, S. Upadhyaya (J. Inorg. Nucl. **28** [1966] 2719/24). — [5] N. K. Dutt, S. Sanyal (Indian J. Chem. **10** [1972] 1033/4).

[6] N. K. Dutt, S. Upadhyaya (J. Inorg. Chem. **29** [1967] 1368/72). — [7] N. K. Dutt, P. Bandyopadhyay (Sci. Cult. [Calcutta] **27** [1961] 403/4). — [8] N. K. Dutt, K. Nag (J. Inorg. Nucl. Chem. **30** [1968] 2779/83).

5.1.7 Complexes with Derivatives of 2,4-Hexanedione

5.1.7.1 With 1,1,1-Trifluoro-5,5-dimethyl-2,4-hexanedione

$CF_3C(O)CH_2C(O)C(CH_3)_3$ (= Pivaloyltrifluoroacetone = $C_8H_{11}F_3O_2$)

5.1.7.1.1 Tris Chelates $M(C_8H_{10}F_3O_2)_3 \cdot nH_2O$ (n = 0, 2)

Each compound that is recorded in Table 5/15 was prepared according to a procedure similar to that described by Sievers and coworkers for preparation of the complexes $M(C_{10}H_{10}F_7O_2)_3 \cdot nH_2O$ with heptafluorodimethyloctanedione (see p. 155) [1, 2]. A solution of pivaloyltrifluoroacetone in methanol was added, with stirring, to an aqueous solution of the rare earth nitrate maintained at pH 4 to 5 with sodium hydroxide solution. The precipitated products were collected by filtration, air dried, and recrystallized from benzene. The compounds were finally dried under vacuum over phosphorus(V) oxide. The anhydrous nature of the chelates was confirmed by the absence of bands attributable to OH stretching near 3400 cm^{-1} in the infrared spectra [1]. Tanaka, Shono, Shinra [2] isolated the monohydrated chelates initially, and showed that they were readily dehydrated under vacuum over phosphorus(V) oxide. The neodymium, gadolinium, and erbium chelates were also prepared by adding methanolic solutions (20 ml) of pivaloyltrifluoroacetone to solutions of the respective rare earth nitrates (2 mmol) in 1:1 (v/v) methanol/water (5 ml), with stirring, while the pH was kept between 5 and 6 with aqueous ammonia. Each reaction mixture was then diluted with a large excess of water, and the resulting precipitate was collected by filtration, dried, and recrystallized from benzene or methanol. The melting points (°C) of the anhydrous chelates thus obtained were: Nd, 126 to 128; Gd, 162 to 164; Er, 158 to 160 [3].

The cerium chelate (n = 2) was prepared by adding 2 N aqueous ammonia to a solution of cerium(III) nitrate hexahydrate (1 mmol) and the ligand (3.5 mmol) in methanol. The mixture was stirred vigorously while it was cooled. The product first separated as an oil, but quickly solidified. The solid was collected, washed with a small amount of a methanol-aqueous ammonia mixture, and then air dried. The yellow-colored solid melted at 60°C and exhibited a magnetic moment of 2.50 μ_B [4].

Thermograms of the chelates revealed weight losses approaching 100% (except for the neodymium chelate) in the range ≈200 to 300°C, indicating that the anhydrous pivaloyltrifluoroacetonate chelates are thermally stable in the volatilization process. Gas chromatograms were obtained for the chelates and mixtures of the chelates. A plot of retention time versus the metal ion radius revealed that retention time tends to decrease with decreasing ionic radius [1, 2]. The data indicated that although it was not possible to separate all the rare earth metals from one another in the form of the pivaloyltrifluoroacetonato chelates, there is a possibility for group separations or for separation of mixtures of rare earth elements which do not contain neighboring elements [1]. For use of tris chelates as chemical shift reagents, see [7].

Table 5/15

Colors and Melting Points of the Tris(pivaloyltrifluoroacetonato) Chelates $M(C_8H_{10}F_3O_2)_3$.

Shifematsu, Matsui, Utsunomiya [1]			Tanaka, Shono, Shinra [2]		
M	color	m.p. in °C	M	color	m.p. in °C
Sc	white	56.5 to 57.0			
Y	white	160.8 to 161.2	Y	white	181 to 183
Nd	lavender	126.5 to 128.0			
Sm	white	113.1 to 118.0	Sm	white	115 to 119
Eu	yellow	113.8 to 114.0	Eu	white	154 to 162
Gd	white	162.1 to 164.0	Gd	white	115 to 119
Tb	white	141.0 to 144.0	Tb	white	169 to 172
Dy	white	150.0 to 151.4	Dy	white	147 to 152
			Ho	beige	117 to 120
Er	pink	158.0 to 161.2	Er	pink	166 to 169
			Tm	white	163 to 165
Yb	white	160.2 to 161.0	Yb	white	162 to 164
Lu	white	160.0 to 163.2	Lu	white	164 to 166

5.1.7.1.2 Lewis Base Adducts of Tris Chelates

The mass spectra and the volatility of $Eu(C_8H_{10}F_3O_2)_3 \cdot 2L$ with L = trimethyl phosphate or tributyl phosphate and of $Eu(C_8H_{10}F_3O_2)_3 \cdot L'$ with L' = triphenylphosphine oxide were determined [8].

5.1.7.1.3 Tetrakis Chelates

$M'[M(C_8H_{10}F_3O_2)_4]$. Belcher and coworkers [5] prepared alkali metal salts with M = La, Ce, Pr, Nd, Sm, Eu, Gd, Tb, Dy, Ho, Er, Yb for M' = Na^+; M = Ho for M' = K^+, Rb^+, Cs^+ by treating solutions of a rare earth nitrate in 1:1 (v/v) water/ethanol with an excess of pivaloyltrifluoroacetone. In each case, the bulk of the acetone was removed from the reaction mixture, and the residue was extracted with chloroform. The chloroform extract was shaken with an excess of the appropriate alkali metal carbonate and allowed to stand overnight. The mixture was filtered, the filtrate evaporated to dryness, and the residue sublimed under vacuum, yielding the purified product [5].

Ismail, Lyle, Newbery [6] also prepared a series of alkali metal salts (Table 5/16) by adding solutions of a rare earth nitrate (2 mmol) in water to 100 ml of aqueous 0.088 M alkali hydroxide solution containing pivaloyltrifluoroacetone. In each case, the mixture was warmed and stirred for about 1 h, after which the product was collected, sublimed at 200 to 220°C under reduced pressure (5 Torr), and then recrystallized from a mixture of chloroform and petroleum ether (boiling range 40 to 60°C). The tetraphenylarsonium (M = Nd) and tetraethylammonium (M = Nd) salts were obtained by treating a mixture of neodymium nitrate and M'Cl (excess) with aqueous ammonia. The products were recrystallized from a chloroform-petroleum ether mixture [6].

The complexes are fairly soluble in both polar and nonpolar solvents. They are monomeric in benzene, as shown by their molecular weights, and behave as 1:1 electrolytes in acetone (Table 5/16). Their high melting points (≈250°C) demonstrate thermal stability. The alkali metal salts were sufficiently volatile to allow purification by sublimation and passage through a gas chromatographic column. The tetraphenylarsonium salt is nonvolatile [6]. The compounds were characterized further by elemental analyses, and by their infrared, emission, ultraviolet absorption, and NMR spectra [5, 6]. Mass spectral data were reported for $Na[Ho(C_8H_{10}F_3O_2)_4]$ [5]. For mass spectrum and volatility of $Na[Eu(C_8H_{10}F_3O_2)_4]$, see [8].

Table 5/16
Properties of the Tetrakis(pivaloyltrifluoroacetonato) Chelate Salts, $M'[M(C_8H_{10}F_3O_2)_4]$ [6].

complex	color	decomposition point in °C	molecular weight*) found	calc.	molar conductance**) in $\Omega^{-1} \cdot cm^2 \cdot mol^{-1}$
$Na[M(C_8H_{10}F_3O_2)_4]$					
M = Y	white	225 to 230	1144	892	89.7
= La	white	252 to 256	1020	942	87.0
= Nd	pale blue	254 to 256	1000	947	87.0
= Eu	white	245 to 257	940	955	94.3
= Tb	white	237 to 240	1131	962	95.7
= Yb	white	214 to 217	1156	976	89.0
$Li[Nd(C_8H_{10}F_3O_2)_4]$	pale blue	252 to 255	1002	931	92.5
$Cs[Nd(C_8H_{10}F_3O_2)_4]$	pale blue	213 to 215	1040	1057	92.5
$(C_6H_5)_4As[Nd(C_8H_{10}F_3O_2)_4]$	pale blue	214 to 216	1473	1307	92.2
$(C_2H_5)_4N[Nd(C_8H_{10}F_3O_2)_4]$	pale blue	285.5	1109	1040	118.6

*) 1 to 2×10^{-3} molar solution in benzene. — **) 10^{-3} molar solution in acetone.

$Ce(C_8H_{10}F_3O_2)_4$ was prepared by adding 2 N aqueous ammonia to an aqueous-methanolic solution of cerium(III) nitrate hydrate (1 mmol) and pivaloyltrifluoroacetone and then passing air through the resulting mixture for 1 h. Precipitation of the chelate was effected by adding water, and the product was collected and recrystallized from ethanol. The red-colored compound melted at 138°C. The compound was diamagnetic, verifying the +4 oxidation state of cerium [4].

5.1.7.2 With Other Derivatives of 2,4-Hexanedione RC(O)CH$_2$C(O)R' (= HL)

$R = CH_3$, $R' = CH(CH_3)_2$ $(= C_7H_{12}O_2)$
$R = CH_3$, $R' = C(CH_3)_3$ $(= C_8H_{14}O_2)$
$R = CF_3$, $R' = C_2H_5$ $(= C_6H_7F_3O_2)$
$R = CF_3$, $R' = CH(CH_3)_2$ $(= C_7H_9F_3O_2)$

$ML_3 \cdot nH_2O$ and $ML_2OH \cdot nH_2O$ (n = 0 to 3). Compounds derived from 5-methyl- and 5,5-dimethyl-2,4-hexanedione were prepared by adding an ethanolic solution (20 ml) of the ligand (6 mmol), with stirring, to an aqueous rare earth nitrate (2 mmol in 5 ml) solution maintained at pH 5 to 6 with 1 molar aqueous ammonia. The resulting precipitate was collected by filtration, dried, and recrystallized from ethanol or methanol. The complexes derived from the fluorinated β-diketones were prepared by adding a methanolic solution (20 ml) of the ligand (6 mmol), with stirring, to a solution of the rare earth nitrate (2 mmol) in 5 ml of 1:1 (v/v) methanol/water, while the pH was maintained at 5 to 6 with ammonia. The solution was then diluted with a large excess of water, and the resulting precipitate was collected by filtration, dried, and recrystallized from benzene or methanol. All the chelates were finally dried under vacuum in a desiccator over phosphorus(V) oxide. The results of carbon and hydrogen analyses of the nonfluorinated β-diketone chelates indicated that the chelates precipitated from solution as the hydrates (n = 2, 3), and on drying under vacuum over phosphorus(V) oxide, the chelates were decomposed to the basic chelate. Melting points of the chelates with M = Nd, Gd, Er are recorded [3].

The monohydrated cerium(III) chelate was prepared by carefully adding 2 N aqueous ammonia to a solution containing 1 mmol of cerium nitrate hexahydrate and 3.5 mmol of 5,5-dimethyl-2,4-hexanedione (pivaloylacetone). The resulting precipitate was collected, washed with a small amount of a methanol/water/ammonia mixture, and dried under a flow of nitrogen. The yellow-colored solid melted at 80°C and exhibited a magnetic moment of 2.38 μ_B [4].

Infrared spectra (Nujol mulls) of the chelates derived from nonfluorinated diketones revealed broad peaks at about 3400 cm^{-1} due to water and a sharp peak at 3650 cm^{-1} assigned to the hydroxyl group. Elemental analyses were not obtained for the fluorinated diketonate metal chelates, but the infrared spectra indicated that these compounds also contained hydroxo groups [3].

Thermogravimetric analyses were reported for erbium chelates. The anhydrous compounds with 5,5-dimethyl- and 1,1,1-trifluoro-5-methyl-2,4-hexanedione, which were analyzed, produced a maximum weight loss of 70 to 80% at temperatures between 150 and 400°C, indicating partial decomposition and incomplete volatilization. Incomplete volatilization was attributed to the presence of a hydroxyl group. Compounds with 5-methyl- and 1,1,1-trifluoro-2,4-hexanedione, all of which are hydrated or contain hydroxyl groups, decomposed above 300°C, instead of subliming [3].

References: to 5.1.7:

[1] T. Shigematsu, M. Matsui, K. Utsunomiya (Bull. Chem. Soc. Japan **42** [1969] 1278/81). — [2] M. Tanaka, T. Shono, K. Shinra (Anal. Chim. Acta **43** [1968] 157/8). — [3] K. Utsunomiya, T. Shigematsu (Anal. Chim. Acta **58** [1972] 411/9). — [4] E. Uhlemann, F. Dietze (Z. Anorg. Allgem. Chem. **386** [1971] 329/34). — [5] R. Belcher, J. Majer, R. Perry, W. I. Stephen (J. Inorg. Nucl. Chem. **31** [1969] 471/8).

[6] M. Ismail, S. J. Lyle, J. E. Newbery (J. Inorg. Nucl. Chem. **31** [1969] 1715/24). — [7] T. Iida, M. Kikuchi, T. Tamura, T. Matsumoto (Yukagaku **27** [1978] 390/3). — [8] V. S. Khomenko, T. A. Rashinina, V. M. Suboch (Vestsi Akad. Navuk Belarusk.SSR Ser. Khim. Navuk **1979** No. 3, pp. 36/40).

5.1.8 Complexes with Derivatives of 2,4-Heptanedione

5.1.8.1 With 1,1,1,5,5,6,6,7,7,7-Decafluoro-2,4-heptanedione $CF_3CF_2CF_2C(O)CH_2C(O)CF_3$ (= $C_7H_2F_{10}O_2$)

5.1.8.1.1 Tris Chelates $M(C_7HF_{10}O_2)_3 \cdot 2H_2O$

Each compound (M = Y, La to Lu except for Ce, Pm) was prepared by extracting an aqueous solution of the rare earth chloride with an ether solution of decafluoroheptanedione. (The rare earth chlorides were used in 10 to 50% excess in order to prevent formation of the tetrakis chelate.) The ether extracts were dried over calcium sulfate and evaporated. Oils were usually obtained, which subsequently crystallized when allowed to stand in air. The compounds oiled out of every solvent from which recrystallization was attempted (for example, benzene, hexane, methylcyclohexane, and heptane). The compounds are too soluble in oxygen donor solvents to allow for crystallization. The hydrated chelates became oils when dried over magnesium perchlorate (1 d under vacuum, several weeks at atmospheric pressure). These oils picked up water when exposed to the atmosphere to reform the dihydrated chelates [1]. Typically, the amount of residual water after drying the europium chelate for 3 days under vacuum over phosphorus(V) oxide was 0.5 mol/mol chelate [2].

Infrared data (Nujol and Fluorolube mulls) for the samarium chelate are tabulated in [1], but specific assignments of all the vibrational modes were not presented. The ν(O-H) stretching mode due to water was observed at 3510 cm^{-1}. The NMR spectrum of the lutetium chelate in deuteriochloroform revealed peaks at 3.58 ppm (water) and 6.47 ppm (=C-H) [1]. Thermogravimetric analytical data showed that the hydrated tris chelates are quite volatile, and that only the lighter rare earth chelates (M = La, Pr, Nd) decompose to any considerable extent on heating, although there is slight decomposition for the heavier rare earth chelates. The half-temperatures (defined as temperatures corresponding to 50% weight loss on the TGA curve) ranged from 165°C for the lutetium chelate to 200°C for the praseodymium chelate [1].

5.1.8.1.2 Lewis Base Adducts of the Tris Chelates

With Dimethylformamide (= C_3H_7NO = DMF)

$M(C_7HF_{10}O_2)_3 \cdot 2DMF$. Each adduct (M = Pr, Yb) was prepared by adding a stoichiometric amount of dimethylformamide to the solid hydrated tris chelate and heating the mixture gently. Vacuum drying yielded the anhydrous 2:1 adducts. The ν(C=O) stretching mode of dimethylformamide shifted to lower wave numbers (1640 to 1650 cm^{-1}) in the infrared spectra of the adducts (Nujol and Fluorolube mulls), relative to the free amide value (1685 cm^{-1}), indicating that ligation of both dimethylformamide molecules occurs through their carbonyl groups. Eight-coordination in the bis adducts was supported by molecular weight measurements in benzene, which showed that neither dissociation of the adduct nor association to form oligomers occurred to any significant extent [1]. The adducts volatilized completely without decomposition on heating. The half-temperatures (defined as the temperature corresponding to 50% weight loss on the TGA curve) were 204 and 179°C, respectively, for the lanthanum and ytterbium chelates [1].

With 2,2-Dimethoxypropane $(CH_3)_2C(OCH_3)_2$, (= $C_5H_{12}O_2$)

$M(C_7HF_{10}O_2)_3 \cdot (C_5H_{12}O_2)$. Each adduct (M = La, Pr, Er) was obtained by dissolving the respective hydrated chelate in ≈5 ml of dimethoxypropane, heating the reaction mixture at reflux for a few seconds, and removing excess solvent under reduced pressure. The adducts are highly volatile, with half-temperatures (defined as temperatures corresponding to 50% weight loss in the TGA curve) ranging from 158°C for the erbium chelate to 205°C for the lanthanum compound. The compounds were slightly decomposed upon volatilization [1].

With Tributyl Phosphate $(C_4H_9O)_3PO$ (= $C_{12}H_{27}O_4P$)

$M(C_7HF_{10}O_2)_3 \cdot 2(C_4H_9O)_3PO$. The overall formation constants, β ($\beta = [M(C_7HF_{10}O_2)_3 \cdot 2(C_4H_9O_3)PO] \cdot [M(C_7HF_{10}O_2)_3]^{-1} \cdot [(C_4H_9O)_3PO]^{-2}$), for the adducts in cyclohexane were determined at 25°C by a solvent extraction method: lg β_2 = 9.96, 10.00, and 10.20 for M = Nd, Eu, and Tm, respectively. Detection of the mixed ligand complexes by gas chromatography at column temperatures of about 200°C is reported in [4].

5.1.8.1.3 Tetrakis Chelates

Ammonium Salts $NH_4[M(C_7HF_{10}O_2]$

Each compound (M = Y, Nd to Lu) was prepared by adding an aqueous solution of ammonium decafluoroheptanedionate, slowly, with stirring, to an aqueous solution of rare earth chloride. The resulting precipitate was collected by filtration, washed several times with water, and recrystallized twice from benzene. The compounds tended to oil out of benzene, but crystallization occurred readily if the solutions were not too concentrated. The lanthanum and praseodymium chelates could not be prepared following this procedure [1].

Infrared absorption bands for the samarium compound were tabulated in [1], but specific assignments of the vibrational modes were not presented. The NMR spectrum of the yttrium chelate in chloroform revealed a broad peak at δ 5.26 ppm (TMS internal reference) due to the NH_4^+, and a peak at δ 6.57 ppm corresponding to the =C-H resonance [1].

The compounds are thermally stable and volatile, with half-temperatures (defined as the temperature corresponding to 50% weight loss on the TGA curve) ranging from 167°C for the lutetium chelate to 191°C for the neodymium compound. The volatilities of the chelates increased with a decrease in metal ion radius [1].

5.1.8.2 With 1,1,1-Trifluoro-5 (or 6)-methyl-2,4-heptanediones

For preparation and properties of tris chelates with neodymium, gadolinium, and erbium according to [5] see analogous chelates with derivatives of 2,4-hexanediones in Section 5.1.7.

References to 5.1.8:

[1] M. F. Richardson, R. E. Sievers (Inorg. Chem. **10** [1971] 498/504). — [2] R. E. Sievers, J. J. Brooks, J. A. Cunningham, W. E. Rhine (Advan. Chem. Ser. **150** [1976] 222/31; C.A. **85** [1976] No. 85111). — [3] J.W. Mitchell, C.V. Banks (Talanta **19** [1972] 1157/69). — [4] R. F. Sieck, C. V. Banks (Anal. Chem. **44** [1972] 2307/12). — [5] K. Utsunomiya, T. Shigematsu (Anal. Chim. Acta **58** [1972] 411/9).

5.1.9 Complexes with 3,5-Heptanedione $CH_3CH_2C(O)CH_2C(O)CH_2CH_3$ and Its Derivatives

5.1.9.1 With 3,5-Heptanedione, Monomethyl- and Dimethyl-3,5-heptanediones
(= $C_7H_{12}O_2$, $C_8H_{14}O_2$, $C_9H_{16}O_2$, respectively)

The preparation and properties for tris chelates of neodymium, gadolinium, and erbium [1] are analogous to those of the chelates with the derivatives of 2,4-hexanediones, which are described in Section 5.1.7.2, see p. 121.

5.1.9.2 With 2,2,6-Trimethyl-3,5-heptanedione
($(CH_3)_2CHC(O)CH_2C(O)C(CH_3)_3{=}C_{10}H_{18}O_2$)

$M(C_{10}H_{17}O_2)_3$. Each chelate with M = Sc, Y, La to Lu ≠ Ce, Pm was prepared by adding a stoichiometric amount of isobutyrylpivaloylmethane, with stirring, to a solution of the rare earth nitrate in a small volume of 1:1 (v/v) water/ethanol. The pH of the reaction mixture was maintained at approximately 6 with 1M aqueous ammonia. The solution was diluted with a large amount of water, and the resulting precipitate was collected by filtration, washed, and dried under vacuum [1, 2]. The colors and melting points of the chelates are recorded below [2].

M	color	m.p. in °C
Sc	white	88.0
Y	white	180 to 184
La	white	116 to 118
Pr	green	166 to 167
Nd	blue	188 to 190
Sm	yellow	176 to 177
Eu	yellow	182 to 183
Gd	white	192 to 193
Tb	white	195 to 198
Dy	white	190 to 193
Ho	yellow	190 to 192
Er	pink	142 to 144
Tm	white	180 to 181
Yb	white	178 to 179
Lu	white	162 to 165

Thermogravimetric analyses revealed that most of the chelates, with the exception of the lanthanum and praseodymium compounds, sublimed almost quantitatively at temperatures between 150 and 260°C. The volatilities of the metal chelates increased with decreasing ionic radius. The lanthanum and praseodymium chelates are also volatile, but they leave a small residue on volatilization, indicating partial thermal decomposition. The chelates were sufficiently volatile to allow chromatography in the vapor phase. The retention times of the chelates increased with increasing ionic radius. The lanthanum and praseodymium chelates were hardly eluted under the conditions used (stainless steel column filled with 5% Dow Corning high vacuum silicone grease on Chromosorb-WAW DMCS [60 to 80 mesh]; column temperature programmed from 210 to 250°C at 4° min^{-1}; injection temperature 300°C, helium flow rate 46 ml · min^{-1}) [2].

References to 5.1.9:

[1] K. Utsunomiya, T. Shigematsu (Anal. Chim. Acta **58** [1972] 411/9). — [2] K. Utsunomiya (Anal. Chim. Acta **59** [1972] 147/51).

5.1.9.3 With 2,2,6,6-Tetramethyl-3,5-heptanedione $(CH_3)_3C(O)CH_2C(O)C(CH_3)_3$ (= Dipivaloylmethane = $C_{11}H_{20}O_2$)

5.1.9.3.1 $M(C_{11}H_{19}O_2)_n^{3-n}$ Complexes. Formation in Solution

Formation constants for the 1:1, 2:1, and 3:1 (ligand : metal) chelates are recorded in Table 5/17. The differences between formation constants for members of the lanthanide series are small, with lg K_1 values varying only by 0.97 lg units [1].

5.1.9.3.2 Tris Chelates $M(C_{11}H_{19}O_2)_3 \cdot nH_2O$ (n = 0, 1)

Preparation

Eisentraut, Sievers [4, 5] effected synthesis of a series of anhydrous chelates (M = Sc, Y, La to Lu except for Ce, Pm) according to the following procedure. Sodium hydroxide (2.4 g) dissolved in 50 ml of 50% aqueous ethanol was added to a solution of dipivaloylmethane (60 mmol) in 30 ml of 95% ethanol. A hydrated rare earth nitrate (20 mmol) dissolved in 50 ml of 50% ethanol was added while the reaction mixture was stirred. Immediately, the head space of the reaction flask was evacuated, the flask was sealed, and the mixture was stirred for 2 h. The volume of the solution was reduced by 50% by distillation under reduced pressure, and 350 ml of water was added. The precipitated product was collected by filtration, washed with water, and air dried. The crude product was then dried under vacuum (0.1 Torr) at 95°C for 1 h and then sublimed by heating (100 to 200°C) under vacuum. The sublimed crystals were recrystallized from n-hexane in vacuo and dried under vacuum. Selbin, Ahmad, Bhacca [6] prepared the series of tris chelates (M = Pr to Lu) by the procedure of Eisentraut, Sievers [4, 5], with the difference that the compounds were purified by recrystallization three times from n-hexane, rather than by sublimation. The melting points of the compounds thus obtained are recorded in Table 5/20.

The cerium(III) chelate was isolated as the monohydrate by adding 2 N ammonia to an ethanolic solution of $Ce(NO_3)_3 \cdot 6H_2O$ (1 mmol) and dipivaloylmethane (3.5 mmol). The resulting precipitate was collected by filtration, washed with a small amount of a deoxygenated methanol/water/ammonia solution, and dried in a nitrogen atmosphere. The yellow solid decomposed at 193°C on heating. The oxidation state of cerium was determined by magnetic susceptibility measurements (μ_{eff} = 2.10 μ_B) [8].

The synthesis of the promethium chelate is reported in [17]. The tris chelates (M = La, Sm, Ho, Er) reported by Hammond, Nonhebel, and Wu [18] were shown later to be actually dimethylformamide adducts of the tris chelates [19].

Crystallographic Properties

Single-crystal X-ray diffraction studies revealed that the anhydrous chelates formed two structurally distinct, isomorphous series [9 to 14]. The chelates of lanthanum to gadolinium crystallized from n-hexane and the vapor phase in the monoclinic system, whereas the chelates of holmium to lutetium were obtained as orthorhombic crystals under the same conditions. The compounds of both terbium and dysprosium were dimorphic, crystallizing in the monoclinic system from hexane and in the orthorhombic system from the vapor phase. Triclinic crystals of $Eu(C_{11}H_{19}O_2)_3 \cdot H_2O$ and $Dy(C_{11}H_{19}O_2)_3 \cdot H_2O$ were obtained from hexane exposed to the atmosphere. The crystallographic data for all compounds studied are summarized in Table 5/18. Crystal and molecular structures, representative of the three crystal types, are discussed below.

X-ray analysis (R = 0.133 for 5373 intensities) revealed that the praseodymium compound (monoclinic system) is dimeric (**Fig. 5-10**, p. 128). Dimerization occurs through equal sharing of two of the diketonate oxygen atoms by praseodymium atoms. As a consequence, each praseodymium ion is seven-coordinate, surrounded by six oxygen atoms at the vertices of a distorted trigonal prism and one oxygen atom at a face. The relevant bond lengths and bond angles in the Pr_2O_{12} system are given in the original paper. The average Pr-O distance is 2.446 Å. The two bridging oxygen atoms are each involved in one long Pr-O bond (2.59 Å). The average intraligand O-O distance (chelate bite) for ligands not involved in bridging is 2.83 Å [9].

Table 5/17

Formation Constants for Rare Earth Complexes with 2,2,6,6-Tetramethyl-3,5-heptanedione $M(C_{11}H_{19}O_2)_n^{3-n}$ (n = 1 to 3).

metal ion	method*)	temp. in °C	medium**)	lg K_1	lg K_2	lg β_2	lg K_3	Ref.
Y^{3+}	gl	30	aq. dioxane, I = 0.1 (KNO_3)	8.57	7.76	16.33		[1]
La^{3+}	gl	30	aq. dioxane, I = 0.1 (KNO_3)	7.60	6.88	14.48		[1]
Pr^{3+}	gl	30	aq. dioxane, I = 0.1 (KNO_3)	8.00	7.18	15.18		[1]
Pr^{3+}	dist	24		8.20		16.30		[2]
Nd^{3+}	gl	30	aq. dioxane, I = 0.1 (KNO_3)	8.06	7.30	15.36		[1]
Nd^{3+}	gl	21	methanol	12.7	9.8		4.4	[3]
Sm^{3+}	gl	30	aq. dioxane, I = 0.1 (KNO_3)	8.26	7.53	15.79		[1]
Sm^{3+}	dist	24		8.30		16.50		[2]
Eu^{3+}	gl	30	aq. dioxane, I = 0.1 (KNO_3)	8.39	7.66	16.05		[1]
Eu^{3+}	dist	24		8.35		16.60		[2]
Gd^{3+}	gl	30	aq. dioxane, I = 0.1 (KNO_3)	8.30	7.59	15.89		[1]
Gd^{3+}	dist	24		8.30		16.53		[2]
Tb^{3+}	gl	30	aq. dioxane, I = 0.1 (KNO_3)	8.62	7.92	16.54		[1]
Tb^{3+}	dist	24		8.46		16.86		[2]
Dy^{3+}	gl	30	aq. dioxane, I = 0.1 (KNO_3)	8.66	7.90	16.56		[1]
Ho^{3+}	gl	30	aq. dioxane, I = 0.1 (KNO_3)	8.68	7.92	16.60		[1]
Er^{3+}	gl	30	aq. dioxane, I = 0.1 (KNO_3)	8.77	7.95	16.72		[1]
Er^{3+}	gl	21	methanol	13.6	10.4		5.5	[3]
Tm^{3+}	gl	30	aq. dioxane, I = 0.1 (KNO_3)	8.90	8.10	17.00		[1]
Yb^{3+}	gl	30	aq. dioxane, I = 0.1 (KNO_3)	9.01	8.14	17.15		[1]
Lu^{3+}	gl	30	aq. dioxane, I = 0.1 (KNO_3)	8.91	8.09	17.00		[1]

*) Abbreviations: gl = glass electrode (potentiometric titration), dist = distribution between two phases. — **) aq. dioxane = 3:1 (v/v) dioxane/water.

Table 5/18

Crystallographic Data for the Rare Earth Chelates with 2,2,6,6-Tetramethyl-3,5-heptanedione $M(C_{11}H_{19}O_2)_3 \cdot nH_2O$ (n = 0, 1)

M	n	system	space group	unit cell dimensions a in Å	b in Å	c in Å	β	c sin β in Å	density in g/cm³ calc.	obs.	Ref.
La	0	monoclinic	$P2_1/n-C_{2h}^5$ (No. 14)	22.32	28.59	12.59	105°		1.179	1.19	[9]
Pr	0	monoclinic	$P2_1/n-C_{2h}^5$ (No. 14)	22.28	28.51	12.56	105°		1.190	1.20	[9]
Nd	0	monoclinic	$P2_1/n-C_{2h}^5$ (No. 14)	22.23	28.45	12.51	105°		1.206	1.22	[9]
Nd	0	monoclinic	$P2_1/c-C_{2h}^5$ (No. 14)	12.40	22.38	28.03	106°35′				[10]
Sm	0	monoclinic	$P2_1/n-C_{2h}^5$ (No. 14)	22.20	28.29	12.49	105°		1.227	1.23	[9]
Sm	0	monoclinic	$P2_1/c-C_{2h}^5$ (No. 14)	12.35	27.83			21.17	1.27	1.26	[11]
Eu	0	monoclinic	$P2_1/n-C_{2h}^5$ (No. 14)	22.16	28.21	12.45	105°		1.240	1.25	[9]
Eu	0	monoclinic	$P2_1/c-C_{2h}^5$ (No. 14)	12.36	27.99			21.28	1.26		[11]
Eu	1	triclinic	$P\bar{1}-C_i^1$ (No. 2)	14.23	14.84	11.66	*)		1.214	1.23	[9]
Gd	0	monoclinic	$P2_1/n-C_{2h}^5$ (No. 14)	22.15	28.12	12.44	105°		1.255	1.25	[9]
Gd	0	monoclinic	$P2_1/c-C_{2h}^5$ (No. 14)	12.31	27.98			21.18	1.28	1.27	[11]
Tb	0	monoclinic	$P2_1/n-C_{2h}^5$ (No. 14)	22.16	28.03	12.45	105°		1.260	1.26	[9]
Tb	0	monoclinic	$P2_1/c-C_{2h}^5$ (No. 14)	12.39	27.85			21.14	1.29	1.29	[11]
Tb	0	orthorhombic	$Pmn2_1-C_{2v}^7$ (No. 31)	18.25	10.05	10.72			1.196	1.20	[9]
Dy	0	monoclinic	$P2_1/n-C_{2h}^5$ (No. 14)	22.11	28.04	12.43	105°		1.271	1.26	[9]
Dy	0	monoclinic	$P2_1/c-C_{2h}^5$ (No. 14)	12.30	27.72			21.08	1.31	1.25	[11]
Dy	0	orthorhombic	$Pmn2_1-C_{2v}^7$ (No. 31)	18.12	10.09	10.71			1.208	1.21	[9]
Dy	1	triclinic	$P\bar{1}-C_i^1$ (No. 2)	14.21	14.88	11.60	**)		1.238	1.24	[9]
Ho	0	orthorhombic	$Pmn2_1-C_{2v}^7$ (No. 31)	9.94	17.83			10.61	1.26	1.25	[11]
Er	0	orthorhombic	$Pmn2_1-C_{2v}^7$ (No. 31)	17.825	10.627	9.950			1.26	1.26	[12]
Er	0	orthorhombic	$Pmn2_1-C_{2v}^7$ (No. 31)	9.94	17.82			10.63	1.26	1.26	[11]
Lu	0	orthorhombic	$Pmn2_1-C_{2v}^7$ (No. 31)	17.713	10.638	9.973			1.286		[15]

*) $\alpha = 99.9°$, $\beta = 109.8°$, $\gamma = 113.9°$. — **) $\alpha = 99.8°$, $\beta = 109.9°$, $\gamma = 114.1°$.

Fig. 5-10

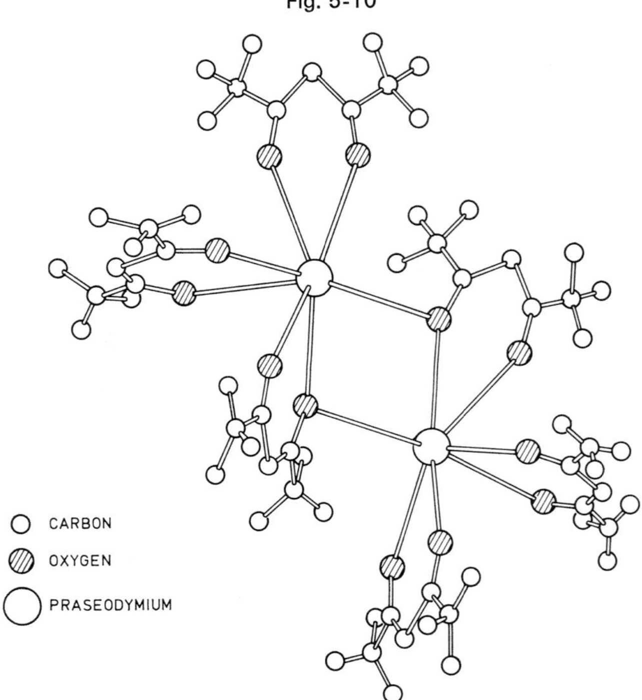

Structural formula of
the dimer praseodymium complex with 2,2,6,6-tetra-
methyl-3,5-heptanedione, $Pr_2(C_{11}H_{19}O_2)_6$.

The orthorhombic modification of the anhydrous erbium chelate consists of monomeric $Er(C_{11}H_{19}O_2)_3$ molecules arranged in almost trigonal fashion within layers lying approximately perpendicular to [010] in the crystal (structure refined to a conventional R = 0.041 over 1637 intensities). The erbium ion is coordinated by six oxygen atoms and the coordination geometry is almost ideally trigonal prismatic. The dimensions of the coordination polyhedron are depicted in **Fig. 5-11**. The average metal-oxygen bond distance is 2.212 Å (range 2.156 to 2.263 Å) with an intraligand O-O distance (chelate bite) of 2.67 Å [12].

The crystal and molecular structures of the lutetium chelate (orthorhombic system) show the compound to be isomorphous with the orthorhombic form of the erbium chelate (R = 0.053 for 2360 reflections). Six oxygen atoms of the three diketonate ligands form a distorted trigonal prism around the central lutetium ion, with an average bond distance of 2.19 Å (range = 2.174 to 2.216 Å). The average chelate bite is 2.66 Å [15].

The crystal structure (R = 0.058 for 3015 intensities) of the monohydrated dysprosium chelate (triclinic system) revealed two $[Dy(C_{11}H_{19}O_2)_3H_2O]$ units linked via hydrogen bonds through the water molecule and in effect forming a dimer (**Fig. 5-12**). Each monomeric unit consists of the dysprosium ion coordinated to six oxygen atoms of the three diketonate ligands and the oxygen atom of the water molecule. The coordination polyhedron was depicted as a monocapped trigonal prism, with the water molecule located at one of the vertices of the trigonal prism rather than capping the rectangular face. Three of the Dy-O bond distances are considerably longer (average = 2.36 Å, range = 2.35 to 2.36 Å) than the other four (average = 2.25 Å, range = 2.22 to 2.26 Å). The longer bond distances include the oxygen of water (Dy-O = 2.36 Å) and the two diketonate oxygen atoms involved in the hydrogen bonds to the water molecule, which link the monomeric units into pairs. The chelate bite ranges from 2.68 to 2.78 Å [13].

Fig. 5-11

Trigonal prismatic coordination of oxygen atoms around the central erbium ion in $Er(C_{11}H_{19}O_2)_3$.

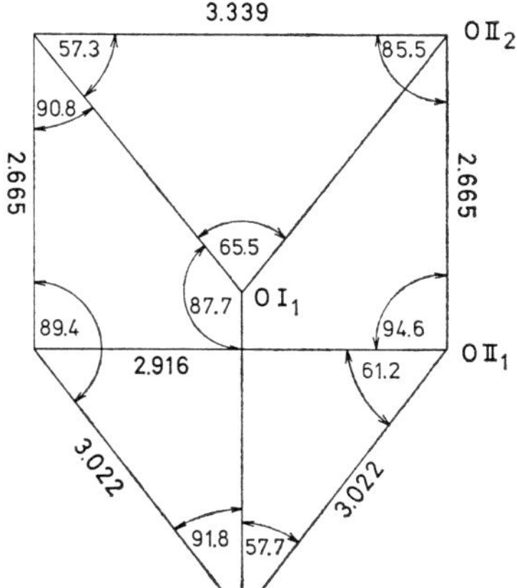

Fig. 5-12

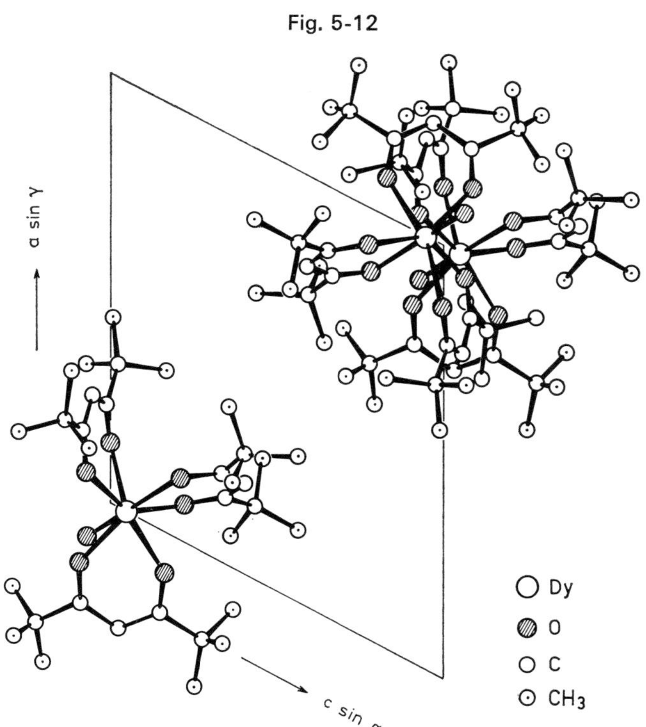

Molecular structure of $[Dy(C_{11}H_{19}O_2)_3H_2O]$ in projection along [010]. One of the overlapping pair around the origin is left out for clarity.

Although the chelates crystallizing in the monoclinic phase (La to Dy) were found to be dimeric in the solid state, molecular weight measurements on the lanthanum, samarium, gadolinium, terbium, thulium, and ytterbium complexes indicated that the compounds are monomeric in benzene [4]. The results obtained for molecular weight measurements (vapor phase osmometry) on solutions of the praseodymium and europium chelates in chloroform, carbon tetrachloride, and n-hexane indicate, as well, that the compounds exist as monomers in these solvents [16].

Magnetic Properties

The magnetic moments (in μ_B) of the complexes are within the ranges expected for the free ions. (The magnetic susceptibilities were measured by an NMR method at ambient temperatures and corrected for the diamagnetism) [6]:

metal	Pr	Nd	Sm	Eu	Gd	Tb	Dy	Ho	Er	Tm	Yb	Lu
μ_B	3.6	3.5	2.1	3.6	7.8	9.6	10.3	10.0	9.3	7.2	4.3	diamag.

Kjekshus, Ledaal [37] noted that the $Pr_2(C_{11}H_{19}O_2)_6$ and $Eu_2(C_{11}H_{19}O_2)_6$ compounds obeyed the Curie-Weiss law over the approximate temperature ranges 100 to 500 K and 50 to 500 K, respectively (the upper temperature limit is imposed by the melting points of the compounds). The minima in $\chi^{-1}(T)$ at $T=50\pm10$ and 100 ± 10 K for the $Pr_2(C_{11}H_{19}O_2)_6$ and $Eu_2(C_{11}H_{19}O_2)_6$ compounds, respectively, were attributed to magnetic coupling between the lanthanide ions in the dimers [37]. Magnetic susceptibility measurements from 4.2 K to room temperature for the praseodymium [38, 39] and europium [39] chelates have also been reported.

The NMR spectrum of $Eu(C_{11}H_{19}O_2)_3$ in carbon tetrachloride (0.03 M) at 30°C revealed singlets at 3.63 and 9.13 τ (TMS = 10.00 τ), assigned to the CH and tert-butyl protons of the ligand, respectively. In addition, a signal at 7.63 τ was observed, which had about one-twelfth of the intensity of the strongest singlet. The extra peak was assigned to the tert-butyl protons associated with the dimeric form of the chelate. In deuterated chloroform solutions (0.03 M), the more intense monomer peak was observed at 9.54 τ and that assigned to the dimer as a doublet centered at 6.49 τ, with an intensity ratio of 2.01 : 1. In methanol-d_4, the tert-butyl protons of $Eu(C_{11}H_{19}O_2)_3 \cdot CD_3OD$ were observed at 10.77 τ, with a very weak signal at 8.86 τ, assigned to a trace of undissociated dimer. Similar results were obtained for the praseodymium chelate [29].

Schwendiman, Zink [41] observed that the isotropic shifts of the ligand tert-butyl protons are approximately +50 Hz (spectra recorded at 60 MHz) in noncoordinating solvents, approximately +110 Hz in weakly coordinating solvents and +175 Hz in pyridine (see Table 5/19). In non-

Table 5/19

Magnetic Gram Susceptibilities of $Eu(C_{11}H_{19}O_2)_3$ and the Isotropic NMR Shifts of the Complexed Tetramethylheptanedionato Ligands in Various Solvents [41].

solvent	$10^6 \chi_g$ in cgsu*)	temp. in °C	isotropic shift in Hz**)
chloroform-d_1	5.41±0.17	36.3	57
carbon tetrachloride	5.34±0.13	30.2	40
methanol-d_4	5.25±0.07	30.0	112
ethyl acetate	5.10±0.11	29.4	125
ethanol	5.03±0.10	37.2	94
acetone-d_6	4.77±0.03	29.4	116
benzene-d_6	4.56±0.09	29.0	48
pyridine-d_5	3.03±0.10	30.5	182

*) Obtained by Evans' NMR method, D. F. Evans (J. Chem. Soc. **1959** 2003). — **) The shift was measured from the proton resonance of the uncomplexed ligand in the appropriate solvent. Spectra were recorded at 60 MHz.

coordinating solvents, three additional resonances attributable to the tert-butyl protons in the dimer (three magnetically nonequivalent tert-butyl groups are present in the dimer) were observed. The integrated intensity of all dimer peaks was smaller by a factor of 10 than the monomer tert-butyl peak. No dimer signals were observed in methanol, acetone, or pyridine. Furthermore, when Lewis bases were added to the chloroform solution, the dimer resonances disappeared [41].

A plot of the isotropic shifts of the tert-butyl protons against the magnetic susceptibility of the complex revealed three groups of points which qualitatively corresponded to noncomplexing, weakly complexing, and strongly complexing solvents. The magnetic susceptibility of the chelate is lower in solution than in the solid state (7.88×10^{-6} cgsu at 22°C) and decreases with increasing coordinating ability of the solvent. The data indicate that the magnetic properties are dependent on the presence or absence of dimers in solution [41].

Optical Properties

Infrared spectra of the $M(C_{11}H_{19}O_2)_3$ chelates (M = La, Ce, Pr, Nd, Gd) were obtained at 300 K in the region 4000 to 300 cm^{-1} for samples dispersed in KBr discs, and at 300 and 77 K, in the region 660 to 76 cm^{-1} for samples of the lanthanum, neodymium, and gadolinium compounds dispersed in polyethylene discs. Raman spectra of the lanthanum, praseodymium, and gadolinium chelates were measured at 300 K, with powdered samples in the region 3000 to 100 cm^{-1}. The vibrational spectral data and tentative band assignments are tabulated in [20]. Assignments were made on the basis of group theory selection rules and group frequencies, and these were compared with the assignments of tris(acetylacetonato) rare earth(III) chelates. In the region of 1500 to 1600 cm^{-1}, bands near 1586, 1545, and 1531 cm^{-1} were assigned to the ν(C=O) stretching modes, and bands near 1565 and 1497 cm^{-1} to the ν(C=C) stretching mode. Bands near 402 cm^{-1} and 246 cm^{-1} were assigned to the ν(M-O) stretching and (O-M-O) out-of-plane bending vibrations, respectively [20].

The emission spectrum of $Eu_2(C_{11}H_{19}O_2)_6$ (Nujol mull) at −196°C revealed five lines in the region of the $^5D_0 \rightarrow {}^7F_2$ transition, consistent with an assignment of the chelate to the trivial C_1 (no symmetry) point group. The spectrum remained characteristic of the dimeric structure in a $CH_2Cl_2:CHCl_3:CCl_4$ (5:5:1) glass solution at −196°C and in a mobile solution at −78°C [21]. The fluorescence spectrum of the europium chelate is also discussed in [22].

The electronic absorption spectrum of each chelate (M = La to Lu except for Ce, Pm) in methanol revealed a strong charge-transfer band (ε = 25000 to 45000 $l \cdot mol^{-1} \cdot cm^{-1}$) in the region 33000 to 36000 cm^{-1}. This band is rather broad, covering a spread of about 10000 cm^{-1}, with a shoulder at about 32250 cm^{-1}. Application of an electrostatic charge-transfer model, whereby the ligand nonbonding electrons are transferred to empty 5d orbitals of the rare earth, produced agreement between the observed and calculated spectra [23]. The absorption spectrum of the neodymium chelate is reported in [24].

The absorption spectra of several $M(C_{11}H_{19}O_2)_3$ chelates (M = Pr, Nd, Sm, Eu, Dy, Ho, Er, Tm) were measured in the vapor phase in the range 4000 to 30000 cm^{-1}. Absorption maxima for each chelate are tabulated in [25], and the oscillator strengths for each of the hypersensitive transitions ($\Delta J = \pm 2$, $\Delta L \leqq 2$, $\Delta S = 0$) are calculated. The results were discussed in terms of the Judd-Ofelt [26, 27] theory. The intensity of the hypersensitive transitions measured in the gaseous phase, as well as the environmental sensitivity and the magnitude of the Judd-Ofelt τ_2 parameter, were accounted for in terms of a vibronic mechanism [25].

Solubility

The solubilities of the ytterbium chelate in several organic solvents are presented below. The compounds are insoluble in water [5].

solvent	methyl-cyclohexane	benzene	chloroform	methanol
temperature in °C	21	23.5	24	24.5
solubility in $g \cdot ml^{-1}$	0.19	0.20	0.42	0.16

solvent	ethyl acetate	ethanol (anhydrous)	carbon tetrachloride	n-hexane
temperature in °C	24	24	23	22
solubility in $g \cdot ml^{-1}$	0.70	0.51	0.36	0.12

Thermal Stability

The most striking properties of these chelates are their volatility and thermal stability. The complexes readily undergo sublimation under vacuum without apparent decomposition at temperatures between 100 to 200°C [4]. Thermogravimetric analyses revealed that the compounds sublimed between 200 to 300°C under an atmosphere of helium [30]. Differential thermal analyses and visual observations of melting behavior indicated that the complexes can be heated above 300°C, allowed to cool and resolidify, then remelted at the original melting point [4].

Eisentraut and Sievers [4] have shown that the tetramethylheptanedionato chelates can be chromatographed in the vapor phase at 200°C without decomposition. The chromatogram of each complex exhibited a single peak, well separated from the solvent peak. The gas chromatographic retention behavior revealed appreciable and significant differences as a function of ionic radius, with the larger rare earth ions showing the largest retention times. The retention time of the yttrium chelate was about the same as that of the erbium chelate, as expected on the basis of their comparable ionic radii [4]. Additional gas chromatographic and thermogravimetric data are given in [31] and [32]. The use of the terbium chelate in the stationary phase (0.131 m solution of the complex in squalane) of a chromatographic column is reported in [33].

Vapor pressure measurements revealed that complexes of the lanthanides with higher atomic numbers are considerably more volatile than those of the lighter, larger members of the series. For example, at 200°C, vapor pressures ranged from 5 Torr for the ytterbium chelate down to 0.2 Torr for the lanthanum complex. Furthermore, the temperature required to produce a vapor pressure of 1 Torr varied from 170°C for the lutetium chelate to 225°C for the lanthanum complex. A plot of lg p (p is the vapor pressure) versus the reciprocal of the absolute temperature for each chelate (except lanthanum) revealed two linear segments, with a pronounced break at the temperature corresponding to the melting point of the chelate (the lanthanum compound did not melt over the temperature range studied, and therefore, no break was observed in the vapor pressure plot). The heats of vaporization and sublimation for each chelate were calculated from the Clausius-Clapeyron equation at the mean temperature of the ranges indicated (Table 5/20). The values obtained further corroborate the observation that the chelates of the smaller rare earths are more volatile than those of the larger rare earths. A plot of heat of vaporization of the chelates against the atomic number of the lanthanide, in general, reflected the effects of the lanthanide contraction, except for a small but distinct discontinuity at gadolinium [7]. (Boeyens [36] noted, however, that a plot of heat of vaporization against ionic radius of the rare earth appeared smoother than the plot of heat of vaporization versus atomic number, and contended that the deviation from the curve at gadolinium seemed barely significant.) As the data in Table 5/20 indicate, the heats of sublimation appear to fall into two groups [7]. Boeyens [36] notes that the transition between the two groups occurs in the same region as the change in structural type (dimeric to monomeric) and contends that the larger heats of sublimation observed for the lighter elements correspond to the energy required to dissociate the dimers into monomers. The heats of fusion, obtained by difference, revealed a similar pattern, with a general increase from 13.4 to 18.6 $kcal \cdot mol^{-1}$ for praseodymium through gadolinium, whereas for dysprosium through lutetium the values are essentially constant at 11 to 12 $kcal \cdot mol^{-1}$ [7].

Berg and Acosta [34] studied the sublimation of the chelates in a thermal gradient fractional vacuum sublimator. The sublimation time was 2 h at 0.1 Torr using deoxygenated nitrogen as the carrier gas, and a thermal gradient of 60 to 190°C. The sublimation recrystallization temperature zones for the various chelates are recorded below.

M	Sc	Y	La	Pr	Nd
temp. zone in °C	74 to 49	92 to 46	132 to 92	124 to 90	120 to 91

M	Sm	Eu	Gd	Tb	Dy
temp. zone in °C. . .	116 to 66	116 to 65	128 to 66	97 to 61	94 to 57
M	Ho	Er	Tm	Yb	Lu
temp. zone in °C. . .	92 to 50	93 to 53	86 to 49	87 to 48	92 to 50

The data revealed that the chelates are divided into two distinct groups according to their volatility. The less volatile group consists of the chelates from lanthanum through gadolinium, whereas the more volatile group consists of the chelates of scandium, yttrium, and the heavier lanthanides [34]. Quantitative separation of an element of the first group from one of the second group by fractional sublimation was demonstrated [4, 34]. Fractional distillation of chelates from solutions of transformer oil has also been demonstrated [35]. For vapor pressure and thermal stability of the europium(III) complex, see also [43], of the chelates with M = Nd, Gd, and Yb, see [44]. The cocrystallization method was used to determine the saturated vapor pressure of the promethium tris chelate [45].

Mass Spectra and Photochemical Reactions

The mass spectra of the series of $M(C_{11}H_{19}O_2)_3$ (M = Sc, Y, La to Lu except for Ce, Pm) chelates were obtained with 50 V electrons. The appearance potentials and intensities of the ions $M(C_{11}H_{19}O_2)_3^+$, $[M(C_{11}H_{19}O_2)_3\text{-}C_4H_9]^+$, and $[M(C_{11}H_{19}O_2)_2]^+$ were depicted graphically. The largest peak in each spectrum, except for the europium chelate, was that of $M(C_{11}H_{19}O_2)_2^+$. The samarium, europium, and ytterbium chelates also revealed a prominent $M(C_{11}H_{19}O_2)^+$ peak, which was attributed to the stability of the +2 oxidation state of these metals [42]. Chemical ionization mass spectra of the series of chelates (M = La to Lu except for Ce, Pm) were also obtained with isobutane at 1 Torr as the reagent gas. The calculated relative peak intensities for each of the $M(C_{11}H_{19}O_2)_3$ chelates are tabulated in [40] and were discussed in terms of the naturally abundant isotopes of each element [40].

Irradiation of the praseodymium, europium, and holmium chelates in coordinating solvents such as pyridine, ethanol, and acetone with an argon laser (100 to 200 mW, tunable over ten transitions in the blue to green region of the spectrum) resulted in a photochemical reaction, with formation of 2,2,6,6-tetramethyl-3,4,5-heptanetrione. No reaction was observed in poorly coordinating solvents such as hexane or carbon tetrachloride. The authors proposed that the production of the triketone was initiated when energy absorbed in the f-f transition (e.g., $^7F_0 \rightarrow {}^5D_0$ in the case of europium) transferred intramolecularly to a ligand triplet level, resulting in photo-substitution of the ligand by solvent. The free diketonate anion, apparently in an excited state, reacted with oxygen in solution, forming a peroxide, which decomposed, yielding the 3,4,5-triketone. The rate of formation of the triketone was observed to be linear in both laser power and irradiation time [28]. For photodecomposition of the terbium(III) chelate in various alcohol solvents, see [46].

Table 5/20

Physical Properties of the Tris Chelates of the Rare Earths with 2,2,6,6-Tetramethyl-3,5-heptanedione.

M	color [6]	m.p. [4]	m.p. [6]	heat of vaporization and sublimation*) [7]			
				temperature range in K	ΔH_{sub} in $kcal \cdot mol^{-1}$	temperature range in K	ΔH_{vap} in $kcal \cdot mol^{-1}$
Sc		152 to 155					
Y		169 to 172.5					
La		238 to 248**)		450 to 520	34.31		
Pr	green	222 to 224**)	222 to 224	450 to 495	39.52	495 to 530	26.10
Nd	violet	215 to 218**)	218 to 219	430 to 491	37.86	491 to 510	23.69
Sm	white	195.5 to 198.5	200 to 201	430 to 468	36.00	468 to 500	21.22
Eu	yellow	187 to 189	190 to 191	430 to 466	39.54	466 to 490	20.88
Gd	white	182 to 184	183 to 184	420 to 456	38.55	456 to 500	21.55
Tb	white	177 to 180	150 to 152	420 to 454	33.81	454 to 500	20.80
Dy	white	180 to 183.5	182 to 183	410 to 456	31.90	456 to 500	20.61
Ho	yellowish white	180 to 182.5	178 to 180	420 to 458	31.40	458 to 500	20.22
Er	pink	179 to 181	179 to 180	410 to 454	31.83	454 to 490	20.49
Tm	white	171.5 to 173.5	170 to 173	410 to 446	31.39	446 to 490	20.09
Yb	white	166 to 169	165 to 167	410 to 444	31.86	444 to 495	19.80
Lu	white	172 to 174	173 to 174	420 to 448	32.08	448 to 490	19.99

*) Calculated from the Clausius-Clapeyron equation at the mean temperature of the ranges indicated. — **) Melting point taken in a sealed, evacuated capillary tube.

References to 5.1.9.3.1 and 5.1.9.3.2:

[1] A. S. Berlyand, L. I. Martynenko, N. P. Potapova, V. I. Spitsyn (Dokl. Akad. Nauk SSSR **189** [1969] 563/5; Dokl. Phys. Chem. Proc. Acad. Sci. USSR **184/189** [1969] 736/8). — [2] T. R. Sweet, H. W. Parlett (Anal. Chem. **40** [1968] 1885/8). — [3] A. S. Berlyand, L. I. Martynenko (Vestn. Mosk. Univ. Khim. **12** [1971] 729/31; Moscow Univ. Chem. Bull. **12** No. 6 [1971] 77/8). — [4] K. J. Eisentraut, R. E. Sievers (J. Am. Chem. Soc. **87** [1965] 5254/6). — [5] K. J. Eisentraut, R. E. Sievers (Inorg. Syn. **11** [1968] 94/8).

[6] J. Selbin, N. Ahmad, N. Bhacca (Inorg. Chem. **10** [1971] 1383/7). — [7] J. E. Sicre, J. T. Dubois, K. J. Eisentraut, R. E. Sievers (J. Am. Chem. Soc. **91** [1969] 3476/81). — [8] E. Uhlemann, F. Dietze (Z. Anorg. Allgem. Chem. **386** [1971] 329/34). — [9] C. S. Erasmus, J. C. A. Boeyens (Acta Cryst. B **26** [1970] 1843/54). — [10] M. D. Danford, J. H. Burns, C. E. Higgins, J. R. Stokely, W. H. Baldwin (Inorg. Chem. **9** [1970] 1953/5).

[11] V. A. Mode, G. S. Smith (J. Inorg. Nucl. Chem. **31** [1969] 1857/9). — [12] J. P. R. DeVilliers, J. C. A. Boeyens (Acta Cryst. B **27** [1971] 2335/40). — [13] C. S. Erasmus, J. C. A. Boeyens (J. Cryst. Mol. Struct. **1** [1971] 83/91). — [14] Wei-Gwo Chen, L. M. Trefonas, E. A. Boudreaux (Decompos. Organometal. Compounds Refract. Ceram. Metals Metal Alloys Proc. Intern. Symp., Dayton, Ohio, 1967, p. 65/75 from C.A. **72** [1970] No. 71832). — [15] S. Onuma, H. Inoue, S. Shibata (Bull. Chem. Soc. Japan **49** [1976] 644/7).

[16] J. F. Desreux, L. E. Fox, C. N. Reilley (Anal. Chem. **44** [1972] 2217/9). — [17] M. Sakanoue, Y. Shirota, H. Shimozawa, K. Yoshihara (Radiochim. Acta **24** [1977] 77/80 according to C.A. **89** [1978] No. 154010). — [18] G. S. Hammond, D. C. Nonhebel, Chin-Hua S. Wu (Inorg. Chem. **2** [1963] 73/6). — [19] J. E. Schwarberg, D. R. Gere, R. E. Sievers, K. J. Eisentraut (Inorg. Chem. **6** [1967] 1933/5). — [20] Hsien-Yu Lee, F. F. Cleveland, J. S. Ziomek, F. Jarke (Appl. Spectrosc. **26** [1972] 251/6).

[21] G. A. Catton, F. A. Hart, G. P. Moss (J. Chem. Soc. Dalton Trans. **1975** 221/6). — [22] P. Procher, P. Caro (Semin. Chim. Etat Solide No. 5 [1970] 141/63 from C.A. **77** [1972] No. 68201). — [23] E. A. Boudreaux, Wei Gwo Chen (J. Inorg. Nucl. Chem. **39** [1977] 595/8). — [24] A. S. Berlyand, L. I. Martynenko, V. I. Spitsyn (Izv. Akad. Nauk SSSR Ser. Khim. **1971** 1100/1; Bull. Acad. Sci. USSR Div. Chem. Sci. **1971** 1016/7). — [25] D. M. Gruen, C. W. Dekock, R. L. McBeth (Advan. Chem. Ser. **71** [1967] 102/21).

[26] B. R. Judd (Phys. Rev. [2] **127** [1962] 750/61). — [27] G. S. Ofelt (J. Chem. Phys. **37** [1962] 511/20). — [28] T. Donohue (J. Am. Chem. Soc. **100** [1978] 7411/3). — [29] M. K. Archer, D. S. Fell, R. W. Jotham (Inorg. Nucl. Chem. Letters **7** [1971] 1135/40). — [30] K. J. Eisentraut, R. E. Sievers (J. Inorg. Nucl. Chem. **29** [1967] 1931/6).

[31] K. Utsunomiya, T. Shigematsu (Anal. Chim. Acta **58** [1972] 411/9). — [32] R. E. Sievers, K. J. Eisentraut, C. S. Springer, D. W. Meek (Advan. Chem. Ser. **71** [1967] 141/54). — [33] B. Feibush, M. F. Richardson, R. E. Sievers, C. S. Springer (J. Am. Chem. Soc. **94** [1972] 6717/24). — [34] E. W. Berg, J. J. C. Acosta (Anal. Chim. Acta **40** [1968] 101/3). — [35] N. P. Kuz'mina, G. G. Rybnikova, V. V. Panin, S. D. Moiseev, L. I. Martynenko, V. I. Spitsyn (Zh. Prikl. Khim. **48** [1975] 2159/62 from C.A. **84** [1976] No. 76441).

[36] J. C. A. Boeyens (J. Chem. Phys. **54** [1971] 75/8). — [37] A. Kjekshus, T. Ledaal (Acta Chem. Scand. **27** [1973] 2665/6). — [38] B. Kanellakopulos, C. Aderhold (Inorg. Nucl. Chem. Letters **9** [1973] 121/6). — [39] S. Jaakkola, R. W. Jotham (Inorg. Nucl. Chem. Letters **8** [1972] 639/42). — [40] T. H. Risby, P. C. Jurs, F. W. Lampe, A. L. Yergey (Anal. Chem. **46** [1974] 161/4).

[41] D. Schwendiman, J. I. Zink (Inorg. Chem. **11** [1972] 3051/4). — [42] J. D. McDonald, J. L. Margrave (J. Less-Common Metals **14** [1968] 236/9). — [43] R. Amano, A. Sato, S. Suzuki (Radiochem. Radioanal. Letters **39** [1979] 451/50). — [44] A. K. Baev, V. Ya. Mishin, I. L. Gaidym, E. M. Rubtsov (Radiokhimiya **22** [1980] 149/51 from C.A. **92** [1980] No. 157042). — [45] S. Yu. Lebedev, S. S. Berdonosov, I. V. Melikhov, N. B. Mikheev, I. A. Rumer (Radiokhimiya **21** [1979] 470/2 from C.A. **91** [1979] No. 146146).

[46] H. G. Brittain (J. Phys. Chem. **84** [1980] 840/2).

5.1.9.3.3 Lewis Base Adducts of the Tris Chelates

Following Hinckley's [1] communication in 1969, several hundred publications have appeared, which are documented in extensive review articles [2 to 12], demonstrating the ability of the tris(dipivaloylmethanato) rare earth chelates to form adducts with a variety of Lewis bases and to induce chemical shifts in the proton NMR spectra of the organic substrates. Theoretical analysis of the induced shifts requires knowledge of the stoichiometry and geometry of the shift reagent adduct, prompting studies aimed at the isolation and structural characterization of the complexes. This section concentrates mainly on the preparation, characterization, and physiochemical properties of these adducts, and includes only those NMR studies most closely associated with the structural aspects and magnetic properties of the complexes.

$M(C_{11}H_{19}O_2)_3 \cdot B$ (B = Lewis Base). **Formation Constants**

Equilibrium constants, corresponding to formation of 1:1 adducts, $M(C_{11}H_{19}O_2)_3 + B \rightleftharpoons M(C_{11}H_{19}O_2)_3 \cdot B$, are recorded in Tables 5/21 to 5/23. Sasaki and coworkers [13, 14] noted, that in general, the values of K_c increased with the dielectric constant of the solvent. The data in Table 5/22 revealed that although methyl substitution in pyridine generally increases the value of K_c, methyl substitution at the α-position decreased K_c due to steric hindrance [14]. Brittain [15] noted that the formation constants for the adducts derived from a series of phosphate esters (Table 5/22) were approximately the same, suggesting that the steric requirements of the alkyl substituents did not play a significant role in the formation of these adducts. He maintained that the relative Lewis acidity of the metal chelate and basicity of the substrate dominate the extent of adduct formation in these systems. On the other hand, triphenyl phosphate apparently presents a more severe steric problem, since no apparent evidence for adduct formation was obtained [15]. The binding strengths of tertiary amides to $Eu(C_{11}H_{19}O_2)_3$, relative to the binding strength of dimethylformamide, were determined for formamides, acetamides, and benzamides. The authors concluded that the equilibrium constants for the amide-$Eu(C_{11}H_{19}O_2)_3$ formation are subject to a combination of steric and electronic effects and depend on the configuration of the amide [16].

From nuclear magnetic resonance studies of acid-base association in solution, thermodynamic parameters of the association between the europium tris chelate and pyridine bases or aliphatic alcohols, respectively, were determined in CCl_4 [45]. The influence of phenyl, 1-adamantyl, and tert-butyl substituents on equilibrium constants was investigated. Association constants of $Eu(C_{11}H_{19}O_2)_3 \cdot B$ with B = ester, ketones, alcohols, or amines were determined at 27°C in $CDCl_3$ to compare hindering effects of the substituents. Donor strength decreases in the order $NH_2 > OH$ > ester-CO, ketone-CO. For alcohols the 1-adamantyl and tert-butyl groups led to higher equilibrium constants than the phenyl group [46]. The intermolecular exchange kinetics was studied by NMR for systems of $(CH_3)_2SO$, $C_6H_5N(CH_3)_2$, or CH_3CN with the tris chelates of europium and praseodymium over wide temperature ranges and at 40, 60, 80, and 100 MHz [47]. For adducts with phenyl carbinols and florenols, see [41].

Isolated Adducts

With Pyridine (= C_5H_5N = py)

$M(C_{11}H_{19}O_2)_3 \cdot py$. Each compound (M = Pr to Lu except for Pm) was prepared by dissolving a tris(dipivaloylmethanato)chelate in absolutely dry pyridine. The solution was left overnight in a tightly stoppered flask, then the excess pyridine was allowed to evaporate under reduced pressure in a desiccator. The dried product was dissolved in absolutely dry benzene and filtered, and the filtrate was allowed to evaporate to dryness under vacuum in a desiccator. All operations were carried out in a dry, inert-atmosphere box. The compounds were stored in a desiccator over sodium hydroxide [22].

The physical properties of the adducts are recorded in Table 5/23. A plot of the magnetic moment versus atomic number revealed a curve characteristic of the free ions [22].

Table 5/21

Equilibrium Constants Corresponding to Formation of 1:1 Adducts Between Tris(dipivaloylmethanato) Rare Earth Chelates and Alcohols or Camphor $M(C_{11}H_{19}O_2)_3 \cdot B$.

alcohol = B	formula	M	solvent	temp. in °C	method*)	K_c in l/mol	Ref.
1-propanol	C_3H_8O	Eu	CCl_4	35	NMR	623	[13]
1-butanol	$C_4H_{10}O$	Eu	CCl_4	35	NMR	818	[13]
		Eu	$CDCl_3$	35	NMR	7.0	[13]
		Eu	CD_2Cl_2	35	NMR	88	[13]
		Eu	C_6H_{12}	35	NMR	778	[13]
2-methyl-1-propanol	$C_4H_{10}O$	Eu	CCl_4	35	NMR	608	[13]
		Eu	$CDCl_3$	35	NMR	71	[13]
		Eu	CD_2Cl_2	35	NMR	129	[13]
		Eu	C_6H_{12}	35	NMR	811	[13]
2-butanol	$C_4H_{10}O$	Eu	CCl_4	35	NMR	482	[13]
		Eu	$CDCl_3$	35	NMR	54	[13]
		Eu	CD_2Cl_2	35	NMR	104	[13]
		Eu	C_6H_{12}	35	NMR	609	[13]
2-methyl-2-propanol	$C_4H_{10}O$	Eu	CCl_4	35	NMR	351	[13]
		Eu	$CDCl_3$	35	NMR	44	[13]
		Eu	CD_2Cl_2	35	NMR	44	[13]
		Eu	C_6H_{12}	35	NMR	571	[13]
2,2-dimethyl-1-propanol	$C_5H_{12}O$	Eu	$CHCl_3$	41	NMR	200	[17]
		Eu	$CDCl_3$	39	NMR	9.8	[18]
1-adamantanol	$C_{10}H_{16}O$	Eu	CCl_4	30	IR	525	[19]
		Eu	CCl_4	37	NMR	692	[19]
		Eu	$CDCl_3$	37	NMR	178	[19]
2-methyl-2-adamantanol	$C_{11}H_{18}O$	Eu	CCl_4		NMR	250	[17]
borneol	$C_{10}H_{18}O$	Ho	CCl_4	22	sp	457	[20]
cedrol	$C_{15}H_{26}O$	Ho	CCl_4	22	sp	457	[20]
cholestanol	$C_{29}H_{53}OH$	Eu	$CDCl_3$	27	NMR	61	[21]
camphor	$C_{10}H_{16}O$	Ho	CCl_4	22	sp	56	[20]

*) Abbreviations: NMR = K_c determined from chemical shift data, IR = infrared absorption spectrophotometry, sp = electronic absorption spectrophotometry.

Table 5/22

Equilibrium Constants Corresponding to Formation of 1:1 Adducts $M(C_{11}H_{19}O_2)_3 \cdot B$ Between Tris (dipivaloylmethanato) Europium(III) and Amines (calculated from chemical shift data in various organic solvents at 38°C) and with Organophosphate or Triphenyl Phosphine Oxide (estimated by electronic luminescence titration).

amine = B	formula	solvent	K in l/mol	Ref.
propylamine	C_3H_9N		37.8	[18]
pyridine	C_5H_5N	CCl_4	1291	[14]
2-methylpyridine (=α-picoline)	C_6H_7N	CCl_4	345	[14]
		$CDCl_3$	31	[14]
		CD_2Cl_2	76	[14]
		C_6H_{12}	558	[14]

Table 5/22 (Continued)

amine = B	formula	solvent	K in l/mol	Ref.
3-methylpyridine (= β-picoline)	C_6H_7N	CCl_4	3223	[14]
		$CDCl_3$	194	[14]
		CD_2Cl_2	1359	[14]
		C_6H_{12}	2523	[14]
4-methylpyridine (= γ-picoline)	C_6H_7N	CCl_4	2390	[14]
		$CDCl_3$	196	[14]
		CD_2Cl_2	1015	[14]
		C_6H_{12}	2865	[14]
3,5-dimethylpyridine (= β,β′-lutidine)	C_7H_9N	CCl_4	2737	[14]
		$CDCl_3$	209	[14]
		CD_2Cl_2	1877	[14]
		C_6H_{12}	3536	[14]
trimethyl phosphate	$C_3H_9PO_4$	CCl_4	1.407	[15]
triethyl phosphate	$C_6H_{15}PO_4$	CCl_4	1.660	[15]
tributyl phosphate	$C_{12}H_{27}PO_4$	CCl_4	1.762	[15]
triphenyl phosphine oxide	$C_{18}H_{15}PO$	CCl_4	4.48*)	[20]

*) At 22°C.

Table 5/23
Physical Properties of the Mono Pyridine Adducts of the Tris (dipivaloylmethanato) Rare Earth Chelates.

M	color	melting point in °C	magnetic moment in μ_B*)	M	color	melting point in °C	magnetic moment in μ_B*)
Pr	green			Dy	white	135 to 137	10.7
Nd	violet			Ho	white	134 to 135	10.3
Sm	white	133 to 135	2.4	Er	pink	131 to 133	9.5
Eu	white	135 to 138	3.7	Tm	white	134 to 136	
Gd	white	134 to 135	7.8	Yb	white	130 to 132	4.5
Tb	white	135 to 137	9.7	Lu	white	125 to 127	diamagnetic

*) Magnetic susceptibilities were measured by an NMR method at ambient temperature. The solvent used was a 1:1 (v/v) mixture of carbon tetrachloride and chloroform. The data were corrected for the diamagnetism of the ligands.

$Eu(C_{11}H_{19}O_2)_3 \cdot 2py$. The compound was isolated in the form of regular parallelopipedon crystals from concentrated solutions of the tris chelate in pyridine. The crystal and molecular structures of the adduct were determined by single-crystal X-ray analysis (R = 0.039 for 6840 reflections) [23, 24]. The crystals are in the triclinic space group $P\bar{1}$-C_i^1 (No. 2) with two molecules in a unit cell of dimensions a = 13.611, b = 17.306, c = 10.493 Å, α = 93.33°, β = 105.20°, and γ = 94.75°. The structure consists of discrete $[Eu(C_{11}H_{19}O_2)_3(py)_2]$ molecules, with each europium ion bonded to six oxygen atoms from the diketonate groups and two nitrogen atoms of the pyridine groups (**Fig. 5-13**). The average Eu-N bond length, 2.649 Å (2.647, 2.651 Å), is significantly longer than the average Eu-O bond distance of 2.347 Å (range = 2.312 to 2.372 Å). The coordination polyhedron was best described as a square antiprism. The average lengths of the square (s) and lateral (l) edges of the antiprism are 2.869 and 3.078 Å, respectively, yielding an l/s ratio of 1.07, which is close to the ideal value of 1.06. The positions of the ligated atoms are depicted in **Fig. 5-14**, p. 141. One chelating group, O(4)-O(5), spans a lateral edge, whereas the other two chelating groups, O(6)-O(7) and O(2)-O(3), span edges in the opposing square faces. The molecule has an approximate twofold symmetry axis, which passes through the europium ion and bisects the O(4)-O(5) chelate ring. Rotation about the C_2 axis carries one pyridine ring into the other and carries the O(2)-O(3) chelate into the O(6)-O(7) chelate. Although the geometry of the pyridine molecules is normal, the europium ion does not lie in either of the pyridine planes (the distortion, as measured by the Eu-N-para c angle, amounts to 8°). The pyridine ring is severely constrained by many short intramolecular contacts and has little freedom for rotation or choice of position [24].

The NMR spectrum of $Eu(C_{11}H_{19}O_2)_3 \cdot 2py$ in carbon disulfide at ambient temperature was studied, and the europium induced isotropic shifts for all protons, including those of the chelate, were calculated using the general equation for dipolar shifts, $\Delta\nu/\nu_0 = D_1(3\cos^2\Theta - 1) \cdot r^{-3} + D_2(\sin^2\Theta\cos 2\,\Omega) \cdot r^{-3}$ (eq. 1). In this equation, $\Delta\nu$ is the dipolar shift at probe frequency ν_0; r is the length of a vector which joins the europium ion and nucleus being examined; Θ is the angle this vector makes with the z magnetic axis; Ω is the angle which the projection of r into the xy plane makes with the x magnetic axis; and D_1 and D_2 are functions of the magnetic anisotropy of the complex. The results (Table 5/24) revealed that the isotropic shifts induced by the europium ion were best accounted for in terms of the complete two term equation, rather than the single term equation, $\Delta\nu/\nu_0 = D_1(3\cos^2\Theta - 1) \cdot r^{-3}$ (eq. 2). The latter equation applies to axially symmetric molecules (or those exhibiting effective axial symmetry in solution) and generally assumes that the magnetic axis passes through the donor atom of the Lewis base. Cramer and coworkers [25] noted the importance of the second term of equation 1, which comprises 29 to 80% of the total shift.

The twofold symmetry apparent in the structure of $Eu(C_{11}H_{19}O_2)_3 \cdot 2py$ produces two types of ortho protons, two types of meta protons, and a single type of para proton for the pyridine ring. The molecular structure also revealed two types of chelate rings which produce a total of three types of tert-butyl groups and two types of methine protons. However, in the temperature range studied (+30 to −105°C), fast exchange between the free and coordinated pyridine molecules must be averaging the signals due to the two coordinated ortho environments and a single free ortho environment, producing the single observed peak; and similarly for the meta position. Fast exchange of pyridine also provides a mechanism by which the chelate rings are interchanged, resulting in averaging of the chelate protons [25]. In solutions in Freon, (CCl_2F_2) at −125°C, the intermolecular exchange was slowed sufficiently to reveal separate resonance signals for both $Eu(C_{11}H_{19}O_2)_3 \cdot 2py$ and free pyridine. Integration of the resonance signals confirmed the existence of the bis adduct [26].

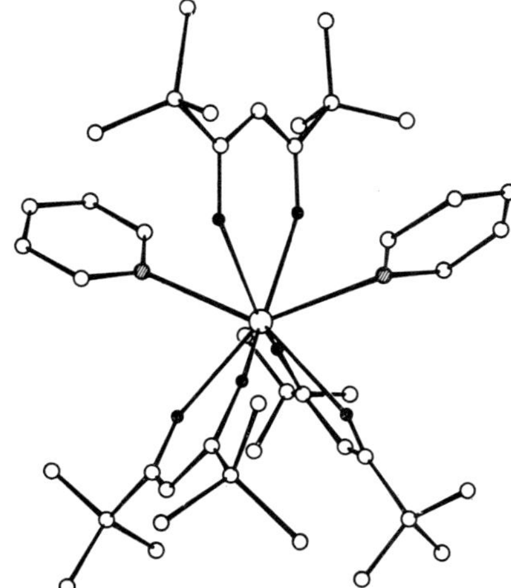

Fig. 5-13

Molecular structure of $Eu(C_{11}H_{19}O_2)_3 \cdot 2py$.

Table 5/24

Chemical Shifts, Isotropic Shifts, Geometric Factors, and Calculated Shifts for $Eu(C_{11}H_{19}O_2)_3 \cdot 2py$ at 30°C [25].

proton	geometric factors [a)]		chemical [b)] shift in ppm	$(\Delta\nu/\nu_0)_{obs}$ in ppm	$(\Delta\nu/\nu_0)_{calc}^{c)}$ in ppm			$(\Delta\nu/\nu_0)_{calc}^{d)}$ in ppm	$(\Delta\nu/\nu_0)_{calc}^{e)}$ in ppm
	$\frac{3\cos^2\Theta-1}{r^3}$	$\frac{\sin^2\Theta\cdot\cos 2\Omega}{r^3}$			$D_1(3\cos^2\Theta-1)\cdot r^{-3}$	$D_2(\sin^2\Theta\cdot\cos 2\Omega)\cdot r^{-3}$	total shift		
methine	0.0090	−0.0028	3.24	8.77	4.99	3.70	8.69	9.15	7.27
tert-butyl	0.0030	−0.0005	2.08	3.08	1.66	0.63	2.29	3.03	3.16
para	−0.0019	+0.0028	−12.65	−5.07	−1.05	−3.67	−4.72	−1.91	−5.43
meta	−0.0032	+0.0030	−12.71	−5.47	−1.83	−3.89	−5.72	−3.28	−6.35
ortho	−0.0167	+0.0048	−24.05	−15.48	−9.26	−6.39	−15.65	−17.08	−15.89

a) The z magnetic axis was assumed to be coincident with the twofold axis of the complex, while the best fit of the data was obtained with the x axis rotated 10° from the N-N vector toward the c crystallographic axis. Values of Θ, Ω, and r were obtained from the X-ray crystallographic results.— b) Measured with respect to internal TMS in carbon disulfide solution.— c) Calculated using equation 1, with $D_1 = 555 \pm 49$ and $D_2 = -1327 \pm 188$ VVk · mol^{-2}; R = 0.047.— d) Calculated using equation 2, with $D_1 = 1022 \pm 98$ VVk · mol^{-1} and the magnetic axis taken along the C_2 axis; R = 0.214.— e) Calculated using equation 2, with $D_1 = -1141 \pm 48$ VVk · mol^{-1} and the magnetic axis taken along the Eu-N line; R = 0.094.

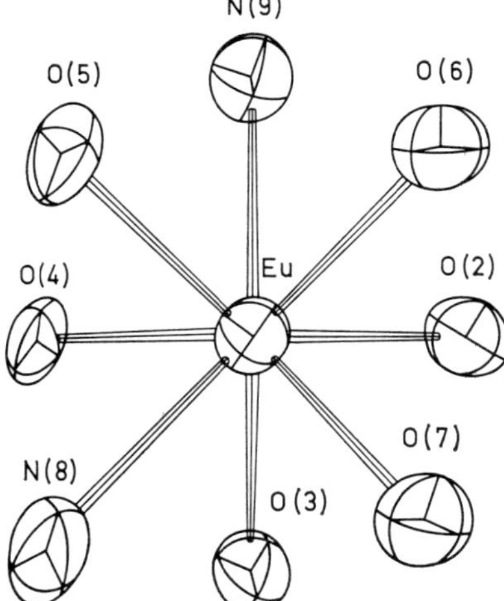

Fig. 5-14

The europium(III) coordination sphere in $Eu(C_{11}H_{19}O_2)_3 \cdot 2py$ viewed along a line approximately normal to the square faces of the antiprism. Ellipsoids of 50% probability are shown.

With 3-Methylpyridine 3-H_3C-C_5H_4N (= β-Picoline = C_6H_7N)

$M(C_{11}H_{19}O_2)_3 \cdot nC_6H_7N$ (n = 1, 2). Crystals of $Lu(C_{11}H_{19}O_2)_3 \cdot C_6H_7N$ were grown from a solution of 3-methylpyridine and also obtained by adding an excess of 3-picoline to a solution of the chelate in n-hexane. The crystal and molecular structures of the compound were determined by single-crystal X-ray analysis (R = 0.079 for 6385 intensities). The crystal system is monoclinic, space group $P2_1/c$-C^5_{2h} (No. 14) with four molecules in a unit cell of dimensions a = 17.404, b = 10.118, c = 24.268 Å, and β = 96.31°. The lutetium ion is seven coordinate (**Fig. 5-15**, p. 144) with the six oxygen atoms of the three diketonate groups and the nitrogen atom of 3-picoline occupying vertices of a distorted, capped trigonal prism. (The nitrogen atom is located at a vertex of the trigonal prism, with an oxygen atom capping one of the rectangular faces.) The Lu-N bond length, 2.492 Å, is significantly longer than the average Lu-O bond distance of 2.238 Å (range = 2.218 to 2.252 Å), accounting for the distortion observed in the coordination polyhedron. The average chelate ring bite (O-O distance) is 2.74 Å. The five atoms O-C-C-C-O in each chelate ring are coplanar, with the lutetium ion lying out of these mean planes by 0.236, 0.470, and 0.827 Å. There is no intermolecular hydrogen bonding in the crystal [27].

The NMR spectrum of a carbon disulfide solution at −110°C containing $Eu(C_{11}H_{19}O_2)_3$ and excess 3-picoline revealed separate resonance signals (slow exchange region) corresponding to free picoline and an adduct of the europium chelate. Comparison of the integrated areas of the chelate peaks and the coordinated 3-picoline peaks established that the species present in solution was the 1:2 adduct, $Eu(C_{11}H_{19}O_2)_3 \cdot 2C_6H_7N$ [25, 28]. The observed and calculated isotropic shifts for all proton resonances of a series of adducts (M = Pr, Sm, Eu, Dy, Ho, Er, Yb) in carbon disulfide are recorded in Table 5/25. Cramer and coworkers [25, 28, 29] observed that the isotropic shifts induced in the ligated picoline groups by the paramagnetic rare earth ions are best accounted for by the general equation, $\Delta\nu/\nu_0 = D_1(3\cos^2\Theta - 1) \cdot r^{-3} + D_2(\sin^2\Theta \cdot \cos 2\Omega) \cdot r^{-3}$ (terms are defined on p. 139). The values of the geometric factors were obtained from X-ray crystallographic results for the analogous $Eu(C_{11}H_{19}O_2)_3 \cdot 2py$ adduct (see p. 138). The values for the magnetic anisotropy parameters D_1 and D_2, which are recorded in Table 5/25, and the location of the x-magnetic axis, were obtained by a least squares fit of the observed data. The z-magnetic axis was assumed to be coincident with the twofold axis based on the structure of the complex $Eu(C_{11}H_{19}O_2)_3 \cdot 2py$. Thermodynamic parameters for the conformational equilibrium from slow-exchange NMR spectra are recorded in [48]. Solvation is important in the equilibrium in CS_2 but less so in CCl_2F_2.

Table 5/25

Observed and Calculated Isotropic Shifts, Magnetic Anisotropies, and Magnetic x-Axis Orientation for the $M(C_{11}H_{19}O_2)_3 \cdot 2B$ Adducts with 3-Methylpyridine and 3,5-Dimethylpyridine.

M	ortho		meta		para		methyl		tert-butyl	
	observed and calculated H isotropic shifts in ppm at 30°C									
	obs	calc	obs	calc	obs	calc	obs	calc	obs	calc
3-methylpyridine[c]										
Pr	23.97	24.16	8.48	8.59	6.74	7.03	5.06	4.47	−3.53	−3.73
Sm	2.35	2.36	0.85	0.81	0.54	0.66	0.47	0.41	−0.35	−0.35
Eu[d]	−62.0	−62.0	−24.5	−25.9	−22.7	−21.4	−16.2	−14.2	10.9	10.4
	−80.0	−79.5								
Dy	218.09	216.65	76.13	75.95	59.54	61.79	44.09	42.03	−23.88	−32.35
Ho	103.04	102.13	37.01	37.17	29.63	30.64	20.32	19.34	−9.74	−14.78
Er	−50.78	−50.64	−17.49	−18.31	−15.73	−15.06	−10.55	−9.35	6.84	7.31
Yb	−59.23	−58.65	−21.53	−22.06	−18.16	−18.36	−12.35	−11.99	4.74	8.50
3,5-dimethylpyridine[f]										
Pr	29.95	29.96			8.64	8.94	6.23	5.85	−3.71	−4.39
Sm	2.63	2.64			0.73	0.75	0.57	0.48	−0.38	−0.39
Eu	−15.3	−15.4			−5.08	−5.08	−3.48	−3.21	2.77	2.19
Dy	209.30	208.20			60.96	62.54	42.89	41.32	−24.12	−30.53
Ho	111.77	110.72			34.55	36.03	24.26	22.81	−9.71	−15.76
Er	−42.19	−42.41			−12.91	−13.00	−8.93	−7.67	6.52	6.00
Yb	−44.34	−44.03			−13.95	−14.24	−9.48	−8.72	4.80	6.21

[a] η is the angle between the crystallographic c-axis and the x-magnetic axis. The c-crystallographic axis is located 34° from the N-N bond vector in $Eu(C_{11}H_{19}O_2)_3 \cdot 2py$ (see p. 138).—[b] Agreement factor for the observed and calculated isotropic shifts.—[c] The diamagnetic references for the 3-picoline protons were taken from the spectrum of free 3-picoline, and for the chelate they were taken as $CH_3 = -116$ Hz and $CH = -575$ Hz with respect to internal TMS, as found for $Lu(C_{11}H_{19}O_2)_3$.

With 4-Methylpyridine $4\text{-}H_3C\text{-}C_5H_4N$ (= γ-Picoline = C_6H_7N)

$M(C_{11}H_{19}O_2)_3 \cdot 2C_6H_7N$. Crystals of the bis adducts (M = Pr, Nd, Sm, Eu, Tb, Dy, Ho, Er, Tm, Yb) were obtained by slow evaporation of 4-picoline solutions of the tris(dipivaloylmethanato) chelates [30 to 32]. X-ray analysis revealed that these series of compounds are isomorphous [30, 31].

The crystal and molecular structures of the holmium compound were determined by single-crystal X-ray analysis (R = 0.077 for 1121 reflections). The adduct crystallizes in the orthorhombic system, space group Pbcn-D_{2h}^{14} (No. 60), with four molecules in a unit cell of dimensions a = 10.260, b = 23.08, and c = 20.23 Å. The holmium ion is eight-coordinate, with six oxygen atoms of the diketonate groups and two nitrogen atoms of the 4-methylpyridine groups located at the vertices of a square antiprism (**Fig. 5-16**, p. 144). The picoline ligands are situated at apices of opposite square faces as far removed from one another as possible. The square edges (s) of the antiprism average 2.74 Å (range = 2.68 to 2.82 Å), while the lateral edges (l) average 2.96 Å (range = 2.71 [constrained by the chelate bite] to 3.14 Å), yielding an l/s ratio of 1.08, which compares well with the most favorable value of 1.06. The Ho-O bond distances range from 2.24 to 2.30 Å and average 2.27 Å, while the Ho-N bond length is 2.65 Å [32].

M	methine		D_1 in VVk · mol^{-1}	D_2 in VVk · mol^{-1}	η°[a]	R[b]	Ref.
	obs	calc					
Pr	−17.88	−17.65	−1189± 55	2937± 167	8	0.023	[29]
Sm	−2.04	−2.02	−141± 24	375± 31	2	0.032	[29]
Eu[d]	34.8	36.5	2305± 16	−5757± 239	29	0.030	[25]
Dy	e)	e)	−7696±417	16143±1914	28	0.037	[29]
Ho	e)	e)	−4724± 95	12022±1623	10	0.046	[29]
Er	39.14	39.39	2615± 82	−7021± 274	6	0.024	[29]
Yb	30.08	29.77	1790±257	−5013± 837	30	0.053	[29]
Pr	−17.85	−17.63	−1145±48	2657± 174	20	0.023	[29]
Sm	−2.00	−2.00	−138± 21	340± 27	6	0.028	[29]
Eu	8.37	8.33	473± 45	−1460± 160	26	0.035	[26]
Dy	e)	e)	−7174±387	17163±1835	26	0.031	[29]
Ho	e)	e)	−3582±341	10572±1768	24	0.054	[29]
Er	38.43	38.49	2525± 72	−7845± 145	2	0.023	[29]
Yb	30.35	30.59	1861±110	−5426± 452	12	0.030	[29]

Calculated isotropic shifts from $\Delta\nu/\nu_0 = D_1\ (\cos^3\Theta - 1) \cdot r^{-3} + D_2\ (\sin^2\Theta \cdot \cos 2\Omega) \cdot r^{-3}$.—[d] Spectrum obtained in slow exchange region at −105°C.—[e] Not observed.—[f] Experimental isotropic shift obtained using $La(C_{11}H_{19}O_2)_3 \cdot 2\,C_7H_9N$ as the diamagnetic reference, measured relative to an internal standard of TMS. Calculated isotropic shifts from $\Delta\nu/\nu_0 = D_1\ (\cos^2\Theta - 1) \cdot r^{-3} + D_2\ (\sin^2\Theta \cdot \cos 2\Omega) \cdot r^{-3}$ (terms defined on p. 139).

The complex has C_2 crystallographic symmetry, with the twofold axis passing through the holmium atom and bisecting one of the chelate rings. This symmetry operation takes the picoline and other chelate ligand into their symmetry related partners. The holmium atom lies in the plane of the chelate ring bisected by the C_2 axis, while the other diketonate ligands are folded about the O-O internuclear axis, such that the plane of the chelate ring makes an angle of 12.0° with the plane defined by the holmium ion and two oxygen atoms [32].

The principal crystal magnetic susceptibilities, χ_a, χ_b, χ_c, recorded in Table 5/26, were measured at 298 K for single crystals of the adducts using a critical torque method [3]. The principal molecular susceptibilities calculated for $\eta = 18°$, where η is the angle the molecular z axis makes with the crystal a axis, are also recorded in Table 5/26 [30]. The data indicate that the susceptibility tensors are highly anisotropic and nonaxial. Dipolar nuclear magnetic resonance shifts for the methyl group of 4-picoline were evaluated from the crystal magnetic and structural data, using the equation $\Delta\nu/\nu_0 = D_1\ (3\cos^2\Theta - 1) \cdot r^{-3} + D_2\ (\sin^2\Theta \cdot \cos 2\Omega) \cdot r^{-3}$, where $D_1 = {}^1/_3 N[\chi_z - {}^1/_2(\chi_x + \chi_y)]$, $D_2 = {}^1/_2 N[\chi_x + \chi_y]$, N = Avogadro's number, with the other terms defined on p. 139, for η in the range 2° to 22°. The calculated values were compared with the observed isotropic shifts as depicted in **Fig. 5-17**, p. 145, along with the experimental NMR results. (The value of $\eta = 18°$ produced the best fits between the calculated and observed values for the dipolar shifts of the methyl group for the entire series of adducts). Horrocks and Sipe [30] noted that the dipolar shifts calculated from the solid state data are in satisfactory agreement with the solution results.

Fig. 5-15

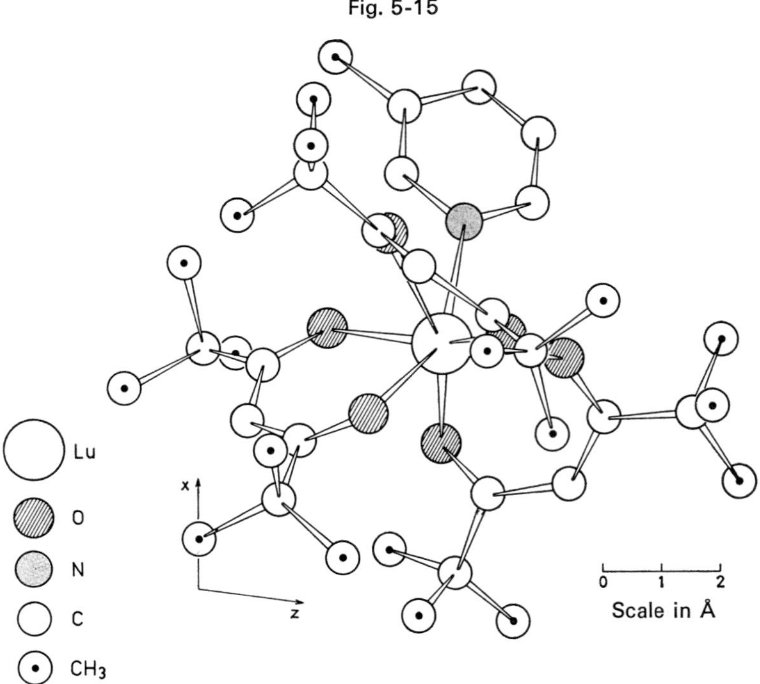

One molecule of $Lu(C_{11}H_{19}O_2)_3 \cdot 3\text{-}H_3C\text{-}C_5H_4N$ projected onto [010].

Fig. 5-16

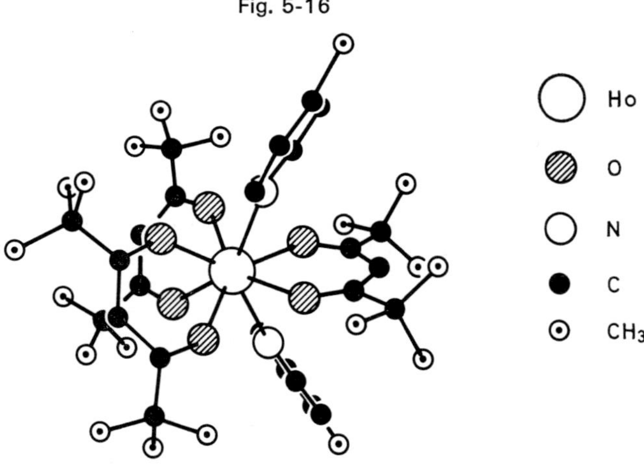

Molecular structure of $Ho(C_{11}H_{19}O_2)_3 \cdot 2(4\text{-}H_3C\text{-}C_5H_4N)$ viewed down the crystallographic c-axis, roughly perpendicular to the square faces of the square antiprism coordination polyhedron. The a-axis is vertical.

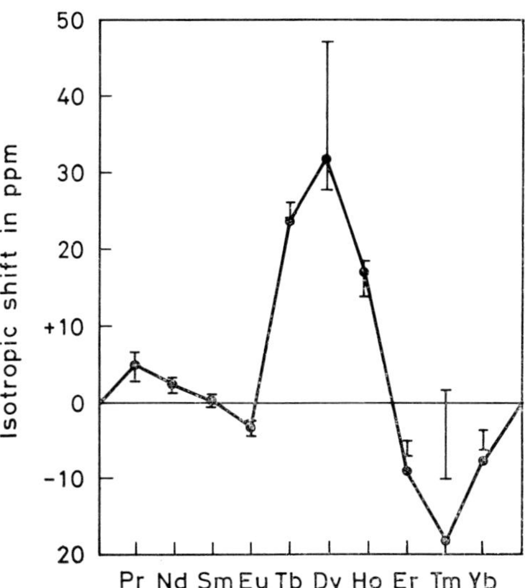

Fig. 5-17

Comparison of the observed isotropic shifts at 298 K for the 4-CH_3 resonance of 4-picoline (points connected by solid line) in the $M(C_{11}H_{19}O_2)_3 \cdot 2(4\text{-}H_3C\text{-}C_5H_4N)$ systems with the dipolar shifts evaluated from single-crystal magnetic anisotropy data for values of the parameter η in the range 2° to 22° (see text). With the exception of the Tb, Er, and Tm systems, for which the opposite is true, the shift of maximum absolute magnitude represents the extreme $\eta = 20°$.

Table 5/26

Principal Crystal and Molecular Magnetic Susceptibilities at 298 K for the $M(C_{11}H_{19}O_2)_3 \cdot 2C_6H_7N$ Adducts with 4-Methylpyridine.

M	principal crystal susceptibilities in VVk · mol^{-1} [3]			principal molecular susceptibilities in VVk · mol^{-1} *) $\eta = 18°$ **) [30]		
	χ_a	χ_b	χ_c	χ_x	χ_y	χ_z
Pr	2779	6315	4865	5111	6315	2533
Nd	3849	5903	4337	4394	5903	3791
Sm	931	1249	973	978	1249	926
Eu	5606	3610	4861	4773	3610	5694
Tb	28268	62209	26589	26390	62210	28470
Dy	33033	65868	45114	46540	65870	31610
Ho	38811	54796	41768	42120	54800	38460
Er	35984	27572	36801	36900	27570	35890
Tm	22688	9072	30051	30920	9072	21820
Yb	9102	3563	8093	7974	3563	9221

*) $\chi_a = \chi_z\cos^2\eta + \chi_x\cos^2\ (90°-n)$; $\chi_b = \chi_y$; $\chi_c = \chi_z\cos^2\ (90°-\eta) + \chi_x\cos^2\eta$. — **) η is the angle between the principal molecular z-axis and the a crystallographic axis. The best fit between calculated dipolar and observed nuclear magnetic resonance shifts was obtained with $\eta = 18°$.

With 3,5-Dimethylpyridine (=ββ'-Lutidine = C_7H_9N)

The NMR spectrum of a Freon solution at about −125°C, containing $Eu(C_{11}H_{19}O_2)_3$ and excess β,β'-lutidine (1:10 mol ratio), revealed separate resonance signals corresponding to coordinated and free lutidine molecules. Comparison of the integrated areas of the chelate peaks and the coordinated lutidine peaks established that the species present in solution was the 1:2 adduct, $Eu(C_{11}H_{19}O_2)_3 \cdot 2C_7H_9N$ [26]. The observed and calculated isotropic shifts for all proton resonances for the series of adducts (M = Pr, Sm, Eu, Dy, Ho, Er, Yb) in carbon disulfide at 30°C are recorded in Table 5/25.

Cramer and coworkers [26, 29] noted that the isotropic shifts induced in the ligated groups by the paramagnetic rare earth ions are best accounted for the general equation, $\Delta\nu/\nu_0 = D_1 \cdot (\cos^2\Theta - 1) \cdot r^{-3} + D_2 \cdot (\sin^2\Theta \cdot \cos 2\Omega) \cdot r^{-3}$ (terms defined on p. 139). The values for the magnetic anisotropy parameters D_1 and D_2, recorded in Table 5/25, and location of the magnetic x-axis were obtained by a least squares fit of the observed data. The magnetic z-axis was assumed to be concident with the twofold axis (based on the structure of $Eu(C_{11}H_{19}O_2)_3 \cdot 2py$) of the complex.

With 1,4-Diazine (= Pyrazine = $C_4H_4N_2$)

$M(C_{11}H_{19}O_2)_3 \cdot C_4H_4N_2$. Each compound (M = Y, La to Yb except for Ce, Pm) was obtained in crystalline form by mixing the tris(dipivaloylmethanato) chelate with pyrazine (1:1 mol ratio) in n-hexane. The adducts can be recrystallized from n-hexane, chloroform, or carbon tetrachloride and exposed to atmospheric moisture without decomposition. The physical properties of the adducts are recorded in Table 5/27. The compounds are thermally stable and do not decompose before melting. They are insoluble in water but soluble in almost all organic solvents. The molar conductance values of millimolar solutions of the adducts in methanol (Table 5/27) are indicative of nonelectrolytes [33]. Changes in the regions of pyrazine absorption in the infrared spectra on adduct formation were attributed to ligation of the pyrazine group.

Table 5/27
Physical Properties of the Mono Pyrazine Adducts of the Tris(dipivaloylmethanato) Rare Earth Chelates, $M(C_{11}H_{19}O_2)_3 \cdot C_4H_4N_2$.

M	color	melting point in °C	magnetic moment in μ_B *)	molar conductance**) in $\Omega^{-1} \cdot cm^2 \cdot mol^{-1}$
Y	white	185	diamagnetic	0.35
La	white	218	diamagnetic	0.54
Pr	light green	217 to 220	3.6	0.47
Nd	violet	220	3.65	0.38
Sm	white	217	2.15	0.34
Eu	white	211 to 214	3.77	0.33
Gd	white	207 to 211	7.51	0.33
Tb	white	200 to 203	10.12	0.24
Dy	white	195 to 198	10.73	0.20
Ho	light yellow	183 to 187	10.73	0.37
Er	pink	168 to 171	7.56	0.21
Tm	white	168 to 171	7.56	0.21
Yb	white	165	4.67	0.15

*) At ambient temperatures. Calculated from magnetic susceptibilities corrected for diamagnetic contributions from the ligand.—**) 10^{-3} M in methanol, temperature not specified.

With 1-Azabicyclo[2,2,2]octane 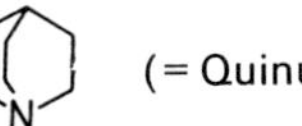 (= Quinuclidine = $C_7H_{13}N$)

$Eu(C_{11}H_{19}O_2)_3 \cdot C_7H_{13}N$. Single crystals of the 1:1 adduct were obtained by slow evaporation of a benzene solution containing equimolar amounts of quinuclidine and $Eu(C_{11}H_{20}O_2)_3$. The crystals are colorless, transparent parallelopipedes.

The crystal and molecular structures of the compound were determined by single-crystal X-ray analysis (R = 0.046 for 4151 reflections). The crystals are monoclinic, space group $P2_1/n$-C_{2h}^5 (No. 14) with four molecules in a unit cell of dimensions a = 10.718, b = 20.417, c = 21.368 Å, β = 106.94°

($D_{osb} = 1.20\ g \cdot cm^{-3}$; $D_{calc} = 1.21\ g \cdot cm^{-3}$). The europium ion is seven-coordinate, surrounded by six chelate-oxygen atoms at an average distance of 2.325 Å (range = 2.316 to 2.337) and the quinuclidine nitrogen atom at a distance of 2.603 Å. The molecular structure is depicted in **Fig. 5-18**. The coordination polyhedron (**Fig. 5-19**) was best described as a distorted octahedron with the quinuclidine nitrogen atom located above the center of one of the faces. Some of the dimensions of the coordination polyhedron are presented in Fig. 5-19. The europium ion is situated 0.51 Å and 1.46 Å from the faces defined by O(2), O(3), O(6) and O(1), O(4), O(5). The compound is axially symmetric, with the threefold axis passing through the europium and nitrogen atoms [34].

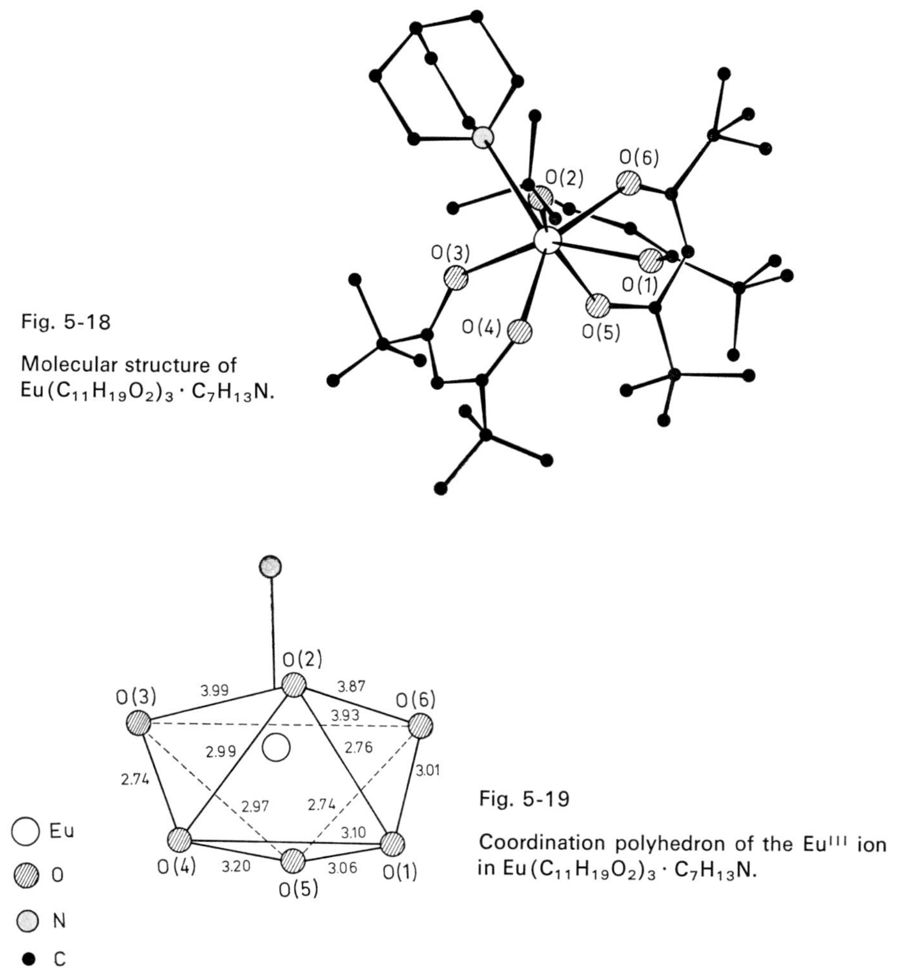

Fig. 5-18

Molecular structure of $Eu(C_{11}H_{19}O_2)_3 \cdot C_7H_{13}N$.

Fig. 5-19

Coordination polyhedron of the Eu^{III} ion in $Eu(C_{11}H_{19}O_2)_3 \cdot C_7H_{13}N$.

With 2,2′-Bipyridine (= $C_{10}H_8N_2$ = bpy) and **1,10-Phenanthroline** (= $C_{12}H_8N_2$ = phen)

$M(C_{11}H_{19}O_2)_3 \cdot$ bpy and **$M(C_{11}H_{19}O_2)_3 \cdot$ phen.** Each series of compounds (M = Pr to Yb except for Pm) was prepared by mixing the tris (dipivaloylmethanato) chelate and the neutral nitrogen donor (1:1 mol ratio) in carbon tetrachloride. Evaporation of the solvent yielded the 1:1 adduct, which was recrystallized three times from n-hexane. The physical properties of the compounds are recorded in Table 5/28.

Enthalpy changes corresponding to the formation of the 1:1 adducts with bipyridine in benzene were determined at 25°C by a calorimetric titration (recorded below). The increase in values of $-\Delta H$ ($kj \cdot mol^{-1}$) observed for adducts of lanthanum through samarium was attributed to increasing electronegativity of the tris chelate resulting from the decrease in metal ion radius. The leveling effect observed for the heavier members of the series was attributed to steric crowding, which opposes the increasing electrostatic effect [35].

M = ...	Y	La	Pr	Nd	Sm	Eu	Gd
$-\Delta H$ . . .	63.7±3.0	29.8±3.0	42.4±3.0	49.2±3.0	63.4±3.0	70.5±3.5	64.9±5.0
M = ...	Tb	Dy	Ho	Er	Tm	Yb	
$-\Delta H$ . . .	66.6±3.0	63.0±3.0	65.5±3.0	63.0±3.0	65.5±3.0	52.9±3.0	

The NMR spectra of the $M(C_{11}H_{19}O_2)_3 \cdot bpy$ and $M(C_{11}H_{19}O_2)_3 \cdot phen$ (M = Eu, Pr, Yb) adducts in carbon tetrachloride or deuterated chloroform were recorded and analyzed in terms of the McConnell-Robertson equation (equation 2, p. 139). The observed isotropic shifts were not in accord with the dipolar shifts predicted from the McConnell-Robertson expression, which assumes axial symmetry. Whereas the spectra of the phenanthroline adducts were rather insensitive to variations in temperature, a conformational process involving biphenyl-type rotational isomerization was deduced from the temperature dependency observed in the spectrum of $Eu(C_{11}H_{19}O_2)_3 \cdot bpy$ [36].

Table 5/28

Physical Properties of $M(C_{11}H_{19}O_2)_3 \cdot bpy$ and $M(C_{11}H_{19}O_2)_3 \cdot phen$.

M	complex with 2,2'-bipyridine			complex with 1,10-phenanthroline		
	color	melting point in °C	magnetic*) moment in μ_B	color	melting point in °C	magnetic*) moment in μ_B
Pr	green	191 to 193	3.6	green	221 to 222	3.6
Nd	violet	189	3.7	violet	222	3.7
Sm	white	189 to 191	2.4	white	225 to 226	2.2
Eu	white	190 to 191	3.6	light yellow	217 to 219	3.6
Gd	white	192 to 193	7.9	white	225 to 228	8.0
Tb	white	195	9.7	greenish white	241 to 242	9.7
Dy	white	193	10.2	white	250	10.0
Ho	light yellow	188 to 190	10.5	light yellow	246 to 248	10.3
Er	pink	183	9.4	pink	254 to 256	9.5
Tm	light yellow	189 to 193	—	light yellow	235 to 239	7.2
Yb	white	167 to 168	4.3	white	—	4.3

*) Magnetic susceptibilities measured by an NMR method at ambient temperature. Solvent employed was a 1:1 (v/v) carbon tetrachloride/chloroform mixture. The data were corrected for diamagnetism of the ligands.

With 1,8-Naphthyridine ($= C_8H_6N_2$)

$M(C_{11}H_{19}O_2)_3 \cdot C_8H_6N_2$. Enthalpy changes $-\Delta H$ in $kj \cdot mol^{-1}$ corresponding to formation of 1:1 naphthyridine-$M(C_{11}H_{19}O_2)_3$ adducts in benzene were determined at 25°C by a calorimetric titration. The data were discussed in terms of two trends which have opposite effects on the enthalpy values, namely, increases in (1) electrostatic attraction and (2) steric crowding, with decreasing ionic radius [35].

M = ...	Y	Sm	Gd	Tb	Ho	Er	Tm	Yb
$-\Delta H$ . . .	39.8±2.5	45.0±3.0	51.0±3.0	52.0±3.0	39.8±3.5	38.1±3.5	37.3±3.0	39.6±2.5

With Phthalazine 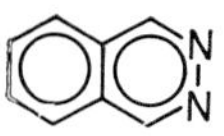 (= $C_8H_6N_2$)

$M(C_{11}H_{19}O_2)_3 \cdot C_8H_6N_2$. The adducts (M = La to Yb except for Ce, Pm) were prepared by mixing the tris(dipivaloylmethanato) rare earth (III) complexes and phthalazine in 1:1 molar ratio in n-hexane. The complexes are quite stable in air. They are insoluble in water but soluble in about all organic solvents. The molar conductance Λ in $\Omega^{-1} \cdot cm^2 \cdot mol^{-1}$ of these complexes at 10^{-3} M dilution in methanol proves their being nonelectrolytes. The IR spectra of the adducts show shifting of all important bands to higher frequencies as compared to their position in the tris(dipivaloylmethanato) rare earth complexes. The important peaks of the noncoordinated phthalazine ligand disappear in the complex. The phthalazine molecule is coordinated to the rare earth metal ions, raising their coordination number from six to seven [37].

M	color	m.p. in °C	μ_{eff} in B.M.	Λ
La	white	132	diamag.	32.23
Pr	dirty white	133	3.59	26.89
Nd	violet	125 to 128	3.61	27.64
Sm	white	125	1.85	22.41
Eu	white	200	3.76	22.41
Gd	white	115 to 118	8.11	22.41
Tb	white	112 to 115	9.87	17.67
Dy	white	120	10.3	14.11
Ho	light yellow	125	10.73	13.87
Er	pink	116 to 120	9.58	13.87
Tm	white	110 to 115	7.59	13.87
Yb	white	120 to 124	4.68	11.45

With Dimethylformamide (= C_3H_7NO = DMF)

$M(C_{11}H_{19}O_2)_3 \cdot DMF$. The compounds (M = Nd to Lu except for Pm) were obtained by recrystallizing the respective tris(dipivaloylmethanato) chelate from hot dimethylformamide [38, 39]. According to the procedure described in [39], the excess dimethylformamide was removed using a water aspirator to pull air through a sintered funnel containing the solid for 4 to 5 h, followed by exposure to air overnight. The melting points (°C) of the compounds thus obtained were: Er, 153 to 154; Sm, 146.5 to 147.5; Ho, 151.5 to 154.5. The melting point of the lanthanum compound (range 118 to 235°C) depended on the length of time it was exposed to air [39]. According to the procedure described in [38], the adducts were suction-dried for several hours, then placed in a vacuum desiccator for at least 12 h [38]. The melting points were tabulated in [38] without clearly identifying the particular adduct to which they belong, but presuming the compounds are listed in order of increasing atomic weight of the rare earth, the values (°C) are: Nd, 134.5; Sm, 143.2; Eu, 145.4, 146.5; Gd, 144, 146.1; Tb, 146.9, 148.8; Dy, 149.8, 150.1; Ho, 150.2; Er, 152.3; Lu, 147.6, 147.9 (multiple entries are data from different sample lots) [38]. Thermogravimetric analyses of the adducts revealed a loss in weight, commencing at 115 to 120°C and ending at 150 to 155°C, prior to the temperature at which sublimation of the tris chelates commenced [39].

$M(C_{11}H_{19}O_2)_3 \cdot 2DMF$. The lanthanum compound was prepared by recrystallizing $La(C_{11}H_{19}O_2)_3$ from hot dimethylformamide, followed by removal of excess dimethylformamide by passing a stream of dry nitrogen over the sample for 72 h. Large colorless crystals of the europium compound were obtained by saturating a dimethylformamide solution with $Eu(C_{11}H_{19}O_2)_3$, and allowing the solvent to evaporate slowly at ambient laboratory temperatures over a period of 1 week. These crystals rapidly decomposed with loss of dimethylformamide if they were not kept in intimate contact with the mother liquor [40, 42].

The crystal and molecular structures of the europium adduct were determined by single-crystal X-ray analysis (R = 0.0621 for 12780 reflections). Four discrete molecules of the adduct crystallized in a triclinic unit cell, space group $P\bar{1}$-C_i^1 (No. 2) with dimensions a = 14.527, b = 18.565, c = 17.583 Å, α = 89.60°, β = 89.03°, γ = 87.01° (D_{obs} = 1.187 g/cm³; D_{calc} = 1.198 g/cm³). Each asymmetric unit within the unit cell consists of two discrete nonsymmetry related molecules of the complex, designated the α and β isomers. The molecular structures of the two isomers are depicted in **Fig. 5-20**, p. 150. The two crystallographically independent molecules have the same overall composition and gross stereochemistry but differ significantly in detail. In both isomers, the europium ion is eight-coordinate, surrounded by the six oxygen atoms of the chelate groups and two oxygen atoms of the dimethylformamide groups. The coordination polyhedra of both isomers were described as distorted square antiprisms. **Fig. 5-21** depicts a schematic representation of the square antiprism showing placement of the monodentate and chelating groups. In each isomer, the dimethylformamide groups are cis on the same square face, and the orientations of the bidentate ligands are such that no edges joining the square faces of the antiprism are spanned. Inspection of the distances and angles (tabulated in [40]) observed in the coordination polyhedra of the α and β isomers, revealed significant differences, which were attributed to crystal packing forces acting upon the sterochemically nonrigid complex. The largest difference, 0.19 Å, is between the two similar edges defined by the lines joining O_{A1} and O_{C2}. Differences of 0.16, 0.11, and 0.11 Å exist between the pairs of similar edges defined by O_{DMF2}-O_{B2}, O_{DMF2}-O_{C1}, and O_{A1}-O_{B1}, respectively. The average lengths of the lateral (l) and square (s) edges for the antiprism of the α-isomer are 1.266 and 1.186 Å, respectively, yielding an l/s value of 1.068, whereas the corresponding l and s values for the β-isomer are 1.275 and 1.182 Å, respectively, yielding an l/s ratio of 1.079 (ideal value = 1.06 Å). The M-O (diketonate) bond distance, averaged over both the α and β isomers, is 2.361 Å (range [α-isomer] = 2.336 to 2.413 Å; range [β-isomer] = 2.326 to 2.371 Å). The average Eu-O(DMF) bond distance of 2.471 Å (2.494 and 2.442 Å for the α-isomer; 2.496 and 2.451 Å for the β-isomer) is significantly longer than the average bond between a europium ion and an oxygen atom of the diketonate ligand [40].

Fig. 5-20

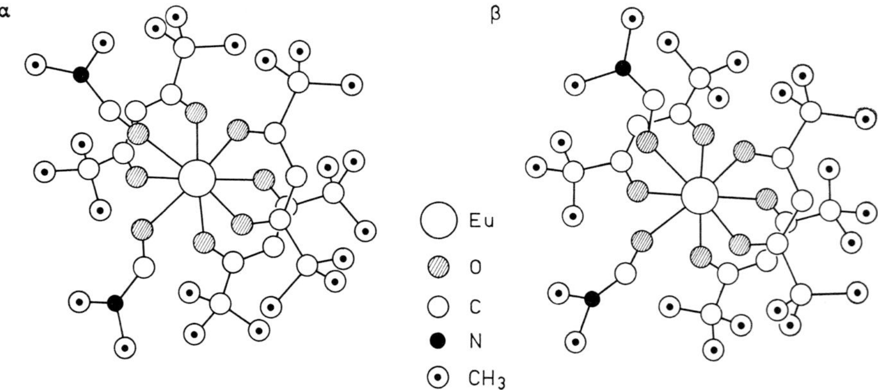

Structures of the two isomers of $Eu(C_{11}H_{19}O_2)_3 \cdot 2\,DMF$.

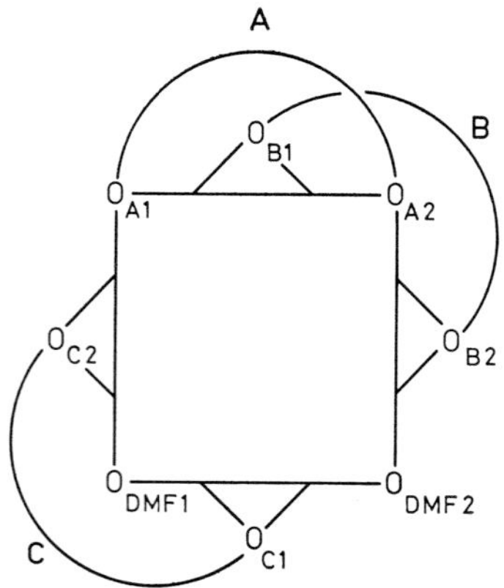

Fig. 5-21

Schematic representation of a square antiprism, showing the ligand connectivity exhibited in crystalline $Eu(C_{11}H_{19}O_2)_3 \cdot 2\,DMF$.

With Dimethyl Sulfoxide (= C_2H_6OS = DMSO)

$Eu(C_{11}H_{19}O_2)_3 \cdot DMSO$. Crystals of the compound were obtained by slowly evaporating at ambient laboratory temperature, a solution of $Eu(C_{11}H_{19}O_2)_3$ in dimethyl sulfoxide. The compound is unstable and loses the dimethyl sulfoxide molecule upon prolonged exposure to ambient laboratory conditions [42].

X-ray analysis revealed that the compound crystallized in the centro-symmetric triclinic unit cell, space group $P\bar{1}$-C_i^1 (No. 2) with four discrete molecules (two nonequivalent pairs) in a unit cell of dimensions a = 21.320, b = 13.291, c = 16.545 Å, cos α = −0.0715, cos β = 0.1926, cos γ = −0.3543 (D_{calc} = 1.204 g/cm³; D_{obs} = 1.206 g/cm³). The europium ion in each crystallographically distinct molecule is seven-coordinate, surrounded by six oxygen atoms of the diketonate groups and the oxygen atom of dimethyl sulfoxide. The coordination polyhedron was described in terms of an idealized trigonal based/tetragonal based geometry that is derived from the square antiprismatic geometry displayed by $Eu(C_{11}H_{19}O_2)_3 \cdot 2DMF$ (Fig. 5-20, p. 150) by removing one unidentate ligand and positioning the remaining unidentate ligand midway along the edge [42].

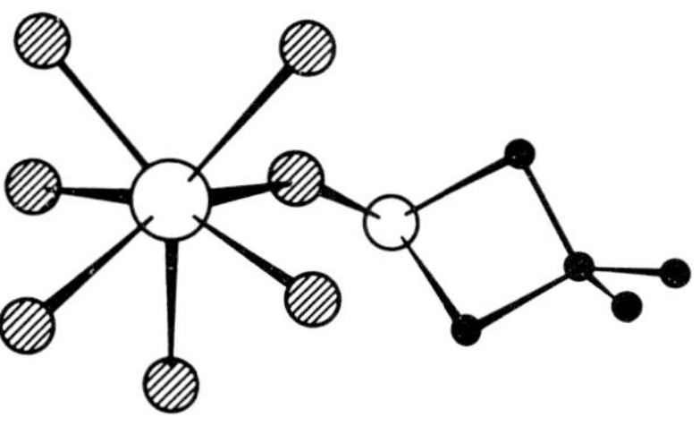

Fig. 5-22

The europium(III) coordination sphere in $Eu(C_{11}H_{19}O_2)_3 \cdot DMSO$. The chelate rings which bridge oxygens A-A, B-B, and C-C have been ommitted for clarity.

With 3,3-Dimethylthietane 1-Oxide $(H_3C)_2C(CH_2)_2SO$ (= $C_5H_{10}OS$)

$Eu(C_{11}H_{19}O_2)_3 \cdot C_5H_{10}OS$. The complex was obtained by recrystallizing $Eu(C_{11}H_{19}O_2)_3$ from an acetonitrile solution containing excess 3,3-dimethylthietane 1-oxide. The compound is an unstable, white solid, which melted at 141 to 142°C [43].

The crystal and molecular structures of the adduct were determined by single-crystal X-ray analysis (R = 0.052 for 2890 reflections). The crystals are monoclinic, space group $P2_1n$-C_{2h}^5 (No. 14), with four molecules in a unit cell of dimensions a = 14.412, b = 20.23, c = 15.660 Å, and β = 98.55° (D_{calc} = 1.21 g/cm³; D_{obs} = 1.19 g/cm³). The europium ion is seven-coordinate, surrounded by six oxygen atoms of the diketonate groups and the oxygen atom of the sulfoxide group (**Fig. 5-22**). The coordination polyhedron was described as a wedged octahedron with the sulfoxide oxygen atom occupying one of the four positions in the equatorial plane. The Eu-O (diketonate) bonds range from 2.32 to 2.37 Å, while the Eu-O (sulfoxide) distance is 2.40 Å [43].

With Tributyl Phosphate $(C_4H_9O)_3PO$ (= $C_{12}H_{27}O_4P$)

$M(C_{11}H_{19}O_2)_3 \cdot (C_4H_9O)_3PO$. Each adduct (M = Nd, Er) was obtained by dissolving equimolar amounts of the tris (dipivaloylmethanato) chelate and tributylphosphate in a minimum volume of n-hexane and then allowing the solvent to evaporate. The product was a vitreous mass with a color characteristic of the particular rare earth ion. The ν(P=O) stretching mode, observed at 1295 cm^{-1} in the spectrum of free tributylphosphate, shifted 60 to 65 cm^{-1} to lower wave numbers on adduct formation (samples of adducts dispersed in liquid paraffin and hexachlorobutadiene mulls), indicating ligation of the phosphoryl oxygen atom [44].

References see p. 152/3

With Triphenylphosphine Oxide $(C_6H_5)_3PO$ (= $C_{18}H_{15}OP$)

$M(C_{11}H_{19}O_2)_3 \cdot (C_6H_5)_3PO$. Each adduct (M = Sm, Eu, Tb) was obtained by heating a mixture of $M(C_{11}H_{19}O_2)_3$ and triphenylphosphine oxide under reflux, while stirring, in a minimum volume of light petroleum ether (b. p. 60 to 80°C) until dissolved. The product crystallized on allowing the reaction mixture to cool. The colors and melting points (°C), under vacuum, of the compounds thus prepared were: Sm, cream, 261 to 269; Eu, colorless with pink fluorescence, 266 to 274; Tb, pale green, 255 to 270 [21]. Crystallization of the praseodymium compound is reported in [49].

The crystal and molecular structures of the praseodymium compound were determined by single-crystal X-ray analysis (R = 0.09). The crystals are monoclinic, in space group $P2_1/n\text{-}C_{2h}^5$ (No. 14), with four molecules in a unit cell of dimensions a = 11.551, b = 22.201, c = 21.786 Å, and γ = 104.92°. The praseodymium ion is seven-coordinate, surrounded by the six diketonate oxygen atoms and the oxygen atom of triphenylphosphine oxide. The Pr-O (diketonate) bond distances range from 2.30 to 2.39 Å, while the Pr-O (phosphine oxide) distance is 2.35 Å [49].

The emission spectrum of the europium adduct (Nujol mull) at −196°C, depicted in [50], revealed 3 lines in the region of the $^5D_0 \to {}^7F_2$ transition, with the short wavelength component showing a further splitting of approximately 2 Å. The $^5D_0 \to {}^7F_0$ transition was observed as a single line and the $^5D_0 \to {}^7F_1$ as two lines. These observations are consistent with an assignment of C_3 or C_{3v} symmetry at the europium(III) site. The spectrum of the europium adduct in a 5:5:1 (v/v) $CH_2Cl_2\text{-}CHCl_3\text{-}CCl_4$ mixture at −196°C (glass) was somewhat broadened relative to that observed in the solid state, but did not differ from the solid state spectrum in any essential features [50].

With Trioctylphosphine Oxide $(C_8H_{17})_3PO$ (= $C_{24}H_{51}OP$)

$Eu(C_{11}H_{19}O_2)_3 \cdot 2(C_8H_{17})_3OP$. The compound was obtained by solvent extraction according to the following procedure. An aqueous solution of europium nitrate was equilibrated with an ether solution of dipivaloylmethane and trioctylphosphine oxide. The mole ratio of chelating agent to trioctylphosphine oxide in the ether solution was fixed at 3:2, and the volumes of solutions used were adjusted to give a mole ratio of 1:3:2 for metal ion : dipivaloylmethane : trioctylphosphine oxide. The complex was formed in almost quantitative yield in a short time, and collected in the ether phase, from which it was isolated. The compound melted at 5°C [51].

References to 5.1.9.3.3:

[1] C. C. Hinckley (J. Am. Chem. Soc. **91** [1969] 5160/2). — [2] R. E. Sievers (Nuclear Magnetic Resonance Shift Reagents, Academic Press, New York 1973). — [3] W. D. W. Horrocks (Lanthanide Shift Reagents and Other Analytical Applications, in: G. N. La Mar, W. D. W. Horrocks, R. H. Holm, Nuclear Magnetic Resonance of Paramagnetic Molecules, Principles and Applications, Academic Press, New York 1973, p. 479/519). — [4] B. C. Mayo (Chem. Soc. Rev. **2** [1973] 49/74). — [5] R. von Ammon, R. D. Fischer (Angew. Chem. Intern. Ed. Engl. **11** [1972] 675/92).

[6] S. P. Sinha (J. Mol. Struct. **19** [1973] 387/401). — [7] J. Grandjean (Ind. Chim. Belge [2] **37** [1972] 220/32). — [8] M. Holik (Chem. Listy **66** [1972] 449/57 from C.A. **77** [1972] No. 54194). — [9] G. Giacometti (Relaz. Corso Teor. Prat. Risonanza Magn. Nucl. **1973** 207/29 from C.A. **80** [1974] No. 53963). — [10] B. Danieli, G. Palmisano (Relaz. Corso Teor. Prat. Risonanza Magn. Nucl. **1973** 231/90 from C.A. **80** [1974] No. 53964).

[11] J. Reuben (Progr. Nucl. Magn. Resonance Spectrosc. **9** [1973] 1/70). — [12] A. F. Cockerill, G. L. O. Davies, R. C. Harden, D. M. Rackham (Chem. Rev. **73** [1973] 553/88). — [13] Y. Sasaki, H. Fujiwara, H. Kawaki, Y. Okazaki (Chem. Pharm. Bull. [Tokyo] **26** [1978] 1066/70). — [14] Y. Sasaki, H. Fujiwara, H. Kawaki, Y. Okazaki (Chem. Pharm. Bull. [Tokyo] **25** [1977] 3177/81). — [15] H. G. Brittain (Inorg. Chem. **19** [1980] 640/3).

[16] A. H. Lewin, R. Alekel (J. Am. Chem. Soc. **98** [1976] 6919/22). — [17] J. Bouquant, J. Chuche (Bull. Soc. Chim. France **1977** 959/66). — [18] I. Armitage, G. Dunsmore, L. D. Hall, A. G. Marshall (Can. J. Chem. **50** [1972] 2119/29). — [19] K. Volka, M. Suchanek, J. Karhan, M. Hajek (J. Mol. Struct. **46** [1978] 329/38). — [20] G. A. Catton, F. A. Hart, G. P. Moss (J. Chem. Soc. Dalton Trans. **1976** 208/10).

[21] H. Huber, J. Sellig (Helv. Chim. Acta **55** [1972] 135/8). — [22] J. Selbin, N. Ahmad, N. Bhacca (Inorg. Chem. **10** [1971] 1383/7). — [23] R. E. Cramer, K. Seff (J. Chem. Soc. Chem. Commun. **1972** 400/1). — [24] R. E. Cramer, K. Seff (Acta Cryst. B **28** [1972] 3281/93). — [25] R. E. Cramer, R. Dubois, K. Seff (J. Am. Chem. Soc. **96** [1974] 4125/31).

[26] R. E. Cramer, R. Dubois, C. K. Furuike (Inorg. Chem. **14** [1975] 1005/7). — [27] S. J. S. Wasson, D. E. Sands, W. F. Wagner (Inorg. Chem. **12** [1973] 187/90). — [28] R. E. Cramer, R. Dubois (J. Am. Chem. Soc. **95** [1973] 3801/2). — [29] R. E. Cramer, R. B. Maynard (J. Magn. Resonance **31** [1978] 295/300). — [30] W. D. W. Horrocks, J. P. Sipe (Science **177** [1972] 994/6).

[31] W. D. W. Horrocks, J. P. Sipe, D. R. Sudnick (Proc. 10th Rare Earth Res. Conf., Carefree, Ariz., 1973, p. 398/404). — [32] W. D. W. Horrocks, J. P. Sipe, J. R. Luber (J. Am. Chem. Soc. **93** [1971] 5258/60). — [33] M. S. Ansari, N. Ahmad (J. Inorg. Nucl. Chem. **37** [1975] 2099/101). — [34] E. Bye (Acta Chem. Scand. A **28** [1974] 731/9). — [35] D. R. Dakternieks (J. Inorg. Nucl. Chem. **38** [1976] 141/3).

[36] N. S. Bhacca, J. Selbin, J. D. Wander (J. Am. Chem. Soc. **94** [1972] 8719/22). — [37] M. S. Ansari, N. Ahmad (Acta Chim. [Budapest] **91** [1976] 429/32). — [38] V. A. Mode, D. H. Sisson (Inorg. Nucl. Chem. Letters **8** [1972] 357/65). — [39] J. E. Scharberg, D. R. Gere, R. E. Sievers, K. J. Eisentraut (Inorg. Chem. **6** [1967] 1933/5). — [40] J. A. Cunningham, R. E. Sievers (Inorg. Chem. **19** [1980] 595/604).

[41] D. M. Rackham, J. Binks (Spectrosc. Letters **12** [1979] 17/23 from C.A. **91** [1979] No. 38348). — [42] J. A. Cunningham, R. E. Sievers (Proc. 10th Rare Earth Res. Conf., Carefree, Ariz., 1973, p. 334/41). — [43] J. J. Uebel, R. M. Wing (J. Am. Chem. Soc. **94** [1972] 8910/2). — [44] N. G. Dzyubenko, L. I. Martynenko, V. I. Spitsyn (Zh. Neorgan. Khim. **21** [1976] 2276/7; Russ. J. Inorg. Chem. **21** [1976] 1253). — [45] H. Kawaki, Y. Okazaki, H. Fujiwara, Y. Sasaki (Chem. Pharm. Bull. **28** [1980] 871/5 from C.A. **92** [1980] No. 221826).

[46] D. M. Rackham (Spectrosc. Letters **12** [1979] 603/7). — [47] Sh. S. Bikew, R. N. Ikhsanov, Yu. Yu. Samitov (Teor. Eksperim. Khim. **15** [1979] 684/90 from C.A. **92** [1980] No. 100096). — [48] R. E. Cramer, R. B. Maynard, R. Dubois (J. Chem. Soc. Dalton Trans. **1979** 1350/5). — [49] L. A. Aslanov, V. M. Ionov, V. B. Rybakov, E. F. Korytnyi, L. I. Martynenko (Koord. Khim. **4** [1978] 1427/9 from C.A. **89** [1978] No. 207560). — [50] G. A. Catton, F. A. Hart, G. P. Moss (J. Chem. Soc. Dalton Trans. **1975** 221/6).

[51] F. Halverson, J. S. Brinen, J. R. Leto (J. Chem. Phys. **41** [1964] 157/63).

5.1.9.3.4 Tetrakis Chelates

$(C_2H_5)_4N[Ce(C_{11}H_{19}O_2)_4]$. The cerium(III) complex was prepared under an atmosphere of pure nitrogen using deoxygenated solvents according to the following procedure. A solution of cerium(III) chloride heptahydrate (2 mmol) and 2,2-dimethoxypropane ($10\ cm^{-3}$) in absolute ethanol ($15\ cm^3$) was added to a mixture of dipivaloylmethane (8 mmol), sodium hydroxide (8 mmol), and tetraethylammonium chloride (2 mmol) in absolute ethanol ($25\ cm^3$). The solvent was removed under reduced pressure, the residue extracted with ethyl acetate ($30\ cm^3$), and the suspended solid removed by filtration. The solution was concentrated under vacuum to approximately $10\ cm^3$. Precipitation of the product was effected by slowly adding n-hexane. The product was collected, washed with n-hexane, and dried under vacuum. The yellow-colored compound was characterized by elemental analysis and its electronic absorption spectrum in acetonitrile (bands at $25000\ cm^{-1}$ [$\varepsilon = 214\ dm^3 \cdot mol^{-1} \cdot cm^{-1}$] and $27100\ cm^{-1}$ [$\varepsilon = 224\ dm^3 \cdot mol^{-1} \cdot cm^{-1}$]) [1].

$Ce(C_{11}H_{19}O_2)_4$. Selbin, Ahmad, Bhacca [2] obtained the cerium(IV) chelate from a cerium(III) salt according to the procedure reported by Eisentraut and Sievers [3] for preparation of the tris chelates (see p. 125). The compound was purified by recrystallization from n-hexane. It was red in color and melted at 276°C. The compound is diamagnetic, confirming that cerium is present in the tetravalent state [2]. Uhlemann and Dietze [3] effected synthesis of the compound by passing air through an aqueous methanolic solution containing cerium(III) nitrate, dipivaloylmethane, and ammonia. The red-brown solid was recrystallized from xylene (decomposed on heating at 180°C).

The absorption spectrum of the complex in acetonitrile revealed bands at 26900 cm^{-1} ($\varepsilon=7100$ $dm^3 \cdot mol^{-1} \cdot cm^{-1}$), 36400 cm^{-1} ($\varepsilon=41000$ $dm^3 \cdot mol^{-1} \cdot cm^{-1}$), and 47600 cm^{-1} ($\varepsilon=23000$ $dm^3 \cdot mol^{-1} \cdot cm^{-1}$) [1].

References to 5.1.9.3.4:

[1] M. Ciampolini, F. Mani, N. Nardi (J. Chem. Soc. Dalton Trans. **1977** 1325/8). — [2] J. Selbin, N. Ahmad, N. Bhacca (Inorg. Chem. **10** [1971] 1383/7). — [3] K. J. Eisentraut, R. E. Sievers (J. Am. Chem. Soc. **87** [1965] 5254/6). — [4] E. Uhlemann, F. Dietze (Z. Anorg. Allgem. Chem. **386** [1971] 329/34).

5.1.9.3.5 Other Complexes with 2,2,6,6-Tetramethyl-3,5-heptanedione

$Ho(C_{11}H_{19}O_2)_2NO_3 \cdot 2H_2O$ was prepared according to the following procedure. Sodium hydroxide (4.44 mmol) in 17 ml of 50% aqueous ethanol was added to an ethanolic solution (10 ml) of dipivaloylmethane (4.44 mmol), and the mixture was stirred for 20 min in a sealed flask. A solution of holmium nitrate in 25 ml of 50% ethanol was added and the mixture stirred for 1.5 to 2 h. Then half the volume of solvent was removed by distillation under reduced pressure. Water was added, while stirring, to effect precipitation of the product. The precipitate was collected by filtration, dried in a desiccator over phosphorus(V) oxide, and recrystallized from n-hexane. Prismatic, amber yellow crystals of the compound were formed during slow evaporation of the solvent at room temperature. The compound was reportedly somewhat more soluble than the tris chelate in n-hexane, dioxane, and ethanol [1].

$Er_8O(C_{11}H_{19}O_2)_{10}(OH)_{12}$. The basic erbium diketonate was isolated as a minor byproduct in the synthesis of $Er(C_{11}H_{19}O_2)_3$. The crystal and molecular structures of the compound (**Fig. 5-23**)

Fig. 5-23

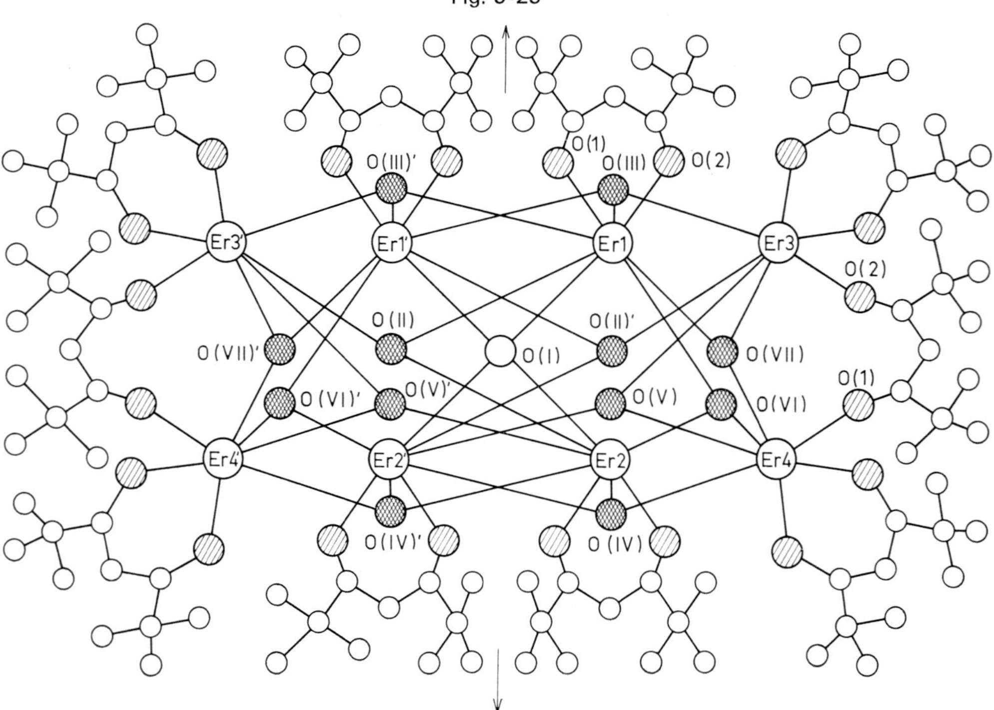

Schematic drawing of the $Er_8O(C_{11}H_{19}O_2)_{10}(OH)_{12}$ molecule, illustrating the atomic numbering. The hydroxyl groups are numbered O(II) to O(VII) and the atoms of the diketonate groups are shown for the chelate ring on Er(1).

were determined by single-crystal X-ray analysis (R = 0.116 for 3103 intensities). The space group is Pbcn-D_{2h}^{14} (No. 60) with four molecules in a unit cell of dimensions a = 18.376, b = 31.625, c = 27.914 Å (calculated density = 1.38 g/cm³, measured density = 1.37 g/cm³). The erbium ions cluster around a central oxygen atom in a dodecahedral configuration. Four erbium ions at an average distance of 2.34 Å (range 2.287 to 2.399 Å) form an inner sphere around the central oxygen atom in a distorted tetrahedral configuration. Each inner-sphere erbium ion is bonded, in addition, to five hydroxo groups and to two oxygen atoms of a bidentate dipivaloylmethanato group and is eight-coordinate. The other four erbium ions in the molecular unit, which formally complete the dodecahedron, form an outer sphere around the central oxygen atom at an average distance of 3.62 Å (range = 3.586 to 3.650 Å). The distance between an outer erbium ion and the central oxygen atom was considered too long to represent coordination, and thus the central oxygen atom is formally four-coordinate. Each outer-sphere erbium ion is bonded to four hydroxo groups and two oxygen atoms of a chelating and one oxygen atom of a bridging dipivaloylmethanato group and is seven-coordinate. The inner and outer spheres of erbium ions are connected through hydroxo bridges. Each hydroxo group is bonded to three erbium ions, two of which belong to either the inner or outer sphere. The coordination geometries about the Er(1) and Er(2) inner-sphere ions were described as a bicapped trigonal prism and a dodecahedron, respectively, whereas the coordination geometry about each of the seven-coordinate outer-sphere ions was described as a monocapped trigonal prism [2].

References to 5.1.9.3.5:

[1] N. G. Dzyubenko, L. I. Martynenko, V. I. Spitsyn (Zh. Neorgan. Khim. **20** [1975] 63/5; Russ. J. Inorg. Nucl. Chem. **20** [1975] 34/6). — [2] J. C. A. Boeyens, J. P. R. DeVilliers (J. Cryst. Mol. Struct. **2** [1972] 197/211).

5.1.10 Complexes with 6,6,7,7,8,8,8-Heptafluoro-2,2-dimethyl-3,5-octanedione $CF_3CF_2CF_2C(O)CH_2C(O)C(CH_3)_3$ (= $C_{10}H_{11}F_7O_2$ = Hfod)

5.1.10.1 $M(fod)_n^{3-n}$ (n = 2, 3). Formation in Solution

Overall formation constants for the 1:2 and 1:3 (metal:ligand) complexes were determined at 24°C in aqueous media (I = 0.1 M $(C_2H_5)_4NClO_4$) by a solvent extraction method [18].

M	Pr	Sm	Eu	Gd	Tb
lg β_2	—	—	11.9	12.7	13.4
lg β_3	18.0	18.4	18.4	18.7	19.4

5.1.10.2 $M(fod)_3 \cdot nH_2O$ (= 0, 0.5, 1)

Each monohydrate (M = Y, La to Lu except for Ce, Pm) was prepared by mixing a solution of the ligand (0.033 mol) in methanol (neutralized with 8.01 ml of 4.12 M sodium hydroxide solution) with a solution of the hydrated rare earth nitrate (0.011 mol) in a minimum amount of absolute methanol. The resulting solution was added dropwise, with vigorous stirring, over a period of 2 h, to approximately 400 ml of distilled water to effect precipitation of the complex. The suspension was stirred until the precipitate was in the form of fine granules. A stirring rod was used during the precipitation to crush the granules and prevent coagulation into a tar. The final methanol concentration was kept very low to minimize the possibility of oiling. The precipitate was collected by filtration and air dried for approximately 1 h. The crude product was recrystallized twice from a minimum amount of methylene chloride (the solution was cooled to < 0°C to effect crystallization), collected, and dried under vacuum at room temperature for approximately 12 h between recrystallizations. The melting points of the compounds thus obtained are recorded in Table 5/29 [1, 2]. Karraker [3], following a procedure similar to that described above, obtained hemihydrates for the lighter rare earths, $M(fod)_3 \cdot 0.5H_2O$ (M = La, Pr, Nd, Eu, Gd, Dy), and monohydrates for the heavier elements (M = Ho, Er, Yb).

All of the monohydrated complexes were dehydrated by storing under vacuum (1 Torr) in a desiccator over phosphorus(V) oxide at room temperature for a few days, yielding the anhydrous chelates [1, 2]. The melting points of the anhydrous chelates are recorded in Table 5/29. The anhydrous scandium chelate was obtained in the form of an oil, directly, according to the procedure described above for the hydrated chelates [1]. A similar preparation of the anhydrous scandium chelate is described in [4]. The anhydrous complexes are hygroscopic, those of the heavier rare earths adding water instantaneously upon exposure to the atmosphere, whereas the lighter metal chelates pick up water more slowly [1].

The crystal and molecular structures of the hydrated praseodymium chelate with empirical composition $Pr(fod)_3 \cdot H_2O$ were determined by single-crystal X-ray analysis ($R = 0.085$ for 3706 intensities). The compound crystallized in the monoclinic space group $P2_1/n\text{-}C_{2h}^5$ (No. 14) with unit cell dimensions $a = 14.280$, $b = 25.589$, $c = 23.182$ Å, and $\beta = 100.62°$ ($\rho_{obs} = 1.65$ g/cm³; $\rho_{calc} = 1.646$ g/cm³ for two crystallographically independent formula units, i.e., $Z = 8$). The two independent formula units were found to constitute a dimer (**Fig. 5-24**) which is formed through bridging across two carbonyl oxygen atoms and one water molecule. The other water molecule was found to be situated between two perfluoro side chains. The oxygen atom of this water molecule, as well as the neighboring $-CF_3$ group, was not well defined in the electron density maps, a fact attributed to rupture of the hydrogen bonds upon X-ray irradiation, leading to considerable disorder. The two praseodymium atoms are 4 Å apart at the apices of a trigonal bipyramid defined by oxygen atoms O(III 2), O(VI 2), and $H_2O(1)$ (Fig. 5-24). Each praseodymium atom is surrounded by eight oxygen atoms at an average distance of 2.42 Å. Neither of the two distinct polyhedra has crystallographic symmetry or corresponds exactly to any of the regular geometries. The polyhedron around Pr(1) was best described as a bicapped trigonal prism (C_{2v} symmetry), whereas that about Pr(2) was described as a dodecahedron [6].

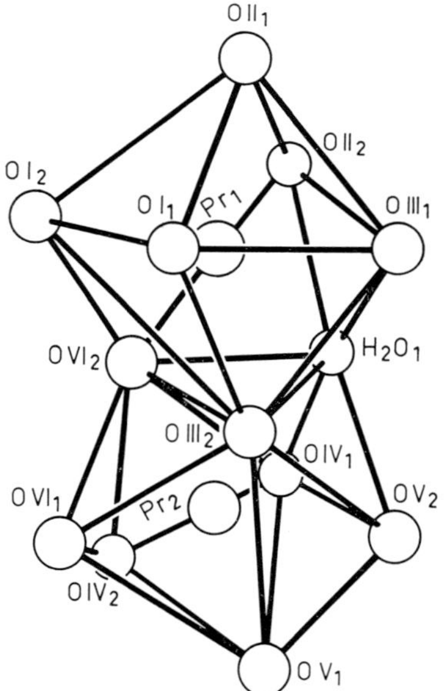

Fig. 5-24

Coordination sphere in the dimeric complex $Pr_2(fod)_6 \cdot 2H_2O$.

Single-crystal X-ray analysis revealed that the monohydrated lutetium complex crystallized in the orthorhombic space group $P\bar{1}\text{-}C_i^1$ (No. 2) with four molecules in a unit cell of dimensions $a = 16.392$, $b = 22.492$, $c = 13.366$ Å, $\alpha = 91.83°$, $\beta = 119.93°$ and $\gamma = 90.00°$ ($R = 0.128$, $\rho_{obs} = 1.66$ g/cm³, $\rho_{calc} = 1.667$ g/cm³). The relatively high R factor was attributed to crystal decomposition, to the prohibitively large asymmetric unit which necessitated refinement in parts, and to the large thermal vibrations in the structure. These factors caused large uncertainties, especially in the positions of the atoms at the ends of the long side chains, but the geometry of the coordination polyhedron, nevertheless, was firmly established. The lutetium ion is seven-coordinate, surrounded by the six oxygen atoms of the three bidentate diketonato groups, and the oxygen atom of the coordinated water molecule at the vertices of a monocapped trigonal prism. The two crystallographically distinct formula units are dimerized through water molecules, with hydrogen bonds across the center of symmetry [7].

Table 5/29
Physical Properties of Tris(heptafluorodimethyloctanedionato) Rare Earth Chelates $M(fod)_3 \cdot nH_2O$ (n = 0, 1).

M	color [1]	m.p. in °C [1] (n = 1)	m.p. in °C [1] (n = 0)	ΔH_{sub} [5] in kcal · mol^{-1}	ΔS_{sub}^{exp} [5] in cal · $mol^{-1} \cdot K^{-1}$	ΔS_{sub}^{est} [5]*) in cal · $mol^{-1} \cdot K^{-1}$
Sc	light yellow		<25			
Y	white	108 to 112	162 to 167			
La	white	215 to 230**)	215 to 230**)	34.7±0.7	57.0±1.72	
Pr	green	218 to 225**)	218 to 225**)			61.6
Nd	blue	210 to 215**)	210 to 215**)	37.1±0.7	64.0±1.68	
Sm	white	63 to 67	208 to 218**)	37.9±0.4	67.9±1.15	
Eu	white***)	59 to 67	205 to 212**)			70.0
Gd	white	60 to 65	203 to 213**)	37.0±0.2	66.8±0.53	
Tb	white***)	92 to 97	190 to 196			67.2
Dy	white	103 to 107	180 to 188	37.4±0.7	67.9±1.82	
Ho	beige	103 to 111	172 to 178			68.2
Er	pink	104 to 112	158 to 164	37.0±1.0	72.3±2.87	
Tm	white	110 to 115	140 to 146			73.8
Yb	white	112 to 115	125 to 132	37.0±0.8	75.2±2.43	
Lu	white	111 to 115	118 to 125			76.7

*) Estimated values obtained by interpolating from a plot of ΔS_{sub} against atomic number. — **) Sample melted with decomposition. — ***) The europium compound fluoresces with an intense red color and the terbium compound with a green color upon irradiation with ultraviolet light.

All of the anhydrous and hydrated chelates are highly soluble in many organic solvents and insoluble in water. The extent of aggregation of the chelates in various organic solvents depends on the degree of hydration of the chelate [8], as well as the nature of the solvent [9]. The association quotients for the formation of dimers and trimers in carbon tetrachloride solutions at 37°C are 140 ± 8 $mol^{-1} \cdot l$ and 45 ± 5 $mol^{-1} \cdot l$, respectively, for the anhydrous praseodymium chelate and 367 ± 22 $mol^{-1} \cdot l$ and 12 ± 2 $mol^{-1} \cdot l$, respectively, for the anhydrous europium chelate [8]. The corresponding values for the hydrated chelates, $M(fod)_3 \cdot 0.5H_2O$, are 314 ± 20 and 15 ± 2 $mol^{-1} \cdot l$ (M = Pr), and 68 ± 5 and 6.9 ± 1 $mol^{-1} \cdot l$ (M = Eu) [8]. The aggregate concentrations of $M(fod)_3$ (M = Eu, Pr) were observed to increase in the order chloroform < carbon tetrachloride < n-hexane (e.g., for M = Eu, lg $\beta_2 = 0.2$, 2.00, and 3.70 for the three solvents, respectively, and lg $\beta_3 = 4.78$ in n-hexane, where $\beta_n = [\text{n-mer}] \cdot [\text{monomer}]^{-n}$) [9].

The physical properties of the various forms of the tris chelates are recorded in Table 5/29. The hydrated compounds are easily transformed into the anhydrous compounds without hydrolysis [1]. The anhydrous compounds are among the most volatile rare earth compounds known, although they are susceptible to thermal decomposition above 200°C [1, 2]. The anhydrous chelates were chromatographed in the vapor phase. A plot of retention time versus the radius of the metal ion revealed a smooth curve, with retention times increasing with the ionic radius. The retention data cannot be correlated with the mass of the complex, as best illustrated by the fact that the yttrium value fits smoothly on the ionic radius curve but is anomalous in either an analogous atomic number or mass curve. Furthermore, thermogravimetric analyses revealed that the chelates of the smaller ions are volatilized at lower temperatures than those of the larger ions [1].

Swain and Karraker [5] observed that the vapor pressures of the $M(fod)_3$ chelates increased with decreasing size of the rare earth ion (a typical pressure is 10^{-4} Torr at 100°C for the gadolinium chelate). Heats of sublimation, ΔH_{sub}, and the entropy of sublimation, ΔS_{sub}, were determined from plots of lg P (P = vapor pressure of the chelate) versus the reciprocal of the absolute temperature. The values thus obtained are recorded in Table 5/29. The ΔH_{sub} values for the $M(fod)_3$ chelates, with the exception of the lanthanum compound, are essentially constant within the accuracy of the measurement. The change in vapor pressure across the rare earth series was therefore attributed entirely to differences in ΔS_{sub}. The ΔS_{sub} values were divided into three sets (La, Nd, Sm; Gd, Dy; and Er, Yb) which coincided with three different crystal structures, as deduced from X-ray diffraction patterns. The entropy differences between two crystal structures were attributed to molecular packing rather than molecular structure [5].

The oscillator strengths (P) of the hypersensitive transitions for the neodymium (${}^4I_{9/2} \rightarrow {}^4G_{5/2}, {}^2G_{7/2}$ in the region 17700 to 16400 cm^{-1}) and erbium (${}^4I_{15/2} \rightarrow {}^2H_{11/2}$ in the region 19600 to 18500 cm^{-1}) chelates, $M(fod)_3 \cdot nH_2O$, were determined in a variety of organic solvents. The oscillator strengths for solutions of anhydrous $Nd(fod)_3$ were considered high relative to values observed for other neodymium diketonato chelates (20 to 40×10^{-6}). Solutions of $Nd(fod)_3 \cdot 0.5H_2O$ showed lower oscillator strengths and appreciably different band shapes relative to the anhydrous chelates in petroleum ether or benzene solutions [3].

M	n	solvent	$P \cdot 10^6$	M	n	solvent	$P \cdot 10^6$
Nd	0	benzene	69	Er	0	petroleum ether	42
Nd	0.5	benzene	61	Er	0	benzene	41
Nd	0	petroleum ether	70	Er	1	petroleum ether	33
				Er	1	benzene	31
Nd	0	100% ethanol	53	Er	0	100% ethanol	26
Nd	0	99% ethanol	53	Er	0	99% ethanol	27

Chemical ionization mass spectra of the $M(fod)_3$ chelates (M = La to Lu except for Ce, Pm) were obtained using isobutane as the reagent gas. The relative peak intensities calculated for each of the $M(fod)_3$ chelates are tabulated in [10] and were discussed in terms of the naturally abundant isotopes of each element present in the complex [10].

5.1.10.3 Lewis Base Adducts of the Tris Chelates

The ability of certain paramagnetic $M(fod)_3$ chelates to interact with a variety of Lewis bases and induce chemical shifts in the proton NMR spectrum of the substrate has been well documented in several extensive review articles [11 to 16]. This section concentrates mainly on the thermodynamic stability of the $M(fod)_3$-substrate interaction in solution and on the kinetics of substrate exchange in adducts of composition $M(fod)_3 \cdot nB$ (n = 1, 2; B = Lewis Base).

The equilibrium constants K_1 and K_2, between a series of ortho- and N-substituted anilines and $Eu(fod)_3$ in carbon tetrachloride, recorded in Table 5/30, were evaluated from NMR chemical shift data using a 4-parameter equation. Hirayama and Ishida [17] noted that the order of magnitude of K_n, corrected for the effect of the pK_a value, corresponded to the accessibility of the nitrogen lone pair to the europium ion in complexation. The order aniline < 2-methyl- < 2-ethyl- < 2-phenyl-substituted aniline holds for steric hindrance [17]. Brittain [27] observed, on the other hand, that steric factors did not play a major role in the adduct formation with a series of primary aliphatic amines (Table 5/31) [27]. An approximate linear correlation was found between the basicity of the substrate and the logarithm of the equilibrium constants for 1:1 adducts of Eu(fod)$_3$ with p-substituted anilines in $CDCl_3$ [29]. The intermolecular exchange kinetics were studied by NMR for systems of $(CH_3)_2SO$, $C_6H_5N(CH_3)_2$, or CH_3CN with Eu(fod)$_3$ at wide temperature ranges and at 40, 60, 80, and 100 MHz [30].

The relative binding abilities of ketones, secondary cyclohexanols, and tertiary cyclohexanols were examined by determining the equilibrium constants corresponding to the interaction of 28 cyclohexane and cyclohexanol substrates with Eu(fod)$_3$ in carbon tetrachloride (Table 5/32). The authors noted that the K_1 values provided the more generally useful and immediately more apparent trends. First, there is a trend related to the degree of substitution on the carbon atom bound to oxygen: K_1 (tertiary alcohols) < K_1 (ketones) < K_1 (secondary alcohols). In addition, for alcohols, the stereochemical disposition of the hydroxyl group (axial or equatorial) has a marked effect: K_1 (axial alcohols) < K_1 (equatorial alcohols). The ordering of K_n values according to compound types is not exceptional, with the tertiary alcohols demonstrating the weakest binding ability [19]. Equilibrium constants for the association of adamantane derivatives with different functional groups to Eu(fod)$_3$ in CCl_4, see [31]. For kinetic investigations of the system Pr(fod)$_3$ · L with L = N,N-dimethylacetamide, see [32]. Stability constants of M(fod)$_3$ · L with M = Eu, Pr, and L = macrocyclic polyethers are given in [33].

Table 5/30

Equilibrium Constants Corresponding to Formation of 1:1 and 2:1 Adducts of Eu(fod)$_3$ with Aniline and Aniline Derivatives in Carbon Tetrachloride at 34°C*) [17].

no.	substrate	formula	K_1	K_2
1	aniline	C_6H_7N	5230	582
2	2-methylaniline	C_7H_9N	1188	109
3	2-ethylaniline	$C_8H_{11}N$	282	65
4	2,3-dimethylaniline	$C_8H_{11}N$	1081	95
5	2,4-dimethylaniline	$C_8H_{11}N$	1101	128
6	2,5-dimethylaniline	$C_8H_{11}N$	740	105
7	2,6-dimethylaniline	$C_8H_{11}N$	48.3	9.3
8	N-methyl-3-methylaniline	$C_8H_{11}N$	11.3	1.6
9	N-ethyl-3-methylaniline	$C_9H_{13}N$	9.2	9.0
10	2-aminobiphenyl	$C_{12}H_{11}N$	9.5	5.5
11	N,N-dimethylaniline	$C_8H_{11}N$	<0.1	—
12	N,N-diethylaniline	$C_{10}H_{15}N$	**)	**)

*) Determined by an NMR method. — **) Analysis was not possible due to small induced shifts.

Brittain [22] noted that the formation constants for Eu(fod)$_3$ adducts derived from a series of phosphate esters (Table 5/31) were approximately the same, suggesting that the steric requirements of the alkyl substituents do not play a significant role in formation of the adducts. He maintained that the relative Lewis acidity of the metal chelate and basicity of the substrate dominate the extent of adduct formation in these systems. The triphenyl phosphate substrate apparently presents a more severe steric problem for the chelate, since its formation constant is measureably smaller [22]. Formation constant for Pr(fod)$_3$ · B with B = phosphoric tris(dimethylamide) = $C_6H_{18}N_3OP$, calculated from chemical shift data: $K_1 = 25 \pm 1\ l \cdot mol^{-1}$ [23]. Furthermore, the alcohol adducts have formation constants which tend to increase as the steric hindrance about the functional group of the substrate increases.

The kinetics of substrate exchange for several $M(fod\text{-}d_9)_3$ adducts with triethylamine, dimethyl sulfoxide, tetramethyl urea, phosphoric tris(dimethylamide) were determined by complete line-shape analysis of the NMR spectrum as a function of temperature. For each compound, at room temperature, a single substrate resonance was observed, which broadened on cooling the solution and finally split into two peaks, corresponding to free and complexed substrate. The number (n) of substrate molecules bound to the M(fod) chelate in solution, determined by integrating the relative areas of the free and complexed substrate peaks, was generally two [24]. Bidzilya and coworkers [25, 26] determined the rate of exchange between phosphoric tris (dimethylamide) = $C_6H_{18}N_3OP$ and $M(fod)_3 \cdot 2C_6H_{18}N_3OP$ (M = Pr, Eu) from +4 to −54°C in $CDCl_3$.

Table 5/31
Equilibrium Constants for the Interaction of $Eu(fod)_3$ with Alcohols, Aliphatic Amines, and Other Donor Ligands in CCl_4.

substrate	temp. in °C	method*)	K_1 in $mol^{-1} \cdot l$	lg K_1	lg β_2**)	Ref.
methanol, (= CH_3OH)		sp			5.17	[27]
ethanol, (= C_2H_5OH)		sp			5.08	[27]
1-propanol, (= C_3H_7OH)		sp			5.01	[27]
2-propanol, (= C_3H_7OH)		sp			5.34	[27]
2-propanol, (= C_3H_7OH)	32	NMR	97±8			[21]
1-butanol, (= C_4H_9OH)		sp			5.25	[27]
1-butanol, (= C_4H_9OH)	22	sp	440			[20]
2-butanol, (= C_4H_9OH)		sp			5.36	[27]
2-butanol, (= C_4H_9OH)	22	sp	340			[20]
2-methyl-2-propanol, (= C_4H_9OH)		sp			5.49	[27]
2-methyl-2-propanol, (= C_4H_9OH)	22	sp	46			[20]
propylamine, (= $C_3H_7NH_2$)		sp			5.66	[27]
isopropylamine, (= $C_3H_7NH_2$)		sp			5.64	[27]
butylamine, (= $C_4H_9NH_2$)		sp			5.64	[27]
sec-butylamine, (= $C_4H_9NH_2$)		sp			5.62	[27]
tert-butylamine, (= $C_4H_9NH_2$)		sp			5.60	[27]
hexylamine, (= $C_6H_{13}NH_2$)***)	37	NMR	535.5	13.9		[28]
2-butanone, (= C_4H_8O)	32	NMR	32±3			[21]
tetrahydrofuran, (= C_4H_8O)	32	NMR	57±3			[21]
allyl acetate, (= $C_5H_8O_2$)	32	NMR	26±3			[21]
isopropenyl acetate, (= $C_5H_8O_2$)	32	NMR	27±3			[21]
trimethyl phosphate, (= $C_3H_9O_4P$)		sp		3.302		[22]
triethyl phosphate, (= $C_6H_{15}O_4P$)		sp		3.467		[22]
tributyl phosphate, (= $C_{12}H_{27}O_4P$)		sp		3.432		[22]
triphenyl phosphate, ($C_{18}H_{15}O_4P$)		sp		2.940		[22]

*) sp = Emission-titration spectroscopy; NMR-formation constants calculated from chemical shitf data. — **) Formation constant has units of $dm^6 \cdot mol^{-2}$. — ***) In $CDCl_3$.

Table 5/32

Equilibrium Constants for Substituted Cyclohexanones and Cyclohexanols Interacting with Eu(fod)$_3$ in CCl_4 at 31°C [19]*).

ligand	type	K_1 in mol^{-1}	K_2 in mol^{-1}
3,3,5-trimethyl-5-(4-chlorophenyl)cyclohexanone (= $C_{15}H_{19}ClO$)	I	260± 11	60± 4
3,3,5-trimethyl-5-(1-naphthyl)cyclohexanone (= $C_{19}H_{22}O$)	I	536± 80	23± 8
3,3,5-trimethyl-5-(2-naphthyl)cyclohexanone (= $C_{19}H_{22}O$)	I	254± 12	10± 2
3,3-dimethyl-5-(1-naphthyl)cyclohexanone (= $C_{18}H_{20}O$)	I	546± 120	35± 8
4-tert-butylcyclohexanone (= $C_{10}H_{18}O$)	I	3103± 433	75± 10
3(a),3(e)-dimethyl-5(e)-(1-naphthyl)cyclohexan-(a)-ol (= $C_{18}H_{22}O$)	II	2500	500
4(e)-tert-butylcyclohexan-(a)-ol (= $C_{10}H_{20}O$)	II	10061±3365	1201±401
3(a),3(e),5(e)-trimethyl-5(a)-(1-naphthyl)cyclohexan-(e)-ol (= $C_{19}H_{24}O$)	III	117000	9000
3(a),3(e),5(a)-trimethyl-5(e)-(1-naphthyl)cyclohexan-(e)-ol (= $C_{19}H_{24}O$)	III	140000	15400
3(e)-methyl-5(e)-(1-naphthyl)cyclohexan-(e)-ol (= $C_{17}H_{20}O$)	III	16311	1279
4(e)-tert-butylcyclohexan-(e)-ol (= $C_{10}H_{20}O$)	III	5345	1024
1(e),3(a),3(e),5(e)-tetramethyl-5(a)-phenylcyclohexan-(a)-ol (= $C_{16}H_{24}O$)	II	168± 10	<0.1
1(e),3(a),3(e),5(e)-tetramethyl-5(a)-(2-methoxyphenyl)cyclohexan-(a)-ol (= $C_{17}H_{26}O_2$)	II	233	<0.5
1(e),3(a),3(e),5(e)-tetramethyl-5(a)-(3-methoxyphenyl)cyclohexan-(a)-ol (= $C_{17}H_{26}O_2$)	II	172	<4.5
1(e),3(a),3(e),5(e)-tetramethyl-5(a)-(4-methoxyphenyl)cyclohexan-(a)-ol (= $C_{17}H_{26}O_2$)	II	149	<0.5
1(e),3(a),3(e),5(e)-tetramethyl-5(a)-(1-naphthyl)cyclohexan-(a)-ol (= $C_{20}H_{26}O$)	II	63	<0.1
1(e),3(a),3(e),5(e)-tetramethyl-5(a)-(2-naphthyl)cyclohexan-(a)-ol (= $C_{20}H_{26}O$)	II	196	<0.1
1(e),3(a),3(e),5(a)-tetramethyl-5(e)-phenylcyclohexan-(a)-ol (= $C_{16}H_{24}O$)	II	77	<0.5
1(e),3(a),3(e),5(a)-tetramethyl-5(e)-(2-methoxyphenyl)cyclohexan-(a)-ol (= $C_{17}H_{26}O_2$)	II	123	4
1(e),3(a),3(e),5(a)-tetramethyl-5(e)-(3-methoxyphenyl)cyclohexan-(a)-ol (= $C_{17}H_{26}O_2$)	II	86	10
1(e),3(a),3(e),5(a)-tetramethyl-5(e)-(4-methoxyphenyl)cyclohexan-(a)-ol (= $C_{17}H_{26}O_2$)	II	83	8
1(e),3(a),3(e),5(a)-tetramethyl-5(e)-(1-naphthyl)cyclohexan-(a)-ol (= $C_{20}H_{26}O$)	II	56	6
1(e),3(a),3(e),5(a)-tetramethyl-5(e)-(2-naphthyl)cyclohexan-(a)-ol (= $C_{20}H_{26}O$)	II	65	<0.2

Table 5/32 (Continued)

ligand	type	K_1 in mol^{-1}	K_2 in mol^{-2}
1(e),3(a),3(e)-trimethyl-5(e)-(1-naphthyl)cyclohexan-(a)-ol (= $C_{19}H_{24}O$)	II	38	< 0.1
4(e)-tert-butyl-1(e)-methylcyclohexan-(a)-ol (= $C_{11}H_{22}O$)	II	209± 11	2.3±0.8
1(a),3(a),3(e)-trimethyl-5(e)-(1-naphthyl)cyclohexan-(e)-ol (= $C_{19}H_{24}O$)	III	207± 40	19±5
8,9,10,11-tetrahydro-9,9,11-trimethyl-7,11-methano-7H-cyclohepta[a]naphthalen-7-ol (= $C_{19}H_{22}O$)	IV	536± 106	64±14
4(e)-tert-butyl-1(a)-methylcyclohexan-(e)-ol (= $C_{11}H_{22}O$)	III	614	31

*) The equilibrium constants were determined by analyzing the concentration dependence of the isotropic shifts induced by $Eu(fod)_3$ in carbon tetrachloride at 31°C.

References to 5.1.10:

[1] C. S. Springer, D. W. Meek, R. E. Sievers (Inorg. Chem. **6** [1967] 1105/10). — [2] R. E. Sievers, K. J. Eisentraut, C. S. Springer, D. W. Meek (Advan. Chem. Ser. **71** [1967] 141/54). — [3] D. G. Karraker (J. Inorg. Nucl. Chem. **33** [1971] 3713/18). — [4] R. E. Sievers, J. W. Connolly (Inorg. Syn. **12** [1970] 72/7). — [5] H. A. Swain, D. G. Karraker (J. Inorg. Nucl. Chem. **33** [1971] 2851/6).

[6] J. P. R. De Villiers, J. C. A. Boeyens (Acta Cryst. B **27** [1971] 692/702). — [7] J. C. A. Boeyens, J. P. R. De Villiers (J. Cryst. Mol. Struct. **1** [1971] 297/306). — [8] A. H. Bruder, S. R. Tanny, H. A. Rockefeller, C. S. Springer (Inorg. Chem. **13** [1974] 880/5). — [9] J. F. Desreux, L. E. Fox, C. N. Reilley (Anal. Chem. **44** [1972] 2217/9). — [10] T. H. Risby, P. C. Jurs, F. W. Lampe, A. I. Yergey (Anal. Chem. **46** [1974] 726/8).

[11] R. E. Sievers (Nuclear Magnetic Resonance Shift Reagents, Academic Press, New York 1973). — [12] W. D. W. Horrocks (Lanthanide Shift Reagents and Other Analytical Applications, in: G. N. La Mar, W. D. W. Horrocks, R. H. Holm, Nuclear Magnetic Resonance of Paramagnetic Molecules, Principles and Applications, Academic Press, New York 1973, p. 479/519). — [13] J. Reuben (Paramagnetic Lanthanide Shift Reagents in NMR Spectroscopy, in: Progress in Nuclear Magnetic Resonance Spectroscopy, Vol. 9, Pergamon Press, New York 1973, p. 1/70). — [14] A. F. Cockerill, G. L. O. Davies, R. C. Harden, D. M. Rackham (Chem. Rev. **73** [1973] 553/88). — [15] R. von Ammon, R. D. Fischer (Angew. Chem. Intern. Ed. Engl. **11** [1972] 675/92).

[16] B. C. Mayo (Chem. Soc. Rev. **2** [1973] 49/74). — [17] M. Hirayama, N. Ishida (Bull. Chem. Soc. Japan **50** [1977] 779/82). — [18] T. R. Sweet, D. Brengartner (Anal. Chim. Acta **52** [1970] 173/81). — [19] M. D. Johnston, B. L. Shapiro, M. J. Shapiro, T. W. Proulx, A. D. Godwin, H. L. Pearce (J. Am. Chem. Soc. **97** [1975] 542/54). — [20] H. G. Brittain, F. S. Richardson (J. Chem. Soc. Dalton Trans. **1976** 2253/7).

[21] D. R. Kelsey (J. Am. Chem. Soc. **94** [1972] 1764/6). — [22] H. G. Brittain (Inorg. Chem. **19** [1980] 640/3). — [23] V. A. Bidzilya, N. K. Davidenko, L. P. Golovkova, K. B. Yatsimirskii (Teor. Eksperim. Khim. **11** [1975] 388/92; Theor. Exptl. Chem. [USSR] **11** [1975] 326/30). — [24] D. F. Evans, M. Wyatt (J. Chem. Soc. Dalton Trans. **1974** 765/72). — [25] V. A. Bidzilya, N. K. Davidenko, L. P. Golovka (Teor. Eksperim. Khim. **11** [1975] 687/90; Theor. Exptl. Chem. [USSR] **11** [1975] 576/8).

[26] V. A. Bidzilya, N. K. Davidenko, L. P. Golovkova, K. B. Yatsimirskii (Dokl. Akad. Nauk SSSR **225** [1975] 842/5 from C.A. **84** [1976] No. 80300). — [27] H. G. Brittain (J. Chem. Soc. Dalton Trans. **1979** 1187/91). — [28] T. Inagaki, M. Tasumi, T. Miyazawa (Bull. Chem. Soc. Japan **48** [1975] 1427/30). — [29] M. Hirayama, M. Owada (Bull. Chem. Soc. Japan **52** [1979] 1786/9). — [30] Sh. S. Bikew, R. N. Ikhsanov, Yu. Yu. Samitov (Teor. Eksperim. Khim. **15** [1979] 684/90 from C.A. **92** [1980] No. 100096).

[31] D. J. Raber et al. (Monatsh. Chem. **111** [1980] 43/52). — [32] H. N. Cheng, H. S. Guitowsky (J. Phys. Chem. **84** [1980] 1039/43). — [33] K. B. Yatsimirskii, N. K. Davidenko, V. A. Bidzilya, L. P. Golokova (Str. Svoistva Primen. Beta Diketonatov Metal. Mater. 3rd Vses. Semin., Moscow 1977 [1978], pp. 19/28).

Table 5/33

Substrate Exchange Kinetics for M(fod-d_9)$_3$ · 2 B Adducts in Solution [24].

substrate	M	solvent*)	mechanism (order)	k_1 in s^{-1}	T_1 in K	$\Delta H_1^{\ddagger}$ in kJ · mol^{-1}	$\Delta S_1^{\ddagger}$ in J · mol^{-1} · K^{-1}	k_2 in mol · l^{-1} · s^{-1}	T_2 in K	$\Delta H_2^{\ddagger}$ in kJ · mol^{-1}	$\Delta S_2^{\ddagger}$ in J · mol^{-1} · K^{-1}
dimethylsulfoxide	Pr	mixed	1st + 2nd	143	186.2	37.6	3	612	186.2	27.6	−39
(= C_2H_6OS)	Nd	mixed	1st + 2nd	84.9	186.2	38.1	0	328	186.2	24.7	−60
	Eu	mixed	1st	29.7	186.2	39.7	0				
	Tb	mixed	1st	131	186.2						
	Ho	mixed	1st	223	186.2						
	Er	mixed	1st	245	186.2						
	Yb	mixed	1st	549	186.2						
tetramethyl urea	Pr	$C_6D_5CD_3$	1st	6.4	186.2	41.5	− 3				
(= $C_5H_{12}N_2O$)	Eu	$C_6D_5CD_3$	1st	4.1	186.2	41.9	− 2				
	Er	$C_6D_5CD_3$	1st	141	186.2						
	Yb	$C_6D_5CD_3$	1st	247	186.2						
phosphoric	Pr	$C_6D_5CD_3$	1st	38.0	273.2	55.1	−12				
tris(dimethylamide)	Eu	$C_6D_5CD_3$	1st	65.9	273.2	54.6	−10				
(= $C_6H_{18}N_3OP$)	Tb	$C_6D_5CD_3$	1st	167	273.2	54.2	− 3				
	Ho	$C_6D_5CD_3$	1st	426	273.2	50.3	−10				
	Er	$C_6D_5CD_3$	1st	702	273.2	52.5	2				
	Yb	$C_6D_5CD_3$	1st	2570	273.2	50.5	3				
	Eu*)	mixed	1st	91.9	273.2	53.0	−13				
	Tb	mixed	1st	180	273.2						
	Er**)	mixed	2nd					419	273.2	25.8	−80
	Yb**)	mixed	2nd					349	273.2	24.6	−87
triethylamine**)	Eu	$C_6D_5CD_3$	1st + 2nd	≈160	243			≈460	243		
(= $C_6H_{15}N$)	Pr	$C_6D_5CD_3$	1st + 2nd	≈330	203			≈320	203		

*) Mixed solvent is $CCl_4/CDCl_3/C_6D_5CD_3$ 1.5:1.8:1 (v/v). — **) These adducts were M(fod-d_9)$_3$ · B.

5.1.11 Complexes with Other Aliphatic Diketones

5.1.11.1 With 2,4-Decanedione $CH_3C(O)CH_2C(O)(CH_2)_5CH_3$ (= $C_{10}H_{18}O_2$)

Formation in Solution

Formation constants (lg β_n) for the $M(C_{10}H_{17}O_2)_n^{3-n}$ (M = Nd, Eu) species in aqueous media at 25°C, I = 0.1 $(C_2H_5)_4NClO_4$, were determined by a solvent extraction method (for M = Nd: lg β_1 = 6.9, lg β_2 = 12.9, lg β_3 = 17.45, lg β_4 = 22.5, lg β_5 = 26.7; for M = Eu: lg β_1 = 6.30, lg β_2 = 12.5, lg β_3 = 19.38) [1].

$M(C_{10}H_{17}O_2)_3$

Each compound (M = Nd, Sm, Gd, Tb, Ho, Er, Yb, Lu) was prepared by adding a methanolic solution of the ligand (neutralized with methanolic sodium hydroxide), dropwise with stirring, to a methanolic solution of the rare earth nitrate (pH adjusted to 5 with methanolic sodium hydroxide). The initial addition produced a precipitate which redissolved with stirring, but at the end some precipitate remained undissolved. The suspension was added dropwise, with stirring, to approximately 400 ml of water. After 2 h of stirring, the fine, granular precipitate was collected by filtration and air dried for 1 h. The precipitate was suspended in acetone and then collected by filtration. This operation was repeated twice. The precipitate was first air dried for 24 h, then dried under vacuum over phosphorus(V) oxide for 7 days. Each chelate was purified by dissolution in a minimum amount of chloroform, effecting precipitation by adding acetone. After standing for 24 h at 0°C, the suspension was filtered, and the precipitate collected was dried under vacuum over phosphorus(V) oxide for 24 h. This operation was repeated twice. Thermogravimetric and elemental analyses, as well as infrared spectral data, confirmed the anhydrous nature of the chelates. The melting points, recorded below, revealed a decreasing trend on going from neodymium to lutetium. Thermogravimetric analyses showed that initial weight losses occurred after the compounds had melted, indicating that decomposition starts in the liquid phase. There are two decomposition regions, one going from 150 to 280°C, the other from this temperature to about 450°C. After 780°C there is a step corresponding to the stoichiometric formation of the oxide, indicating that the chelates are not appreciably volatile. X-ray powder patterns revealed a very low order of crystallinity [2].

M	Nd	Sm	Gd	Tb	Ho	Er	Yb	Lu
color	blue	beige	beige	beige	beige	pink	beige	beige
m.p. in °C . . .	123	126.0	110.5	117.0	110.0	112.0	118	96

References to 5.1.11.1:

[1] M. De Jesus Tavares, M. A. Gouveia, R. G. De Carvalho (J. Inorg. Nucl. Chem. **38** [1976] 1363/5). — [2] M. A. Gouveia, M. De Jesus Tavares, R. G. De Carvalho (J. Inorg. Nucl. Chem. **33** [1971] 817/22).

5.1.11.2 With Stearoylacetone $CH_3(CH_2)_{16}C(O)CH_2C(O)CH_3$ (= 2,4 Heneicosandione $C_{21}H_{40}O_2$)

$Ce(C_{21}H_{39}O_2)_4$

The compound was prepared by adding 60 ml of 2 N ammonia to an ethanolic solution containing cerium(III) nitrate hexahydrate (1 mmol) and stearoylacetone (4.5 mmol). Air was then passed though the reaction mixture for 4 h. The red-brown precipitate was collected and recrystallized from an acetonitrile/ethyl acetate mixture. The wax-like substance melted at 36.0°C. The composition of the complex was verified by elemental analysis. The compound is a nonelectrolyte in isooctane, K. H. Jahr, R. Gelius (Z. Chem. [Leipzig] **15** [1975] 280/1).

5.1.11.3 With β-Diketones Containing Alicyclic Terminal Groups, $R_1COCH_2COR_2 = HL$ ($R_1 = (CH_3)_3C$, R_2 = Cyclopropyl, Cyclobutyl, Cyclopentyl, and Cyclohexyl; $R_1 = CF_3$, R_2 = Cyclopropyl, Cyclohexyl)

ErL_3

In a search for chelating agents which give volatile and thermally stable metal chelates for gas chromatography, Kito, Nakane, and Miyake [1 to 4] prepared a series of erbium complexes derived from β-diketones with a tert-butyl or trifluoromethyl group constituting one side chain, and an alicyclic substituent at the other end of the molecule. Thermogravimetric analyses revealed partial decomposition of all chelates on vaporization, with increased decomposition occurring upon substitution of the trifluoromethyl for the tertiary butyl group [2, 4]. Thermal stability and volatility of the alicyclic β-diketone chelates were intermediate between those with acyclic terminal groups and those with aromatic groups [1, 3].

References to 5.1.11.3:

[1] A. Kito, M. Nakane, Y. Miyake (Bunseki Kagaku **27** [1978] 1/6 from C.A. **89** [1978] No. 35891). — [2] A. Kito, M. Nakane, Y. Miyake (Bunseki Kagaku **27** [1978] 7/11 from C.A. **89** [1978] No. 52744). — [3] A. Kito, M. Nakane, Y. Miyake (Osaka Kogyo Gijutsu Shikensho Kiho **29** [1978] 231/7 from C.A. **90** [1979] No. 161587). — [4] A. Kito, M. Nakane, Y. Miyake (Osaka Kogyo Gijutsu Shikensho Kiho **29** [1978] 238/43 from C.A. **90** [1979] No. 145265).

5.1.12 Complexes with 1-Phenyl-1,3-butanedione $C_6H_5C(O)CH_2C(O)CH_3$ (= $C_{10}H_{10}O_2$ = Benzoylacetone)

Mixed complexes derived from amino-N-polycarboxylic acids and benzoylacetone are described in "Rare Earth Elements" D 1, 1980, pp. 234 and 236/7.

5.1.12.1 $M(C_{10}H_9O_2)_n^{3-n}$ Complexes (n = 1 to 4). Formation in Solution

Formation constants for benzoylacetonato rare earth(III) chelates have been determined potentiometrically in nonaqueous and aqueous-organic solutions, and the results are recorded in Table 5/34 [1, 6, 7]. Potentiometric evaluation of formation constants in pure aqueous media was precluded due to the difficultly soluble nature of the chelates in water. For values determined in 4:1 (v/v) methanol/water, plots of lg K_n (n = 1 to 3) versus ionic radius revealed maxima at europium and erbium, with a dip in the curve at gadolinium [6]. Stability constants for the complexes in aqueous media, determined by a solvent extraction method, are: La, lg K_1 = 5.2, lg K_2 = 4.1, lg K_3 = 2.8 [2]; Ce, lg K_1 = 6.17, lg K_2 = 5.93, lg K_3 = 4.89 [3]; Gd, $K_1 = 2.4 \times 10^7$, $K_2 = 3.2 \times 10^6$, $K_3 = 2.2 \times 10^5$ [4]; Y, lg K_1 = 6.55, lg K_2 = 4.85, lg K_3 = 3.00 [5] (values from [2] determined at I = 0.1; values from [3] determined at 20°C, I = 0.1). Samelson, Brecher, and Lempicki [8] determined the stepwise instability constants for the $Eu(C_{10}H_9O_2)_n^{3-n}$ (n = 1 to 4) species in 3:1 (v/v) ethanol/methanol by measuring the relative intensities of the peaks in the emission spectrum corresponding to the $^5D_0 \rightarrow {}^7F_0$ transition ($K_1 = 1.5 \times 10^{-10}$, $K_2 = 2.9 \times 10^{-8}$, $K_3 = 1.3 \times 10^{-6}$, $K_4 = 1.2 \times 10^{-3}$).

Table 5/34
Formation Constants for the (Benzoylacetonato) Rare Earth(III) Chelates, $M(C_{10}H_9O_2)_n^{3-n}$ (n = 1 to 4).

metal ion	methanol/water [6] a) (4:1 volume ratio)			acetone/water [1] b) (3:1 volume ratio)			methanol [7] c)			
	lg K_1	lg K_2	lg K_3	lg K_1	lg K_2	lg K_3	lg K_1	lg K_2	lg K_3	lg K_4
Y^{3+}	8.29	6.35	4.42	8.21	6.68	5.68				
La^{3+}	7.02	5.25	3.36	6.24	5.21	4.76				
Pr^{3+}	7.76	6.06	4.06	6.84	6.25	5.21	10.7	8.4	4.4	2.6

Table 5/34 (Continued)

metal ion	methanol/water [6][a) (4 : 1 volume ratio)			acetone/water [1][b) (3 : 1 volume ratio)			methanol [7][c)			
	lg K_1	lg K_2	lg K_3	lg K_1	lg K_2	lg K_3	lg K_1	lg K_2	lg K_3	lg K_4
Nd^{3+}	7.83	6.13	4.13	6.94	6.61	5.58	10.8	8.5	4.4	2.6
Sm^{3+}	8.03	6.28	4.30							
Eu^{3+}	8.17	6.30	4.36				11.1	8.7	4.6	2.9
Gd^{3+}	8.09	6.21	4.18							
Tb^{3+}	8.27	6.44	4.41							
Dy^{3+}	8.36	6.47	4.45							
Ho^{3+}	8.41	6.50	4.50							
Er^{3+}	8.47	6.55	4.55							
Yb^{3+}	8.44	6.49	4.46							
Lu^{3+}	8.37	6.44	4.45							

a) Determined potentiometrically at 25°C, I = 0.1 (NaCl).—b) Determined potentiometrically at 30°C, I = 0.1 ($NaClO_4$).—c) Determined potentiometrically at 22°C, I = 0.1 (NaCl).

Stepwise and overall formation constants of $M(C_{10}H_9O_2)_n^{3-n}$ complexes in dioxane/water mixture (75% v/v) at 30°C and ionic strength I = 0.1 M ($NaClO_4$) have been determined by Mehrotra et al. [9].

References to 5.1.12.1:

[1] N. K. Dutt, P. Bandyopadhyay (J. Inorg. Nucl. Chem. **26** [1974] 729/36). — [2] J. Stary (Collection Czech. Chem. Commun. **25** [1960] 86/92). — [3] I. P. Efimov, L. S. Vorontsev, L. G. Makarova, V. M. Peshkova (Vestn. Mosk. Univ. Ser. II Khim. **24** No. 4 [1969] 121/3; Moscow Univ. Chem. Bull. **24** No. 4 [1969] 83/4). — [4] M. I. Gromova, G. N. Perelyaeva (Vestn. Mosk. Univ. Ser. II Khim. **18** [1963] 58/60; C.A. **58** [1963] 13203). — [5] J. Stary, N. P. Rudenko (Zh. Neorgan. Khim. **4** [1959] 2405/9; Russ. J. Inorg. Chem. **4** [1959] 1100/3; C.A. **1960** 13925).

[6] N. K. Davidenko, A. A. Zholdakov (Zh. Neorgan. Khim. **12** [1967] 1195/8; Russ. J. Inorg. Chem. **12** [1967] 633/5). — [7] A. A. Zholdakov, N. K. Davidenko (Zh. Neorgan. Khim. **12** [1967] 3066/71; Russ. J. Inorg. Chem. **12** [1967] 1622/5). — [8] H. Samelson, C. Brecher, A. Lempicki (J. Mol. Spectrosc. **19** [1966] 349/71). — [9] R. C. Mehrotra, B. P. Bachlas, B. P. Gupta (Indian J. Chem. A **18** [1979] 370/2).

5.1.12.2 Tris Chelates

5.1.12.2.1 $M(C_{10}H_9O_2)_3$

The preparation of the anhydrous europium chelate was effected by heating the dihydrate under vacuum (25 to 30 Torr) at 70°C for 1 h [1]. A similar procedure for preparation of the anhydrous europium chelate is reported in [2]. Charles [1] attributed the low carbon analysis (56.7, calc; 55.5, found) to the hydrolysis reaction $Eu(C_{10}H_9O_2)_3 \cdot 2H_2O \rightarrow Eu(C_{10}H_9O_2)_2OH + C_{10}H_{10}O_2 + H_2O$, which may occur during dehydration, even under mild conditions. The light yellow colored tris chelate softened in a melting point capillary at ca. 70°C, before melting at 90°C to a viscous liquid [1]. According to Murav'eva et al. [61] anhydrous chelates $M(C_{10}H_9O_2)_3$ were not obtained by heating of the hydrates $M(C_{10}H_9O_2)_3 \cdot nH_2O$ with M = Pr, Nd, Gd, Er, Yb in He atmosphere to 250°C.

The anhydrous praseodymium, neodymium, and samarium chelates were prepared by allowing the respective anhydrous rare earth isopropoxide salts to react with solutions of benzoylacetone in benzene (1:3 mol ratio) heated at reflux. The isopropanol liberated during the reaction was removed from the reaction mixture by distillation as the azeotrope with benzene. The products were isolated by removing the solvent from the clear solution under reduced pressure [3, 4]. The molecular weights of the praseodymium and neodymium chelates, determined ebullioscopically in boiling benzene, corresponded to dimeric formulations [3].

The emission spectrum at 93 K of the anhydrous europium chelate (sample prepared according to the procedure described in [2]) in 3:1 (v/v) ethanol/methanol (standard solvent) and 4:1 (v/v) standard solvent/dimethylformamide, as well as a powdered crystalline sample, revealed a multiplicity of lines that could not be accounted for in terms of a single chemical species. For the powdered sample: $^5D_0 \rightarrow {}^7F_2$ = 6240, 6202, 6168, 6138, 6125, 6114, 6040 Å; $^5D_0 \rightarrow {}^7F_1$ = 5960, 5928, 5898 Å; and $^5D_0 \rightarrow {}^7F_0$ = 5799, 5788 Å [2].

5.1.12.2.2 $M(C_{10}H_9O_2)_3 \cdot nH_2O$ (n = 1, 2)

Sacconi and Ercoli [5] prepared the dihydrated chelates (M = La, Pr, Nd, Sm, Gd) by adding saturated ethanolic solutions of ammonia to solutions containing a hydrated rare earth nitrate and benzoylacetone in aqueous ethanol. The colors and melting points (°C) of the compounds are recorded in Table 5/35. Purushottam and coworkers [6, 7] effected precipitation of the tris chelates (M = Sc, Y, La to Yb except for Ce, Pm) according to a procedure similar to that described by Sacconi, Ercoli [5], but adjusted the pH of each reaction mixture to an optimum value by adding ammoniacal ethanol (optimum pH for precipitation of each chelate, together with their physical properties, are recorded in Table 5/35). In each case, the precipitate formed was collected by filtration, washed with alcohol, and then dissolved in benzene. The solution was filtered. The filtrate was allowed to evaporate at room temperature, and the compound obtained was dried under vacuum over concentrated sulfuric acid [6, 7].

Table 5/35
Properties of the Tris (benzoylacetonato) Rare Earth(III) Hydrates.

$M(C_{10}H_9O_2)_3 \cdot 2H_2O$ [5]			$M(C_{10}H_9O_2)_3 \cdot nH_2O$*) [6, 7]							
M	color	melting point in °C	M	pH of pptn.	color	melting point in °C	M	pH of pptn.	color	melting point in °C
La	straw-yellow	108 to 109	La	6.5	yellow	120	Dy	5.8	straw-yellow	148 to 150
Pr	green-yellow	106 to 108	Pr	6.5	yellow-green	108	Ho	6.5	straw-yellow	144
Nd	pale-lilac	106 to 108	Nd	6.5	pink	107	Er	6.5	yellow	142
Sm	straw-yellow	103 to 105	Sm	6.5	straw-yellow	104	Tm	6.4	red	118
			Eu	6.5	yellow	155	Yb	6.2	red	125
Gd	white	100 to 101	Gd	6.5	pink	100	Sc	6.2	pink	205
			Tb	6.4	yellow-green	140	Y	6.4	yellow	145

*) M = La to Yb from [6], M = Sc, Y from [7]. Value of n not given in [6] or [7].

Whan, Crosby [8] effected synthesis of each dihydrate (M = La to Lu except for Ce, Pm) by adding, slowly and with stirring, an aqueous solution of the rare earth chloride to a solution of sodium benzoylacetonate in water. (The sodium benzoylacetonate was prepared by allowing benzoylacetone to react with sodium (1:1 mol ratio) in p-xylene heated at reflux.) The resulting flocculent precipitate was collected by filtration, washed several times with distilled water, and dried under vacuum at ambient temperature for 10 h. Filipescu, Hurt, McAvoy [9] found that the europium chelate prepared by this method was dimeric in benzene, chloroform, and acetonitrile (molecular weight determined by vapor pressure osmometry). Brecher, Samelson, Lempicki [2] obtained the monohydrated europium chelate following the procedure described by Whan and Crosby [8]. Kuznetsova, Sevchenko, Khomenko [10] effected precipitation of the tris chelates of europium and terbium by adding either ammonium or sodium benzoylacetonato to an aqueous solution of the rare earth nitrate.

Lyle, Witts [11] obtained the dihydrate, $Eu(C_{10}H_9O_2)_3 \cdot 2H_2O$, by adding a solution of sodium benzoylacetonate (3 mmol) in ethanol (25 ml), dropwise over a period of about 2 h, to a stirred aqueous solution (25 ml) of europium(III) chloride (1 mmol). The product precipitated during the addition of the ligand, but the mixture was kept overnight before filtration. The precipitate was air dried.

Cotton, Legzdins [12] effected synthesis of crystalline $Y(C_{10}H_9O_2)_3 \cdot H_2O$ by allowing sodium benzoylacetonate and yttrium chloride (3:1 mol ratio) to react in aqueous ethanol. The product was recrystallized from the same solvent. The compound was air sensitive and melted at 114 to 115°C. It was soluble in most nonpolar organic solvents and was found to exist as a monomer in benzene (molecular weight determined osmometrically).

The preparation of $Eu(C_{10}H_9O_2)_3 \cdot 2H_2O$ was reported by Charles [1] in "Inorganic Syntheses": To a solution (200 ml) of europium chloride (5 mmol) in water was added, with stirring, a solution of 4.0 g (an excess) of benzoylacetone in 50 ml of 95% ethanol. The resulting suspension was stirred while 15 ml of 1 M aqueous ammonia was added dropwise over a period of 2 h. The solid mixture (product and excess benzoylacetone) was collected by filtration, washed with water, and dried in a vacuum desiccator (25 to 30 Torr—too low a pressure resulted in partial dehydration and decomposition). The solid was then dissolved in acetone, and the solution was filtered. The addition of 50 ml of water to the filtrate effected separation of a yellow oil which solidified on shaking the solution. After a few minutes, the solid was collected by filtration, washed with water, and dried in the open air at room temperature overnight. Excess benzoylacetone was removed by stirring the product with 25 ml of petroleum ether for 0.5 h and filtering. The solid collected was washed with petroleum ether and dried in a vacuum desiccator (25 to 30 Torr) or in the open air at room temperature (74% yield). The compound is a yellowish-white solid, insoluble in water, readily soluble in acetone, ethanol, benzene and chloroform, and slightly soluble in cyclohexane. In a capillary, the compound softened when heated in the range 90 to 100°C and melted at 100 to 104°C [1].

Khalmurzaev and coworkers [13], employing the procedure described by Charles, Perotto [14] for preparation of the tris(dibenzoylmethanato) rare earth(III) chelates, effected syntheses of compounds of composition $M(C_{10}H_9O_2)_3 \cdot nH_2O$ (n = 2 for M = Y, La, Pr, Nd, Gd; n = 1 for M = Sm, Er, Yb). To a solution of the rare earth chloride in water was added an ethanolic solution of benzoylacetone, followed by the addition of an aqueous ammonia solution. The resulting precipitate was collected by filtration and dried under vacuum in a desiccator at room temperature. In the case of the terbium, dysprosium, and holmium chelates, apparent hydrolysis of the product occurred during preparation, resulting in formation of compounds formulated as $M_2(C_{10}H_9O_2)_5OH$. Thermogravimetric analyses revealed that only the monohydrated erbium and ytterbium chelates were completely dehydrated when heated (90 to 110°C), whereas the majority of compounds were only partially dehydrated (90 to 110°C), losing 1.3 to 1.7 mol of water. The authors proposed that partial dehydration is accompanied by hydrolysis, with formation of polymeric products [13].

Anufrieva and Zaidel [15] prepared the samarium chelate by allowing freshly precipitated samarium hydroxide to react with an aqueous suspension of benzoylacetone. The solid was collected and dried, then extracted with chloroform. White and light-yellow crystals of the product were obtained from chloroform.

The hydrolytic decomposition of the tris chelates $M(C_{10}H_9O_2)_3 \cdot nH_2O$ was studied by a gas-liquid chromatographic method. Complexes with M = Pr, Nd, Gd, n = 2; M = Er, Yb, n = 1.5 decomposed during heating in He atmosphere with the loss of 1 or 0.5 mol H_2O (1st stage of heating, 130 to 140°C), 0.5 mol acetophenone, and 0.5 mol CH_3COOH (2nd stage of heating, 250°C) [61].

The crystal and molecular structures of $Y(C_{10}H_9O_2)_3 \cdot H_2O$ (compound prepared by method described by Cotton and Legzdins [12]) were determined by single-crystal X-ray analysis (R = 0.059 for 4820 independent reflections). The compound crystallizes in the triclinic space group $P\bar{1}$-C_i^1 (No. 2) with two molecules in a unit cell of dimensions a = 6.214, b = 12.462, c = 19.299 Å, α = 95°37′, β = 104°53′, and γ = 96°20′, density(observed) = 1.36 g · cm^{-3}; density(calculated) = 1.37 · cm^{-3}. The crystal structure consists of staggered molecular chains, each chain containing only one enantiomer of $Y(C_{10}H_9O_2)_3(H_2O)$, running parallel to the crystallographic a axis. The molecules are monomeric, with the yttrium ion surrounded by a distorted octahedral array of chelate oxygen atoms, and the oxygen atom of the coordinated water molecule capping one face of the octahedron, see **Fig. 5-25**. The octahedron is distorted mainly by a spreading apart of the O_1, O_4, O_6 atoms (see original paper) defining the capped face. The average Y-O (benzoylacetonate) bond length is 2.28 Å (range 2.248 to 2.325 Å) and the Y-O (water) bond length is 2.341 Å. The benzoylacetonate ligands wrap asymmetrically (two phenyl groups up, one down) about the coordination polyhedron and exhibit characteristic folding to aid packing. All intermolecular distances are indicative of normal non-bonded contacts, except in a direction parallel to the crystallographic a axis, where hydrogen bonding is indicated [12].

Fig. 5-25

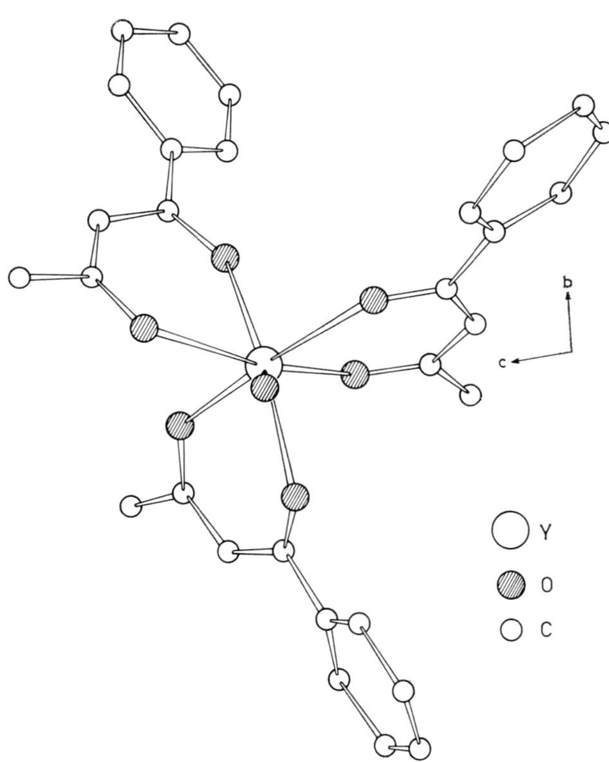

The molecular structure of $Y(C_{10}H_9O_2)_3 \cdot H_2O$ projected along the bc plane.

Infrared spectral data for $Er(C_{10}H_9O_2)_3 \cdot 2H_2O$ are recorded in Table 5/36 (spectrum recorded on a sample dispersed in a KBr disc) [16]. The band assignments were based on those reported [17 to 21] for the tris (acetylacetonato) rare earth(III) chelates. Similar assignments for the benzoyl-acetonato chelates of lanthanum [22], praseodymium [22], neodymium [22, 23], and samarium [22] also have been reported. Stretching force constants, on the order of 2.7×10^5 dyn·cm^{-1} for the M-O bonds (M = La, Pr, Nd, Sm) were determined by normal coordinate analysis [24]. Additional infrared studies are reported in [25] and [26].

Brecher, Samelson, Lempicki [2] noted that the emission spectrum at 93 K of a sample of $Eu(C_{10}H_9O_2)_3 \cdot H_2O$ (prepared according to the procedure described in [8] and dried under vacuum), in the crystalline state and in 3:1 (v/v) ethanol/methanol (standard solvent) and 4:1 (v/v) standard solvent/dimethylformamide, revealed a multiplicity of lines that was greater than the number allowed for a single chemical species. For the powdered sample: ${}^5D_0 \to {}^7F_2$ = 6240, 6202, 6168, 6138, 6125, 6114 Å, ${}^5D_0 \to {}^7F_1$ = 6040, 5960, 5928, 5898 Å, and ${}^5D_0 \to {}^7F_0$ = 5799, 5788 Å. (As noted by Charles [1], drying the dihydrate under vacuum at too low a pressure results in partial dehydration and decomposition of the sample.) On the other hand, Bhaumik [27] interpreted the emission spectrum for a sample of the microcrystalline tris chelate (method of sample preparation not given) at 77 K in terms of a pure chelate with D_3 symmetry at the europium(III) site: ${}^5D_0 \to {}^7F_2$, 6132, 6125 Å; ${}^5D_0 \to {}^7F_1$, 5920, 5907 Å; ${}^5D_0 \to {}^7F_0$, 5801 Å (very low intensity, not allowed in D_3 symmetry). Korol'kov, Kuznetsova [28] also interpreted the emission spectrum of the europium chelate in terms of D_3 symmetry at the metal ion site. Luminescence studies of the hydrated tris chelates also have been reported for: the europium compound in acetonitrile [29], ethanol (−150°C) [30], 3:1 (v/v) ethanol/methanol [29], polymethylacrylate [31, 32], and the crystalline state (20 K) [33]; terbium in ethanol (−150°C) [30] and polymethylacrylate [31]; thulium and holmium in a hydrocarbon matrix (77 K) [34]; and samarium [35]. The frequencies of the electronic vibrational and vibrational transitions observed in the spectrum of $Eu(C_{10}H_9O_2)_3 \cdot 2H_2O$ are tabulated in [36], and the relative intensities of the vibrational lines were calculated and are tabulated in [37].

Table 5/36
Assignments of Absorption Bands in the Infrared Spectrum of $Er(C_{10}H_9O_2)_3 \cdot 2H_2O$ (KBr Disc).

absorption band in cm⁻¹ *)	assignment**)	absorption band in cm⁻¹ *)	assignment**)
3065 (w)	ν_{as}(C-H)	1032 (s)	$\rho(CH_3)$
2915 (s)	$\nu_{as}(CH_3)$	1005 (s)	ϕ ring breathing
2850 (sh)	ν_s (CH_3)	965 (ms)	π(C-H)
1600 (vs)	ν_{as}(C=O)	925 (w)	ν(C=O) + $\nu(C\text{-}CH_3)$
1578 (vs)	$\phi\,\nu$(C-C)	850 (s)	$\phi\,\pi$(C-H)
1530 (vs)	ν_{as}(C=C)	805 (w)	$\nu(C\text{-}\phi)$
1490 (vs)	$\phi\,\nu$(C-C)	760 (s)	π(C-H)
1458 (vs)	ν_s (C=O) + δ(C-H)	716 (vs)	$\phi\,\pi$(C-H)
1395 (s)	$\delta_{as}(CH_3)$	692 (m)	5 adjacent H on ϕ
1360 (m)	δ_s (CH_3)	682 (w)	$\nu(C\text{-}CH_3)$
1285 (s)	ν_s(C=C) + $\nu(C\text{-}CH_3)$	618 (w)	$\nu(C\text{-}\phi)$
1210 (w)		560 (m)	
1182 (m)	δ(C-H) + ν(C=O) (1210, 1182, 1160)	520 (m)	$\delta(C\text{-}\phi)$
1160 (w)		410 (m)	ν Er-O
1112 (m)		375 (w)	ring deformations
1075 (m)	$\phi\,\delta$(C-H)	315 (w)	

*) Relative intensity: vs = very strong, s = strong, m = medium, sh = shoulder, w = weak, vw = very weak.— **) Assignment: ν = stretching, δ = in-plane bending, π = out-of-plane bending, ρ = rocking, s = symmetric, as = asymmetric, ϕ = phenyl ring.

The values of the Slater-Condon (F_2=308.2 cm⁻¹, F_4=42.55 cm⁻¹, F_6=4.657 cm⁻¹), Racah interelectronic repulsion (E^1=4436 cm⁻¹, E^2=23.68 cm⁻¹, E^3=457.50 cm⁻¹), and Lande (645.5 cm⁻¹) parameters for $Pr(C_{10}H_9O_2)_3 \cdot 2H_2O$ (sample prepared according to the procedure described in [8]) were calculated from the diffuse reflectance spectrum [38]. The diffuse reflectance spectra of the hydrated tris chelates of lanthanum, cerium, and praseodymium [39], neodymium [39, 40], and samarium [39] also have been reported. The nephelauxetic ratio, β, evaluated from reflectance spectra, has values of 0.9568 and 0.9849, respectively, for the $Pr(C_{10}H_9O_2)_3 \cdot 2H_2O$ [38] and $Nd(C_{10}H_9O_2)_3 \cdot 2H_2O$ [40] complexes.

Crystal field calculations have been reported for $Tm(C_{10}H_9O_2)_3 \cdot 2H_2O$ [42] and $Yb(C_{10}H_9O_2)_3 \cdot 2H_2O$ (samples prepared according to the procedure given in [8]) [41]. The energy calculated for the $^3H_6$ and $^3H_5$ terms of the thulium chelate agreed with the experimental values obtained from the absorption spectrum recorded for carbon tetrachloride solutions, whereas the calculated and experimental values obtained for the $^3H_4$ term were not in agreement [42]. Good agreement was observed between the calculated and experimental term energies obtained in the case of the ytterbium chelate [41].

Oscillator strengths for the absorption bands observed in the spectra of the hydrated praseodymium [43] and neodymium [44] chelates (samples prepared according to the procedure described in [8]) recorded for methanol, dimethylformamide, 3:1 (v/v) ethanol/methanol, and 1:1:1 (v/v) methanol/ethanol/dimethylformamide solutions were evaluated and compared with values obtained from the Judd-Ofelt theory—see B. R. Judd (Phys. Rev. **127** [1962] 750/61) and G. S. Ofelt (J. Chem. Phys. **37** [1962] 511/20). Good correlation between the experimental and calculated oscillator strengths was observed for the neodymium chelates [44], but poor correlation was observed in the case of the praseodymium chelates [43]. Values for the oscillator strengths of the hypersensitive transitions in the neodymium, holmium, and erbium chelates (spectra were recorded for solutions of the chelates in carbon tetrachloride, chloroform, isoamyl alcohol, and acetone/water) are tabulated in [45]. Mason, Peacock, and Stewart [46] attribute the oscillator strengths of the hypersensitive transitions to "coulombic correlation" between transient induced electric dipoles in the ligands and the f-electron quadrupole moment of the rare earth ion.

The effects of the solvent medium on the spectrum of the neodymium chelate in water/methanol mixtures are discussed in [47 to 52].

Fry and Pirie [53] observed that exposure of a solution of $Eu(C_{10}H_9O_2)_3 \cdot H_2O$ in 3:1 (v/v) ethanol-methanol to ultraviolet light or heat resulted in decomposition of the chelate. The reaction yielded ethyl acetate and acetophenone as the principal products, suggesting that the ligand decomposition is the reverse of a Claisen condensation. Kinetic experiments carried out at 70°C revealed first-order plots for the disappearance of the europium chelate [53]. Upon irradiation with ultraviolet light, the diamagnetic lanthanum chelate produced an ESR signal, indicating formation of an organic radical upon decomposition of the complex [54].

5.1.12.3 Adducts of the Tris(benzoylacetonato) Complexes

With Ammonia (= NH_3)

$Eu(C_{10}H_9O_2)_3 \cdot 2NH_3$. The compound was prepared by adding, at ambient temperature, a solution of europium chloride (0.00387 mol) in 25 ml of absolute ethanol to an absolute ethanolic solution (25 ml) of benzoylacetone saturated with ammonia. An immediate precipitate resulted, which was collected by filtration, washed with cold absolute ethanol, and air dried (yield = 56%). The melting range of the compound was 120 to 122°C. The authors proposed that both ammonia molecules are coordinated to the europium ion, and that the coordination number of the central ion is eight [55].

With Diethylamine and Benzoylacetone $M(C_{10}H_9O_2)_3 \cdot C_{10}H_{10}O_2 \cdot (C_2H_5)_2NH$

The crystal and molecular structures of a compound having the same empirical composition as the diethylammonium salt of the anionic tetrakis europium chelate, see p. 178, were determined by single-crystal X-ray analysis (R = 0.150). The structural analysis revealed that the compound was a diethylamine adduct of the neutral tetrakis chelate, and not the salt of the anionic complex. Furthermore, a series of compounds of analogous compositions derived from the lighter rare earths (M = La to Gd) were shown by X-ray analyses to be isomorphous. The detailed procedures for preparation of the compounds were not presented, although the authors noted that the compounds were obtained upon rapid crystallization, presumably from solutions containing a europium salt, benzoylacetone, and diethylamine [60].

The europium compound crystallizes in space group B 2/b-C_{2h}^{6} (No. 15),with four molecules in a unit cell of dimensions a = 24.912, b = 18.200, c = 10.206 Å, γ = 115.6°. The metal chelate molecules are packed such that channels 5.8 Å in diameter are formed along the [001] direction. The diethylamine molecules are located in these channels in a disordered manner. The europium ion is coordinated to eight oxygen atoms from four bidentate benzoylacetone groups. The average Eu-O interatomic bond distance is 2.38 Å. The coordination polyhedron was described as a tetragonal antiprism (*s s s s* isomer). The chelated edges of the retangular faces are shorter than the unchelated edges (2.74 versus 2.94 Å). Other evidences which support the conclusion that this compound is a crystal solvate are: (1) the packing density is lower in this compound than in piperidinium tetrakis(benzoylacetonato)europate(III); (2) hydrogen bonds of the O-N-H type, analogous to those found in the piperidinium salt of the anionic chelate and which fixed the position of the outer-sphere ions are absent; (3) the diethylamine groups are disordered within the channels; (4) the diethylamine can be eliminated from the compound without destroying its structure [60].

With 2,2′-Bipyridine (= $C_{10}H_8N_2$ = bpy)

$M(C_{10}H_9O_2)_3 \cdot$ bpy. The neodymium and erbium compounds were prepared by recrystallizing the respective hydrated tris chelates from an ethanolic solution containing bipyridine [56].

With 1,10-Phenanthroline (= $C_{12}H_8N_2$ = phen)

$M(C_{10}H_9O_2)_3 \cdot$ phen. To a hot solution of 6 mmol of benzoylacetone and 2 mmol of europium chloride in 50 ml of ethanol and 5 ml of water were added 3.0 ml of 2.0 N sodium hydroxide solution. The solution was filtered while hot to effect separation of the precipitated sodium chloride, and to the filtrate was added 2 mmol of 1,10-phenanthroline monohydrate in 10 ml of ethanol. After the reaction mixture had cooled, the precipitated product was collected by filtration, washed with water, and dried under vacuum. The europium compound was recrystallized from a 1:1 acetonitrile-methanol solution. The melting range of the compound thus obtained was 192 to 194°C [57]. The europium compound was also obtained by dissolving the hydrated tris chelate and phenanthroline in benzene and adding petroleum ether to effect separation of the brightly yellow colored product (melting range = 186 to 188°C) [58].

Butter, Kreher [58] noted that the fluorescence intensity of the adduct was enhanced relative to that observed for the tris chelate of europium(III). The neodymium and erbium compounds were prepared by recrystallizing the respective hydrated tris chelates from alcoholic solutions containing 1,10-phenanthroline [56]. The isomer shift (−0.58±0.05 mm · s^{-1} relative to the source, ^{151}Sm in Sm_2O_3) obtained for $Eu(C_{10}H_9O_2)_3 \cdot$ phen from the Mössbauer spectrum provided no evidence for participation of 4f-electrons in covalent bonds [59].

With Tributylphosphate $(C_4H_9O)_3PO$ (= $C_{12}H_{27}O_4P$) or Triphenylphosphine Oxide $(C_6H_5)_3PO$ (= $C_{18}H_{15}OP$)

$M(C_{10}H_9O_2)_3 \cdot$ L. Each series of adducts (M = Pr, Nd, Gd, Er, Yb) was prepared by allowing the respective hydrated tris chelate to react directly with the neutral O-donor ligand (1:1 mol ratio). The shifts of νP=O (60 cm^{-1} for L = tributylphosphate, 20 to 35 cm^{-1} for L = triphenylphosphine oxide) to lower wave numbers in the infrared spectra of the adducts relative to the free ligand values suggested introduction of the neutral ligand into the inner coordination sphere of the central ion [56].

With Diantipyrylmethane (= $C_{23}H_{24}N_4O_2$)

$La(C_{10}H_9O_2)_3 \cdot C_{23}H_{24}N_4O_2$. The complex was prepared from an aqueous acetone solution of $LaCl_3$, diantipyrylmethane, and benzoylacetone. On the basis of the IR spectrum, the La atom is assumed to have a coordination number of eight [62].

References to 5.1.12.2 and 5.1.12.3:

[1] R. G. Charles (Inorg. Syn. **9** [1967] 37/40). — [2] C. Brecher, H. Samelson, A. Lempicki (J. Chem. Phys. **42** [1965] 1081/96). — [3] M. Hasan, K. Kumar, S. Dubey, S. N. Misra (Bull. Chem. Soc. Japan **41** [1968] 2619/23). — [4] B. S. Sankhla, R. N. Kapoor (Australian J. Chem. **20** [1967] 685/8). — [5] L. Sacconi, R. Ercoli (Gazz. Chim. Ital. **79** [1949] 731/8).

[6] D. Purushottam, V. Ramachandra Rao, Bh. S. V. Raghava Rao (Anal. Chim. Acta **33** [1965] 182/97). — [7] D. Purushottam, Bh. S. V. Raghava Rao (Indian J. Chem. **4** [1966] 109/10). — [8] R. E. Whan, G. A. Crosby (J. Mol. Spectrosc. **8** [1962] 315/27). — [9] N. Filipescu, C. R. Hurt, N. McAvoy (J. Inorg. Nucl. Chem. **28** [1966] 1753/6). — [10] V. V. Kuznetsova, A. N. Sevchenko, V. S. Khomenko (Zh. Prikl. Spektrosk. **2** [1965] 147/53; J. Appl. Spectrosc. [USSR] **2** [1965] 98/103).

[11] S. J. Lyle, A. D. Witts (Inorg. Chim. Acta **5** [1971] 481/4). — [12] F. A. Cotton, P. Legzdins (Inorg. Chem. **7** [1968] 1777/83). — [13] N. K. Khalmurzaev, I. A. Murav'eva, L. I. Martynenko, V. I. Spitsyn, A. I. Byrke (Zh. Neorgan. Khim. **20** [1975] 2062/5; Russ. J. Inorg. Chem. **20** [1975] 1148/50). — [14] R. G. Charles, A. Perotto (J. Inorg. Nucl. Chem. **26** [1964] 373/6). — [15] E. V. Anufrieva, A. N. Zaidel (Zh. Eksperim. Teor. Fiz. **24** [1953] 114/6; C.A. **1955** 2188/9).

[16] R. C. Mathur, S. S. L. Surana, S. P. Tandon (Z. Naturforsch. **30b** [1975] 207/9). — [17] S. Misumi, N. Iwasaki (Bull. Chem. Soc. Japan **40** [1967] 550/4). — [18] C. Y. Liang, E. J. Schimitschek, J. A. Trias (J. Inorg. Nucl. Chem. **32** [1970] 811/31). — [19] M. F. Richardson, W. F. Wagner, D. E. Sands (Inorg. Chem. **7** [1968] 2495/500). — [20] P. C. Mehta, S. S. L. Surana, S. P. Tandon (Indian J. Pure Appl. Phys. **7** [1969] 767/8 from C.A. **72** [1970] No. 26664).

[21] S. P. Tandon, P. C. Mehta, R. N. Kapoor, S. N. Mishra (Z. Naturforsch. **25b** [1970] 472). — [22] P. C. Mehta, S. S. L. Surana, S. P. Tandon (Can. J. Spectrosc. **18** [1973] 55/60)). — [23] A. I. Byrke, N. N. Magdesieva, L. I. Martynenko, V. I. Spitsyn (Zh. Neorgan. Khim. **12** [1967] 666/71; Russ. J. Inorg. Chem. **12** [1967] 348/51). — [24] P. C. Mehta, S. P. Tandon (Z. Naturforsch. **26a** [1971] 759/62). — [25] K. I. Gur'ev (Nekot. Vop. Nelinein. Opt. Teor. Spektrosk. Kvant. Khim. [1972] 32/40; C.A. **79** [1973] No. 141234).

[26] G. A. Domrachev, V. P. Ippolitova (Tr. Khim. Tekhnol. **1966** No. 2, p. 227/40; C.A. **68** [1968] No. 73665). — [27] M. L. Bhaumik (J. Mol. Spectrosc. **11** [1963] 323/4). — [28] V. S. Korol'kov, V. V. Kuznetsova (Opt. Spektroskopiya **19** [1965] 764/70; Opt. Spectrosc. [USSR] **19** [1965] 423/6). — [29] L. D. Derkacheva, A. D. Kudryavtseva, G. V. Peregudov, A. I. Sokolovskaya (Zh. Prikl. Spektrosk. **6** [1967] 346/9; J. Appl. Spectrosc. [USSR] **6** [1967] 233/5). — [30] P. A. Bazhulin, L. D. Derkacheva, B. G. Dustanov, G. V. Peregudov, A. M. Prokhorov, A. I. Sokolovskaya, D. N. Shigorin (Opt. Spektroskopiya **18** [1965] 526/9; Opt. Spectrosc. [USSR] **18** [1965] 298/300).

[31] V. A. Voloshin, A. G. Goryushko, V. A. Kul'chitskii (Ukr. Fiz. Zh. **9** [1964] 192/5; C.A. **61** [1964] 192). — [32] V. A. Voloshin, A. G. Goryushko, V. A. Kul'chitskii (Opt. Spektroskopiya **15** [1963] 286/7; Opt. Spectrosc. [USSR] **15** [1963] 154/5). — [33] V. A. Voloshin, A. G. Goryushko, N. K. Davidenko, G. V. Klimusheva, V. A. Kul'chitskii (Opt. Spektroskopiya **18** [1965] 628/30; Opt. Spectrosc. [USSR] **18** [1965] 357/8). — [34] G. A. Crosby, R. E. Whan (Naturwissenschaften **47** [1960] 276/7). — [35] A. N. Sevchenko, A. K. Trofimov (Zh. Eksperim. Teor. Fiz. **21** [1951] 220/9; C.A. **1952** 2403).

[36] K. I. Gur'ev, N. I. Davydova, I. A. Zhigunova, V. F. Zolin, M. A. Kovner, V. A. Kudryashova, V. I. Tsaryuk (Opt. Spektroskopiya **28** [1970] 921/5; Opt. Spectrosc. [USSR] **28** [1970] 499/501). — [37] K. I. Gur'ev (Issled. Nelinein. Opt. Spektroskopii [1973] 52/7; C.A. **83** [1975] No. 18166). — [38] S. P. Tandon, R. C. Govil (Z. Naturforsch. **26a** [1971] 1357/9). — [39] S. P. Tandon, R. C. Govil, K. Tandon (Spectrosc. Letters **6** [1973] 125/9). — [40] S. P. Tandon, R. C. Govil (Spectrosc. Letters **4** [1971] 73/7).

[41] W. G. Perkins, G. A. Crosby (J. Chem. Phys. **42** [1965] 407/14). — [42] W. G. Perkins, G. A. Crosby (J. Chem. Phys. **42** [1965] 2621/3). — [43] S. P. Tandon, P. C. Mehta (J. Chem. Phys. **52** [1970] 4313/5). — [44] P. C. Mehta, S. P. Tandon (J. Chem. Phys. **53** [1970] 414/7). — [45] N. S. Poluektov, L. I. Kononenko, S. V. Bel'tyukova, V. N. Drobyazko, E. V. Melent'eva (Zh. Fiz. Khim. **49** [1975] 1379/81; Russ. J. Phys. Chem. **49** [1975] 814/6; C.A. **83** [1975] No. 170268).

[46] S. F. Mason, R. D. Peacock, B. Stewart (Mol. Phys. **30** [1975] 1829/41). — [47] N. K. Davidenko, A. A. Zholdakov (Zh. Neorgan. Khim. **14** [1969] 408/11; Russ. J. Inorg. Chem. **14** [1969] 210/2). — [48] N. K. Davidenko, K. B. Yatsimirskii, L. N. Lugina (Zh. Neorgan. Khim. **13** [1968] 138/42; Russ. J. Inorg. Chem. **13** [1968] 70/3). — [49] A. G. Goryushko, N. K. Davidenko (Zh. Neorgan. Khim. **19** [1974] 1772/7; Russ. J. Inorg. Chem. **19** [1974] 965/8). — [50] A. G. Goryushko, N. K. Davidenko (Zh. Prikl. Spektrosk. **13** [1970] 60/5; J. Appl. Spectrosc. [USSR] **13** [1970] 901/5; C.A. **73** [1970] No. 125336).

[51] N. K. Davidenko, A. A. Zholdakov (Zh. Neorgan. Khim. **14** [1969] 83/9; Russ. J. Inorg. Chem. **14** [1969] 44/7). — [52] N. K. Davidenko, A. G. Goryushko, Yu. V. Naboikin, A. A. Zholdakov (Zh. Prikl. Spektrosk. **8** [1968] 74/9; J. Appl. Spectrosc. [USSR] **8** [1968] 44/8; C.A. **69** [1968] No. 6827). — [53] F. H. Fry, W. R. Pirie (J. Chem. Phys. **43** [1965] 3761/2). — [54] S. N. Kisel'-nikova, E. L. Sorin, V. B. Tsaregradskii, G. A. Domrachev (Zh. Obshch. Khim. **37** [1967] 2608/10; J. Gen. Chem. [USSR] **37** [1967] 2482/4). — [55] S. M. Lee, L. J. Nugent (J. Inorg. Nucl. Chem. **26** [1964] 2304/6).

[56] I. A. Murav'eva, L. I. Martynenko, V. I. Spitsyn (Zh. Neorgan. Khim. **22** [1977] 3009/12; Russ. J. Inorg. Chem. **22** [1977] 1636/8). — [57] H. Bauer, J. Blanc, D. L. Ross (J. Am. Chem. Soc. **86** [1964] 5125/31). — [58] E. Butter, K. Kreher (Z. Naturforsch. **20a** [1965] 408/12). — [59] M. F. Taragin, J. C. Eisenstein (J. Inorg. Nucl. Chem. **35** [1973] 3815/9). — [60] A. L. Il'inskii, M. A. Porai-Koshits, L. A. Aslanov, P. I. Lazarev (Zh. Strukt. Khim. **13** [1972] 277/86; J. Struct. Chem. [USSR] **13** [1972] 254/62).

[61] I. A. Murav'eva, L. I. Martynenko, N. Denliev (Str. Svoistva Primen. Beta-Diketonatov Metal. Mater. 3rd Vses. Semin., Moscow 1977 [1978], pp. 84/7 from C.A. **91** [1979] No. 150507). — [62] M. A. Tishchenko, G. I. Gerasimenko, Yu. N. Anisimov, N. S. Poluektov (Dokl. Akad. Nauk SSSR **250** [1980] 122/5; Dokl. Chem. Proc. Acad. Sci. USSR **250** [1980] 16/9).

5.1.12.4 Tetrakis Chelates $M'[M(C_{10}H_9O_2)_4]$

5.1.12.4.1 Alkali Metal Salts (M' = Na, K)

Melby and coworkers [1] effected synthesis of the alkali metal salts (M' = Na, K) of the tetrakis chelate of europium by adding solutions of europium chloride (2 mmol) in 10 ml of water to a solution, heated at reflux, prepared by mixing benzoylacetone (10 mmol) in 50 ml of acetone with 10 ml of 1 N alkali hydroxide. Each solution was gravity filtered while hot, and the filtrate was allowed to cool to room temperature. The product, in the form of crystalline needles, was collected by filtration and washed with 1:1 (v/v) acetone/water. The melting point of the sodium salt was 220°C [1].

Verron, Meyer [2] effected synthesis of $Na[Eu(C_{10}H_9O_2)_4]$ by three methods. In the first method, an ethanolic solution containing 1.52 g of europium chloride was added, dropwise, to a solution prepared by adding sodium hydroxide (1.20 g) to 4.86 g of benzoylacetone in 30 ml of ethanol. The resulting crystals were collected by filtration, washed with ethanol, and air dried. In the second method, an ethanolic solution (50 ml) of $Eu(C_{10}H_9O_2)_3 \cdot 2H_2O$ (3.36 g) was mixed with 4.62 g of sodium benzoylacetonate (prepared by dissolving 2.3 g of sodium in a solution containing 16.2 g benzoylacetone in 100 ml p-xylene). The precipitated product was collected by filtration, washed with alcohol, and air dried. Alternatively, an aqueous solution of europium chloride (1.52 g) was added to 1.48 g of sodium benzoylacetonate in 10 ml of absolute ethanol. The resulting precipitate was collected, washed, and dried as described above [2]. Meyer [3], following the latter procedure, also effected syntheses of several sodium salts (M = La, Eu, Tb). The latter compounds decomposed without melting at 250°C on heating. Their solubility in dimethylformamide at 20°C is about 10^{-3} mol · 1^{-1} [3].

Crozet, Meyer [4] recorded the emission spectrum of $Na[Eu(C_{10}H_9O_2)_4]$ at 77 K in pure dimethylformamide and 1:3 (v/v) dimethylformamide/ethanol. The appearance of two peaks (at 5801 and 5798 Å in the mixed solvent) in the region of the $^5D_0 \rightarrow {}^7F_0$ transition was attributed to partial dissociation of the tetrakis chelate in solution, with formation of a dimethylformamide adduct of the tris chelate (a peak at 5798 Å was attributed to the latter species). These investigators interpreted conductance data (10^{-3} M solutions in both solvents at 295 K) in terms of the existence of two forms of the tetrakis chelate in solution, a molecular form, $Na[Eu(C_{10}H_9O_2)_4]$, and an anionic form, $Eu(C_{10}H_9O_2)_4^-$. Using the intensity of the 5798 Å line to determine the relative concentration of the tris chelate, together with the limiting conductance values calculated from the Onsager equation for solutions of the chelate in both dimethylformamide and 1:3 dimethylformamide-ethanol, the mole per cent of each species in solution was calculated. For dimethylformamide: $Eu(C_{10}H_9O_2)_3 \cdot DMF$ = 7%, $Eu(C_{10}H_9O_2)_4^-$ = 33%, $Na[Eu(C_{10}H_9O_2)_4]$ = 60%; for 1:3 (v/v) dimethylformamide/ethanol: $Eu(C_{10}H_9O_2)_3 \cdot DMF$ = 30%, $Eu(C_{10}H_9O_2)_4^-$ = 17%, $Na[Eu(C_{10}H_9O_2)_4]$ = 53% [4].

5.1.12.4.2 Piperidinium Salts $(C_5H_{12}N)[M(C_{10}H_9O_2)_4]$

General. Bhaumik and coworkers [5] first isolated a substance with an empirical composition corresponding to that of the piperidinium salt of the tetrakis chelate. However, their data were insufficient to allow for an unequivocal assignment of the molecular structure. Later it was shown that the two isomeric forms of the species were obtained, depending on the method of preparation [6]. Single-crystal X-ray analyses [7 to 12, 34] revealed that one isomer was the true piperidinium salt of the anionic tetrakis chelate, whereas the other form was a piperidine solvate of the neutral tetrakis chelate. In each isomer, however, the rare earth ion is bonded to eight benzoylacetonate oxygen atoms. Inasmuch as the preparative methods for the isomeric series of chelates are similar and in some instances, the particular isomer obtained was not identified, the preparation and properties of both forms of the chelate are discussed in this section.

Preparation and Properties. Addition of an ethanolic [5, 33] or aqueous [1] solution of europium chloride to a hot solution of benzoylacetone and piperidine in ethanol yielded, on standing, a crystalline precipitate, with a composition corresponding to that of the piperidinium salt of the anionic tetrakis chelate. The precipitate was collected, washed with cold absolute ethanol, and air dried [5]. The compound isolated by Bhaumik and coworkers [5] melted at 125.5 to 126.5°C, whereas the compound obtained by Melby and coworkers [1] melted at 132°C. Workman, Burns [6] observed that rapid recrystallization from methanol or butanol of the initial product, obtained by allowing europium chloride to react with benzoylacetone in the presence of piperidine, produced one crystalline modification of the chelate, designated the β-form, which, if left standing in the recrystallization liquor, redissolved, and the other isomer, designated the α-form, crystallized slowly. Preparation of the two isomeric forms of the tetrakis neodymium chelate was effected in the same manner. X-ray diffraction studies revealed that the corresponding forms of the europium and neodymium chelates are isomorphous [6]. Ippolitova and coworkers [14] prepared the tetrakis europium and gadolinium chelates according to the procedure described by Bhaumik and coworkers [5], then effected separation of the two isomeric forms of the chelate by fractional crystallization. Khomenko, Kuznetsova [15] prepared the α-form of the europium and terbium chelates by adding piperidine (0.15 ml) to a solution of $M(NO_3)_3 \cdot 6H_2O$ (100 mg) and benzoylacetone (250 mg) in 5 ml of ethanol. In each case, rectangular crystals formed within a few hours, which were collected, washed with ethanol, and dried in air (melting point of both chelates = 110 to 112°C). The β-form of the europium and terbium chelates was prepared according to two procedures: 1) 300 mg of the α-form was dissolved in 8 ml of hot methanol, and within a few hours, clumps of small needles were produced, which were washed with a little methanol and dried in air; 2) 300 mg of $M(NO_3)_3 \cdot 6H_2O$ and 700 mg of benzoylacetone were dissolved in 5 ml of methanol, and 0.45 ml of piperidine was added to the mixture, together with a few crystals of the β-form, effecting crystallization of the product (m.p. in °C: M = Eu, 124 to 126; M = Tb, 125). Lastovskii, Temkina, and Tsirul'nikova [16, 17] prepared a compound they formulated as the piperidinium salt of the europium chelate by allowing the ligand and europium nitrate to react in aqueous ethanol in the presence of piperidine.

Crystal and Molecular Structures. The crystal and molecular structures of the europium chelate (obtained according to a procedure similar to that described by Bhaumik and coworkers [5]) were determined by single-crystal X-ray analysis [8, 10, 12, 34] (R = 0.200 [34]). The compound crystallizes in the space group $P2_1/c\text{-}C_{2h}^5$ (No. 14) with four molecules in a unit cell of dimensions a = 11.786, b = 19.643, c = 18.867 Å, β = 110°47.7′ [10]. These crystallographic data correspond to those obtained by Workman, Burns [6] for the α-form of the europium chelate (space group $P2_1/c$; a = 11.91, b = 19.71, c = 19.06, β = 110°43′). The crystal consists of discrete tetrakis(benzoylacetonato)europate(III) anions (**Fig. 5-26**) and piperidinium cations, linked by hydrogen bonds ($N\text{-}O_2$ = 2.60 Å, $N\text{-}O_6$ = 2.88 Å) [10, 34]. The europium(III) ion is eight-coordinate, with oxygen atoms from four bidentate benzoylacetonato groups as donors. The average Eu-O bond distance is 2.39 Å (range = 2.33 to 2.42 Å). The coordination polyhedron was described as a tetragonal antiprism (*s s s s* isomer), with the chelate rings located on opposing sides of the quadrangular faces of the antiprism. Distortions from the square antiprism are relatively small and were attributed to constraints imposed by the ligands. The faces of the antiprism remain planar but are transformed into rectangles (the edges spanned by the chelate rings are 2.74 Å, whereas the unchelated edges are 2.94 Å). All four phenyl groups are located on one side of a plane passing through the central ion and perpendicular to the two-fold symmetry axis [34]. The complex belongs to the C_2 symmetry point group [12]. However, if the phenyl and methyl substituents are ignored, the symmetry at the europium(III) site is D_2 [34].

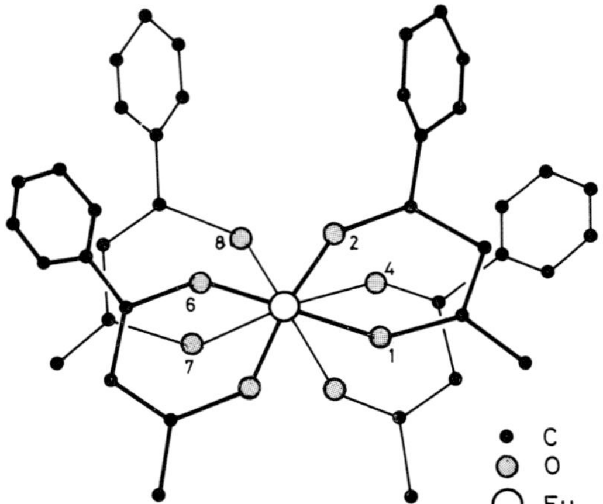

Fig. 5-26

Schematic representation of the tetrakis-(benzoylacetonato) europate(III) ion.

While preparing single crystals of the gadolinium chelate for X-ray analysis, Butman and coworkers [11] obtained two crystalline modifications of the compound which had different crystallographic characteristics. The crystals of one isomer belonged to space group $B2/b\text{-}C_{2h}^6$ (No. 15) and the other to space group $P2_1/c\text{-}C_{2h}^5$ (No. 14). The latter phase was isostructural with the α-form of the europium chelate (structure discussed above). The compound which crystallized in space group B 2/b was shown to be more properly formulated as a piperidine adduct of the neutral tetrakis gadolinium chelate, rather than as a piperidinium salt of the anionic tetrakis chelate. This compound crystallizes with four molecules in a unit cell of dimensions a = 31.568, b = 20.214, c = 10.692 Å, γ = 138°10′. The chelate molecules are packed such that channels of cross section 7 × 9 Å are formed along the crystallographic c-axis. Piperidine molecules, not piperidinium ions, are contained in the channels. The coordination polyhedron is similar to that described for piperidinium tetrakis(benzoylacetonato) europate(III), with the gadolinium atom located on the two-fold axis, surrounded by a tetragonal antiprism of eight oxygen atoms. The range of Gd-O bond distances is 2.38 to 2.42 Å. The chelated edge of the rectangular face of the antiprism is 0.3 Å shorter than the unchelated edge. The phenyl rings are located on one side of the plane perpendicular to the two-fold axis and passing through the gadolinium atom [11].

On the basis of X-ray diffraction data, Il'inskii and coworkers [34] reported that compounds derived from lanthanum through lutetium were isomorphous, although the method of preparation was not presented.

Spectral Data. Emission spectra recorded for microcrystalline samples of the europium chelate at 77 K [18] and 93 K [19] are recorded in Table 5/37. Good agreement between the two sets of data was observed in the region of the $^5D_0 \rightarrow {}^7F_0$ transition and the strongest emissions in the regions of the $^5D_0 \rightarrow {}^7F_1$ and $^5D_0 \rightarrow {}^7F_2$ transitions. Nugent and coworkers [18] concluded on the basis of their data that the symmetry at the europium site in the crystalline sample is C_1, C_{1h}, C_2, or C_{2v}. Brecher, Samelson, Lempicki [19], on the other hand, proposed that the europium ion is located in a site of C_{4v} symmetry. They noted that the intensity of the $^5D_0 \rightarrow {}^7F_0$ transition, forbidden under C_{4v} symmetry, was fully three orders of magnitude less intense than the $^5D_0 \rightarrow {}^7F_1$ and $^5D_0 \rightarrow {}^7F_2$ emissions [19]. Goryushko and coworkers [20] also interpreted the emission spectrum of the europium chelate (microcrystalline solid) in terms of C_{4v} symmetry.

Khomenko and Kuznetsova [15] noted that the fluorescence spectra of the α- and β-forms of the crystalline europium chelates revealed a single line in the region of the $^5D_0 \rightarrow {}^7F_0$ transition (580 nm), indicating that each chelate contains only one type of emission center, i.e., the crystals of each compound are uniform in composition. The $^5D_0 \rightarrow {}^7F_0$ transition in the β-form is much stronger than in the α-form, suggesting a lower symmetry at the metal ion site in the β-form. The $^5D_0 \rightarrow {}^7F_2$ transition for the α-form revealed two closely spaced components (612.7 and 613.3 nm), whereas the β-form revealed four lines (611.5, 614.4, 618.0, 619.9 nm). These results were interpreted in terms of C_{4v} or C_4 symmetry at the europium site in the α-isomer versus C_{2v} symmetry for the β-isomer.

Table 5/37
Emission Spectrum of Microcrystalline Piperdinium Tetrakis(benzoylacetonato) Europate(III) in the Regions of the $^5D_0 \rightarrow {}^7F_J$ (J = 0, 1, 2) Transitions.

transition	Nugent, Bhaumik, George, Lee[a] [18] wavelength (Å)	relative intensity[c]	Brecher, Samelson, Lempicki[b] [19] wavelength (Å)[d]
$^5D_0 \rightarrow {}^7F_0$	5801.7		5803
$^5D_0 \rightarrow {}^7F_1$	5907.6	10	5909 (s)
			5916
	5923.4	10	5923 (s)
	5951.4	5	5952 (s)
			5980
$^5D_0 \rightarrow {}^7F_2$			6038
			6114
	6125.1	180	6126 (s)
	6132.6	180	6132 (s)
	6173.6	6	6173
	6184.1	6	6184
	6245.5		

[a] Spectrum recorded at 77 K. — [b] Spectrum recorded at 93 K. — [c] The relative intensities are consistent only among ligand field transitions with the same values of J. — [d] The symbol (s) indicates the strongest lines within each individual transition region.

Brecher and coworkers [19] recorded the spectrum of the europium chelate (compound prepared by adding a stoichiometric amount of europium chloride dissolved in 95% ethanol to an ethanolic solution containing equimolar amounts of benzoylacetone and piperidine) in 3:1 (v/v) ethanol/methanol (standard alcohol solvent) and 4:1 (v/v) standard alcohol solvent/dimethylformamide. In each case, two distinct lines were observed in the region of the $^5D_0 \rightarrow {}^7F_0$ transition. One of the lines corresponded to that observed in the spectrum of the tris chelate, see p. 167, indicating a dissociative equilibrium of the type $Eu(C_{10}H_9O_2)_4^- \rightleftharpoons Eu(C_{10}H_9O_2)_3 + C_{10}H_9O_2^-$. In the standard alcohol solvent, the degree of dissociation of the tetrakis chelate is 37%, compared with 82% in the dimethylformamide/alcohol mixture (the degree of dissociation was calculated from the relative intensities of the $^5D_0 \rightarrow {}^7F_0$ transitions corresponding to the tris and tetrakis chelates) [19]. In the alcohol solvent, the spectrum of the tetrakis chelate was interpreted in terms of D_{2d} symmetry at the europium site [19, 21], whereas in the alcohol-DMF solvent, the spectrum was interpreted in terms of C_{4v} symmetry at the metal ion site. This lower symmetry was attributed to coordination of dimethylformamide, resulting in formation of a nine-coordinate species [19].

The emission spectrum of the europium chelate in a polymethylmethacrylate host has also been reported [22]. Luminescence polarization studies are reported in [23] and analysis of vibrational fine structure is reported in [24].

Molecular weight (vapor pressure osmometry) and conductance measurements for solutions of the europium chelate in benzene, chloroform, and acetonitrile indicated that the complex is monomeric, and undergoes 100 per cent ionic dissociation in each solvent [25].

The ESR spectrum of the gadolinium ion doped (10 per cent) into a single crystal of piperidinium tetrakis(benzoylacetonato)europate(III) showed all the lines in fields up to 14 kOe ($g_z = g_x = 1.992$) [13].

The Mössbauer spectrum of the europium chelate (prepared by allowing europium chloride, benzoylacetone, and piperidine to react in ethanol [26]), taken with a ^{151}Eu source in an Sm_2O_3 lattice, was obtained over the temperature range 79 to 220 K. The mean value ($-0.44\ mm \cdot s^{-1}$) of the isomer shift is positive relative to the predominantly ionic $EuF_3 \cdot 0.5H_2O$ [27].

Alkylammonium Salts

Diethylammonium tetrakis(benzoylacetonato)europate(III) was prepared according to the following procedure: To a hot solution (90°C) containing 0.0155 mol of benzoylacetone and 0.0155 mol of diethylamine in 25 ml of absolute ethanol, was added a warm solution (25 ml) of europium chloride in the same solvent. The mixture was maintained at ambient temperature for four days in a dry box and then filtered. The resulting crystalline precipitate was collected, washed with cold absolute ethanol and air dried. The melting point of the compound was 128 to 130°C [18]. For the crystal and molecular structures of a compound having the same composition, but which is formulated as adduct of $M(C_{10}H_9O_2)_3$, see p. 171.

The tetrapropylammonium salt $(n\text{-}C_3H_7)_4N[Eu(C_{10}H_9O_2)_4]$ was prepared in the following way: To a solution of 8 mmol of benzoylacetone and 2 mmol of europium chloride in 75 ml of hot ethanol were added 8 mmol of 10% aqueous tetrapropylammonium hydroxide. The compound separated in crystalline form upon allowing the solution to cool. The product, recrystallized from 2-propanol, melted at 158 to 166°C [28].

Morpholinium Salt $(C_4H_{10}NO)[Eu(C_{10}H_9O_2)_4]$

The compound was prepared according to the procedure described in Section 1.1.3 for preparation of the corresponding diethylammonium salt. The melting point of the product was 127 to 128°C [18].

5.1.12.5 Cerium(IV) Compound $Ce(C_{10}H_9O_2)_4$

Yoshimura, Miyake, and Imoto [29] prepared the cerium(IV) compound by adding a solution of ligand in ethanol to an aqueous solution of ammonium cerium(IV) sulfate tetrahydrate. The product was extracted from the resulting precipitate with toluene or benzene and allowed to crystallize from the resulting solution. The complex was recrystallized from benzene and dried under vacuum [29].

Sacconi and Ercoli [30] prepared the compound by adding an ethanolic ammonia solution to a solution of cerium(III) nitrate hexahydrate and benzoylacetone in the same solvent, then heating the mixture on a steam bath. The brown-red precipitate which formed was collected and recrystallized from benzene or chloroform/petroleum ether (melting point = 100 to 101°C).

Purushottam and coworkers [31] also obtained the compound by adding an ethanolic solution of ligand to a solution of cerium(IV) nitrate in absolute ethanol, then adjusting the pH of the reaction mixture to 5.6 by the addition of alcoholic ammonia. The precipitated product was collected by filtration, washed with ethanol, and dissolved in benzene. The benzene solution was filtered and the filtrate allowed to evaporate at room temperature. The product was dried under vacuum over concentrated sulfuric acid. The brown colored substance melted at 185°C [31].

Infrared spectral data (samples dispersed in Nujol and hexachlorobutadiene mulls) are tabulated and discussed in [29]. The νCe-O stretching mode was assigned to a band at 242 cm^{-1} [29]. The electronic absorption spectrum of the red-brown compound in acetonitrile revealed bands at 32500 (76700), and 40400 cm^{-1} (35700) (absorption coefficient, $\varepsilon/dm^3 \cdot mol^{-1} \cdot cm^{-1}$, given in parentheses) [32].

References to 5.1.12.4 and 5.1.12.5:

[1] L. R. Melby, N. J. Rose, E. Abramson, J. C. Caris (J. Am. Chem. Soc. **86** [1964] 5117/25). — [2] M. Verron, Y. Meyer, Centre National de la Recherche Scientifique (Fr. 1453474 [1966] from C.A. **67** [1967] No. 16686). — [3] Y. H. Meyer (J. Phys. [Paris] **27** [1966] 415/21 from C.A. **66** [1967] No. 24073). — [4] P. Crozet, Y. H. Meyer (Nature **213** [1967] 1115/6). — [5] M. L. Bhaumik, P. C. Fletcher, L. J. Nugent, S. M. Lee, S. Higa, C. L. Telk, M. Weinberg (J. Phys. Chem. **68** [1964] 1490/3).

[6] M. O. Workman, J. H. Burns (Inorg. Chem. **8** [1969] 1542/4). — [7] L. A. Butman, L. A. Aslanov, M. A. Porai-Koshits (Zh. Neorgan. Khim. **11** [1966] 2400/1; Russ. J. Inorg. Chem. **11** [1966] 1287). — [8] L. A. Aslanov, A. L. Il'inskii, L. A. Butman, M. A. Porai-Koshits (Zh. Strukt. Khim. **8** [1967] 705/7; Russ. J. Struct. Chem. **8** [1967] 634/5). — [9] L. A. Butman, L. A. Aslamov, M. A. Porai-Koshits (Zh. Strukt. Khim. **9** [1968] 332/3; Russ. J. Struct. Chem. **9** [1968] 274). — [10] L. A. Aslanov, A. L. Il'inskii, P. I. Lazarev, M. A. Porai-Koshits (Zh. Strukt. Khim. **10** [1969] 345/7; Russ. J. Struct. Chem. **10** [1969] 321/2).

[11] L. A. Butman, L. A. Aslanov, M. A. Porai-Koshits (Zh. Strukt. Khim. **11** [1970] 46/53; Russ. J. Struct. Chem. **11** [1970] 40/46). — [12] A. V. Aristov, Yu. S. Maslyukov, M. I. Gryaznova, G. A. Domrachev, L. A. Aslanov, A. L. Il'inskii (Teor. Eksperim. Khim. **6** [1970] 61/6; Theor. Exptl. Chem. [USSR] **6** [1970] 53/7). — [13] A. G. Anders, A. I. Zvyagin, V. A. Moiseev, A. I. Otko, Yu. K. Khudenskii (Teor. Eksperim. Khim. **5** [1969] 282/4; Theor. Exptl. Chem. [USSR] **5** [1969] 182/3). — [14] V. P. Ippolitova, G. A. Domrachev, M. I. Gryaznova, L. A. Aslanov, Yu. K. Khudenskii (Zh. Neorgan. Khim. **14** [1969] 723/5; Russ. J. Inorg. Chem. **14** [1969] 377/9). — [15] V. S. Khomenko, V. V. Kuznetsova (Zh. Prikl. Spektrosk. **7** [1967] 850/3; J. Appl. Spectrosc. [USSR] **7** [1967] 564/7; C.A. **69** [1968] No. 6836).

[16] R. P. Lastovskii, V. Ya. Temkina, N. V. Tsirul'nikova (Tr. Vses. Nauchn. Issled. Inst. Khim. Reaktivov Osobo Chist. Khim. Veshchestv No. 29 [1966] 299/303 from C.A. **67** [1967] No. 87321). — [17] R. P. Lastovskii, V. Ya. Temkina, N. V. Tsirul'nikova (Zh. Vses. Khim. Obshchestva **11** [1966] 587 from C.A. **67** [1967] No. 3132). — [18] L. J. Nugent, M. L. Bhaumik, S. George, S. M. Lee (J. Chem. Phys. **41** [1964] 1305/12). — [19] C. Brecher, H. Samelson, A. Lempicki (J. Chem. Phys. **42** [1965] 1081/96). — [20] A. G. Goryushko, V. A. Kul'chitskii, Yu. V. Naboikin, Yu. K. Khudenskii (Spektrosk. Krist. Mater. 2nd Simp., 1967 [1970], pp. 227/32 from C.A. **74** [1971] No. 17664).

[21] C. Brecher, A. Lempicki, H. Samelson (J. Chem. Phys. **41** [1964] 279/80). — [22] A. G. Goryushko, Yu. V. Naboikin, L. A. Ogurtsova, A. P. Podgornyi (Opt. Spektroskopiya **24** [1968] 732/9; Opt. Spectrosc. [USSR] **24** [1968] 392/5). — [23] R. A. Puko, N. N. Mit'kina, V. V. Kuznetsova, V. N. Kotlo, V. S. Khomenko (Zh. Prikl. Spektrosk. **21** [1974] 736/8; J. Appl. Spectrosc. [USSR] **21** [1974] 1405/6; C.A. **82** [1975] No. 36947). — [24] A. Baczynski, W. Orzeszko (Acta Phys. Polon. A **48** [1975] 127/34 from C.A. **83** [1975] No. 170316). — [25] N. Filipescu, C. R. Hurt, N. McAvoy (J. Inorg. Nucl. Chem. **28** [1966] 1753/6).

[26] S. J. Lyle, A. D. Witts (Inorg. Chim. Acta **5** [1971] 481/4). — [27] S. J. Lyle, A. D. Witts (J. Chem. Soc. Dalton Trans. **1975** 185/8). — [28] H. Bauer, J. Blanc, D. L. Ross (J. Am. Chem. Soc. **86** [1964] 5125/31). — [29] T. Yoshimura, C. Miyake, S. Imoto (Bull. Chem. Soc. Japan **46** [1973] 2096/101). — [30] L. Sacconi, R. Ercoli (Gazz. Chim. Ital. **79** [1949] 731/8).

[31] D. Purushottam, V. Ramachandra Rao, Bh. S. V. Raghava Rao (Anal. Chim. Acta **33** [1965] 182/97). — [32] M. Ciampolini, F. Mani, N. Nardi (J. Chem. Soc. Dalton Trans. **1977** 1325/8). — [33] S. M. Lee, L. J. Nugent (J. Inorg. Nucl. Chem. **26** [1964] 2304/6). — [34] A. L. Il'inskii, M. A. Porai-Koshits, L. A. Aslanov, P. I. Lazarev (Zh. Strukt. Khim. **13** [1972] 277/86; Russ. J. Struct. Chem. **13** [1972] 254/62).

5.1.12.6 Other Benzoylacetonato Complexes

Hydroxide $M(C_{10}H_9O_2)_2OH \cdot H_2O$

Each compound (M = Y, La, Pr, Nd, Sm, Eu) was prepared by adding dilute aqueous ammonia to an aqueous ethanolic solution of benzoylacetone and the rare earth nitrate, until the pH of the reaction mixture was 7 to 7.5. The resulting granular precipitate was collected, recrystallized from acetone, and air dried. The colors and melting points (°C) of the compounds are: Y, white, 142; La, white, 155; Pr, green-yellow, 162; Nd, pink, 164; Sm, white, 165. The compounds were characterized by infrared [1], electronic absorption [1], and emission [2] spectral measurements.

Acetate $Eu(C_{10}H_9O_2)_2(C_2H_3O_2)$

The compound was prepared by adding, with vigorous shaking, an aqueous solution (25 ml) containing sodium acetate (0.03 mol) and europium chloride (0.005 mol) to a solution of benzoylacetone (0.0165 mol) in absolute ethanol (100 ml). The reaction mixture, containing a precipitate, was heated on a water bath for 15 to 20 min. After standing at room temperature overnight, the solution was filtered, and the precipitate collected was washed with absolute ethanol and dried in air. The compound melted at 53 to 54°C [3]. The emission spectrum of the product in 3:1 (v/v) ethanol/methanol at 77 K revealed bands at 16371 Å ($^5D_0 \rightarrow {}^7F_2$), 16840, 16865, 16931, 16984 Å ($^5D_0 \rightarrow {}^7F_1$), and 17331 Å ($^5D_0 \rightarrow {}^7F_0$) [4].

2-Propanolato Salts $M(C_{10}H_9O_2)_n(C_3H_7O)_{3-n}$ (n = 1, 2)

Each compound (M = Pr, Nd, Sm) was prepared by allowing stoichiometric amounts of the rare earth 2-propanolate to react with benzoylacetone in benzene. The 2-propanol formed during the course of the reaction was removed by distillation as the azeotrope with benzene. Molecular weight measurements in benzene (ebullioscopically) revealed that the n = 1 compounds exist as trimers, whereas the n = 2 compounds exist as dimers in this solvent [5, 6].

Mixed Chelates with 2-Pyridinecarboxylic Acid (= $C_6H_5NO_2$)
8-Hydroxyquinoline (= C_9H_7NO) and **2-Quinolinecarboxylic Acid** (= $C_{10}H_7NO_2$) (= HL)

$M(C_{10}H_9O_2)_2L$. Each series of mixed chelates (M = Y, La, Pr, Nd, Sm) was prepared by heating at reflux a solution (solvent not specified) containing $M(C_{10}H_9O_2)_2OH \cdot H_2O$ (see above) and the heterocyclic ligand. The resulting precipitates were collected, washed with boiling solvent, and allowed to dry. The colors and melting points of the compounds are recorded in Table 5/38 [60].

Table 5/38
Physical Properties of the Mixed Chelates Derived from Benzoylacetone and N-heterocyclic Ligands, $M(C_{10}H_9O_2)_2L$ [8].

M	HL = 2-pyridinecarboxylic acid		HL = 8-hydroxyquinoline		HL = 2-quinolinecarboxylic acid	
	color	m.p. in °C	color	m.p. in °C	color	m.p. in °C
Y	white	215	yellow	206	white	208
La	white	252	yellow	—	white	255
Pr	light yellow	295	yellow	262	light yellow	252
Nd	light pink	248	yellow	253	light pink	252
Sm	white	209	yellow	259	white	205

Mixed Complex with N,N'-Bis(salicylidene)ethylenediamine
$HOC_6H_4CH{=}NCH_2CH_2N{=}CHC_6H_4OH$ (= $C_{16}H_{16}N_2O_2$)

$M(C_{10}H_9O_2)(C_{16}H_{14}N_2O_2)$. Each mixed ligand complex (M = Y, La, Pr, Nd, Sm) was prepared by adding 30 ml of an ethanolic solution of bis(salicylidene)ethylenediamine (1 mmol) to 50 ml of an ethanolic solution of the tris(benzoylacetonato) rare earth chelate (1 mmol), then heating the mixture at reflux on a steam bath. After 0.5 h, the mixed chelates began to precipitate, but heating at reflux was continued for 2 h. The solution was filtered while hot, and the compound was washed several times with ethanol and dried under vacuum at room temperature over sulfuric acid. The colors and melting points (°C) of the compounds were: Y, light yellow, 266 to 270; La, yellow, 268 to 270; Pr, greenish yellow, 273 to 275; Nd, light pink, 270 to 272; Sm, yellow, 274 to 276. The compounds were further characterized by elemental analyses and their infrared spectra (KBr discs) [7].

References to 5.1.12.6:

[1] N. K. Dutt, S. Sur (J. Inorg. Nucl. Chem. **31** [1969] 3171/6). — [2] N. K. Dutt, S. Sur, S. Rahut (J. Inorg. Nucl. Chem. **33** [1971] 1717/24). — [3] V. Janardhanarao, A. P. B. Sinha (Indian J. Chem. **4** [1966] 196/7). — [4] J. V. Rao, D. R. Rao, A. P. B. Sinha (Indian J. Chem. **8** [1970] 270/4). — [5] M. Hasan, K. Kumar, S. Dubey, S. N. Misra (Bull. Chem. Soc. Japan **41** [1968] 2619/23).

[6] B. S. Sankhla, R. N. Kapoor (Australian J. Chem. **20** [1967] 685/8). — [7] N. K. Dutt, K. Nag (J. Inorg. Nucl. Chem. **30** [1968] 2779/83). — [8] N. K. Dutt, S. Sur (J. Inorg. Nucl. Chem. **31** [1969] 3171/6).

5.1.13 Complexes with 4,4,4-Trifluoro-1-phenyl-1,3-butanedione
$CF_3C(O)CH_2C(O)C_6H_5$ (= Benzoyltrifluoroacetone = $C_{10}H_7F_3O_2$)

5.1.13.1 Tris Chelates $M(C_{10}H_6F_3O_2)_3 \cdot nH_2O$ (n = 1 to 3)

Ismail, Lyle, Newbery [1] effected syntheses of several hydrated chelates (n = 1 for M = La, Eu; n = 2 for M = Pr, Nd, Eu) by adding sodium hydroxide (6 mmol in 20 ml of 50% (v/v) ethanol/water mixture) dropwise, over 2 h, to well-stirred solutions of the respective rare earth nitrates (2 mmol) and benzoyltrifluoroacetone (6.6 mmol) in 150 ml of 50% (v/v) ethanol/water. The ethanol was allowed to evaporate, and the products were recrystallized from a chloroform/petroleum ether mixture. The physical properties of the compounds are recorded below. The molecular weight data indicate the compounds tend to associate in benzene. The two monohydrates revealed a greater association than the dihydrates, leading effectively to formation of dimeric species [1].

M	n	color	decomposition range in °C	molecular weight in benzene	
				found	calc*)
La	1	white	122 to 129	2200	802
Pr	2	green	112 to 125	1040	822
Nd	2	pale blue	133 to 134	1080	825
Eu	1	white	123 to 130	1470	815
Eu	2	white	152 to 160	1040	833

*) Calculated for $M(C_{10}H_6F_3O_2)_3 \cdot nH_2O$

Livingstone, Zimmermann [2] prepared a series of tris chelates (n = 3 for M = Pr, Sm, Ho; n = 2 for M = Nd, Eu, Gd, Dy, Er; n = 1 for M = Yb) according to the following procedure. Benzoyltrifluoroacetone (1.0 g) was dissolved in alcohol (30 ml), and a stoichiometric amount of 0.2 M ammonia was added slowly to the solution. The resulting solution was added, dropwise (0.5 to

1.0 ml · min^{-1}) and with rapid stirring, to a solution of the rare earth nitrate in 60% (v/v) ethanol/water, while the pH was maintained within the range 6.0 to 6.5 by addition of dilute ammonia or hydrochloric acid. Complete precipitation of each product was effected by adding 100 ml of water, dropwise and with rapid stirring, while the pH was maintained within the range 6.0 to 6.5. The microcrystalline product was collected by filtration, washed with water, and dried under vacuum in a desiccator over silica gel [2].

Charles, Riedel [3] prepared the europium chelate (n = 2) by combining benzoyltrifluoroacetone (0.018 mol), triethanolamine (0.018 mol), and europium chloride (0.004 mol) in a warm ethanol/water mixture. The addition of more water effected separation of an oil which solidified on long standing at room temperature. The crystalline solid was collected by filtration, washed with 50% ethanol/water, and air dried at room temperature. The compound melted at 107 to 110°C.

Brecher, Samelson, Lempicki [4] noted that the emission spectrum at 93 K of a sample of $Eu(C_{10}H_6F_3O_2)_3 \cdot 2H_2O$ in the crystalline state and in 3:1 (v/v) ethanol/methanol and 3:1:1 (v/v) ethanol/methanol/dimethylformamide revealed a multiplicity of lines that was greater than the number allowed for a single chemical species. For the powdered sample: $^5D_0 \to {}^7F_2$ = 6235, 6215, 6205, 6161, 6135, 6114 Å; $^5D_0 \to {}^7F_1$ = 5980, 5960, 5915, 5880, 5838 Å; $^5D_0 \to {}^7F_0$ = 5802, 5793, 5774 Å.

Ismail and coworkers [1] observed that dehydration of the dihydrates at 10°C under reduced pressure (15 Torr) produced a weight loss greater than that expected for the loss of two water molecules. They proposed that a decomposition of the type, $M(C_{10}H_6F_3O_2)_3 \cdot 2H_2O \to M(C_{10}H_6F_3O_2)_2OH + C_{10}H_7F_3O_2 + H_2O$, occurred upon heating the sample. The thermogram of $Eu(C_{10}H_6F_3O_2)_3 \cdot 2H_2O$ [3] revealed that decomposition occurred immediately above the dehydration temperature region, rendering difficult the obtaining of the anhydrous chelate through thermal dehydration.

The mass spectra of the chelates (samples prepared according to the procedure described in [2]) revealed a peak corresponding to the molecular ion (= M) of the anhydrous complex, indicating that the water of hydration was lost under the low pressure conditions obtained in the mass spectrometer. Although peaks for the ion M-L ($L = C_{10}H_6F_3O_2$) occurred in virtually all spectra, peaks for the ions M-2L and M-3L were absent. However, peaks for M-2L+F and M-3L+2F(MF_2^+) were obtained for nearly every chelate, indicating fluorine migration to the metal had taken place. A peak for the ion M-3L+F was obtained for samarium, europium, and ytterbium, corresponding to a valency change from 3 to 2 for these compounds in the mass spectrometer [2].

The Mössbauer spectrum of the europium chelate (n = 2), taken with a ^{151}Eu source in an Sm_2O_3 lattice, was obtained over the temperature range 79 to 220 K. The average isomer shift, −0.43 mm · s^{-1} (range −0.40 to −0.45 mm · s^{-1}) was positive relative to the predominantly ionic compound, $EuF_3 \cdot 0.5H_2O$, indicating some covalency in the europium/oxygen bond [5].

5.1.13.2 Lewis Base Adducts of the Tris Chelates

With Butylamine (= $C_4H_9NH_2$) and Pyridine (= C_5H_5N = py)

$Nd(C_{10}H_6F_3O_2)_3 \cdot H_2O \cdot C_4H_9NH_2$ and $Nd(C_{10}H_6F_3O_2)_3 \cdot H_2O \cdot 2py$. The butylamine and pyridine adducts were obtained by treating solutions of $Nd(C_{10}H_6F_3O_2)_3 \cdot H_2O$ in benzene with the amines at room temperature. The butylamine adduct melted at 205 to 207°C and the pyridine adduct at 87 to 92°C. The compounds were characterized further by elemental and thermogravimetric analyses and by infrared, visible, and ultraviolet spectroscopy [6].

With 1,10-Phenanthroline (= $C_{12}H_8N_2$ = phen)

$Eu(C_{10}H_6F_3O_2)_3 \cdot phen$. To a hot solution of 6 mmol of benzoyltrifluoroacetone and 2 mmol of europium chloride in 50 ml of ethanol and 5 ml of water were added 3.0 ml of 2.0 N sodium hydroxide solution. The solution was filtered, while hot, to effect separation of the precipitated sodium chloride, and to the filtrate was added 2 mmol of 1,10-phenanthroline monohydrate in 10 ml of ethanol. After the reaction mixture had cooled, the precipitated product was collected by filtration, washed with water and dried under vacuum. The product was recrystallized from a 1:1 acetonitrile/methanol mixture (m.p. = 197 to 199°C) [7].

With Hexanol, Tributyl Phosphate and Trioctylphosphine Oxide (= L)

$M(C_{10}H_6F_3O_2)_3 \cdot nL$ (n = 1, 2). Formation in Solution

Equilibrium constants corresponding to formation of several neutral oxygen donor adducts were determined by a solvent extraction method. The authors noted that the stability of the adducts increased with the ionic radius of the rare earth. The general order of stability observed for a particular rare earth ion, hexanol < tributyl phosphate < trioctylphosphine oxide, corresponds to the donating power of lone pair electrons of the oxygen atoms in these compounds [8].

O-donor	La		Eu		Tb		Lu	
	lg β_1	lg β_2	lg β_1	lg β_2	lg β_1	lg β_2	lg β_1	lg β_2
hexanol	1.95	3.30	1.85	3.10	1.85	3.10	1.65	2.70
tributyl phosphate	4.38	7.80	4.55	7.40	4.50	7.30	4.70	6.00
trioctylphosphine oxide	7.00	12.30	6.85	11.70	6.90	11.20	7.50	*)

*) Formation of the bis adduct was not observed.

The adducts $Nd(C_{10}H_6F_3O_2)_3 \cdot H_2O \cdot 2C_2H_5OH$ and $Nd(C_{10}H_6F_3O_2)_3 \cdot 2C_4H_9OH$ were obtained by treating a solution of $Nd(C_{10}H_6F_3O_2)_3 \cdot H_2O$ in benzene with the alcohol at room temperature. The melting points of both the ethanol and butanol adducts were 84 to 88°C. The compounds were characterized further by elemental and thermogravimetric analyses and by infrared, visible and ultraviolet spectroscopy [6].

5.1.13.3 Tetrakis Chelates of Type $M'[M(C_{10}H_6F_3O_2)_4]$ ($M' = NH_4^+$ or an Organic Cation)

The compounds recorded in Table 5/39, p. 184, were prepared according to the following procedures:

A) To 3.89 g (0.018 mol) benzoyltrifluoroacetone dissolved in 60 ml of 95% ethanol was added 0.018 mol of the organic base (ammonium or tetraalkylammonium hydroxide), followed by 0.004 mol hydrated europium chloride dissolved in 20 ml of water. The mixture was heated briefly, nearly to boiling, and then allowed to cool to room temperature overnight. The precipitate was collected by filtration, washed with 70% ethanol/water, and air dried at room temperature [3].

B) A solution of hydrated europium chloride (0.0039 mol) in 25 ml of absolute ethanol was added to a mixture of benzoyltrifluoroacetone (0.0155 mol) and the organic amine (0.0155 mol) in 25 ml of the same solvent. The solution was allowed to stand at room temperature until crystallization of the product was complete. The product was collected by filtration, washed with a small amount of cold absolute ethanol, and air dried at room temperature [3].

C) An aqueous solution (20 ml) of a rare earth chloride (4 mmol) was added to a solution of benzoyltrifluoroacetone (18 mmol) and triethylamine (18 mmol) in 60 ml of boiling 95% ethanol. The solution was filtered while hot, and the filtrate allowed to stand overnight at room temperature, effecting crystallization of an analytically pure chelate [9].

D) A solution (20 ml) of 1 N NaOH was pipetted into a boiling mixture of benzoyltrifluoroacetone (20 mmol) and tetraethylammonium bromide (5 mmol) in 50 ml of 95% ethanol, followed by the addition of a solution of europium nitrate (5 mmol) in 20 ml of warm water. The mixture was allowed to stand at room temperature for 2 days; then the crystalline product was collected by filtration, washed with 50% alcohol, and dried [9].

E) Stoichiometric amounts (1:4:4 mol ratio) of the rare earth chloride, benzoyltrifluoroacetone, and amine were mixed in hot ethanol, and the solution was allowed to cool to room temperature. The product was collected by filtration and washed with water or alcohol [7, 10 to 12]. The 2,4,6-trimethylpyridinium (No. 19) and the isoquinolinium (No. 21) salts were recrystallized from carbon tetrachloride and 2-propanol, respectively [7].

Table 5/39
Salts of Tetrakis (Benzoyltrifluoroacetonato) Rare Earth(III) Chelates $M'[M(C_{10}H_6F_3O_2)_4]$ and Their Physical Properties.

No.	cation M′	formula	M	method of preparation	melting point in °C	molar conductance in $\Omega^{-1} \cdot cm^2 \cdot mol^{-1}$ *) in acetonitrile	in acetone	Ref.
1	ammonium	NH_4	Eu	**)				[14,16]
2	tetramethylammonium	$(CH_3)_4N$	Eu	A	217 to 218	109(23.5°)	102(24.5°)	[3]
3	diethylammonium	$(C_2H_5)_2NH_2$	Eu	B	135 to 136	113(26°)	96(14.5°)	[3]
4***)	triethylammonium	$(C_2H_5)_3NH$	Eu	A	87 to 88	100(26°)	95(25°)	[3]
5	triethylammonium	$(C_2H_5)_3NH$	La Nd Tb	C	130 130 130			[9]
6	tetraethylammonium	$(C_2H_5)_4N$	Eu Eu	A D	152 to 153 152	110(23.5°)	94(24.5°)	[3] [9]
7	2-hydroxyethylammonium	$HOCH_2CH_2NH_3$	Eu	A	171 to 172	99(24.5°)	80(24°)	[3]
8	tetrapropylammonium	$(C_3H_7)_4N$	Eu	A	167 to 168	88(23.5°)	84(24°)	[3]
9	butylammonium	$(C_4H_9)NH_3$	Eu	A	78 to 80	101(26°)	82(24°)	[3]
10	tetrabutylammonium	$(C_4H_9)_4N$	Eu	B	136 to 137	80(23.5°)	78(24.5°)	[3]
11	benzylammonium	$C_6H_5CH_2NH_3$	Eu	A	68 to 70	91(26°)	72(25°)	[3]
12	dibenzylammonium	$(C_6H_5CH_2)_2NH_2$	Eu	A	152 to 153	78(26°)	65(25°)	[3]
13	benzyltrimethylammonium	$C_6H_5CH_2N(CH_3)_3$	Eu	**)				[14,16]
14	imidazolium	$C_3H_5N_2$	Eu	E	194 to 195			[10]
15	pyrrolidinium	$C_4H_{10}N$	Eu, Tb	E				[11]
16	piperazinium	$C_4H_{11}N_2$	Eu	**)				[14,16]
17	morpholinium	$C_4H_{10}NO$	Eu	**)				[14,16]
18	pyridinium	C_5H_5NH	Eu	A	180 to 185	76(26°)	13(26°)	[3]
19	2,4,6-trimethylpyridinium	$C_8H_{11}NH$	Eu	E	168 to 170			[7]

Table 5/39 (Continued)

No.	cation M'	formula	M	method of preparation	melting point in °C	molar conductance in $\Omega^{-1} \cdot cm^2 \cdot mol^{-1}$ *) in acetonitrile	in acetone	Ref.
20	quinolinium	C_9H_7NH	Eu	B	155 to 157	63 (24.5°)	7 (24.5°)	[3]
21	isoquinolinium	C_9H_7NH	Eu	A	153 to 155	72 (24.5°)	14 (25.5°)	[3]
			Eu	E	150 to 151			[7]
22	piperidinium	$C_5H_{10}NH_2$	Eu	A	170 to 172	109 (26°)	88 (24.5)	[3]
			Gd	E				[11]
23	tetramethylguanidinium	$C_5H_8N_2$	Eu	A	154 to 155	82 (24.5°)	89 (24.5°)	[3]
24	tetraphenylphosphonium	$(C_6H_5)_4P$	Eu	**)				[14]
25	tetraphenylarsonium	$(C_6H_5)_4As$	Eu	**)				[14]

*) 0.005 M solutions, temperature indicated in parentheses. — **) Method of preparation not reported. — ***) Monohydrate, $(C_2H_5)_3NH[Eu(C_{10}H_6F_3O_2)_4] \cdot H_2O$, resolidified after melting to anhydrous compound: melting point 134 to 135°C.

The tetraethylammonium salt of the cerium(III) chelate, $(C_2H_5)_4N[Ce(C_{10}H_6F_3O_2)_4]$, was prepared under an atmosphere of pure nitrogen using deoxygenated solvents according to the following procedure: A solution of cerium chloride heptahydrate (2 mmol) and 2,2-dimethoxypropane (10 cm^3) in absolute ethanol (15 cm^3) was added to a mixture of benzoyltrifluoroacetone (8 mmol), sodium hydroxide (8 mmol), and tetraethylammonium chloride (2 mmol) in absolute ethanol (25 cm^3). The solvent was removed under reduced pressure, the residue extracted with ethyl acetate (30 cm^3), and the suspended solid removed by filtration. The solution was concentrated under vacuum to approximately 10 cm^3, and then hexane was added, slowly, to effect crystallization of the product. The orange-colored crystals were collected, washed with hexane and dried under vacuum [13].

The emission spectral data for the europium chelates are tabulated and discussed in [3, 4, 7, 9, 14, and 15]. Charles, Riedel [3] noted that the fluorescence properties of the tetrakis europium chelates (compounds No. 2 to 4, 6 to 12, 18, 20 to 23) in the solid state (ambient temperatures) are dependent, to a significant extent, on the nature of the cation. For example, the $^5D_0 \rightarrow {}^7F_0$ transition observed in the spectra of the crystalline tetraalkylammonium salts showed an increase in wavelength of peak position with increasing length of the alkyl chain : methyl (5788 Å), ethyl (5795 Å), propyl (5799 Å), butyl (5801 Å). Wavelengths (in Å) of the $^5D_0 \rightarrow {}^7F_0$ transition for the remainder of the solid chelates studied: piperidinium (5787), butylammonium (5798), diethylammonium (5798), triethylammonium (5797), benzylammonium (5798), dibenzylammonium (5789), 2-hydroxyethylammonium (5790), pyridinium (5795), quinolinium (5798), and isoquinolinium (5797). Shepherd [14] observed at least five very strong lines at 77 K in the region (6100 to 6200 Å) of the $^5D_0 \rightarrow {}^7F_2$ transition for the ammonium (No. 1), piperazinium (No. 16), and morpholinium salts (No. 17) and proposed that the symmetry at the metal ion site is C_1, C_{1h}, C_2, or C_{2v}. The spectrum of the benzyltrimethylammonium salt (No. 13) revealed only two very strong lines, corresponding to this transition at 6114 and 6133 Å, indicating C_{4v} or C_4 symmetry at the metal ion site [14]. Brecher, Samelson, and Lempicki [4] observed two lines in the region of the $^5D_0 \rightarrow {}^7F_0$ transition (5792, 5788 Å) in the spectrum (93 K) of the solid piperidinium salt (No. 22), suggesting the presence of two molecular species. Two strong lines at 6127 and 6114 Å in the region of the $^5D_0 \rightarrow {}^7F_2$ transition were interpreted in terms of C_{4v} symmetry at the europium site (the less intense set of isomer lines was disregarded in this interpretation).

Molar conductance data (Table 5/39) indicated that the tetrakis europium salts are extensively dissociated to M'^+ and $[Eu(C_{10}H_6F_3O_2)_4]^-$ in acetonitrile. The fluorescence spectra of the acetonitrile solutions (ambient temperatures), therefore, are independent of the cation and characteristic of the $Eu(C_{10}H_6F_3O_2)_4^-$ chelate, except in those instances (pyridinium, quinolinium, and isoquinolinium) where interaction between the cation and anion give equilibrium concentrations of the tris chelate (further discussion of conductivity data is presented below) [3]. Brecher and coworkers [4], on the basis of their fluorescence measurements in solution (glass) at 93 K, reported that the tetrakis europium chelate (piperidinium salt) was 100% dissociated to the tris chelate in a 3:1 (v/v) ethanol/methanol mixture and 47% dissociated in a 3:1:1 (v/v) ethanol/methanol/dimethylformamide mixture. Samelson and coworkers [15] noted that the equilibria established among the various forms of the chelated species are strongly dependent on both the concentration of the chelate and the chemical properties of the solvent (solvents employed were: 3:1 ethanol-methanol, 3:1:1 ethanol/methanol/dimethylformamide, 5 : 5 : 2 ether/isopentane/ethanol, 1 : 1 : 1 propionitrile/butyronitrile/isobutyronitrile, and 8:1:1 toluene/benzene/xylene).

The absorption spectra of the compounds in acetonitrile are nearly identical, as expected, if each salt dissociates to the same absorbing species, namely, $[M(C_{10}H_6F_3O_2)_4]^-$ [3].

The molar magnetic susceptibilities of several salts of the europium chelate, as a function of temperature, are recorded in Table 5/40. Shepherd [16] noted that the differences between magnetic susceptibility values for the various salts are significant, with a maximum spread at a particular temperature of about 10%. With the exception of the benzyltrimethylammonium salt, the differences decreased with temperature and at −150°C, the spread was within experimental error (2%). The effective magnetic moments calculated from these values were plotted as a function of temperature, and compared with theoretical values calculated from the Van Vleck equation. Qualitatively, the magnetic behavior of these compounds is in good agreement with that expected for the free europium(III) ion [16].

Table 5/40

Magnetic Susceptibilities in cgsu · 10^4*) of Tetrakis(Benzoyltrifluoroacetonato) Salts, $M'[Eu(C_{10}H_6F_3O_2)_4]$, as a Function of Temperature.

°C	M'^+ = piperidinium	piperazinium	morpholinium	ammonium	benzyltrimethylammonium
100	47.10	44.10	43.00	44.40	42.90
50	50.20	48.10	46.30	47.50	46.00
0	55.10	53.10	51.50	51.70	50.10
−50	59.90	58.70	57.00	57.10	54.20
−100	64.50	63.50	62.00	62.80	58.30
−150	68.00	67.30	66.30	68.00	61.60

*) Experimental values were corrected for diamagnetism of the samples.

Molar conductance values (Table 5/39) for solutions of the salts in acetonitrile and acetone, in most instances, are large enough to be considered indicative of 1:1 electrolytic dissociation ($M'^+ + [Eu(C_{10}H_6F_3O_2)_4]^-$). Conductance values for the pyridinium (No. 18), quinolinium (No. 20), and isoquinolinium (No. 21) compounds are lower than those of the other salts, particularly in acetone. These results were rationalized in terms of the equilibria $M'^+ + [Eu(C_{10}H_6F_3O_2)_4]^- \rightleftharpoons Eu(C_{10}H_6F_3O_2)_3 + C_{10}H_7F_3O_2 + X$ (X = free amine) or $M'^+ + [Eu(C_{10}H_6F_3O_2)_4]^- \rightleftharpoons Eu(C_{10}H_6F_3O_2)_3 + M'(C_{10}H_6F_3O_2)$, by assuming that the equilibria are displaced to the right on dilution [3].

Thermogravimetric analyses indicated that the compounds are thermally stable, generally showing appreciable decomposition only above their melting points. Exceptions were the pyridinium (No. 18), benzylammonium (No. 11), and dibenzylammonium (No. 12) salts which revealed some decomposition at temperatures below the melting point [3].

The Mössbauer spectrum of piperidinium tetrakis(benzoyltrifluoroacetonato)europate(III), taken with a ^{151}Eu source in an Sm_2O_3 lattice, was obtained over the temperature range 79 to 220 K. The average isomer shift, −0.46 mm · s^{-1} (range −0.39 to −0.52 mm · s^{-1}), was positive relative to the predominantly ionic compound, $EuF_3 \cdot 0.5\,H_2O$, indicating some covalency in the europium-oxygen bonds [5].

Cerium(IV) Compound $Ce(C_{10}H_6F_3O_2)_4$

The complex was prepared by adding an ethanolic solution of benzoyltrifluoroacetone to an aqueous solution of cerium(IV) ammonium sulfate tetrahydrate, and gradually neutralizing the mixture with concentrated ammonia. The crude precipitate was extracted with benzene or toluene, and the crystalline product obtained from the extract solution was recrystallized from benzene and dried under vacuum [17].

The absorption spectrum of the red-brown colored product revealed bands at 22500 cm^{-1} (sh), 29900 cm^{-1} (ε = 73200 $dm^3 \cdot mol^{-1} \cdot cm^{-1}$), 36000 cm^{-1} (ε = 27000 $dm^3 \cdot mol^{-1} \cdot cm^{-1}$), and 47000 cm^{-1} (sh) [13]. Absorption bands in the region 1500 to 1700 cm^{-1} in the infrared spectrum (Nujol and hexachlorobutadiene mulls) are tabulated and discussed. The ν(Ce-O) stretching mode was assigned to a band at 230 cm^{-1} [17].

References to 5.1.13:

[1] M. Ismail, S. J. Lyle, J. E. Newbery (J. Inorg. Nucl. Chem. **31** [1969] 1715/24). — [2] S. E. Livingstone, W. A. Zimmermann (Australian J. Chem. **29** [1976] 1845/50). — [3] R. G. Charles, E. P. Riedel (J. Inorg. Nucl. Chem. **28** [1966] 3005/18). — [4] C. Brecher, H. Samelson, A. Lempicki (J. Chem. Phys. **42** [1965] 1081/96). — [5] S. J. Lyle, A. D. Witts (J. Chem. Soc. Dalton Trans. **1975** 185/8).

[6] T. Shigematsu, T. Honjyo (Bull. Inst. Chem. Res. Kyoto Univ. **45** [1967] 282/9 from C.A. **69** [1968] No. 24116). — [7] H. Bauer, J. Blanc, D. L. Ross (J. Am. Chem. Soc. **86** [1964] 5125/31). — [8] T. Shigematsu, M. Tabushi, M. Matsui, T. Honjyo (Bull. Chem. Soc. Japan **40** [1967] 2807/12). — [9] L. R. Melby, N. J. Rose, E. Abramson, J. C. Caris (J. Am. Chem. Soc. **86** [1964] 5117/25). — [10] E. J. Schimitschek, R. B. Nehrich, J. A. Trias (J. Chem. Phys. **42** [1965] 788/90).

[11] S. Sato, M. Wada (Bull. Chem. Soc. Japan **43** [1970] 1955/62). — [12] S. J. Lyle, A. D. Witts (Inorg. Chim. Acta **5** [1971] 481/4). — [13] M. Ciampolini, F. Mani, N. Nardi (J. Chem. Soc. Dalton Trans. **1977** 1325/8). — [14] T. M. Shepherd (J. Inorg. Nucl. Chem. **29** [1967] 2551/9). — [15] H. Samelson, C. Brecher, A. Lempicki (J. Mol. Spectrosc. **19** [1966] 349/71).

[16] T. M. Shepherd (J. Phys. Chem. **71** [1967] 4137/9). — [17] T. Yoshimura, C. Miyake, S. Imoto (Bull. Chem. Soc. Japan **46** [1973] 2096/101).

5.1.14 Complexes with Aryl-substituted β-Diketones $ArC(O)CH_2C(O)R$ (=HL)

5.1.14.1 $M(L)_n^{3-n}$ (n = 1, 2, 3). Formation in Solution

The stepwise and overall formation constants of La^{3+}, Pr^{3+}, Nd^{3+}, and Sm^{3+} with 4-fluoro-, 4-chloro-, 4-bromo-, and 4-methyl-benzoylacetone as well as the ligand-proton stability constants of the ligands, lg K_H, were determined potentiometrically in 75 vol% aqueous dioxane at 30 ± 0.1°C and ionic strength I = 0.1 M ($NaClO_4$).

lg K_H	formation constant	metalion La^{3+}	Pr^{3+}	Nd^{3+}	Sm^{3+}
		$4\text{-}FC_6H_4COCH_2COCH_3$			
9.70	lg K_1	6.61	6.95	7.08	7.34
	lg K_2	5.62	6.06	6.15	6.28
	lg K_3	4.16	4.74	4.81	4.50
	lg β_3	16.39	17.75	18.04	18.12
		$4\text{-}ClC_6H_4COCH_2COCH_3$			
9.475	lg K_1	6.41	6.93	7.16	7.35
	lg K_2	5.38	5.98	6.23	6.16
	lg K_3	4.04	4.78	4.98	4.58
	lg β_3	15.83	17.69	18.37	18.09
		$4\text{-}BrC_6H_4COCH_2COCH_3$			
9.43	lg K_1	6.48	7.00	7.13	7.28
	lg K_2	5.40	6.03	6.13	6.08
	lg K_3	4.06	4.75	4.85	4.35
	lg β_3	15.94	17.78	18.11	17.71
		$4\text{-}CH_3C_6H_4COCH_2COCH_3$			
10.20	lg K_1	6.80	7.25	7.34	7.54
	lg K_2	5.74	6.23	6.53	6.21
	lg K_3	4.13	4.58	4.76	4.34
	lg β_3	16.67	18.06	18.43	18.09

Reference to 5.1.14.1:

R. C. Mehrotra, B. P. Bachlas, B. P. Gupta (Indian J. Chem. A **18** [1979] 370/2).

5.1.14.2 Isolated Tris Chelates

With 4,4,4-Trifluoro-1-(4-methylphenyl)-1,3-butanedione $CH_3C_6H_4C(O)CH_2C(O)CF_3$ ($=C_{11}H_9F_3O_2$) **and 4,4,4-Trifluoro-1-(4-methoxyphenyl)-1,3-butanedione** $CH_3OC_6H_4C(O)CH_2C(O)CF_3=C_{11}H_9F_3O_3$)

$M(C_{11}H_8F_3O_2)_3\cdot nH_2O$ (n = 2, 3) and $M(C_{11}H_8F_3O_3)_3\cdot 2H_2O$. Each compound (n = 3 for M = Pr, Dy, Ho; n = 2 for M = Nd, Sm, Eu, Gd, Er, in the trifluoromethylphenyl series; M = Pr, Sm, Eu, Gd, Dy, Er, Yb in the trifluoromethoxyphenyl series) was prepared according to the following procedure. The fluorinated β-diketone (1.0 g) was dissolved in alcohol (30 ml) and neutralized with a stoichiometric amount of 0.2 M ammonia. The β-diketone solution was then added dropwise (0.5 to 1.0 ml/min), with rapid stirring, to a solution of the rare earth nitrate (0.5 mol equivalent) in 2:3 (v/v) water/ethanol, while the pH was maintained within the range 6.0 to 6.5 with dilute ammonia or hydrochloric acid. Complete precipitation of the rare earth complex was effected by adding 100 ml of water, dropwise, with rapid stirring, while the pH was maintained within the range 6.0 to 6.5. The microcrystalline product was collected by filtration, washed with water, and dried under vacuum in a desiccator over silica gel.

The mass spectra of the chelates were essentially the same. The base peak was usually the molecular ion, M, but in a few cases, peaks corresponding to other ions, e.g., M-2L+F-CF_2, were the most intense. Although peaks for the ion, M-L, occurred in virtually all the spectra, peaks for the ions M-2L and M-3L were absent. However, peaks for the ions M-2L+F and M-3L+2F (LnF_2^+ ion) were obtained for nearly every complex, indicating that fluorine migration to the metal ion had occurred. Peaks corresponding to the ions M-3L+F (LnF^+) were observed in the spectra of the samarium, europium, and ytterbium chelates [1].

With 4,4-Dimethyl-1-(2-nitrophenyl)-1,3-pentanedione $NO_2C_6H_4C(O)CH_2C(O)C(CH_3)_3$ ($=C_{13}H_{15}NO_4$)

$Ce(C_{13}H_{14}NO_4)_3\cdot 2H_2O$. The compound was prepared by adding 2 N aqueous ammonia to a solution of cerium(III) nitrate hexahydrate (1 mmol) and o-nitrobenzoylpivaloylmethane (3.5 mmol) in methanol. The mixture was cooled while it was stirred vigorously. The product first separated as an oil but quickly solidified. The solid was collected, washed with a small amount of a methanol-aqueous ammonia mixture, and then air dried. The yellow colored solid melted at 95°C and had a magnetic moment of 2.24 μ_B [2].

With 1-(2-Naphthyl)-1,3-butanedione $C_{10}H_7C(O)CH_2C(O)CH_3$ (=β-Naphthoylacetone = $C_{14}H_{12}O_2$)

$Eu(C_{14}H_{11}O_2)_3$. The compound was prepared by adding a solution of β-naphthoylacetone in 95% ethanol to an aqueous solution of europium chloride and then neutralizing the resulting mixture with aqueous ammonia. The precipitate was collected, recrystallized from a benzene-petroleum ether mixture, and dried under vacuum at 100°C. The yellow colored compound melted at 156°C [3].

$Eu(C_{14}H_{11}O_2)_3\cdot phen$. The adduct was obtained by treating a solution of the tris chelate in benzene with 1,10-phenanthroline and then effecting precipitation by adding petroleum ether. The yellow colored compound melted at 142 to 146°C. It was shown that the fluorescence of the tris chelate was enhanced upon ligation of 1,10-phenanthroline [3].

References to 5.1.14.2:

[1] S. E. Livingstone, W. A. Zimmermann (Australian J. Chem. **29** [1976] 1845/50). — [2] E. Uhlemann, F. Dietze (Z. Anorg. Allgem. Chem. **386** [1971] 329/34). — [3] E. Butter, K. Kreher (Z. Naturforsch. **20a** [1965] 408/12).

5.1.14.3 Tetrakis Chelates

The piperidinium salts of tetrakis chelates recorded in Table 5/41 were prepared according to the following procedures:

A) A mixture of the β-diketone (0.018 mol) and piperidine (1.53 g) was dissolved, with warming, in 60 ml of 95% ethanol. An aqueous solution (12 ml) of europium chloride (0.004 mol) was added, and the mixture was allowed to stand overnight. The precipitate was collected by filtration, washed with ethanol-water, and air dried at room temperature [1].

B) Similar to procedure A, except that the piperidinium salt of the chelating agent was used [1].

C) A solution of the β-diketone (0.018) in 95% ethanol (60 ml) was treated with piperidine (1.53 g), followed by 0.004 mol of the rare earth chloride dissolved in water (20 ml). The mixture was heated at reflux for 2 h and then allowed to cool to room temperature overnight. The complexes were recrystallized from a suitable solvent such as benzene, chloroform, or methanol [2, 3].

The isoquinolinium salt $C_9H_7NH[Eu(C_{14}H_8F_3O_2)_4]$ derived from naphthoyltrifluoroacetone $C_{10}H_7C(O)CH_2C(O)CF_3$ (= $C_{14}H_9F_3O_2$) was prepared from a solution containing 8 mmol of ligand and 2 mmol of europium chloride in 70 ml of hot ethanol to which 8 mmol of isoquinoline were added. The solution was allowed to stand and cool to room temperature. The resulting precipitate was collected by filtration, washed with water, and dried under vacuum. The product was recrystallized from 2-propanol (m.p. = 111 to 114°C). The compound was characterized by elemental analysis and its emission spectrum [4].

Table 5/41
Piperidinium Salts of Tetrakis Rare Earth Chelates Derived from Arylsubstituted β-diketones, $C_5H_{10}NH_2[M(ArCOCHCOR)_4]$.

No.	Ar =	R =	formula of ligand	M	method of synthesis	melting point in °C	Ref.
1	4-fluorophenyl	CF_3	$C_{10}H_6F_4O_2$	Eu	A	157 to 160	[1]
2	2,5-difluorophenyl	CF_3	$C_{10}H_5F_5O_2$	Sm	C	330	[2]
3	2,5-difluorophenyl	CF_3	$C_{10}H_5F_5O_2$	Eu	C	121	[2]
4	2,5-difluorophenyl	CF_3	$C_{10}H_5F_5O_2$	Tb	C	115	[2]
5	2-chloro-4-fluorophenyl	CF_3	$C_{10}H_5ClF_4O_2$	La	C	210 to 212	[3]
6	2-chloro-4-fluorophenyl	CF_3	$C_{10}H_5ClF_4O_2$	Pr	C	196 to 198	[3]
7	2-chloro-4-fluorophenyl	CF_3	$C_{10}H_5ClF_4O_2$	Nd	C	145 to 147	[3]
8	2-chloro-4-fluorophenyl	CF_3	$C_{10}H_5ClF_4O_2$	Eu	C	249 to 250	[3]
9	3-chloro-4-fluorophenyl	CF_3	$C_{10}H_5ClF_4O_2$	Pr	C	66 to 68	[3]
10	3-chloro-4-fluorophenyl	CF_3	$C_{10}H_5ClF_4O_2$	Nd	C	87 to 88	[3]
11	3-chloro-4-fluorophenyl	CF_3	$C_{10}H_5ClF_4O_2$	Tb	C	142	[2]
12	4-fluoro-2-methylphenyl	CF_3	$C_{11}H_8F_4O_2$	La	C	>360 d	[3]
13	4-fluoro-2-methylphenyl	CF_3	$C_{11}H_8F_4O_2$	Pr	C	173 to 174	[3]
14	4-fluoro-2-methylphenyl	CF_3	$C_{11}H_8F_4O_2$	Nd	C	135 to 136	[3]
15	4-fluoro-3-methylphenyl	CF_3	$C_{11}H_8F_4O_2$	Pr	C	96 to 97	[3]
16	4-fluoro-3-methylphenyl	CF_3	$C_{11}H_8F_4O_2$	Nd	C	146 to 147	[3]
17	4-fluoro-3-methylphenyl	CF_3	$C_{11}H_8F_4O_2$	Eu	C	128 to 130	[3]
18	2-methoxyphenyl	CF_3	$C_{11}H_9F_3O_3$	Eu	A	117 to 120	[1]
19	4-methoxyphenyl	CF_3	$C_{11}H_9F_3O_3$	Eu	A	217 to 219	[1]
20	3-fluoro-4-methoxyphenyl	CF_3	$C_{11}H_8F_4O_3$	Gd	C	107	[2]
21	2,5-dimethoxyphenyl	CF_3	$C_{12}H_{11}F_3O_4$	Eu	A	180 to 182	[1]
22	1-naphthyl	CF_3	$C_{14}H_9F_3O_2$	Eu	B	170 to 175	[1]

Table 5/41 (Continued)

No.	Ar =	R =	formula of ligand	M	method of synthesis	melting point in °C	Ref.
23	2-naphthyl	CF_3	$C_{14}H_9F_3O_2$	Eu	A	203 to 205	[1]
24	2-biphenylyl	CF_3	$C_{16}H_{11}F_3O_2$	Eu	A	237 to 238	[1]
25	2-fluorenyl	CF_3	$C_{17}H_{11}F_3O_2$	Eu	A	205 to 208	[1]
26	3-phenanthryl	CF_3	$C_{18}H_{11}F_3O_2$	Eu	A	145 to 149	[1]
27	4-fluorophenyl	CH_3	$C_{10}H_9FO_2$	La	C	158 to 160	[3]
28	4-fluorophenyl	CH_3	$C_{10}H_9FO_2$	Pr	C	173 to 175	[3]
29	4-fluorophenyl	CH_3	$C_{10}H_9FO_2$	Nd	C	134 to 135	[3]
30	2,5-difluorophenyl	CH_3	$C_{10}H_8F_2O_2$	Eu	C	208	[2]
31	2-chloro-4-fluorophenyl	CH_3	$C_{10}H_8ClFO_2$	La	C	175 to 177	[3]
32	3-chloro-4-fluorophenyl	CH_3	$C_{10}H_8ClFO_2$	La	C	86 to 88	[3]
33	3-chloro-4-fluorophenyl	CH_3	$C_{10}H_8ClFO_2$	Pr	C	102 to 105	[3]
34	3-chloro-4-fluorophenyl	CH_3	$C_{10}H_8ClFO_2$	Nd	C	116 to 117	[3]
35	3-chloro-4-fluorophenyl	CH_3	$C_{10}H_8ClFO_2$	Eu	C	>360 d	[3]
36	2-fluoro-5-methylphenyl	CH_3	$C_{11}H_{11}FO_2$	La	C	122 to 125	[3]
37	4-fluoro-2-methylphenyl	CH_3	$C_{11}H_{11}FO_2$	La	C	135 to 136	[3]
38	4-fluoro-2-methylphenyl	CH_3	$C_{11}H_{11}FO_2$	Pr	C	127 to 129	[3]
39	4-fluoro-2-methylphenyl	CH_3	$C_{11}H_{11}FO_2$	Nd	C	101 to 102	[3]
40	4-fluoro-3-methylphenyl	CH_3	$C_{11}H_{11}FO_2$	La	C	130 to 132	[3]
41	4-fluoro-3-methylphenyl	CH_3	$C_{11}H_{11}FO_2$	Pr	C	140 to 141	[3]
42	4-fluoro-3-methylphenyl	CH_3	$C_{11}H_{11}FO_2$	Nd	C	133 to 135	[3]
43	4-fluorophenyl	C_2F_5	$C_{11}H_6F_6O_2$	Eu	C	250 d	[2]
44	2,5-difluorophenyl	C_2F_5	$C_{11}H_5F_7O_2$	Eu	C	113	[2]
45	3-fluoro-4-methoxyphenyl	C_2F_5	$C_{12}H_8F_6O_2$	Eu	C	241 d	[2]
46	4-fluorophenyl	C_2H_5	$C_{11}H_{11}FO_2$	La	C	124 to 125	[3]
47	4-fluorophenyl	C_2H_5	$C_{11}H_{11}FO_2$	Pr	C	114 to 115	[3]
48	4-fluorophenyl	C_2H_5	$C_{11}H_{11}FO_2$	Nd	C	170 to 172	[3]

Infrared spectral data for several chelates (Nos. 2, 3, 5 to 17, 20, 27 to 48) are reported and discussed in [2] and [3]. The $\nu(C{=}O)$ and $\nu(C{=}C)$ stretching modes were assigned to bands in the regions 1620 to 1550 cm^{-1} and 1530 to 1500 cm^{-1}, respectively. The fluorescence properties of the europium chelates were examined at room temperature [1 to 3] and at 77 K [2] in the microcrystalline solid state [1, 2] and in solution [2, 3]. Charles and Riedel [1] noted that the room temperature fluorescence intensity of the solid varied appreciably from compound to compound, an observation they attributed principally to the local symmetry condition about the europium ion, rather than electronic effects. Joshi and coworkers [2, 3] noted that fluorine substitution for hydrogen atoms in the ligand increased fluorescence intensity, and attributed the enhanced fluorescence intensity to decreased vibrational energy. The chelates of praseodymium and neodymium either failed to fluoresce or fluoresced very weakly so that there was no visible fluorescence [3]. The emissions spectra of piperidinium salts $C_5H_{10}NH_2[Eu(ArCOCH_2COR)_4]$ with Ar = 4-methylphenyl, 4-methoxyphenyl, and 4-bromophenyl and $R = CF_3$ were examined. The observed number of lines and their intensities in the 500 to 700 nm region indicate that the ligand field symmetry around the europium ion in the unsubstituted chelate and its 4-methyl and 4-methoxy derivatives belongs to one of the following point groups: C_1, C_{1h}, C_2, or C_{2v}. A higher symmetry, approximately C_4 or C_{4v}, is found for the 4-bromo derivative [5].

For preparation and luminescence properties of an europium tetrakis chelate with benzylidenetrifluoroacetyl acetone, $Ar = C_6H_5CH{=}CH$, $R = CF_3$, $(= C_{12}H_9F_3O_2)$, see [6].

References to 5.1.14.3:

[1] R. G. Charles, E. P. Riedel (J. Inorg. Nucl. Chem. **29** [1967] 715/23). — [2] K. C. Joshi, V. N. Pathak, S. Bhargava, P. G. Seybold (J. Fluorine Chem. **9** [1977] 387/97). — [3] K. C. Joshi, V. N. Pathak (J. Inorg. Nucl. Chem. **35** [1973] 3161/70). — [4] H. Bauer, J. Blanc, D. L. Ross (J. Am. Chem. Soc. **86** [1964] 5125/31). — [5] E. Kwiatkowski, H. Dabrowska, S. Zachara (Acta Phys. Polon. A **55** [1979] 517/20 from C.A. **91** [1979] No. 29943).

[6] M. Tanaka, G. Yamaguchi, J. Shiokawa, C. Yamanaka (Bull. Chem. Soc. Japan **43** [1970] 549/50).

5.1.15 With 1,3-Diphenyl-1,3-propanedione $C_6H_5C(O)CH_2C(O)C_6H_5$ (= Dibenzoylmethane $C_{15}H_{12}O_2$)

5.1.15.1 $M(C_{15}H_{11}O_2)_n^{3-n}$ Complexes (n = 1 to 3). Formation in Solution

Formation constants for the 1:1, 2:1, and 3:1 (ligand:metal) rare earth chelates derived from dibenzoylmethane (M = Pr, Nd, Eu) were determined potentiometrically (glass electrode) at 24°C, in 80% (v/v) methanol-water, I = 0.1 (NaCl) (values are presented below) [1]. Stepwise formation constants for the europium chelate were determined at 25°C by a solvent extraction method (aqueous phase maintained at I = 0.1 (KNO_3), benzene or chloroform used as the organic phase): $K_1 = 3.8 \times 10^7$, $K_2 = 4.6 \times 10^6$, $K_3 = 2.85 \times 10^5$ [2].

metal ion	lg K_1	lg K_2	lg K_3
Pr^{3+}	8.2	6.81	4.6
Nd^{3+}	8.3	6.92	4.7
Eu^{3+}	8.7	7.13	4.9

5.1.15.2 Tris Chelates

5.1.15.2.1 Anhydrous Compounds $M(C_{15}H_{11}O_2)_3$

Whan, Crosby [3] effected synthesis of each chelate (M = La to Lu except for Ce, Pm) by adding slowly and with stirring, a solution (20 ml) of the hydrated rare earth chloride in absolute ethanol to an ethanolic solution (300 ml) containing 3 g of dibenzoylmethane (25% excess) and 2 ml of piperidine. The reaction mixture was evaporated to about two-thirds of its original volume, and the bright yellow precipitate formed was collected by filtration. The precipitate dissolved readily in approximately 50 ml of acetone, and this solution was poured slowly into 200 ml of absolute ethanol. The solution was evaporated to about 50 ml, and the resulting precipitate was collected. After a second recrystallization, the product was air dried. Analysis revealed that this product contained an extra mole of chelating agent per mole of tris chelate (species was later shown to be the piperidinium salt of the anionic tetrakis chelate, see p. 203). The tris chelate was obtained by heating the substance under vacuum at 125 to 150°C for 24 h. The final products melted with decomposition in the range 210 to 245°C [3]. Lyle, Witts [4], in a critical examination of methods used to prepare tris and tetrakis rare earth diketonate chelates, noted that $Eu(C_{15}H_{11}O_2)_3$ was obtained without difficulty according to the Whan and Crosby procedure.

Purushottam and coworkers [5, 6] effected precipitation of the tris chelates (M = Sc, Y, La to Yb, except for Ce, Pm) by adding saturated solutions of ethanolic ammonia to solutions of rare earth nitrates in absolute ethanol (for optimum pH for precipitation of each chelate, see [5], color, and melting points after drying in vacuum over concentrated sulfuric acid see Table 5/42).

The anhydrous tris chelates of praseodymium [7], neodymium [7], and samarium [8] were prepared by allowing the respective rare earth 2-propanolates to react with a solution, heated at reflux, of dibenzoylmethane in benzene (1:3 mol ratio). The 2-propanol liberated during the reaction was removed from the reaction mixture by distillation as the azeotrope with benzene. The products were isolated as crystalline solids by removing the solvent from the clear solution under reduced pressure.

The compounds are soluble in benzene and alcohol. Molecular weight determinations (ebullioscopically) in benzene revealed that the praseodymium and neodymium chelates exist as dimers [7], but that the samarium chelate exists as a monomer in this solvent [8].

The scandium chelate was prepared by adding, slowly and with stirring, an aqueous solution (10 ml) of scandium chloride to a solution of dibenzoylmethane (18 g, 0.080 mol) in 2:1 (v/v) ethanol/ether, followed by the addition of dilute ammonia until no further reaction was evident. The golden yellow precipitate was collected by filtration, washed with water and several small portions of ether, and recrystallized from benzene/hexane (yield 76.7%). The standard heat of combustion, $\Delta H^{\circ}_{298.16}$, of the compound, was determined using a static oxygen bond calorimeter ($\Delta H^{\circ}_{298.16} = -5303.8 \pm 10.0$ kcal · mol^{-1}). The homolytic and heterolytic metal to oxygen bond enthalpies were evaluated from the thermodynamic data, and the values obtained were 65.1 and 231.3 kcal · mol^{-1}, respectively. The lattice energy of the compound was calculated as −1398.0 kcal · mol^{-1} [9].

Perrotto, Charles [10] observed that the anhydrous tris chelates could also be obtained by carefully dehydrating the $M(C_{15}H_{11}O_2)_3 \cdot nH_2O$ (n = 0.5, 1) species. Color and melting points, see Table 5/42. However, the hydrates, particularly those of the lighter rare earths, were susceptible to hydrolysis upon heating, resulting, in some instances, in the formation of impure products.

The infrared spectrum (KBr disc) of the anhydrous tris chelate of samarium was recorded in the region 4000 to 250 cm^{-1}, and the absorption bands, together with their assignments, are tabulated in [11]. Bands at 1588 (s), 1500 (m), and 1470 cm^{-1} (s) were assigned, respectively, to the ν_{asym} (C-O), ν_{asym} (C-C), and ν_{sym} (C-O) stretching modes. Three bands at 655, 510, and 350 cm^{-1} were assigned to M-O vibrational modes, consistent with their assignment of D_3 symmetry for the complex [11].

The emission spectrum at 93 K of the anhydrous europium chelate (compound prepared according to the procedure described by Whan, Crosby [3]) in 3:1 (v/v) ethanol/methanol (standard solvent) and 4:1 (v/v) standard solvent/dimethylformamide, and as a microcrystalline powder, revealed a multiplicity of lines that could not be accounted for in terms of a single chemical species. For the crystalline sample: $^5D_0 \rightarrow ^7F_2$ = 6210, 6154, 6120, 6111 Å; $^5D_0 \rightarrow ^7F_1$ = 6047, 5990, 5984, 5969, 5963, 5957, 5934, 5922, 5906, 5890 Å; $^5D_0 \rightarrow ^7F_0$ = 5806, 5802 Å [12]. The fluorescence lifetime at 77 K of the anhydrous europium chelate (compound prepared according to the procedure of Whan, Crosby [3]) was determined for the microcrystalline solid and for the chelate in 5:5:2 (v/v) ether: 3-methylpentane: ethanol and in several polystyrene and epoxy polymer hosts. Values were in the range of 332 to 568 µs [13, 14]. A discussion of emission spectra of the series of anhydrous tris chelates is presented in [3].

Oscillator strengths for absorption bands observed in the spectrum of the neodymium chelate (compound prepared according to the procedure of Whan, Crosby [3]) in methanol, dimethylformamide, 3:1 (v/v) ethanol/methanol, 1:1:3 (v/v) methanol/ethanol/dimethylformamide, were determined and compared with values calculated from the Judd-Ofelt theory [15, 16]. Good correlation of the experimental and calculated oscillator strengths was generally obtained using only the τ_4 and τ_6 parameters in the Judd-Ofelt equation, except for the hypersensitive transition observed in the region 16250 to 18250 cm^{-1}. The intensity of the hypersensitive transition depended markedly on the parameter τ_2 [17]. The value of the nephelauxetic ratio, $\beta = 0.9836$, for the neodymium chelate was determined from the diffuse reflectance spectrum [18]. The experimental energy levels of ytterbium(III), obtained from the absorption spectrum of $Yb(C_{15}H_{11}O_2)_3$ in carbon tetrachloride solution, were compared with predictions of electrostatic crystal field theory and good agreement was noted [19].

The mass spectra of the scandium and yttrium chelates exhibited the following m/e peaks (relative intensity in percentage): for M = Sc, 715 (25), 714 (56) = $[Sc(C_{15}H_{11}O_2)_3]$, 492 (56), 491 (100) = $[Sc(C_{15}H_{11}O_2)_2]^+$, 413 (0.2), 267 (10), 224 (10), 223 (10), 105 (28), 77 (17), 61 (17); for M = Y, 759 (23), 758 (63) = $[Y(C_{15}H_{11}O_2)_3]$, 536 (32), 535 (100) = $[Y(C_{15}H_{11}O_2)_2]^+$, 457 (0.6), 224 (11), 223 (11), 105 (19), 77 (11) [20].

 References to 5.1.15 see p. 208/10

5.1.15.2.2 Hydrates, $M(C_{15}H_{11}O_2)_3 \cdot nH_2O$ (n = 0.5, 1, 2)

Sacconi, Ercoli [21] effected syntheses of the monohydrated tris chelates (M = La, Pr, Nd, Sm, Gd) by adding saturated solutions of ethanolic ammonia to solutions containing hydrated rare earth nitrates and benzoylacetone in aqueous ethanol. The colors and melting points of the compounds are recorded in Table 5/42. The monohydrated terbium chelate was also prepared according to this procedure and recrystallized from ethanol [22].

Perrotto, Charles [10] prepared a series of hydrated tris chelates (n = 1 for M = La, Pr, Nd, Sm, Eu, Gd, Tb, Dy, Ho, Tm; n = 0.5 for M = Er, Yb, Lu) according to the following procedure: To a suspension prepared by mixing 6.0 g of dibenzoylmethane in 50 ml of 95% ethanol with 5×10^{-3} mol of a rare earth chloride in 200 ml of water was added, over a period of 2 h using a motor driven syringe, 15.0 ml of 1.00 M aqueous ammonia. In each case, the resulting precipitate was collected by filtration and dried under vacuum in a desiccator at room temperature. Recrystallization of the crude product was effected by dissolving the sample in 200 ml of hot 95% ethanol, filtering the solution while hot, adding 100 ml of water to the cooled filtrate, and then allowing the mixture to cool further in a refrigerator. The product was collected by filtration and dried under vacuum. Excess dibenzoylmethane was removed by stirring a suspension of the recrystallized substance in cyclohexane. Melting points of the compounds thus obtained are recorded in Table 5/42.

Table 5/42

Colors and Melting Points of Tris(dibenzoylmethanato) Rare Earth Chelates, $M(C_{15}H_{11}O_2)_3 \cdot nH_2O$ (n = 0, 0.5, 1) According to Various Procedures.

Sacconi, Ercoli [21]				Purushottam and coworkers [5, 6]			Perotto, Charles [10]	Khomenko, Kuznetsova [24]	
M	n	color	melting point in °C	M	color	melting point in °C*)	melting point in °C**)	color	melting point in °C
				Sc	light brown	248			
				Y	light pink	224			
La	1	straw yellow	142 to 145				~120	brown	131 to 133
Pr	1	yellow	149 to 151	Pr	green yellow	230	113 to 115	brown	134 to 135
Nd	1	lilac green	147 to 150	Nd	green yellow	200	122 to 155	greenish red	132 to 133
Sm	1	golden yellow	148 to 149	Sm	yellow	165	227 to 230	yellow	138 to 139
Eu				Eu	yellow	171	230 to 234	yellow	144 to 145
Gd	1	green yellow	214 to 216	Gd	yellow	222	250 to 253	yellow	140 to 143
Tb	1						253 to 257	light yellow	141 to 143
Dy	1			Dy	yellow	212	261 to 265	light yellow	130 to 133
Ho	1						262 to 267		
Er	0.5			Er	yellow	210	267 to 271		
Tm	1						265 to 272		
Yb	1			Yb	yellow	216		light yellow	140 to 142
Lu	0.5						265 to 270		

*) Degree of hydration not given. — **) Except for the lanthanum, praseodymium, and neodymium compounds, dehydration was effected on heating before the sample melted. Thus the melting ranges reported correspond to the anhydrous compounds.

Murav'eva and coworkers [23] prepared each chelate (n = 1 for M = Pr to Tm except for Pm, n = 0.5 for M = Yb, Lu) also by mixing an ethanolic solution of a hydrated rare earth nitrate with a stoichiometric amount of dibenzoylmethane in the same solvent, and then neutralizing by slowly adding an alcoholic ammonia solution. The precipitate formed was collected by filtration, washed with a small quantity of alcohol, and dried in air. The lanthanum compound obtained according to this procedure had the composition $La(C_{15}H_{11}O_2)_2OH$.

Several monohydrates (M = La, Pr, Nd, Sm, Eu, Gd, Tb, Dy, Yb) were also prepared by adding slowly and with stirring, an ethanolic solution (100 ml) containing sodium hydroxide (24 to 28 mg) and dibenzoylmethane (230 mg) to a solution of a rare earth nitrate (100 mg) in the same solvent [24]. The colors and melting points of each compound obtained according to this procedure are recorded in Table 5/42.

$Eu(C_{15}H_{11}O_2)_3 \cdot 2H_2O$ was prepared by combining a stoichiometric amount of europium chloride with sodium dibenzoylmethanate in 50% ethanol. The resulting precipitate was collected, washed with several portions of water, and dried under vacuum at room temperature [36].

The monohydrated cerium chelate was prepared by carefully adding 2 N aqueous ammonia to a solution containing 1 mmol of cerium(III) nitrate hexahydrate and 3.5 mmol of dibenzoylmethane. The resulting precipitate was collected, washed with a small amount of a methanol/water/ammonia mixture, and dried under a flow of nitrogen. The lilac colored product melted at 140°C. The magnetic moment of the compound is 2.00 μ_B [25].

Infrared spectral data for a sample of $Er(C_{15}H_{11}O_2)_3 \cdot 2H_2O$ dispersed in a KBr disc are recorded in Table 5/43 [26]. The assignments were based on those reported [27 to 31] for the hydrated tris acetylacetonato rare earth chelates. Additional infrared data for the dibenzoylmethane chelates are discussed in [32] to [35]. Mehta, Tandon [33] computed the stretching force constants, f M-O, for the M-O bonds, from the ν(M-O) ($\nu_1[A_1]$ = 500 to 510 cm^{-1}; $\nu_2[A_2]$ = 345 to 350 cm^{-1}; $\nu_3[B_2]$ = 600 to 655 cm^{-1}) vibrational modes (M, f M-O in dynes · cm^{-1}): La, 2.87×10^{-5}; Pr, 2.75×10^{-5}; Nd, 2.63×10^{-5}; Sm, 2.96×10^{-5}. The assignment of three νM-O vibrational modes was based on D_3 symmetry [35].

The emission spectrum of $Eu(C_{15}H_{11}O_2)_3 \cdot 2H_2O$ in 3:1 (v/v) ethanol/methanol (standard solvent) and 4:1 (v/v) standard solvent/dimethylformamide, as well as in the crystalline state, revealed a multiplicity of lines that could not be accounted for in terms of a single chemical species. For the crystalline sample: $^5D_0 \rightarrow {}^7F_2$, 6245, 6142, 6128, 6120, 6105 Å; $^5D_0 \rightarrow {}^7F_1$, 6077, 6038, 6012, 5982, 5978, 5974, 5956, 5948, 5940, 5933, 5892, 5886 Å; $^5D_0 \rightarrow {}^7F_0$, 5805, 5802, 5800 Å [36].

Table 5/43

Assignment of Absorption Bands in the Infrared Spectrum of $Er(C_{15}H_{11}O_2)_3 \cdot 2H_2O$.

absorption band in cm^{-1} *)	assignment**)	absorption band in cm^{-1} *)	assignment**)
3080 (m)	ν_{as}(C-H)	1022 (m)	
2920 (sh)	$\phi\nu$(C-H)	1000 (m)	ϕ ring breathing
1589 (vs)	ν_{as}(C=O)	940 (s)	πC-H
1560 (s)	$\phi\nu$(C-C)	810 (m)	νC-ϕ
1540 (w)	$\phi\nu$(C-C)	782 (m)	
1515 (s)	ν_{as}(C=C)	750 (s)	πC-H
1475 (vs)	$\phi\nu$(C-C)	718 (vs)	$\phi\pi$ C-H
1450 (m)	ν_s C=O + δC-H	682 (s)	5 adjacent H on ϕ
1433 (m)		610 (m)	νC-ϕ
1400 (vs)	δ C-H	600 (m)	
1300 (s)	$\phi\nu$C-C	520 (m) }	
1280 (s)	ν_s C=C	500 (w) }	δC-ϕ
1218 (s) }		425 (w) }	
1175 (m) }	δ C-H + νC=C	415 (m)	Er-O
1130 (m) }		370 (w)	ring deformations
1065 (m)	$\phi\delta$C-H	300 (w)	

*) Relative intensity: vs = very strong; s = strong; m = medium; sh = shoulder; w = weak. — **) Assignment: ν = stretching, δ = in-plane bending, π = out-of-plane bending, s = symmetric, as = asymmetric, ϕ = phenyl ring.

The absorption spectra of the tris chelates (compounds prepared according to the procedure described in [5]) exhibited two bands characteristic of the ligand (at 250 to 255 and 342 to 355 nm for methanolic solutions, at 255 to 265 and 350 to 355 nm in ethanol, at 250 and 335 to 355 nm in n-hexane, and at 250 and 335 to 345 nm in cyclohexane) [5].

The oscillator strengths for absorption bands observed in the spectrum of the erbium chelate (degree of hydration not presented) in methanol and dimethylformamide were determined and compared with those calculated using the Judd-Ofelt theory [15, 16]. The study revealed that it was possible to use the Judd-Ofelt parameters (τ_2, τ_4, τ_6) for the evaluation of the spectral intensities. The oscillator strength of the hypersensitive transition $^4I_{15/2} \rightarrow {}^2H_{11/2}$ was attributed to the τ_2 parameter originating from the forced electric dipole transition [37].

Perrotto, Charles [10] noted that the thermograms for samples of the $M(C_{15}H_{11}O_2)_3 \cdot nH_2O$ (n = 0.5, 1) species (compounds prepared according to the procedure described in [10] and presented above) heated in flowing argon revealed weight losses up to 110°C corresponding, in each instance, to evolution of the water of hydration. Further weight losses were observed for the lighter rare earth chelates in the temperature region 110 to 250°C, and were attributed to the hydrolysis reaction $M(C_{15}H_{11}O_2)_3 \cdot H_2O \rightarrow M(C_{15}H_{11}O_2)_2OH + C_{15}H_{12}O_2$. The tendency toward hydrolysis decreases with increasing atomic weight of the lanthanide and becomes negligible for the heavier rare earth dibenzoylmethides. Weight losses observed above 300°C were attributed to decomposition of the anhydrous chelate [10]. Murav'eva and coworkers [23] noted that the temperature of dehydration for the $M(C_{15}H_{11}O_2)_3 \cdot nH_2O$ (n = 0.5, 1) species (compounds prepared according to the procedure described in [23] and presented above) decreased across the series from lanthanum to lutetium. The hydrolysis reaction noted above was observed only for the lanthanum and praseodymium chelates.

5.1.15.3 Lewis Base Adducts of the Tris Chelates

Formation in Solution. With Alcohols and Amines

$Eu(C_{15}H_{11}O_2)_3 \cdot nB$ (n = 1, 2). The formation constants (Table 5/44) for the interaction of $Eu(C_{15}H_{11}O_2)_3$ with several alcohol and amine substrates in carbon tetrachloride were determined by emission-titration spectroscopy. Job's method revealed that 1:2 adducts were formed for all the amines used, but only 1:1 adducts were detected in the case of the alcohol substrates. The values of lg β_2 and lg K_1 did not depend strongly on the steric nature of the substrate for either the amine or alcohol adducts [38].

Table 5/44

Formation Constants for the Interaction of the Tris(dibenzoylmethanato) Europium Chelate with Alcohol and Amine Substrates in CCl_4*) [38].

substrate	lg K_1**)	lg β_2**)
propylamine (= C_3H_9N)	***)	4.40
isopropylamine (= C_3H_9N)	***)	4.42
butylamine (= $C_4H_{11}N$) see p. 199, Table 5/45	***)	4.42
sec-butylamine (= $C_4H_{11}N$)	***)	4.41
tert-butylamine (= $C_4H_{11}N$)	***)	4.38
methanol (= CH_4O)	2.98	
ethanol (= C_2H_6O)	2.98	
propanol (= C_3H_8O)	2.97	
isopropyl alcohol (= C_3H_8O)	2.89	
butanol (= C_4H_9O)	2.97	
sec-butyl alcohol (= $C_4H_{10}O$)	2.78	
tert-butyl alcohol (= $C_4H_{10}O$)	2.70	

*) Formation constants determined by emission-titration spectroscopy. — **) The K_1 formation constants have units of dm^3/mol, and the overall formation constants, β_2, have units of dm^6/mol^2. — ***) Could not be determined.

Isolated Compounds

With 2,2'-Bipyridine (= bpy = $C_{10}H_8N_2$) and 1,10-Phenanthroline (= phen = $C_{12}H_8N_2$)

$M(C_{15}H_{11}O_2)_3 \cdot$ bpy and $M(C_{15}H_{11}O_2)_3 \cdot$ phen. Melby and coworkers [39] effected synthesis of $Eu(C_{15}H_{11}O_2)_3 \cdot$ bpy by adding, dropwise and with stirring, an aqueous solution of a europium salt (2 mmol) to a solution containing dibenzoylmethane (6 mmol), bipyridine (2 mmol), and 6 ml of 1 N sodium (or potassium) hydroxide in 20 ml of hot 95% ethanol. The mixture was allowed to cool, and the product was collected and recrystallized from 95% ethanol. The melting point of the compound thus obtained was 210 to 213°C.

The phenanthroline adduct was also prepared according to the procedure described above, except that the compound was purified by dissolving it in benzene and effecting precipitation with hexane. The melting point of the product was 210 to 213°C [39]. Bauer, Blanc, Ross [40] also prepared the europium compound according to a similar procedure. The precipitated product was collected by filtration, washed with water, and dried under vacuum. The melting point of the adduct thus obtained was 184 to 187°C.

Butter, Kreher [41] prepared the bipyridine and phenanthroline adducts of the europium chelate by adding solid dibenzoylmethane (3×10^{-3} mol) and the respective Lewis base (10^{-3} mol) to solutions of hydrated europium nitrate (10^{-3} mol) in 20 ml of absolute ethanol, followed by the addition of gaseous ammonia, while stirring, until the pH of the reaction mixture reached 7 to 8. Each solution was stirred for an additional 20 min, after which time, the product was collected and dried under vacuum. The compounds were recrystallized from a benzene/petroleum ether mixture. The melting points of the bipyridine and phenanthroline adducts obtained in this manner were 201 to 204°C and 199 to 202°C, respectively.

It was observed that the emission intensity of the tris europium chelate in solution or the solid state (at 77 K and 300 K) was greatly enhanced by the addition of bipyridine or phenanthroline, suggesting a direct interaction between the rare earth and neutral donor [41].

X-Ray diffraction studies established that the $Eu(C_{15}H_{11}O_2)_3 \cdot$ bpy compound crystallizes in space group $P2_1/a = C_{2h}^5$, with four molecules in a unit cell of dimensions a = 16.3, b = 22.5, c = 13.2 Å, $\beta = 109.6°$ (density = 1.43 g/cm) [42].

Denliev, Murav'eva, Martynenko [43] prepared a series of adducts (M = Pr, Nd, Gd, Er, Yb) derived from phenanthroline and bipyridine by treating each respective hydrated tris chelate with a stoichiometric amount of the Lewis base in a minimum amount of ethanol. Each solution was heated at reflux, with stirring, for 10 h. The crystalline products were collected by filtration, washed with ethanol, and dried at room temperature over phosphorus(V) oxide.

Melent'eva, Kononenko, Poluektov [44] prepared a series of phenanthroline adducts (M = Y, Nd, Sm, Eu, Tb) by treating ethanolic solutions containing the respective rare earth chlorides, dibenzoylmethane, and 1,10-phenanthroline with 40% aqueous urotropine solutions. After stirring each mixture for 1 h, the resulting precipitate was collected by filtration, washed with a small quantity of ethanol, and dried in a desiccator over sulfuric acid. The colors and melting points (°C) of the compounds thus obtained were: Y, yellow, 180 to 182; Nd, lilac, 182 to 183; Sm, yellow, 182 to 184; Eu, straw yellow, 184 to 186; Tb, yellow, 185 to 186. The adducts are insoluble in water, sparingly soluble in ether, butanol, and carbon tetrachloride, and readily soluble in acetone, benzene, and methanol.

Thermogravimetric analyses revealed that both series of adducts (M = Pr, Nd, Gd, Er, Yb) began losing mass at 190 to 200°C when heated, with maximum decomposition observed in the range 300 to 350°C [43].

With Other Lewis Bases

The Lewis base adducts of the tris(dibenzoylmethanato) chelates, listed in Table 5/45, were prepared according to the following procedures:

A) A solution of the Lewis base in benzene was added dropwise to a warm solution of the tris chelate (0.5×10^{-3} mol) in the same solvent. Warm petroleum ether was added to the resulting solution, and the mixture was allowed to cool, effecting crystallization. The compound was recrystallized from a benzene/petroleum ether mixture [45].

References to 5.1.15 see p. 208/10

B) The butylamine adduct was prepared by adding 1 g of europium dibenzoylmethanate to a solution of 3 ml of butylamine in 3 ml of benzene. After briefly heating the mixture on a hot plate, excess solvent was removed by evaporation with a stream of nitrogen [46, 47].

C) Anhydrous europium dibenzoylmethanate (1 g) was dissolved in 3 ml of the pure liquid Lewis base at room temperature, then 25 ml of petroleum ether was added to effect precipitation of the adduct. The compound was collected and dried under vacuum in a desiccator [46, 47].

D) Anhydrous europium dibenzoylmethanate (1 g) was dissolved in 3 ml of the liquid Lewis base at room temperature; then excess liquid was removed by evaporation in a stream of dry nitrogen. Final drying was accomplished under vacuum in a desiccator [46, 47].

E) A solution (6 ml) of 1 N sodium (or potassium) hydroxide was added to a solution of 6 mmol of dibenzoylmethane and 2 mmol of the neutral ligand in 20 ml of hot 95% ethanol. The mixture was stirred while 2 mmol of rare earth salt solution was added dropwise. The mixture was allowed to cool, and the product was collected. The compound was purified by dissolving it in benzene and effecting precipitation with hexane [39].

F) A methanolic solution of the Lewis base was added dropwise to a solution of the tris chelate in benzene. Petroleum ether was added to the clear solution to effect separation of the product [45].

G) The compound was prepared by recrystallizing the tris chelate from an ethanolic solution containing the neutral ligand [49].

H) The adducts were obtained by allowing the direct reaction of $M(C_{15}H_{11}O_2)_3 \cdot H_2O$ with the Lewis bases (1:1 mol ratio) [49].

Emission and infrared spectral studies provided evidence for direct bonding between the neutral Lewis base and the central metal ion of the tris chelate. Ohlmann, Charles [47] found that both the lifetime and relative fluorescence intensity at room temperature of microcrystalline samples of compounds 2 to 6 and 8 to 11 (Table 5/45) were markedly dependent on the nature of the Lewis base present. They also observed that the number and position of Stark components in the emission spectra were dependent on the Lewis base [46]. Kreher, Butter, Seifert [45] interpreted the emission spectra at 77 K of compounds 1 and 11 to 13, in the solid state, in terms of C_{2v} symmetry (and in some cases D_{4d} or D_{2d} symmetry to a first approximation) at the europium(III) site. The molecular structure they proposed for these adducts is a face-centered trigonal prism, which provided for ligation of the Lewis base.

Infrared spectra of compounds 14 to 16 (vaseline and hexachlorobutadiene mulls) revealed that the ν(P=O) stretching mode shifted to lower wave numbers on complexation of the neutral ligand ($\Delta\nu$(P=O) is 60 cm^{-1} for the tributylphosphate adducts, 25 to 33 cm^{-1} for the triphenylphosphine oxide adducts, and 30 to 37 cm^{-1} for the trioctylphosphine oxide adducts), indicating that ligation occurs through the phosphoryl or phosphate oxygen atom [49]. Infrared studies of compounds 2 to 6 and 8 to 11 are reported in [46].

X-ray diffraction studies showed that the ethylenediamine adduct of $Eu(C_{15}H_{11}O_2)_3$ (No. 1) crystallizes in the space group $P2_1/c = C_{2h}^5$ (No. 14) with four molecules in a unit cell of dimensions a = 15.1, b = 22.9, c = 12.17 Å, β = 110.5° (density = 1.48 g/cm) [42].

Thermogravimetric analyses (compounds 2 to 6 and 8 to 10) revealed that the Lewis base molecules are lost in a stepwise manner for those adducts which contain more than one mole of Lewis base. In most instances, weight losses corresponding to complete removal of the Lewis base were observed when the sample was heated to 300°C. Weight losses which occurred above 300°C were attributed to decomposition of the tris chelate [46].

With Diantipyrylmethane

$La(C_{15}H_{11}O_2)_3 \cdot C_{23}H_{24}N_4O_2$. The complex was prepared from an aqueous acetone solution containing $LaCl_3$, diantipyrylmethane, and dibenzoylmethane. On the basis of the IR spectrum the La atom is assumed to have a coordination number of eight [82].

Table 5/45
Lewis Base Adducts of the Tris(dibenzoylmethanato) Rare Earth Chelates.

No.	Lewis base	formula of adduct	method of preparation	melting point in °C	Ref.
1	ethylenediamine ($=C_2H_8N_2=$ en)	$Eu(C_{15}H_{11}O_2)_3 \cdot en$	A	170 to 171	[45]
2	butylamine ($=C_4H_{11}N$)	$Eu(C_{15}H_{11}O_2)_3 \cdot 2C_4H_{11}N$	B	145 to 147	[46, 47]
3	pyridine ($=C_5H_5N=$ py)	$Eu(C_{15}H_{11}O_2)_3 \cdot 2py$	C	103 to 106	[46, 47]
4	piperidine ($=C_5H_{11}N$)	$Eu(C_{15}H_{11}O_2)_3 \cdot 2C_5H_{11}N$	C	183 to 185	[46, 47]
5	aniline ($=C_6H_7N$)	$Eu(C_{15}H_{11}O_2)_3 \cdot C_6H_7N$	D	206 to 209	[46, 47]
6	quinoline ($=C_9H_7N$)	$Eu(C_{15}H_{11}O_2)_3 \cdot 2C_9H_7N$	C	109 to 111	[46, 47]
7	2,2',2''-terpyridine ($=C_{15}H_{11}N_3$)	$Eu(C_{15}H_{11}O_2)_3 \cdot C_{15}H_{11}N_3 \cdot C_6H_6$	E	190 to 194	[39]
8	dimethyl sulfoxide ($=C_2H_6SO=$ DMSO)	$Eu(C_{15}H_{11}O_2)_3 \cdot 3DMSO$	D	112 to 115	[46, 47]
9	dimethylformamide ($=C_3H_7NO=$ DMF)	$Eu(C_{15}H_{11}O_2)_3 \cdot DMF$	D	133 to 138	[46, 47]
10	dioxane ($=C_4H_8O_2$)	$Eu(C_{15}H_{11}O_2)_3 \cdot 2C_4H_8O_2$	D	170 to 175	[46, 47]
11	pyridine N-oxide ($=C_5H_5NO$)	$Eu(C_{15}H_{11}O_2)_3 \cdot C_5H_5NO$	A	179 to 180 [45] 189 to 190 [47]	[45, 46, 47]
12	2,2'-bipyridyl N,N'-dioxide ($=C_{10}H_8N_2O_2$)	$Eu(C_{15}H_{11}O_2)_3 \cdot C_{10}H_8N_2O_2$	F	220 to 230	[45]
13	diethyl methylphosphonate ($=C_5H_{13}O_3P$)	$Eu(C_{15}H_{11}O_2)_3 \cdot 2C_5H_{13}O_3P$	A	89 to 90	[45]
14	tributyl phosphate ($=C_{12}H_{27}O_4P=$ TBP)	$M(C_{15}H_{11}O_2)_3 \cdot TBP$; M = Pr, Nd, Gd, Er, Yb	G, H	Pr (275), Nd (285) Gd (295), Er (333) Yb (324)	[12, 49]
15	triphenylphosphine oxide ($=C_{18}H_{15}OP$)	$M(C_{15}H_{11}O_2)_3 \cdot C_{18}H_{15}OP$; M = Pr, Nd, Gd, Er, Yb	G, H	Pr (283), Nd (345) Gd (305), Er (323) Yb (350)	[12, 49]
16	trioctylphosphine oxide ($=C_{24}H_{51}OP$)	$M(C_{15}H_{11}O_2)_3 \cdot C_{24}H_{51}OP$; M = Pr, Nd, Gd, Er, Yb	G, H	Pr (322), Nd (323) Gd (324), Er (338) Yb (314)	[12, 49]

References to 5.1.15 see p. 208/10

5.1.15.4 Tetrakis Chelates of Type $M'[M(C_{15}H_{11}O_2)_4]$

Alkali Metal Salts

Dutt, Bandyopadhyay [50] first effected syntheses of tetrakis chelates $M'[Nd(C_{15}H_{11}O_2)_4]$ (M' = Na, K) by neutralizing an aqueous ethanolic solution containing neodymium nitrate and dibenzoylmethane (1:4 mol ratio) with an ethanolic solution of sodium (or potassium) hydroxide. In each case, the precipitate that formed was collected by filtration and dissolved in a small amount of acetone. The acetone solution was diluted with absolute ethanol and then allowed to evaporate in air to effect crystallization of the product. The crystals were collected by filtration, and recrystallized from absolute ethanol. The compounds thus obtained were highly soluble in acetone and moderately soluble in alcohol.

Melby and coworkers [39] effected synthesis of the series of europium chelates (M' = Na, K, Rb, Cs) according to the following procedure: Dibenzoylmethane (8 mmol) was dissolved in a mixture of 25 ml of ethanol and 16 ml of 0.5 M aqueous base solution. To this solution was added 2 mmol of europium chloride in water. The mixture was cooled in ice water, and the solid formed was collected by filtration. The compounds were recrystallized from 200 ml of o-dichlorobenzene. The melting points (°C) of the compounds thus obtained were (M' =): Na, 167; K, 300; Rb, 286; Cs, 280.

Butter, Seifert [51] reported that the luminescence spectra of alkali tetrakis europate(III) chelates in the crystalline state at 77 K revealed the existence of different modifications of these compounds. The spectra of the α-forms were interpreted in terms of D_4 symmetry at the europium(III) site, whereas the spectra of the β-forms were interpreted in terms of D_2 site symmetry. Generally, the β-modification of each compound was obtained by recrystallizing the product prepared according to the procedure of Melby and coworkers [39] from dichlorobenzene [51]. The α-form of $Na[Eu(C_{15}H_{11}O_2)_4]$ was obtained by recrystallizing the reaction product from methanol. Conversion of the β-form to the α-form was observed for the potassium and caesium salts. The authors proposed that the conversion involves a change from dodecahedral into antiprismatic coordination around the europium ion [51]. For Mössbauer studies of $Na[Eu(C_{15}H_{11}O_2)_4]$, see [83].

Ammonium and Alkylammonium Salts

Ammonium Salts $NH_4[M(C_{15}H_{11}O_2)_4]$. Each compound (M = La, Nd) was prepared by adding an ethanolic solution of dibenzoylmethane to a solution of the rare earth nitrate in aqueous ethanol, then neutralizing the mixture with alcoholic ammonia. The resulting precipitate was collected by filtration and extracted into a small amount of acetone. The acetone solution was diluted with absolute ethanol, and the mixture was allowed to evaporate in air to effect crystallization of the product. The products were recrystallized from absolute ethanol, collected, and dried in air [50].

Tetramethylammonium Salt $(CH_3)_4N[Eu(C_{15}H_{11}O_2)_4]$. Dibenzoylmethane (8 mmol) and tetramethylammonium chloride (3 mmol) in 25 ml of ethanol and 8 ml of 1 N sodium hydroxide solution were heated on a steam bath until all solids had dissolved. To this warm solution was added 2 mmol of europium chloride in 10 ml of water. The precipitate was collected, washed with water, and recrystallized from o-dichlorobenzene (m. p. = 259°C) [39].

Diethylammonium Salt $(C_2H_5)_2NH_2[Eu(C_{15}H_{11}O_2)_4]$. The compound was prepared by adding a solution of europium chloride (0.00387 mol) in 25 ml of absolute ethanol to a hot (90°C) ethanolic solution (25 ml) containing 0.0155 mol of dibenzoylmethane and 0.0155 mol of diethylamine. The mixture was maintained at ambient temperature for 4 days and then filtered. The crystalline product collected was washed with cold absolute ethanol and air dried (yield 100%). The melting point of the compound was 196 to 204°C [52].

The Mössbauer spectrum of the chelate at 77 and 4 K yielded an average isomer shift of $-0.60\ mm \cdot s^{-1}$ relative to a source of ^{151}Sm in Sm_2O_3. The isomer shift data provided no evidence for participation of 4f-electrons in covalent bonds [53].

Triethylammonium Salt $(C_2H_5)_3NH[Eu(C_{15}H_{11}O_2)_4]$. A solution of europium chloride (2 mmol) in 10 ml of water was added to an ethanolic solution containing dibenzoylmethane (8 mmol) and triethylamine (8 mmol). The mixture was cooled in ice water and the resulting precipitate collected by filtration. The product was recrystallized from o-dichlorobenzene (m. p. = 175°C) [39].

The emission spectrum of a powdered microcrystalline sample of the compound at 77 K revealed the following bands: 5799 Å = $^5D_0 \rightarrow {}^7F_0$; 5915, 5920, 5949 Å = $^5D_0 \rightarrow {}^7F_1$; 6120, 6136, 6167, 6180, 6258 Å = $^5D_0 \rightarrow {}^7F_2$; 5264 Å = $^5D_1 \rightarrow {}^7F_0$; 5362, 5387 Å = $^5D_1 \rightarrow {}^7F_1$; 5528, 5542, 5565, 5579, 5638 Å = $^5D_1 \rightarrow {}^7F_2$ [54]. Triboluminescence from the compound was clearly visible in broad daylight at room temperature when crystals were broken with a glass rod in a test tube [57].

Tetraethylammonium Salt $(C_2H_5)_4N[M(C_{15}H_{11}O_2)_4]$. Melby and coworkers [39] prepared the europium compound by adding a solution of europium chloride (2 mmol) in 10 ml of water to a warm ethanolic solution containing dibenzoylmethane (8 mmol), tetraethylammonium chloride (3 mmol), and 8 ml of aqueous 1N sodium hydroxide solution. The resulting precipitate was collected, washed with water, and recrystallized from o-dichlorobenzene (m.p. = 230°C). Bjorklund and coworkers [55] effected synthesis of the europium compound according to a procedure similar to that described by Melby and coworkers [39], except that 8 mmol of tetraethylammonium bromide was employed instead of 3 mmol of tetraethylammonium chloride. After recrystallization from o-dichlorobenzene, the product was washed several times with absolute ethanol and dried under vacuum at 110°C. The chelate decomposed at 222 to 225°C on heating [55].

The cerium(III) compound was prepared under a pure nitrogen atmosphere using deoxygenated solvents according to the following procedure: A solution of cerium(III) chloride heptahydrate (2 mmol) and 2,2-dimethoxypropane (10 cm^3) in absolute ethanol (15 cm^3) was added to a mixture of dibenzoylmethane (8 mmol), sodium hydroxide (8 mmol), and tetraethylammonium chloride (2 mmol) in absolute ethanol (25 cm^3). The solvent was removed under reduced pressure, and the residue was extracted with ethyl acetate (30 cm^3). The ethyl acetate solution was concentrated under reduced pressure to approximately 10 cm^3. Crystallization of the product was effected by slowly adding n-hexane. The red-brown colored solid was collected, washed with n-hexane, and dried under vacuum. The electronic absorption spectrum of the compound in acetonitrile revealed bands at 20800 ($\varepsilon = 264\ dm^3 \cdot mol^{-1} \cdot cm^{-1}$) and 22300 cm^{-1} (shoulder) [56].

The internal Stark splitting observed in the emission spectrum of a microcrystalline sample of the europium chelate at 77 K was analyzed by treating the ligand field as a perturbation on the free ion levels. Assignments of spectral bands, together with energy levels obtained from the spectral data and those obtained from the ligand-field calculations are recorded in Table 5/46. The ligand-field parameters were derived from the splitting in the $^7F_1$ and $^7F_2$ levels of the europium ion without assuming a molecular geometry, and these values were then used to establish the configuration about the emitting ion. The best correlation (see Table 5/46) between energy levels predicted by the ligand-field model and those observed in the spectrum was obtained for square antiprismatic coordination geometry with the chelate rings spanning opposite edges of the square faces of the antiprism. A value of 2.40 Å for the interatomic Eu-O bond distance was determined using the ligand-field model [55].

The Mössbauer spectrum of the europium chelate at 77 and 4 K revealed an average isomer shift of $-0.58\ mm \cdot s^{-1}$ relative to a ^{151}Sm source in Sm_2O_3. The data provided no evidence for participation of 4f electrons in covalent bond formation [53].

Propyl-, Dipropyl-, Tripropyl-, and Tetrapropylammonium Salts $(C_3H_7)_nNH_{4-n}[Eu(C_{15}H_{11}O_2)_4]$ (n = 1 to 4). Filipescu, Degnan, and McAvoy [54] effected preparation of the propylammonium and tripropylammonium salts according to the procedure described on p. 200 for preparation of the triethylammonium salts. Bauer, Blanc, Ross [40] effected synthesis of the dipropylammonium salt by adding 8 mmol of the amine to a solution of 8 mmol of dibenzoylmethane and 2 mmol of europium chloride in 70 ml of hot ethanol. The mixture was allowed to cool to room temperature, and the resulting precipitate was collected, washed with water, and dried under vacuum. The product was recrystallized from carbon tetrachloride (melting range = 195 to 200°C). The tetrapropylammonium salt was prepared by adding 8 mmol of 10% aqueous tetrapropylammonium hydroxide to a hot ethanolic solution (75 ml) containing 8 mmol of dibenzoylmethane and 2 mmol of europium chloride. Crystallization of the product was effected by allowing the solution to cool to room temperature. The compound was recrystallized from 2-butanone (melting range = 203 to 207°C) [40].

References to 5.1.15 see p. 208/10

Table 5/46
Emission Spectrum of Microcrystalline Sample of $(C_2H_5)_4N[Eu(C_{15}H_{11}O_2)_4]$ at 77 K in the Regions of the $^5D_0 \rightarrow {}^7F_n$ (n = 0 to 2) Transitions [55].

transition	bands in cm^{-1} *)	energy level in cm^{-1} **) from spectral data	predicted by ligand-field model
$^5D_0 \rightarrow {}^7F_0$	17248		
$^5D_0 \rightarrow {}^7F_1$	17007	241	241
	16819	429	429
	16792	456	456
$^5D_0 \rightarrow {}^7F_2$	16306	942	940
	16298	950	952
	16199	1049	1051
	16129	1119	1122
	16119	1129	1124

*) E_0 of $^7F_1$ term is 375 cm^{-1} above $^7F_0$; E_0 of $^7F_2$ term is 1038 cm^{-1} above $^7F_0$. — **) Relative to $^7F_0$ set equal to zero.

The emission spectra of the propylammonium and tripropylammonium salts were recorded for microcrystalline powders at 77 K. As revealed in the data presented below, the splitting of the $^7F_1$ and $^7F_2$ levels in the europium ion depends on the steric requirements of the organic cation. The energy levels derived from the spectral data were related to the combined effects of electrostatic and steric interactions between the cation and complex anion, resulting from approach of the cation along the C_2 symmetry axis of the square antiprism, which was the geometry proposed for the coordination polyhedron [54].

complex	emission lines in Å $^5D_0 \rightarrow {}^7F_0$	$^5D_0 \rightarrow {}^7F_1$	$^5D_0 \rightarrow {}^7F_2$
$C_3H_7NH_3[Eu(C_{15}H_{11}O_2)_4]$	5800	5922, 5940	6114, 6144, 6158, 6195, 6243
$(C_3H_7)_3NH[Eu(C_{15}H_{11}O_2)_4]$	5801	5913, 5925, 5944	6116, 6142, 6166, 6188, 6251

Tetrahexylammonium Salt $(C_6H_{13})_4N[Eu(C_{15}H_{11}O_2)_4]$. The compound was prepared according to the procedure described by Bauer and coworkers [40] for preparation of the analogous tetrapropylammonium salt of the chelate. The product was recrystallized from 2-butanone (melting range = 220 to 226°C) [40]. X-ray diffraction studies revealed that the compound crystallizes in the monoclinic space group $P2_1/c$-C_{2h}^5 (No. 14) [58].

Polarized absorption in the $^7F_0 \rightarrow {}^5D_2$ transition (spectrum obtained at room temperature) and polarized emission in the $^5D_0 \rightarrow {}^7F_2$ transition (spectrum obtained at 78 K) of the europium ion were obtained for single crystals (monoclinic) of the chelate. Both sets of transitions consist of three lines, two strongly polarized perpendicular to the crystallographic c axis, and the third polarized parallel to this axis (the emission spectral data are presented below). Applying electric-dipole selection rules, the authors concluded that these results are consistent with S_4 symmetry for the chelate anion [58].

bands in cm^{-1}	relative intensities electric vector $\parallel$ to c	electric vector $\perp$ to c	species under S_4
16355	0.07	1	B
16310	0.06	1	B
16185	0.09	0.06	E

Ethanolammonium Salts $HO(CH_2)_2NH_3[M(C_{15}H_{11}O_2)_4] \cdot nC_2H_5OH$ (n = 1,2). The mono solvated samarium and europium complexes were prepared by adding ethanolamine (0.001 mol), with stirring, to solutions prepared by mixing anhydrous ethanolic solutions (10 cm^3) of dibenzoylmethane (0.004 mol) with ethanolic solutions (5 cm^3) of the respective rare earth chlorides (0.001 mol). The cream colored, crystalline complexes were collected after 24 h and dried under vacuum (10^{-5} Torr) at room temperature. Heating the europium chelate under vacuum (10^{-5} Torr) at 100°C for 24 h produced the tris chelate. Each compound melted at 125°C. They are readily soluble in alcohol and other organic solvents, but insoluble in water [59]. The europium compound (n = 0) was similarly prepared using methanol as the solvent [60]. Baczynski, Rozploch, Zachara [61] reported, however, that homogeneous, well defined crystals of the europium chelate were obtained only when there was a large excess of ethanolamine in the methanolic solution, e.g., 100:1 ethanolamine: metal ion mole ratio.

Infrared spectral date (samples dispersed in Nujol mulls) for each complex (M = Sm, Eu; n = 0, 1) are tabulated in [59] and [60]. Bands at 1515, 1550, and 1600 cm^{-1} were assigned to the ν(C=C) and ν(C=O) vibrational modes, and a band at 520 cm^{-1} was assigned to the ν(M-N) mode [59]. The emission spectrum of the solid europium chelate (n = 0) at 77 K ($^5D_0 \rightarrow {}^7F_0 = 17232\ cm^{-1}$; $^5D_0 \rightarrow {}^7F_1 = 16890, 16800\ cm^{-1}$; $^5D_0 \rightarrow {}^7F_2 = 16350, 16310, 16213\ cm^{-1}$) was interpreted in terms of S_4 symmetry at the site of the eight coordinate europium ion [60]. The emission spectrum of a solution of the europium complex in methanol indicated that the tetrakis chelate dissociated in solution with formation of the solvated tris chelate [61].

Pyridinium and Piperidinium Salts

N-Hexadecylpyridinium Salt $C_5H_5NC_{16}H_{33}[Eu(C_{15}H_{11}O_2)_4]$. The compound was prepared by adding, with stirring, 4.0 ml of 2.0 N sodium hydroxide solution to a solution of 8 mmol of dibenzoylmethane, 2 mmol of europium chloride, and 2 mmol of hexadecylpyridinium chloride in 70 ml of hot ethanol. A dense off-white microcrystalline precipitate separated immediately. After the mixture was allowed to cool, the solid was collected and recrystallized from 1:1 acetone/2-butanone (yield 74%). The melting range of the compound was 220 to 226°C [40].

Piperidinium Salts $C_5H_{11}NH[M(C_{15}H_{11}O_2)_4] \cdot nH_2O$ (n = 0, 1) and $C_5H_{11}NH$-$[M(C_{15}H_{11}O_2)_4] \cdot$ DMF. Bauer, Blanc, Ross [40] effected syntheses of two different modifications (M = Eu, Gd) of the piperidinium chelates, which they designated the α- and β-forms, according to the following procedure: To a solution containing 8 mmol of dibenzoylmethane and 2 mmol of the rare earth chloride in 70 ml of hot ethanol was added 8 mmol of piperidine. The solution was allowed to cool to room temperature, and the well-formed needles which separated were collected, washed with water, and dried under vacuum. In each case, rapid recrystallization of the product from methanol yielded the β-form of the complex as feathery needles (melting points: M = Eu, 182 to 184°C; M = Gd, 183 to 185°C). Within 1 h, the β-form slowly began to redissolve in the mother liquor, and the α-form slowly deposited over a 24 h period as dense blocks (melting points: M = Eu, 184 to 187°C; M = Gd, 185 to 189°C). The emission spectra of the two forms of the europium chelate were recorded on solid samples at 78 and 300 K and analyzed in the region of the $^5D_0 \rightarrow {}^7F_2$ transition (6100 to 6150 Å). The less stable β-form revealed a prominent doublet (wavelength not presented), whereas the α-form showed three emission bands (6120, 6124, 6130 Å) of appreciable intensity. The change from two lines for the β-form to three lines for the α-form was attributed to an isomerization, resulting in a change of symmetry at the europium site from D_{2d} (β-form) to S_4 (α-form) [40].

Workman, Burns [62], following the procedure described by Bauer and coworkers [40], obtained two forms of the europium chelate, which corresponded to those previously reported, both in the physical appearance of the crystals and in their fluorescence spectra. However, the authors noted that elemental analyses and Karl Fischer titrations for water content indicated that the β-form contained one molecule of water per molecule of complex, $C_5H_{11}NH[Eu(C_{15}H_{11}O_2)_4] \cdot H_2O$, whereas, the α-form was anhydrous. The infrared spectrum of the β-form exhibited a characteristic O-H stretching mode as a broad band at 3740 cm^{-1}, whereas this band was absent in the spectrum of the α-form. The hydrated form tends to lose water on standing and converts to the anhydrous form. X-ray diffraction studies revealed that the α-form crystallizes in the monoclinic space group $C2/c\text{-}C^6_{2h}$ (No. 15) or $Cc\text{-}C^4_s$ (No. 9) with eight molecules in a unit cell of dimensions a = 24.0, b = 23.8, c = 19.4Å, β = 90°25' (density = 1.36 g/cm^3). The β-form (monohydrate), on the other hand, crystallizes in monoclinic space group $P2_1/n\text{-}C^5_{2h}$ (No. 14) with four molecules in a unit cell of dimensions a = 18.93, b = 18.75, c = 16.18 Å, β = 92°42' (density = 1.31 g/cm^3) [62].

The dimethylformamide adduct, $C_5H_{11}NH[Eu(C_{15}H_{11}O_2)_4] \cdot DMF$, was prepared by recrystallizing the initial reaction product obtained according to the procedure of Bauer and coworkers [40] from methanol to which some dimethylformamide had been added. This compound crystallizes in the monoclinic space group C2/c or Cc with four molecules in a unit cell of dimensions a = 29.2, b = 9.38, c = 28.6 Å, β = 128°42' (density = 1.31 g/cm^3) [62].

Khomenko, Kuznetsova [63] isolated three modifications of the europium chelate, which they designated as the α-, β-, and γ-forms. The α-form was prepared by adding 0.18 ml of piperidine to a solution of europium nitrate hexahydrate (200 mg) and dibenzoylmethane (450 mg) in 20 ml of hot ethanol. The needle shaped crystals that formed were collected, washed with ethanol and air dried (m.p. = 172°C). The β-form was prepared by adding 0.15 ml of piperidine to 50 ml of an absolute methanolic solution containing 100 mg of europium nitrate hexahydrate and 225 mg of dibenzoylmethane. Rectangular crystals were produced, which were washed with methanol and dried in air (m. p. = 170°C). The β-form was also obtained by recrystallizing 800 mg of the α-form from 5 ml of hot methanol. The γ-form was prepared by adding 0.15 ml of piperidine (the α-form was produced if only 0.09 ml of piperidine was used) to a solution of europium nitrate hexahydrate (100 mg) and dibenzoylmethane (225 mg) in 1 ml of hot, absolute ethanol. The resulting mass of fine crystals was washed with methanol and dried (m. p. = 159°C). The emission spectra of the α- and γ-forms were interpreted in terms of S_4 or D_2 symmetry at the europium site, whereas the spectrum of the β-form was interpreted in terms of C_{4v} or D_{2d} symmetry at the metal ion site.

Baczynski, Zachara [64] also obtained three crystalline forms of the europium chelate by changing the amount of piperidine used in the preparation. They analyzed the $^5D_0 \rightarrow {}^7F_J$ and $^5D_1 \rightarrow {}^7F_J$ (J = 0, 1, 2) emissions of the three crystalline forms and concluded from the nature of the Stark splitting that the symmetry at the metal ion site was different for each compound. The symmetry (S_4) was highest for the chelate obtained using the highest (100:1) piperidine to metal mole ratio, whereas four crystal field symmetries (C_1, C_{1h}, C_2, C_{2v}) were considered possible for the complexes obtained from solutions with the smallest (10:1) and intermediate piperidine : europium mole ratios. The metal ion site symmetry was considered to be greater in the former compound (prepared using 10:1 mol ratio) than in the latter compound [64].

A comparison of the properties of different modifications of the europium chelate prepared by various groups of investigators is presented in [65].

Brecher, Samelson, Lempicki [36] prepared the europium chelate by adding, slowly and with stirring, a solution of europium chloride in 95% ethanol to a solution of dibenzoylmethane and piperidine in the same solvent (metal : ligand : piperidine mole ratio = 1:4:4). A precipitate formed immediately which was collected by filtration, washed with several portions of 95% ethanol, and dried by forcing dry nitrogen through the powder. The emission spectrum of the crystalline powder at 93 K revealed two intense emissions in the region of the $^5D_0 \rightarrow {}^7F_2$ transition at 6136 and 6120 Å, which is consistent with the C_{4v} symmetry they proposed for the europium site. The coordination sphere was depicted as a square antiprismatic array of diketonate oxygen atoms, with the piperidinium ion capping a square face of the antiprism along its C_4 axis [36]. (Two intense lines are also predicted for D_{2d} symmetry as noted by Bauer and coworkers [40].) The emission spectrum of the compound in 3:1 (v/v) ethanol-methanol (standard alcohol solvent) at 93 K revealed three intense peaks at 6147, 6130, and 6118 Å in the region of the $^5D_0 \rightarrow {}^7F_2$ transition. These results were interpreted in terms of dodecahedral geometry with S_4 symmetry at the europium site [36, 66]. The emission spectrum of the chelate in a 4:1 (v/v) standard alcohol/dimethylformamide mixture at 93 K, however, resembled the spectrum of the crystalline powder, exhibiting two intense lines in the region of the $^5D_0 \rightarrow {}^7F_2$ transition at 6142 and 6120 Å. The authors proposed ligation of the dimethylformamide molecule along the C_4 axis of the square antiprismatic tetrakis chelate, with C_{4v} or C_4 symmetry at the europium site [36, 66]. The authors accounted for extraneous lines in the solution spectra of the europium chelate by proposing dissociation of the tetrakis complex. The degrees of dissociation, calculated from the relative intensities of lines corresponding to the tetrakis and tris chelates, were 0.43 and 0.51 in the standard alcohol and 4:1 standard alcohol/dimethylformamide solutions, respectively [36]. The vibronic spectra of europium dibenzoylmethane chelate with piperidine in the neighborhood of the $^5D_0 \rightarrow {}^7F_0$ line at 293 and 77 K and the IR absorption spectra in the region of 20 to 200 cm^{-1} at various temperatures were also studied by Zachara [81]. All vibrational frequencies obtained from the vibronic spectrum at 77 K can be observed in the IR spectrum at the same temperature. For the above frequencies some vibrations were assigned as belonging to irreducible representations of S_4 point group symmetry [81].

Infrared and Raman spectral data for the europium chelate are recorded in Table 5/47. The band assignments are based on a single ring (C_{2v} symmetry), normal coordinate analysis. (The compound was prepared by dissolving europium chloride, dibenzoylmethane, and piperidine in ethanol. After partial evaporation of the solvent, the crystalline precipitate was collected by filtration, washed with ethanol, and air dried.) The authors concluded that the symmetry group of the chelate is either S_4 ($=C_2$), or C_1 (no symmetry). There was no definite evidence to support the symmetry groups C_{4v}, C_{2v}, and D_{2d}, since under these groups, the Raman spectrum is expected to exhibit a larger number of bands than the infrared spectrum. The data in Table 5/47 indicate that this is not the case. The symmetry groups containing a center of inversion were ruled out because no evidence for mutual exclusion of the u and g modes was observed in the infrared and Raman spectra [67].

Table 5/47
Infrared and Raman Spectral Data for Piperidinium Tetrakis(dibenzoylmethanato)Europate(III).

infrared band in cm^{-1} *)	Raman band in cm^{-1} **)	assignment	infrared band in cm^{-1} *)	Raman band in cm^{-1} **)	assignment
1600 vs	1590 vs	ν_2 mainly C=O stretch	940 w		ν_4
1580 vw		ϕ ring stretch	925 vw		
1550 vs		ϕ ring stretch		855 vw	ϕ CH out-of-plane bend
1520 vs	1510 vw	ν_9 mainly C=C stretch	810 w	795 w	ν_{12}
1480 vs	1485 m	ϕ ring stretch	782 w		
1460 s		ν_{10} CH bend and C=O stretch	750 m		CH out-of-plane bend (chelate ring)
1440 vw	1440 w			730 m	
1420 s			720 s		ϕ CH out-of-plane bend
1400 m			690 m	658 w	ν_5
1360 sh	1380 m		620 w		
1310 m	1305 s		610 w	610 w	ν_{13}
1284 m	1275 vs	ν_3	520 w		ϕ ring in-plane deformation
1220 m	1220 w	ν_{11} CH bend and C=O stretch	510 sh	510 w	
1175 vw	1180 m		440 w	460 m	
1155 vw	1130 vw		420 vw		
1070 m	1065 m	ϕ CH in-plane bend	405?		ν_6 complicated mode involving Eu-O stretch
1025 m		ϕ CH in-plane bend	355 w		
1000 vw	1000 s		260?		ν_7
970 vw					

*) KBr and CsI pellets. — **) Powder.

References to 5.1.15 see p. 208/10

Proton magnetic resonance data (recorded below) obtained for solutions of the paramagnetic praseodymium and dysprosium chelates (method for synthesizing compounds not presented) in chloroform, revealed relatively small isotropic shifts for the phenyl rings protons, but large downfield shifts for the −CH proton. Large isotropic shifts were also observed for the $-\overset{+}{N}H_2$ and α-protons of the piperidine ring, suggesting an interaction between the piperidinium cation and complex anion in solution [48].

compound	chemical shift in Hz at 100 MHz relative to $CHCl_3$ *)					
	dibenzoylmethanate		piperidinium ion			
	C_6H_5	CH	NH_2	α-H	β-H	γ-H
$C_5H_{10}NH_2[Pr(C_{15}H_{11}O_2)_4]$	35	− 505(48)	1740(90)	1040(20)	965(20)	836(20)
$C_5H_{10}NH_2[Dy(C_{15}H_{11}O_2)_4]$	80	−1150(50)	4250(250)	1430(100)	870(50)	660(25)

*) Line width (Hz) enclosed in parentheses.

The Mössbauer spectrum of the europium chelate (prepared by allowing europium chloride, dibenzoylmethane, and piperidine to react in ethanol [4]), taken with a ^{151}Eu source in the Sm_2O_3 lattice over the temperature range 79 to 220 K, revealed a mean value for the isomer shift of −0.40 mm/s [68].

Other Salts

Morpholinium Salt $C_4H_9ONH[Eu(C_{15}H_{11}O_2)_4]$. Lee, Nugent [52] prepared the compound by adding 0.00387 mol of europium chloride in 25 ml of absolute ethanol to a hot (90°C) ethanolic solution (25 ml) containing 0.0155 mol of dibenzoylmethane and 0.0155 mol of morpholine. The mixture was maintained at ambient temperature for four days and then filtered. The crystalline product was collected, washed with cold absolute ethanol, and air dried (yield = 89%). The melting point of the compound was 179 to 180°C [52]. Rozploch [60] effected synthesis of the compound according to a similar procedure.

The emission spectrum of the chelate ($^5D_0 \to {}^7F_0$ = 17244 cm^{-1}; $^5D_0 \to {}^7F_1$ = 16927, 16888, 16827 cm^{-1}; $^5D_0 \to {}^7F_2$ = 16353, 16286, 16260, 16167 cm^{-1}) was interpreted in terms of C_{2v} symmetry at the site of the eight-coordinate europium ion [60]. The vibronic spectrum and the $^5D_{J'} \to {}^7F_{J''}$ electronic emissions (J' = 0, 1 and J'' = 0, 1, 2) are tabulated in [69]. A correlation was observed between vibrational frequencies obtained from the infrared spectra and those obtained from the vibronic spectrum.

Azabicyclononane Salt $C_8H_{16}N^+[Eu(C_{15}H_{11}O_2)_4]$. The compound was prepared according to the procedure described by Bauer, Blanc, Ross [40] for preparation of the analogous piperidinium salts of the dibenzoylmethane chelates. The product was recrystallized from cyclohexane (melting point = 164 to 168°C).

5.1.15.5 Cerium(IV) Compound $Ce(C_{15}H_{11}O_2)_4$

Overall formation constant, determined by a solvent extraction method (I = 0.1); lg β_4 = 51.60 [70].

Sacconi, Ercoli [21] effected synthesis of the cerium(IV) chelate by adding an ethanolic ammonia solution to a solution of cerium(III) nitrate hexahydrate (2 g) and dibenzoylmethane (4 g) in 110 cm^3 of ethanol and heating the resulting mixture on a steam bath. The brown red solid product which formed had a melting point of 192 to 193°C. Purushottam and coworkers [5] prepared the compound by adding an ethanolic solution of the ligand to a solution of cerium(IV) nitrate in absolute ethanol and then adjusting the pH of the mixture to 5.6 by the addition of alcoholic ammonia. The precipitate was collected by filtration, washed with ethanol, and then dissolved in benzene. The benzene solution was filtered, and the filtrate was allowed to evaporate at room temperature. The product was dried under vacuum over concentrated sulfuric acid. The brown colored substance melted at 192°C.

The crystal and molecular structures of the compound have been investigated by single-crystal X-ray analysis. The orthorhombic crystals are in space group Pccn = D_{2h}^{10} (No. 56), with four molecules in a unit cell of dimensions a = 10.320, b = 20.109, and c = 23.514 Å [71]. Wolf, Barnighausen [72] interpreted their electron-density projections in terms of a dodecahedral configuration of diketonate oxygen atoms about the eight-coordinate central ion, whereas Shankar, Kunchur [73] interpreted their data in terms of a square antiprismatic geometry.

Infrared spectral data (spectra obtained for samples dispersed in Nujol and hexachlorobutadiene mulls) are tabulated and discussed in [74]. The ν(C-O) stretching mode was assigned to a band at 225 cm^{-1}.—The electronic absorption spectra (in methanol, ethanol, hexane, cyclohexane) revealed two bands in the ranges 250 to 255 and 340 to 350 nm [5].

5.1.15.6 Other Dibenzoylmethanato Complexes

5.1.15.6.1 Bis(dibenzoylmethanato) Salts

Charles [75] effected synthesis of the mixed chelate-acetate $Eu(C_{15}H_{11}O_2)_2(CH_3COO) \cdot nCH_3COOH$ (n = 0, 1) according to the following procedure: A solution of sodium acetate (0.03 mol) in 15 ml of water was added to 10 ml of an aqueous solution of europium chloride (0.005 mol). This mixture was added slowly and with vigorous shaking, to a solution of dibenzoylmethane (0.0165 mol) in 100 ml of 95% ethanol. The resulting mixture, containing a yellow precipitate, was heated on a hot plate for 10 min. After standing at room temperature overnight, the precipitate was collected by filtration, washed with 95% ethanol, and dried under vacuum at room temperature. The compound was purified by heating with 95% ethanol (melting point = 269 to 270°C). Sinha and coworkers [76, 77] prepared this compound according to a similar procedure.

The emission spectrum of the compound (n = 0) in 3:1 (v/v) ethanol-methanol at 77 K revealed bands at 15267 Å ($^5D_1 \rightarrow {}^7F_3$), 16331 Å ($^5D_0 \rightarrow {}^7F_2$), 16909, 16966 Å ($^5D_0 \rightarrow {}^7F_1$), and 17284 Å ($^5D_0 \rightarrow {}^7F_0$) [76]. The compound was further characterized by thermogravimetric and differential thermal analyses and infrared spectroscopy [75].

Baczynski, Rozploch, Orzeszko [78] effected preparation of two forms of the complex by adding a solution of sodium acetate (0.82 g) in methanol (30 cm^3) to a solution containing dibenzoylmethane (0.9 g) and europium chloride (0.26 g) in about 40 cm^3 of methanol, and heating the resulting mixture. The solution was allowed to cool, and after 24 h the crystallization process was complete. A mixture containing both a yellowish crystalline form and a brown form of the complex was obtained. The separation of the two forms was achieved by recrystallization from benzene. The yellow form was found to have the empirical composition, $Eu(C_{15}H_{11}O_2)_2(CH_3COO) \cdot CH_3COOH$, and the brown form the composition, $Eu(C_{15}H_{11}O_2)_2(CH_3COO)$. Both forms exhibited strong fluorescence at liquid air temperature, but only the brown form showed luminescence at room temperature when irradiated with ultraviolet light [78].

The propionate $Eu(C_{15}H_{11}O_2)_2(C_3H_5O_2)$ and benzoate $Eu(C_{15}H_{11}O_2)_2(C_7H_5O_2)$ were prepared according to the procedure described by Charles [75] for preparation of the mixed acetato-dibenzoylmethanato chelates. The propionato salt melted at 268 to 270°C and the benzoato salt at 223 to 225°C. The compounds were characterized by elemental, thermogravimetric, and differential analysis and infrared spectroscopy [75].

For Mössbauer studies on the acetate, benzoate, and 4-nitrobenzoate, see [83].

The 2-propanolates $M(C_{15}H_{11}O_2)_n(C_3H_7O)_{3-n}$ (n = 1, 2) (M = Pr, Nd, Sm) were prepared by allowing stoichiometric amounts of a rare earth 2-propanolate to react with dibenzoylmethane in benzene. The 2-propanol formed during the reaction was removed by distillation as the azeotrope with benzene. The molecular weights of the compounds in benzene (determined ebullioscopically) revealed that the n = 1 compounds exist as trimers and that the n = 2 compounds exist as dimers in this solvent [7, 8].

References to 5.1.15 see p. 208/10

5.1.15.6.2 Chelates with Dibenzoylmethane and β-Ketoesters $CH_3C(O)CH(R)COOC_2H_5$ ($R = H, CH_3, C_2H_5$)

The compounds listed in Table 5/48 were prepared according to the following procedure: A solution of dibenzoylmethane and the appropriate ester in 20 ml of ethanol was added to a solution of hydrated europium nitrate in 20 ml of acetone (mole ratio of reactants given in Table 5/48), followed by addition of alcoholic sodium hydroxide solution, with continuous stirring, until the pH of the reaction mixture reached 6.5 to 7. The sodium nitrate crystals formed were removed by filtration, and the filtrate was allowed to evaporate at room temperature. Yellow crystals appeared when half the solvent had evaporated. These were collected by filtration and washed with hot alcohol. The chelates were purified by repeated recrystallization from acetone. The compounds were characterized by elemental analyses and infrared and electronic emission spectral measurements [79].

Table 5/48
Mixed Ligand Chelates with Dibenzoylmethane and β-Ketoesters $CH_3C(O)CH(R)COOC_2H_5$.

No.	R	formula of β-ketoester	formula of complex	mole ratio of reactants	m. p. in °C
1	H	$C_6H_{10}O_3$	$Eu(C_{15}H_{11}O_2)(C_6H_9O_3)_2$	1:5:1.5	78 to 80
2	CH_3	$C_7H_{12}O_3$	$Eu(C_{15}H_{11}O_2)(C_7H_{11}O_3)(OH) \cdot H_2O$	1:5:1.5	258 to 260
3	CH_3	$C_7H_{12}O_3$	$Eu(C_{15}H_{11}O_2)_2(C_7H_{11}O_3)$	1:5:2.5	247 to 249
4	C_2H_5	$C_8H_{14}O_3$	$Eu(C_{15}H_{11}O_2)(C_8H_{13}O_3)(OH) \cdot H_2O$	1:5:1.5	255
5	C_2H_5	$C_8H_{14}O_3$	$Eu(C_{15}H_{11}O_2)_2(C_8H_{13}O_3)$	1:5:2.5	236 to 238

*) Metal : β-ketoester : dibenzoylmethane mole ratio.

5.1.15.6.3 Chelates with Dibenzoylmethane and N,N'-Bis(salicylidene)ethylenediamine $HOC_6H_4CH{=}NCH_2CH_2N{=}CHC_6H_4OH$ ($= C_{16}H_{16}N_2O_2$)

$M(C_{15}H_{11}O_2)(C_{16}H_{14}N_2O_2)$. Each mixed ligand complex (M = Y, La, Pr, Nd, Sm) was prepared by adding 30 ml of an ethanolic solution of bis(salicylidene) ethylenediamine (1 mmol) to 50 ml of an ethanolic solution of the tris (dibenzoylmethanato) rare earth chelate (1 mmol) and then heating the mixture at reflux on a steam bath. After 0.5 h, the mixed chelates began to precipitate, but heating at reflux was continued for 2 h. The solution was filtered while hot, and the compound was washed several times with ethanol and dried under vacuum at room temperature over sulfuric acid. Similar compounds were obtained from the reaction of the sodium salts of the tetrakis chelates. The colors and melting points of the compounds were: Y, yellow, 284 to 288°C; La, yellow, 280 to 282°C; Pr, yellow brown, 286 to 290°C; Nd, yellowish pink, 275 to 280°C; Sm, yellow, 282 to 284°C. The compounds were further characterized by elemental analyses and their infrared spectra (KBr discs) [80].

References to 5.1.15:

[1] A. A. Zholdakov, N. K. Davidenko (Zh. Neorgan. Khim. **13** [1968] 3223/6; Russ. J. Inorg. Chem. **13** [1968] 1662/5). — [2] A. S. Berlyand, A. I. Byrke, L. I. Martynenko (Zh. Neorgan. Khim. **13** [1968] 2106/10; Russ. J. Inorg. Chem. **13** [1968] 1089/91). — [3] R. E. Whan, G. A. Crosby (J. Mol. Spectrosc. **8** [1962] 315/27). — [4] S. J. Lyle, A. D. Witts (Inorg. Chim. Acta **5** [1971] 481/4). — [5] D. Purushottam, V. Ramachandra Rao, Bh. S. V. Raghava Rao (Anal. Chim. Acta **33** [1965] 182/97).

[6] D. Purushottam, Bh. S. V. Raghava Rao (Indian J. Chem. **4** [1966] 109/10). — [7] M. Hasan, K. Kumar, S. Dubey, S. N. Misra (Bull. Chem. Soc. Japan **41** [1968] 2619/23). — [8] M. Hasan, S. N. Misra, R. N. Kapoor (Indian J. Chem. **7** [1969] 519/20). — [9] J. L. Wood, M. M. Jones (J. Inorg. Nucl. Chem. **29** [1967] 113/22). — [10] A. Perotto, R. G. Charles (J. Inorg. Nucl. Chem. **26** [1964] 373/6).

[11] S. P. Tandon, P. C. Mehta, R. N. Kapoor, S. N. Mishra (Z. Naturforsch. **25b** [1970] 472/5).— [12] I. A. Murav'eva, L. I. Martynenko, V. I. Spitsyn (Zh. Neorgan. Khim. **22** [1977] 3009/12; Russ. J. Inorg. Chem. **22** [1977] 1636/8). — [13] M. Metlay (J. Electrochem. Soc. **111** [1964] 1253/5). — [14] M. Metlay (J. Chem. Phys. **39** [1963] 491/2). — [15] B. R. Judd (Phys. Rev. [2] **127** [1962] 750/61).

[16] G. S. Ofelt (J. Chem. Phys. **37** [1962] 511/20).— [17] P. C. Mehta, S. P. Tandon (J. Chem. Phys. **53** [1970] 414/7). — [18] S. P. Tandon, R. C. Govil (Spectrosc. Letters **4** [1971] 73/7). — [19] W. G. Perkins, G. A. Crosby (J. Chem. Phys. **42** [1965] 407/14). — [20] M. J. Lacey, C. G. Macdonald, J. S. Shannon (Org. Mass Spectrom. **1** [1968] 115/26).

[21] L. Sacconi, R. Ercoli (Gazz. Chim. Ital. **79** [1949] 731/8 from C.A. **1950** 3832). — [22] A. I. Byrke, L. A. Aslanov, E. F. Korytnyi (Vestn. Mosk. Univ. Ser. II Khim. **21** No. 6 [1966] 117/9; Moscow Univ. Chem. Bull. **21** [1966] 533/4).— [23] I. A. Murav'eva, N. Denliev, N. K. Khalmurzaev, A. I. Byrke, A. S. Berlyand, L. I. Martynenko, V. I. Spitsyn (Izv. Akad. Nauk SSSR Ser. Khim. **1975** 1258/62; Bull. Acad. Sci. USSR Div. Chem. Sci. **1975** 1161/4). — [24] V. S. Khomenko, V. V. Kuznetsova (Dokl. Akad. Nauk Belorussk. SSR **7** [1963] 610/3 from C.A. **60** [1964] 2533). — [25] E. Uhlemann, F. Dietze (Z. Anorg. Allgem. Chem. **386** [1971] 329/34).

[26] R. C. Mathur, S. S. L. Surana, S. P. Tandon (Z. Naturforsch. **30b** [1975] 207/9). — [27] S. Misumi, N. Iwasaki (Bull. Chem. Soc. Japan **40** [1967] 550/4). — [28] C. Y. Liang, E. J. Schimitschek, J. A. Trias (J. Inorg. Nucl. Chem. **32** [1970] 811/31).— [29] M. F. Richardson, W. F. Wagner, D. E. Sands (Inorg. Chem. **7** [1968] 2495/500). — [30] P. C. Mehta, S. S. L. Surana, S. P. Tandon (Indian J. Pure Appl. Phys. **7** [1969] 767/8).

[31] S. P. Tandon, P. C. Mehta, R. N. Kapoor, S. N. Mishra (Z. Naturforsch. **25b** [1970] 472).— [32] A. I. Byrke, N. N. Magdesieva, L. I. Martynenko, V. I. Spitsyn (Zh. Neorgan.Khim. **12** [1967] 666/71; Russ. J. Inorg. Chem. **12** [1967] 348/51).— [33] P. C. Mehta, S. P. Tandon (Z. Naturforsch. **26a** [1971] 759/62). — [34] G. A. Domrachev, V. P. Ippolitova (Tr. Khim. Khim. Tekhnol. **1966** No. 2, p. 227/40 from C.A. **68** [1968] No. 73665).— [35] P. C. Mehta, S. S. L. Surana, S. P. Tandon (Can. J. Spectrosc. **18** [1973] 55/60).

[36] C. Brecher, H. Samelson, A. Lempicki (J. Chem. Phys. **42** [1965] 1081/96).— [37] T. Isobe, S. Misumi (Bull. Chem. Soc. Japan **47** [1974] 281/4). — [38] H. G. Brittain (J. Chem. Soc. Dalton Trans. **1979** 1187/91). — [39] L. R. Melby, N. J. Rose, E. Abramson, J. C. Caris (J. Am. Chem. Soc. **86** [1964] 5117/25). — [40] H. Bauer, J. Blanc, D. L. Ross (J. Am. Chem. Soc. **86** [1964] 5125/31).

[41] E. Butter, K. Kreher (Z. Naturforsch. **20a** [1965] 408/12). — [42] W. Schmidt, R. Butter (Z. Chem. [Leipzig] **8** [1968] 117). — [43] N. Denliev, I. A. Murav'eva, L. I. Martynenko (Zh. Neorgan. Khim. **22** [1977] 255/6; Russ. J. Inorg. Chem. **22** [1977] 141/2).— [44] E. V. Melent'eva, L. I. Kononenko, N. S. Poluektov (Zh. Neorgan. Khim. **11** [1966] 369/73; Russ. J. Inorg. Chem. **11** [1966] 200/3). — [45] E. Kreher, E. Butter, W. Seifert (Z. Naturforsch. **22b** [1967] 242/7).

[46] R. G. Charles, R. C. Ohlmann (J. Inorg. Nucl. Chem. **27** [1965] 119/27). — [47] R. C. Ohlmann, R. G. Charles (J. Chem. Phys. **40** [1964] 3131/3). — [48] P. K. Burkert, H. P. Fritz, W. Gretner, H. J. Keller, K. E. Schwarzhans (Inorg. Nucl. Chem. Letters **4** [1968] 31/2). — [49] N. Denliev, I. A. Murav'eva, L. I. Martynenko, V. I. Spitsyn, A. V. Davydov, B. F. Myasoedov (Izv. Akad. Nauk SSSR Ser. Khim. **1976** 1455/8; Bull. Acad. Sci. USSR Div. Chem. Sci. **1976** 1392/5). — [50] N. K. Dutt, P. Bandyopadhyay (Sci. Cult. [Calcutta] **24** [1958] 38).

[51] E. Butter, W. Seifert (Z. Anorg. Allgem. Chem. **384** [1971] 67/80). — [52] S. M. Lee, L. J. Nugent (J. Inorg. Nucl. Chem. **26** [1964] 2304/6). — [53] M. F. Taragin, J. C. Eisenstein (J. Inorg. Nucl. Chem. **35** [1973] 3815/9).— [54] N. Filipescu, J. J. Degnan, N. McAvoy (J. Chem. Soc. A **1968** 1594/8). — [55] S. Bjorklund, N. Filipescu, N. McAvoy, J. Degnan (J. Phys. Chem. **72** [1968] 970/8).

[56] M. Ciampolini, F. Mani, N. Nardi (J. Chem. Soc. Dalton Trans. **1977** 1325/8). — [57] C. R. Hurt, N. McAvoy, S. Bjorklund, N. Filipescu (Nature **212** [1966] 179/80). — [58] J. Blanc, D. L. Ross (J. Chem. Phys. **43** [1965] 1286/9). — [59] A. Lodzinska, A. Rozploch (Rocznicki Chem. **46** [1972] 565/72; C.A. **77** [1972] No. 108899). — [60] A. Rozploch (Acta Phys. Polon. A **48** [1975] 93/103).

[61] A. Baczynski, A. Rozploch, S. Zachara (Acta Phys. Polon. A **42** [1972] 31/7; C.A. **77** [1972] No. 107263). — [62] M. O. Workman, J. H. Burns (Inorg. Chem. **8** [1969] 1542/4). — [63] V. S. Khomenko, V. V. Kuznetsova (Zh. Prikl. Spektrosk. **7** [1967] 850/3; J. Appl. Spectry. **7** [1967] 564/7). — [64] A. Baczynski, S. Zachara (Acta Phys. Polon. A **42** [1972] 55/61). — [65] L. A. Aslanov, M. A. Porai-Koshits (Zh. Strukt. Khim. **15** [1974] 836/40; Russ. J. Struct. Chem. **15** [1974] 737/40).

[66] H. Samelson, C. Brecher, A. Lempicki (J. Mol. Spectrosc. **19** [1966] 349/71). — [67] C. Y. Liang, E. J. Schimitschek, J. A. Trias (J. Inorg. Nucl. Chem. **32** [1970] 811/31). — [68] S. J. Lyle, A. D. Witts (J. Chem. Soc. Dalton Trans. **1975** 185/8). — [69] A. Baczynski, W. Orzeszko (Acta Phys. Polon. A **48** [1975] 127/34). — [70] L. S. Voronets, I. P. Efimov, V. M. Peshkova (Zh. Neorgan. Khim. **15** [1970] 886/7; Russ. J. Inorg. Chem. **15** [1970] 451/2).

[71] L. Wolf, H. Bärnighausen (Acta Cryst. **10** [1957] 605/6). — [72] L. Wolf, H. Bärnighausen (Acta Cryst. **13** [1960] 778/85). — [73] J. Shankar, N. R. Kunchur (Acta Cryst. **12** [1959] 940/1). — [74] T. Yoshimura, C. Miyake, S. Imoto (Bull. Chem. Soc. Japan **46** [1973] 2096/101). — [75] R. G. Charles (J. Inorg. Nucl. Chem. **26** [1964] 2195/9).

[76] V. J. Rao, D. R. Rao, A. P. B. Sinha (Indian J. Chem. **8** [1970] 270/4). — [77] V. Janardhanarao, A. P. B. Sinha (Indian J. Chem. **4** [1966] 196/8). — [78] A. Baczynski, A. Rozploch, W. Orzeszko (Acta Phys. Polon. A **43** [1973] 211/8). — [79] N. K. Dutt, S. Sur, S. Rahut (J. Inorg. Nucl. Chem. **33** [1971] 1717/24). — [80] N. K. Dutt, K. Nag (J. Inorg. Nucl. Chem. **30** [1968] 2779/83).

[81] S. Zachara (Z. Naturforsch. **34a** [1979] 748/51). — [82] M. A. Tishchenko, G. I. Gerasimenko, Yu. N. Anisimov, N. S. Poluektov (Dokl. Akad. Nauk SSSR **250** [1980] 122/5 from C.A. **92** [1980] No. 139945). — [83] O. K. Medhi, U. Agarwal (Z. Naturforsch. **34a** [1979] 625/30).

5.1.16 Complexes with Derivatives of 1,3-Diphenyl-1,3-propanedione

5.1.16.1 With o-Hydroxydibenzoylmethane $HOC_6H_4C(O)CH_2C(O)C_6H_5$ (= $C_{15}H_{12}O_3$)

$M(C_{15}H_{11}O_3)_n^{3-n}$ Complexes in Solution (n = 1 to 3)

Formation constants were determined potentiometrically (glass) electrode at 27.5°C in 75% (v/v) acetone/water, I = 0.1 (NaCl). The thermodynamic formation constants (I = 0) were determined at 27.5°C by interpolating data obtained at I = 0.025, 0.05, and 0.1 (NaCl) [1].

metal ion	formation constants*) lg K_1	lg K_2	lg K_3	lg β_3	$-\Delta H_{total}$**) in kJ·mol⁻¹	$-\Delta S_{total}$ in J·mol⁻¹·K⁻¹
La^{3+}	8.40 (9.48)	7.62 (8.58)	7.46 (7.96)	23.28 (26.02)	343.3	651.0
Pr^{3+}	8.70 (9.56)	7.92 (8.66)	7.52 (8.32)	24.14 (26.64)	329.9	631.8
Nd^{3+}	9.16 (9.58)	8.10 (8.88)	7.70 (8.44)	24.62 (26.90)	217.5	626.6
Sm^{3+}	9.20 (9.70)	8.34 (8.90)	7.98 (8.66)	25.52 (27.28)	326.8	622.0

*) The value in parentheses is the thermodynamic formation constant, I = 0.0. — **) Calculated from the temperature coefficient of the overall formation constant.

Isolated Compounds

$Eu(C_{15}H_{11}O_3)_2(CH_3COO)$. An aqueous solution (15 ml) of sodium acetate (0.03 mol) was added to an aqueous solution (10 ml) of europium chloride (0.005 mol), and the resulting solution was added in small portions, with vigorous shaking, to a solution of the chelating agent (0.0165 mol) in absolute ethanol (100 ml). The reaction mixture which contained a precipitate, was heated on a water bath for 15 to 20 min. After standing at room temperature overnight, the precipitate was collected by filtration, washed with absolute ethanol, and dried in air. The compound melted at 109 to 110°C [2]. The emission spectrum of the mixed ligand chelate in 3:1 (v/v) ethanol/methanol at 77 K revealed the following bands in the regions of the $^5D_0 \rightarrow {}^7F_J$ (J = 0, 1, 2) transitions: $^5D_0 \rightarrow {}^7F_2$ = 16241, 16380 Å; $^5D_0 \rightarrow {}^7F_1$ = 16828, 16867, 16062 Å; $^5D_0 \rightarrow {}^7F_0$ = 17272, 17355 Å [3].

$Eu(C_{15}H_{11}O_3)_3 \cdot bpy$ and $Eu(C_{15}H_{11}O_3)_3 \cdot phen$. To a solution of o-hydroxydibenzoylmethane (6 mmol) and the Lewis base (2 mmol) in hot absolute ethanol (20 ml) was added 6 ml of 1 N sodium (or potassium) hydroxide. The mixture was stirred while a solution of europium chloride (2 mmol) was added dropwise. The mixture was allowed to cool, and the products were collected, recrystallized from absolute ethanol, and dried in air at room temperature. The phenanthroline adduct melted at 144°C (no melting point was reported for the bipyridine adduct) [2]. The emission spectra of the adducts in 3:1 (v/v) ethanol/methanol at 77 K revealed the following bands in the regions of the $^5D_0 \rightarrow {}^7F_J$ (J = 0, 1, 2) transitions: $Eu(C_{15}H_{11}O_3)_3 \cdot bpy$, $^5D_0 \rightarrow {}^7F_2$ = 16260, 16338 Å; $^5D_0 \rightarrow {}^7F_1$ = 17027 Å; $^5D_0 \rightarrow {}^7F_0$ = 17382 Å; $Eu(C_{15}H_{11}O_3)_3 \cdot phen$, $^5D_0 \rightarrow {}^7F_2$ = 16231, 16382 Å; $^5D_0 \rightarrow {}^7F_1$ = 16813, 16935, 17004 Å; $^5D_0 \rightarrow {}^7F_0$ = 17298 Å [3].

$Eu(C_{15}H_{11}O_3)_3 \cdot 2(C_4H_9O)_3PO$. An aqueous europium nitrate solution was extracted with an ether solution of the chelating agent and tributyl phosphate, and the product was recovered from the ether layer. The emission spectrum of the adduct in 3:1 (v/v) ethanol/methanol at 77 K revealed bands at 16369 Å ($^5D_0 \rightarrow {}^7F_2$), 16975 ($^5D_0 \rightarrow {}^7F_1$), and 17298 Å ($^5D_0 \rightarrow {}^7F_0$) [3].

$NH_4[Eu(C_{15}H_{11}O_3)_4]$. The compound was prepared by passing anhydrous ammonia through a solution of europium chloride and o-hydroxydibenzoylmethane in absolute ethanol (1:4 mol ratio). The resulting precipitate was collected by filtration, washed with alcohol, and dried in air. The compound melted at 204 to 205°C [2]. The emission spectrum of the compound in 3:1 (v/v) ethanol/methanol at 77 K revealed the following bands in the regions of the $^5D_0 \rightarrow {}^7F_J$ (J = 0, 1, 2) transitions: $^5D_0 \rightarrow {}^7F_2$ = 16142, 16355 Å; $^5D_0 \rightarrow {}^7F_1$ = 16813, 16894, 16922, 16997 Å; $^5D_0 \rightarrow {}^7F_0$ = 17255 Å [3].

$C_5H_{11}NH[Eu(C_{15}H_{11}O_3)_4]$. A warm, filtered solution of europium chloride (0.00337 mol) in absolute ethanol (25 ml) was added to a solution of o-hydroxydibenzoylmethane (0.0155 mol) and piperidine (0.0155 mol) in absolute ethanol (25 ml) maintained at 90°C. The mixture was kept at this temperature for 4 days, then filtered. The collected precipitate was washed with absolute ethanol, recrystallized from absolute ethanol, and dried in air at room temperature. The compound melted at 104 to 105°C [2]. The emission spectrum of the compound in 3:1 (v/v) ethanol/methanol at 77 K revealed the following bands in the regions of the $^5D_0 \rightarrow {}^7F_J$ (J = 0, 1, 2) transitions: $^5D_0 \rightarrow {}^7F_2$ = 16241, 16381 Å; $^5D_0 \rightarrow {}^7F_1$ = 16921, 16966, 16982 Å; $^5D_0 \rightarrow {}^7F_0$ = 17301 Å [3].

References to 5.1.16.1:

[1] K. P. Menon (J. Indian Chem. Soc. **55** [1978] 11/3). — [2] V. Janardhanarao, A. P. B. Sinha (Indian J. Chem. **4** [1966] 196/8). — [3] V. J. Rao, D. R. Rao, A. P. B. Sinha (Indian J. Chem. **8** [1970] 270/4).

5.1.16.2 With Sulfoderivatives of Dibenzoylmethane

Formation constants (Table 5/49) for the 1:1, 2:1, and 3:1 (ligand:metal) rare earth chelates were determined potentiometrically in aqueous media, I = 0.1 ($NaClO_4$). A compound formulated as $La(C_{15}H_{11}O_5S)_3 \cdot 8H_2O$ was obtained by treating an aqueous solution of lanthanum chloride with the sodium salt of 1-phenyl-3-(p-sulfophenyl)-1,3-propanedione. The colorless, fine crystalline precipitate was collected, washed with water, and dried under vacuum over calcium chloride, C. Troeltzsch (Z. Chem. [Leipzig] **6** [1966] 432/3).

Table 5/49

Formation Constants for Rare Earth Chelates Derived from the Sodium Salt of 1-Phenyl-3-(p-sulfophenyl)-1,3-propanedione and the Disodium Salt of 1,3-Bis(m-sulfophenyl)-1,3-propanedione.

metal ion	with the sodium salt of 1-phenyl-3-(p-sulfophenyl)-1,3-propanedione (= $C_{15}H_{11}NaO_5S$)			with the disodium salt of 1,3-bis(m-sulfophenyl)-1,3-propanedione (= $C_{15}H_{10}Na_2O_8S_2$)		
	lg K_1	lg K_2	lg K_3	lg K_1	lg K_2	lg K_3
Y^{3+}	6.68	5.52	4.21	7.05	5.18	3.28
La^{3+}	5.57	4.57	3.77	6.10	4.34	
Pr^{3+}	5.99	5.00	4.01	6.44	4.65	
Nd^{3+}	6.05	5.13	4.08	6.50	4.79	
Sm^{3+}	6.40	5.41	4.24	6.89	5.06	3.28
Eu^{3+}	6.53	5.48	4.24	6.92	5.12	3.33
Gd^{3+}	6.56	5.43	4.32	6.92	5.09	3.28
Tb^{3+}	6.71	5.56	4.30	7.09	5.25	3.36
Dy^{3+}	6.81	5.60	4.25	7.20	5.27	3.40
Ho^{3+}	6.76	5.59	4.34	7.22	5.27	3.40
Er^{3+} *)	6.80	5.66	4.32	7.13	5.31	3.38
Tm^{3+} **)	6.77	5.67	4.34	7.18	5.34	3.37
Yb^{3+}	6.88	5.69	4.36	7.11	5.33	3.23

*) lg K_4 = 3.4. — **) lg K_4 = 3.5.

5.1.16.3 With Other Dibenzoylmethane Derivatives

5.1.16.3.1 Tris Chelates

Each of the compounds recorded in Table 5/50 was prepared by treating an ethanolic solution containing a rare earth chloride and the respective β-diketone with piperidine. The absorption and phosphorescence spectra of the gadolinium chelates were recorded in order to investigate the effect of substituents on the electronic states of the ligand as related to the efficiency of intramolecular energy transfer [1]. The fluorescence spectra of the europium and terbium chelates were modified significantly on changing substituents in the organic ligand. The relative intensity, spectral distribution, shifting, and splitting of the fluorescence lines were discussed in relation to the nature of substituents, their position, molecular configuration, and overall intramolecular energy transfer [2].

Table 5/50

Tris Chelates $ML_3 \cdot nH_2O$ (n = 0, 1) Derived from Substituted Dibenzoylmethanes.

R-C6H4-C(=O)-CH(R')-C(=O)-C6H4-R''

No.	ligand = HL R	R'	R''	formula	M	Ref.
1	4-F	H	4-F	$C_{15}H_{10}F_2O_2$	Eu, Gd, Tb	[1]
2*)	H	Cl	H	$C_{15}H_{11}ClO_2$	La, Sm, Eu Gd, Tb, Dy	[3]

Table 5/50 (Continued)

No.	ligand = HL R	R′	R″	formula	M	Ref.
3*)	H	Br	H	$C_{15}H_{11}BrO_2$	La, Sm, Eu Gd, Tb, Dy	[3]
4	3-NO_2	H	H	$C_{15}H_{11}NO_4$	Eu, Gd, Tb	[1]
5	4-NO_2	H	H	$C_{15}H_{11}NO_4$	Eu, Gd, Tb	[1]
6	3-NO_2	H	3-NO_2	$C_{15}H_{10}N_2O_6$	Eu, Gd, Tb	[1]
7	4-NO_2	H	4-NO_2	$C_{15}H_{10}N_2O_6$	Eu, Gd, Tb	[1]
8	3-CH_3O	H	H	$C_{16}H_{14}O_3$	Eu, Gd, Tb	[1]
9	4-CH_3O	H	H	$C_{16}H_{14}O_3$	Eu, Gd, Tb	[1]
10	3-CH_3O	H	3-CH_3O	$C_{17}H_{16}O_4$	Eu, Gd, Tb	[1]
11	4-CH_3O	H	4-CH_3O	$C_{17}H_{16}O_4$	Eu, Gd, Tb	[1]
12	4-C_6H_5	H	H	$C_{21}H_{16}O_2$	Eu, Gd, Tb	[1]

*) Isolated as the monohydrates.

5.1.16.3.2 Tetrakis Chelates

The piperidinium salts (No. 1 to 4, recorded in Table 5/51) are prepared in the following way: A solution of the β-diketone (0.018) in 95% ethanol (60 ml) was treated with piperidine (1.53 g), followed by 0.004 mol of the rare earth chloride dissolved in water (20 ml). The mixture was heated at reflux for 2 h and then allowed to cool to room temperature overnight. The complexes were recrystallized from a suitable solvent such as benzene, chloroform, or methanol [4, 5].

Infrared spectral data for the chelates (No. 1 to 4) are reported and discussed in [4] and [5]. The ν(C=O) and ν(C=C) stretching modes were assigned to bands in the regions 1620 to 1550 cm^{-1} and 1530 to 1500 cm^{-1}, respectively. The fluorescence properties of the europium chelates were examined at room temperature [5] and at 77 K [4] in the microcrystalline solid state [4] and in solution [4, 5]. Joshi and coworkers noted that fluorine substitution for hydrogen atoms in the ligand increased fluorescence intensity, and attributed the enhanced fluorescence intensity to decreased vibrational energy. The chelates of praseodymium and neodymium either failed to fluoresce or fluoresced very weakly so that there was no visible fluorescence [3].

For preparation of the tetraethylammonium salt (Table 5/51, No. 5), an ethanolic solution (25 ml) of anhydrous europium chloride (0.25 mmol) was added slowly with stirring, to a hot solution containing the β-diketone (1 mmol) and aqueous 10% tetraethylammonium hydroxide (1.4 ml) in ethanol (25 ml). The solution was filtered hot and left overnight. The crystalline product was collected, washed several times with hot 95% ethanol, air dried, and placed in a desiccator over silica gel. The fluorescence spectrum was analyzed in detail by treating the ligand field as a perturbation on the free-ion levels and calculating the ligand field parameters [6].

The piperidinium compounds No. 6 to 8 (see Table 5/51) were prepared by treating an ethanolic solution of europium chloride and the β-diketone with piperidine. The emission spectra of the chelates in 3:1 (v/v) ethanol/methanol at 77 K revealed the following bands in the regions of the $^5D_0 \rightarrow {}^7F_J$ (J = 0, 1, 2) transitions: $Eu(C_{15}H_{10}NO_5)_4^-$, $^5D_0 \rightarrow {}^7F_2$ = 16351 Å, $^5D_0 \rightarrow {}^7F_1$ = 16901 Å, $^5D_0 \rightarrow {}^7F_0$ = 17340 Å; $Eu(C_{16}H_{13}O_3)_4^-$, $^5D_0 \rightarrow {}^7F_2$ = 16242, 16393 Å, $^5D_0 \rightarrow {}^7F_1$ = 16837, 16891 Å; $^5D_0 \rightarrow {}^7F_0$ = 17340 Å; $Eu(C_{17}H_{15}O_5)_4^-$, $^5D_0 \rightarrow {}^7F_2$ = 16241, 16304, 16330 Å; $^5D_0 \rightarrow {}^7F_1$ = 16760, 16900, 16940, 16970, 17030 Å; $^5D_0 \rightarrow {}^7F_0$ = 17280 Å [7].

Table 5/51
Piperidinium Salts of Tetrakis Chelates Derived from Dibenzoylmethane Derivatives $C_5H_{11}NH[M(RC_6H_4COCH_2COC_6H_4R')_4]$.

No.	R	R'	formula of ligand	M	melting point in °C	Ref.
1	4-F	H	$C_{15}H_{11}FO_2$	La	280 to 282	[5]
2	4-F	H	$C_{15}H_{11}FO_2$	Pr	165 to 166	[5]
3	4-F	H	$C_{15}H_{11}FO_2$	Nd	>360	[5]
4	3-F, 4-CH_3O	H	$C_{16}H_{13}FO_3$	Eu	130	[4]
5*)	4-Br	4-Br	$C_{15}H_{10}Br_2O_2$	Eu	233 to 235	[6]
6	2-OH	4-NO_2	$C_{15}H_{11}NO_5$	Eu	—	[7]
7	2-CH_3O	H	$C_{16}H_{14}O_3$	Eu	—	[7]
8	2-OH, 2-CH_3O, 4-CH_3O	H	$C_{17}H_{16}O_5$	Eu	—	[7]

*) Tetraethylammonium salt.

References to 5.1.16.3:

[1] W. F. Sager, N. Filipescu, F. A. Serafin (J. Phys. Chem. **69** [1965] 1092/100). — [2] N. Filipescu, W. F. Sager, F. A. Serafin (J. Phys. Chem. **68** [1964] 3324/46). — [3] M. Morita, S. Shionoya (Bull. Chem. Soc. Japan **43** [1970] 2404/9). — [4] K. C. Joshi, V. N. Pathak, S. Bhargava, P. G. Seybold (J. Fluorine Chem. **9** [1977] 387/97). — [5] K. C. Joshi, V. N. Pathak (J. Inorg. Nucl. Chem. **35** [1973] 3161/70).

[6] J. J. Degnan, C. R. Hurt, N. Filipescu (J. Chem. Soc. Dalton Trans. **1972** 1158/63). — [7] V. J. Rao, D. R. Rao, A. P. B. Sinha (Indian J. Chem. **8** [1970] 270/4).

5.1.17 Complexes with 1,3-Di(1-naphthyl)- and 1,3-Di(2-naphthyl)-1,3-propanedione
$C_{10}H_7C(O)CH_2C(O)C_{10}H_7$ (= $C_{23}H_{16}O_2$)

$M(C_{23}H_{15}O_2)_3$

Each compound (M = Eu, Gd, Tb) was prepared by treating an ethanolic solution of the rare earth chloride and the appropriate β-diketone with piperidine [1, 2]. The gadolinium chelates were characterized by their absorption (λ_{max} = 346 and 372 nm for the di-1-naphthoylmethanato and di-2-naphthoylmethanato chelates, respectively) and phosphorescence (λ_{max} = 522 and 518 nm, for the 1-naphthoyl and 2-naphthoyl derivatives, respectively) spectra. The europium and terbium chelates were characterized by their fluorescence spectra.

References to 5.1.17:

[1] W. F. Sager, N. Filipescu, F. A. Serafin (J. Phys. Chem. **69** [1965] 1092/100). — [2] N. Filipescu, W. F. Sager, F. A. Serafin (J. Phys. Chem. **68** [1964] 3324/46).

5.1.18 Complexes with Curcumin
(= 1,7-Bis(4-hydroxy-3-methoxyphenyl)-1,6-heptadiene-3,5-dione = $C_{21}H_{20}O_6$)

$$HO-C_6H_3(OCH_3)-CH=CH-C(=O)-CH_2-C(=O)-CH=CH-C_6H_3(OCH_3)-OH$$

$M(C_{21}H_{19}O_6)Cl_2 \cdot 2H_2O$

Each chelate (M = Y, La, Pr, Nd, Sm) was prepared by adding a solution of saturated ethanolic ammonia to an aqueous ethanolic solution containing the ligand and a hydrated rare earth nitrate. The ligand to metal ratio in the resulting product was found to be 1:1. The chloride content of the chelate was determined gravimetrically as silver chloride and the water content by a weight loss at 100°C, which corresponded to dehydration of the compound. The molar conductance values (Λ_M = 42 to 82 $\Omega^{-1} \cdot cm^2 \cdot mol^{-1}$) for solutions of the complexes in ethanol were interpreted in terms of nonelectrolyte behavior. The chelates were characterized further by their infrared and electronic absorption spectra, P. V. Takalkar, V. R. Rao (Indian J. Chem. **7** [1969] 943/4).

5.1.19 Complexes with 1,3-Bis(4-pyridyl)-1,3-propanedione $C_5H_4N\text{-}C(O)CH_2C(O)C_5H_4N$ (= Diisonicotinoylmethane = $C_{13}H_{10}N_2O_2$)

$M(C_{13}H_9N_2O_2)_3$

Each chelate (M = Eu, Gd, Tb) was prepared by treating an ethanolic solution containing the respective rare earth chloride and diisonicotinoylmethane with piperidine [1, 2]. The gadolinium chelate was characterized by its absorption (λ_{max} = 363 nm) and phosphorescence (λ_{max} = 480 nm) spectra [1] and the europium and terbium chelates were characterized by their fluorescence spectra [2].

References to 5.1.19:

[1] W. F. Sager, N. Filipescu, F. A. Serafin (J. Phys. Chem. **69** [1965] 1092/100). — [2] N. Filipescu, W. F. Sager, F. A. Serafin (J. Phys. Chem. **68** [1964] 3324/46).

5.1.20 Complexes with Furylderivatives of 1,3-Propanedione and 1,3-Butanedione

5.1.20.1 With 4,4,4-Trifluoro-1-(2-furyl)-1,3-butanedione $OC_4H_3C(O)CH_2C(O)CF_3$ (= Furoyltrifluoroacetone = $C_8H_5F_3O_3$)

Tris Chelates $M(C_8H_4F_3O_3)_3 \cdot n\,H_2O$

Each compound (M = Nd, Gd, Er) was prepared by adding a methanolic solution (20 ml) of furoyltrifluoroacetonate (6 mmol) with stirring, to a solution of the rare earth nitrate (2 mmol) in a 1:1 (v/v) methanol/water mixture (5 ml). The pH of the reaction mixture was kept between 5 and 6 with aqueous ammonia. The solution was then diluted with a large excess of water, and the resulting precipitate was collected by filtration, dried, and recrystallized from benzene or methanol. The chelates were dried under vacuum in a desiccator over phosphorus(V) oxide. The melting points (°C) of the compounds were: Nd, 110 to 112, Gd, 130 to 131, Er, 105 to 107. Elemental analyses were not obtained, and thus, the degree of hydration was not established with certainty [1].

Tetrakis Chelates $NH_4[M(C_8H_4F_3O_3)_4]$

Each compound (M = La, Ce, Pr, Sm) was prepared by mixing a boiling 95% ethanolic solution of a rare earth salt (either chloride or nitrate) and ammonium furoyltrifluoroacetonate (1:5 mol ratio). The mixture was allowed to cool, and the resulting precipitate was separated by centrifugation. The precipitate was dissolved in hot acetone, together with additional ammonium furoyltrifluoroacetonate (1 mol per mole of complex). The acetone was evaporated with a stream of nitrogen until a small amount of solvent remained. The resulting precipitate was collected, washed with 95% ethanol, dried, and analyzed [2].

5.1.20.2 With 1,3-Bis(2-furyl)-1,3-propanedione $OC_4H_3C(O)CH_2C(O)C_4H_3O$ (= Difuroylmethane = $C_{11}H_8O_4$)

$M(C_{11}H_7O_4)_3$

Each chelate (M = Eu, Gd, Tb) was prepared by treating an ethanolic solution of the rare earth chloride and difuroylmethane with piperidine [3, 4]. The gadolinium chelate was characterized by its electronic absorption (λ_{max} = 372 nm) and phosphorescence (λ_{max} = 493 nm) spectra [3], and the europium and terbium chelates were characterized by their fluorescence spectra [4].

5.1.20.3 With 1-[4-Methoxy-5-benzofuranyl]-3-phenyl-1,3-propanedione (= Pongamol = $C_{18}H_{14}O_4$)

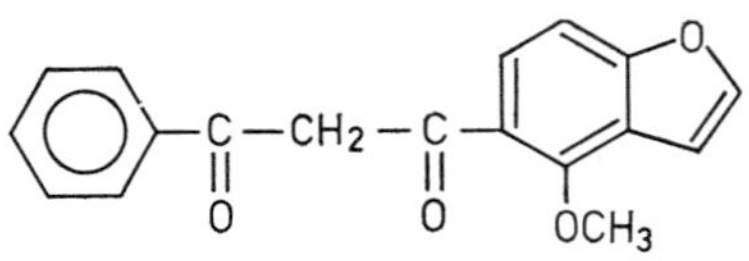

5.1.20.3.1 Tris Chelates $M(C_{18}H_{13}O_4) \cdot n\,H_2O$

The samarium, dysprosium, and ytterbium chelates (degree of hydration not given) were prepared by adding ethanolic solutions of pongamol to solutions of the respective rare earth chlorides in absolute ethanol, then adjusting the pH of the reaction mixtures with alcoholic ammonia to effect precipitation of the chelates [7]. Tris complexes for M = Nd [8], Eu (n = 2) [6], Gd [9], Ho [9] and Er [9] also have been reported. The emission spectrum of the europium chelate in 3:1 (v/v) ethanol/methanol revealed the following bands in the region of the $^5D_0 \rightarrow {}^7F_J$ (J = 0, 1, 2) transitions: $^5D_0 \rightarrow {}^7F_2$ = 16233, 16323, 16380 Å; $^5D_0 \rightarrow {}^7F_1$ = 16809, 16922, 16990 Å; $^5D_0 \rightarrow {}^7F_0$ = 17277 Å [6]. In the ultraviolet region, ethanolic solutions of the gadolinium, holmium, and erbium chelates exhibited bands at 240 and 350 nm [9]. The neodymium chelates in chloroform exhibited bands at 350 nm (ε = 61480) and 250 nm (ε = 62630). Bands at 1530 and 1500 cm^{-1} in the infrared spectrum of the neodymium chelate were assigned to the ν(C=O) and ν(C=C) stretching modes, respectively [8].

5.1.20.3.2 Lewis Base Adducts of the Tris Chelates

The bipyridine and phenanthroline adducts $Eu(C_{18}H_{13}O_4)_3 \cdot bpy$ and $Eu(C_{18}H_{13}O_4)_3 \cdot phen$ were obtained according to the following procedure: To a solution of pongamol (6 mmol) and the neutral ligand (2 mmol) in hot absolute ethanol (20 ml) was added 6 ml of 1 N sodium (or potassium) hydroxide. The mixture was stirred while a solution of europium chloride (2 mmol) was added dropwise. The mixture was allowed to cool; the products were collected, recrystallized from absolute ethanol, and dried in air at room temperature. The bipyridine adduct melted at 123 to 124°C and the phenanthroline adduct at 105 to 111°C [5]. The procedure for preparing the ammonia adduct $Eu(C_{18}H_{13}O_4)_3 \cdot 2NH_3$ was not presented [6]. The emission spectra of the adducts in 3:1 (v/v) ethanol/methanol at 77 K revealed the following bands in the regions of the $^5D_0 \rightarrow {}^7F_J$ (J = 0, 1, 2) transitions: $Eu(C_{18}H_{13}O_4)_3 \cdot 2NH_3$, $^5D_0 \rightarrow {}^7F_2$ = 16320 Å; $^5D_0 \rightarrow {}^7F_1$ = 16871, 16950 Å; $^5D_0 \rightarrow {}^7F_0$ = 17280 Å; $Eu(C_{18}H_{13}O_4)_3 \cdot bpy$, $^5D_0 \rightarrow {}^7F_2$ = 16229, 16317, 16374 Å; $^5D_0 \rightarrow {}^7F_1$ = 16922 Å; $^5D_0 \rightarrow {}^7F_0$ = 17284 Å; $Eu(C_{18}H_{13}O_4)_3 \cdot phen$, $^5D_0 \rightarrow {}^7F_2$ = 16220, 16310, 16320, 16450 Å; $^5D_0 \rightarrow {}^7F_1$ = 16890, 16920, 16970, 17010 Å; $^5D_0 \rightarrow {}^7F_0$ = 17271, 17360 Å [6].

5.1.20.3.3 Tetrakis Chelates

Ammonium Salt $NH_4[Eu(C_{18}H_{13}O_4)_4]$. The compound was prepared by passing anhydrous ammonia through an absolute ethanolic solution of europium chloride and pongamol (1:4 mole ratio). The resulting precipitate was collected by filtration, washed with alcohol, and dried in air. The compound melted at 183 to 184°C [5]. The emission spectrum of the compound at 77 K in 3:1 (v/v) ethanol/methanol revealed the following bands in the regions of the $^5D_0 \rightarrow {}^7F_J$ (J = 0, 1, 2) transitions: $^5D_0 \rightarrow {}^7F_2$ = 16322 Å; $^5D_0 \rightarrow {}^7F_1$ = 16877, 16950 Å; $^5D_0 \rightarrow {}^7F_0$ = 17341, 17370 Å [6].

Piperidinium Salt $C_5H_{11}NH[Eu(C_{18}H_{13}O_4)_4]$. A warm, filtered solution of europium chloride (0.00337 mol) in absolute ethanol (25 ml) was added to a solution of pongamol (0.0155 mol) and piperidine (0.0155 mol) in absolute ethanol (25 ml) maintained at 90°C. The mixture was kept at this temperature for 4 days and then filtered. The collected precipitate was washed with absolute ethanol, recrystallized from absolute ethanol, and dried in air at room temperature. The compound melted at 117 to 119°C [5]. The emission spectrum of the compound in 3:1 (v/v) ethanol/methanol at 77 K revealed the following bands in the regions of the $^5D_0 \rightarrow {}^7F_J$ (J = 0, 1, 2) transitions: $^5D_0 \rightarrow {}^7F_2$ = 16186, 16241, 16318 Å; $^5D_0 \rightarrow {}^7F_1$ = 16813, 16914, 17005 Å; $^5D_0 \rightarrow {}^7F_0$ = 17284 Å [6].

5.1.20.3.4 Acetato Compound $Eu(C_{18}H_{13}O_4)_2(C_2H_3O_2)$

An aqueous solution (15 ml) of sodium acetate (0.03 mol) was added to an aqueous solution (10 ml) of europium chloride (0.005 mol), and the resulting solution was added in small portions with vigorous shaking to a solution of pongamol (0.0165 mol) in absolute ethanol (100 ml). The reaction mixture, containing a precipitate, was heated on a water bath for 15 to 20 min. After standing at room temperature overnight, the solution was filtered, and the precipitate was washed with absolute ethanol and dried in air. The complex melted at 123 to 124°C [5]. The emission spectrum of the compound in 3:1 (v/v) ethanol/methanol at 77 K revealed the following bands in the regions of the $^5D_0 \rightarrow {}^7F_J$ (J = 0, 1, 2) transitions: $^5D_0 \rightarrow {}^7F_2$ = 16191, 16393 Å; $^5D_0 \rightarrow {}^7F_1$ = 16978, 17018 Å, and $^5D_0 \rightarrow {}^7F_0$ = 17301 Å [6].

References to 5.1.20:

[1] K. Utsunomiya, T. Shigematsu (Anal. Chim. Acta **58** [1972] 411/9). — [2] W. O. McSharry, M. Cefola (J. Inorg. Nucl. Chem. **31** [1969] 2777/9). — [3] W. F. Sager, N. Filipescu, F. A. Serafin (J. Phys. Chem. **69** [1965] 1092/100). — [4] N. Filipescu, W. F. Sager, F. A. Serafin (J. Phys. Chem. **68** [1964] 3324/46). — [5] V. Janardhanarao, A. P. B. Sinha (Indian J. Chem. **4** [1966] 196/8).

[6] V. J. Rao, D. R. Rao, A. P. B. Sinha (Indian J. Chem. **8** [1970] 270/4). — [7] P. R. Murty, V. R. Rao (Current Sci. [India] **35** [1966] 12/3). — [8] V. R. Rao, P. R. Murty (Current Sci. [India] **38** [1969] 110). — [9] D. Purushotham, V. R. Rao (Current Sci. [India] **39** [1970] 486).

5.1.21 Complexes with 4,4,4-Trifluoro-1-(2-thienyl)-1,3-butanedione

$C_4H_3S{-}C({=}O){-}CH_2{-}C({=}O){-}CF_3$ (Thenoyltrifluoroacetone = $C_8H_5F_3O_2S$)

5.1.21.1 $M(C_8H_4F_3O_2S)_n^{3-n}$ Complexes (n = 1 to 4). Formation in Solution

Although the solvent extraction of trivalent rare earth ions with thenoyltrifluoroacetone and mixtures of thenoyltrifluoroacetone with neutral N- or O-donors has been well documented [1 to 41], formation constant data for the metal chelates are relatively limited. Ikeda and coworkers [1] determined the cumulative formation constants for the 1:1, 2:1, and 3:1 (ligand:metal) europium chelates ($K_1 = 4.5 \times 10^4$, $K_2 = 4.7 \times 10^9$, $K_3 = 1.1 \times 10^{12}$) by a solvent extraction method at 25°C, I = 0.1. Zolotov, Shakhova, Alimarin [2] obtained the values lg β_1 = 6.34, lg β_2 = 11.76, and lg β_3 = 19.26 for the scandium chelate at 25°C also using a solvent extraction method, I = 0.1 ($NaClO_4$). The pH-potentiometric method has been used by Shilov, Batyaev [96] to determine the acidic dissociation constant of thenoyltrifluoroacetone ($pK_a = 6.80 \pm 0.02$) and the first stability constants of lanthanide complexes in 50 vol% aqueous acetonitrile solution at 25 ± 0.2°C and ionic strength I→0:

metal ion	La	Pr	Nd	Gd	Tb
lg K_1	5.30 ± 0.02	5.56 ± 0.01	5.60 ± 0.01	5.80 ± 0.01	5.85 ± 0.02
metal ion	Ho	Er	Yb	Lu	Eu*)
lg K_1	5.91 ± 0.02	5.97 ± 0.01	5.93 ± 0.01	5.96 ± 0.01	5.81 ± 0.01

*) For the europium complex: lg K_2 = 5.34 ± 0.03, lg K_3 = 5.06 ± 0.04.

For extraction of the 2-aminopyridinium salt of the tetrakis chelate by benzene, see Kononenko, Vitkun [51].

The kinetics of the reaction of scandium ion with thenoyltrifluoroacetone was reported by Taft, Cool [42] under conditions which precluded the formation of higher order complexes. The rate determining step for the formation of the 1:1 chelate was found to be the ionization of thenoyltrifluoroacetone (k = 0.56 ± 0.02 min^{-1} at 25°C). The minimum value of the rate constant for the reactions of scandium(III) with the enolate ion was calculated at $1.0 \times 10^7\ l \cdot mol^{-1} \cdot min^{-1}$ [42].

5.1.21.2 Tris Chelates $M(C_8H_4F_3O_2S)_3 \cdot nH_2O$ (n = 0 to 2.5)

Purushottam and coworkers [58, 59] prepared each chelate (M = Sc, Y, La to Yb except for Ce, Lu) by adding an ethanolic solution of thenoyltrifluoroacetone to a solution of the rare earth nitrate in the same solvent, and then adjusting the pH of the reaction mixture to an optimum value (see Table 5/52) by adding alcoholic ammonia. The abundant precipitate formed was collected by filtration, washed with a little ethanol, and dissolved in benzene. The benzene solution was filtered. The filtrate was allowed to evaporate at room temperature, and the product obtained was dried under vacuum over concentrated sulfuric acid. The compounds were not formulated as hydrates. The gadolinium chelate gave accurate results for carbon, hydrogen, sulfur, and fluorine analyses, based on the formulation $Gd(C_8H_4F_3O_2S)_3$, whereas the carbon analyses were approximately 1% low for the erbium and ytterbium chelates (elemental analyses were not obtained for the other chelates). The physical properties of the compounds thus obtained are recorded in Table 5/52 [58, 59].

Lis, Siekierski [60, 61] prepared the series of hydrates (n = 1 for M = La, Yb, Lu; n = 1.5 for M = Tm; n = 2 for M = Ce to Dy except for Pm; n = 2.5 for M = Er, Ho) by adding aqueous ammonia and an aqueous solution of the respective rare earth chloride to solutions of thenoyltrifluoroacetone in ethanol. Thermogravimetric analyses of the compounds revealed a certain degree of dehydration in the temperature range 100 to 120°C, with decomposition of the ligand occurring in the range 220 to 310°C. When heated at 107°C, the complexes from cerium through dysprosium showed mass changes corresponding to the loss of 1.5 water molecules, whereas the holmium and erbium complexes showed losses corresponding to 2.5 water molecules (complete dehydration). The thulium, ytterbium, and lutetium complexes each showed a continuous loss in mass when heated at 107°C. Elemental analyses of the dehydrated products deviated from the values calculated for the anhydrous compounds, indicating that decomposition had occurred [62].

Table 5/52

Physical Properties and Electronic Absorption Spectral Data for the Thenoyltrifluoroacetato Rare Earth Chelates, $M(C_8H_4F_3O_2S)_3$.

M	precipitation at pH	color	melting point in °C	λ_{max} in nm (ε in $mol^{-1} \cdot l \cdot cm^{-1}$) in methanol*)	Ref.
Sc	6.2	light pink	149	270(7943), 342(31620)	[59]
Y	6.4	gray	132	270(19950), 341(50120)	[59]
La	6.5	brown	135	270(5012), 344(25120)	[58]
Pr	6.5	brown	164	264(12590), 339(25120)	[58]
Nd	6.5	pink	181 to 182	240(8215), 270(17380), 340(50930)	[58]
Sm	6.5	straw yellow	148	266(26730), 344(68050)	[58]
Eu	6.5	pink	180	268(19950), 340(50120)	[58]
Gd	6.5	gray	122	268(25120), 341(79430)	[58]
Tb	6.4	pink	115	265(15850), 340(63100)	[58]
Dy	5.8	yellow	193	269(18220), 344(51770)	[58]
Ho	6.5	brown	135	265(19950), 340(63100)	[58]
Er	6.5	pink	125	265(15850), 340(50120)	[58]
Tm	6.4	brown	115	265(15850), 340(50120)	[58]
Yb	6.2	yellow	145	265(17490), 342(62790)	[58]

*) Absorption spectral data were also tabulated for the compounds in ethanol, hexane, and cyclohexane.

Charles, Ohlmann [63] effected synthesis of the dihydrate, $Eu(C_8H_4F_3O_2S)_3 \cdot 2H_2O$, according to the following procedure: Thenoyltrifluoroacetone (0.03 mol) was dissolved in 75 ml of 95% ethanol, and 30 ml of 1 M aqueous ammonia (0.03 mol) was added. A 0.5 M solution of europium chloride (0.01 mol) in water was added to the resulting solution, and the mixture was allowed to stand several hours until the initially formed oil solidified. The yellow solid was collected by filtration, washed with water, and dried under vacuum. The compound was recrystallized from acetone and again dried under vacuum. Elemental analysis revealed that the compound thus obtained contained excess thenoyltrifluoroacetone, which was removed by stirring the solid with petroleum ether for 0.5 h [63]. Kertes and Kassierer [81], following the procedure described by Charles and Ohlmann [63], effected preparation of the dihydrated tris chelates of lanthanum (m.p. = 148 to 150°C), neodymium (m.p. = 158°C), gadolinium (170°C decomposed), and lutetium (170°C decomposed).

Uhlemann, Dietze [64] obtained the dihydrated cerium chelate by adding 2 N aqueous ammonia to a solution of cerium(III) nitrate hexahydrate (1 mmol) and thenoyltrifluoroacetone (3.5 mmol) in methanol. The mixture was cooled while being stirred vigorously. The product first separated as an oil, but quickly solidified. The solid was collected, washed with a small amount of a methanol/aqueous ammonia mixture, and then air dried. The yellow solid thus obtained melted at 105°C and had a magnetic moment of 2.24 μ_B [64].

The anhydrous europium chelate was prepared from thenoyltrifluoroacetone and europium formate in the form of pale yellow crystals, which melted at 118 to 120°C [72].

The crystal and molecular structures of $Eu(C_8H_4F_3O_2S)_3 \cdot 2H_2O$ were determined by single-crystal X-ray analysis (R = 0.111 for 1444 independent reflections). The compound crystallizes in the space group $P2_1/c$-C^5_{2h} (No. 14) with four molecules in a unit cell of dimensions a = 11.40, b = 22.40, c = 13.72 Å, and β = 105.0° (calculated density = 1.673 g/cm³). The europium ion is coordinated to eight oxygen atoms at the vertices of a square antiprism (**Fig. 5-27**). One thenoyltrifluoroacetate group is ligated with its oxygen atoms at adjacent vertices of one square face, whereas the other two chelates span vertices of the opposing square faces. The two water molecules are located at diagonally opposite corners of one square and complete the coordination sphere. The chelate rings formed by the thenoyltrifluoroacetate groups bridging corners of opposite square faces are nonplanar, folded on the oxygen-oxygen line by 20° to 23°. The chelate ring formed by the thenoyltrifluoroacetate group ligating corners of the same face, on the other hand, is planar. Due to slow decomposition of the crystals, the X-ray data were not sufficiently complete to determine accurate bond distances. However, from best estimates, the average europium-oxygen (thenoyltrifluoroacetate) bond distance is 2.42 Å, and that of the europium-oxygen (water) bond is 2.53 Å. The distance between the two square faces of the antiprism is smaller than an edge so that the distances between oxygen atoms of the same thenoyltrifluoroacetate group are almost equal for all three groups (average 2.88 Å) [65].

Fig. 5-27

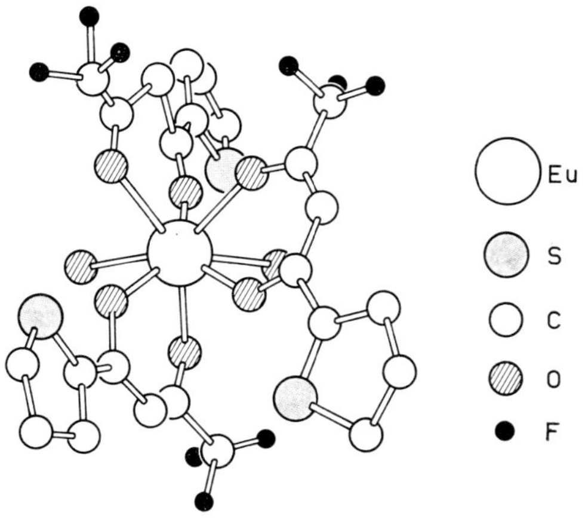

Eu
S
C
O
F

Molecular structure of $Eu(C_8H_4F_3O_2S)_3 \cdot 2H_2O$.

Infrared spectral data are tabulated and discussed in [58] and [66 to 71]. Some representative data in the regions of ligand absorptions for the scandium chelate [67] are: ν(C=O) = 1603, ν(C-C) = 1584, thienyl ring stretching = 1550, 1517, 1411, 1363, 1354, 1319; ν(C-F) = 1235, 1197; ν(C-CF_3) = 850 cm^{-1}. Mehta, Tandon [66] computed the stretching force constants, f(M-O), for the M-O bonds from the ν(M-O) (ν_1 [A_1] = 565 to 580 cm^{-1}, ν_2[A_2] = 430 to 450 cm^{-1}, ν_3[B_2] = 590 to 600 cm^{-1}) vibrational modes (M, f(M-O) in 10^5 dyn · cm^{-1}): La, 2.88; Pr, 2.86; Nd, 2.96; Sm, 2.98. The assignment of three M-O vibrational modes is based on D_3 symmetry [70]. Lis [68] assigned a band near 490 cm^{-1} to the ν(M-O) stretching mode, and noted that the wave numbers of this band increased monotonically with the atomic number of the lanthanide.

The emission spectrum of the anhydrous europium chelate in the crystalline state at 77 K revealed a number of lines corresponding to complete removal of the degeneracy (2 J + 1) associated with the ^7F_J (J = 1, 2) levels ($^5D_0 \rightarrow {}^7F_0$ = 5790 Å, $^5D_0 \rightarrow {}^7F_1$ = 5900, 5930, 5970 Å, $^5D_0 \rightarrow {}^7F_2$ = 6130, 6160, 6175, 6195, 6210 Å). Only three lines are observed for the $^5D_0 \rightarrow {}^7F_2$ transition in the spectra recorded at 77 K for samples dispersed in polymethylmethacrylate (6130, 6150, 6170 Å) and in ethanol (6135, 6150, 6180 Å) [72]. Additional emission spectral studies have been reported for crystalline samples of the europium chelate [63] and samples of the samarium chelate dispersed in polymethylmethacrylate [73] (emission lines were observed at 645, 598, and 562 nm at both 300 and 77 K in the spectrum of the samarium chelate [73]).

The oscillator strengths for absorption bands observed in the spectra of the praseodymium [74] and neodymium [75] chelates in methanol, 1:3 (v/v) methanol/ethanol, dimethylformamide, and 1:3:3 (v/v) dimethylformamide/methanol/ethanol, and the erbium chelate in methanol and in dimethylformamide [76] (degree of hydration of chelates not given), were determined and discussed in terms of the Judd-Ofelt theory—see B. R. Judd (Phys. Rev. **27** [1962] 750/61) and G. S. Ofelt (J. Chem. Phys. **37** [1962] 511/20). Poor correlation between the observed and calculated intensities was observed for the praseodymium chelate [74], whereas the studies revealed that it was possible to use the Judd-Ofelt parameters (τ_2, τ_4, τ_6) for evaluation of band intensities in the spectra of the neodymium [75] and erbium [76] chelates. The intensities of the hypersensitive transitions were best explained by the τ_2 parameter, originating from the forced electric dipole transition [75, 76]. The value of the nephelauxetic ratio, $\beta = 0.9846$, for the neodymium chelate was determined from the diffuse reflectance spectrum [77].

The mass spectra of the samarium, europium, gadolinium, and terbium chelates were obtained and discussed in terms of the fragmentation patterns which were similar for each chelate. Fragmentation was initiated by loss of the CF_3 group from the molecular ion, followed by loss of the first diketonate group. Fluorine atom migration from the -CF_3 group to the metal was observed in subsequent fragmentation, with concomitant loss of :CF_2 groups. Peaks were observed in the spectra corresponding to formation of the $[MF_2]^+$ (M = Sm, Gd, Tb) and $[MF]^+$ (M = Sm, Eu) ions. The occurrence of the $[MF]^+$ ion in the spectra of the samarium and europium chelates was attributed to the tendency for these two lanthanides to be reduced to the bivalent state [78]. For the mass spectrum and the volatility of $Eu(C_8H_4F_3O_2S)_3 \cdot H_2O$, see also [97].

5.1.21.3 Lewis Base Adducts of the Tris Chelates

5.1.21.3.1 With Nitrogen Donor Ligands

With Primary Aliphatic Amines

$Eu(C_8H_4F_3O_2S)_3 \cdot RNH_2$. Formation in Solution. Formation constants for the interaction of tris(thenoyltrifluoroacetonato)europium(III) with primary amine substrates in carbon tetrachloride were determined by emission-titration spectroscopy. Analysis of the emission-titration data by Job's method revealed that only 1:1 adducts were formed [95].

amine	propyl-	isopropyl-	butyl-	sec-butyl-	tert-butyl
lg K_1	4.08	4.35	4.01	4.34	4.39

With 2,2'-Bipyridine (= $C_{10}H_8N_2$ = bpy)

$M(C_8H_4F_3O_2S)_3 \cdot n$ bpy (n=1, 2). **Formation in Solution.** The stepwise equilibrium constants, enthalpy, and entropy changes for the addition of 2,2'-bipyridine to several $M(C_8H_4F_3O_2S)_3 \cdot 2H_2O$ chelates (see p. 218) were determined by direct titration calorimetry in chloroform solution at 25°C. The data were interpreted in terms of formation of both a mono and a bis adduct. Each adduct is stabilized by the large exothermic enthalpy changes. The ΔH_n values become increasingly exothermic as the ionic radius of the metal ion decreases [81].

M	lg β_1	lg β_2	$-\Delta H_1$ in kcal/mol	$-\Delta H_2$ in kcal/mol	$-\Delta S_1$ in cal·mol^{-1}·K^{-1}	$-\Delta S_2$ in cal·mol^{-1}·K^{-1}
La	3.25	5.50	6.92	12.47	7.46	16.65
Nd	3.61	6.87	7.90	18.44	9.97	30.42
Gd	3.71	7.50	9.78	23.18	15.63	43.42
Lu	4.75	7.75	14.32	29.14	26.40	62.28

Isolated Adducts. $M(C_8H_4F_3O_2S)_3 \cdot$ bpy. The europium compound was prepared by adding, dropwise and with stirring, an aqueous solution of a europium salt (2 mmol) to a hot 95% ethanolic solution containing 6 mmol of thenoyltrifluoroacetone, 2 mmol of 2,2'-bipyridine, and 6 ml of 1 N sodium (or potassium) hydroxide. The mixture was allowed to cool, and the product was collected and recrystallized from 95% ethanol. The europium compound was also obtained by extracting an aqueous solution containing europium chloride, thenoyltrifluoroacetone, and 2,2'-bipyridine with benzene, separating the benzene layer, and adding petroleum ether to the extractant solution to effect precipitation of the product. The resulting precipitate was collected by filtration, washed with petroleum ether, and dried in air. The melting point of the compound thus obtained was 220 to 221°C [82]. The neodymium compound was prepared by adding 1 mmol of the tris chelate in 15 cm^3 of ethanol to a solution of 1 mmol of 2,2'-bipyridine in 15 cm^3 of ethanol. The resulting compound was recrystallized from ethanol [83].

The scandium compound precipitated from a mixture of scandium chloride, and 2,2'-bipyridine in 40% (v/v) ethanol/water (pH = 3.5 with acetate buffer). The product was collected by filtration, washed several times with small portions of 40% ethanol, and dried in air. The fine, white, crystalline powder melted at 179°C. The compound is moderately soluble in benzene, acetone, toluene, xylene, ether, cyclohexanone, ethyl acetate, and chloroform; sparingly soluble in isopentyl alcohol, cyclohexanol, and isobutyl alcohol; and almost insoluble in carbon tetrachloride and ethanol [84].

The crystal and molecular structures of the neodymium compound were determined by single-crystal X-ray analysis (R = 0.139% for 2160 independent reflections). The light blue crystals are in the monoclinic system, space group $P2_1/c\text{-}C^5_{2h}$ (No. 14) with four molecules in a unit cell of dimensions a = 10.42, b = 16.93, c = 22.53 Å, and β = 104.0° (D_{calc} = 1.66 g · cm^{-3}; D_{obs} = 1.65 g · cm^{-3}). The neodymium ion is coordinated to six oxygen atoms from the three thenoyltrifluoroacetonate groups and two nitrogen atoms from the bipyridine molecule. The coordination polyhedron was described as a distorted square antiprism with the chelate rings spanning the s edges. The average Nd-O and Nd-N bond distances are 2.45 Å (range 2.397 to 2.515 Å) and 2.71 Å (2.724, 2.689 Å), respectively. The difference in the neodymium-oxygen and neodymium-nitrogen bond lengths causes a relatively large distortion in the individual interatomic distances within the coordination polyhedron (s lengths range from 2.82 to 3.14 Å and the l lengths range from 2.95 to 3.36 Å). The mean value of the angle between the metal-ligand bond and the fourfold symmetry axis is 57.4°. The chelate rings formed by the thenoyltrifluoroacetonate groups are approximately planar, whereas the best planes through the atoms of the two pyridine rings showed that they are planar with an angle of 0.5° between the two planes [83].

With 1,10-Phenanthroline (= $C_{12}H_8N_2$ = phen)

$M(C_8H_4F_3O_2S)_3 \cdot$ phen. Bauer, Blanc, Ross [46] effected preparation of the europium and terbium compounds according to the following procedure: To a hot solution of thenoyltrifluoroacetone (6 mmol) and the rare earth chloride (2 mmol) in 50 ml of ethanol and 5 ml of water was added 3.0 ml of 2.0 N sodium hydroxide solution. The solution was filtered while hot to remove the precipitated sodium chloride, and 10 ml of an ethanolic solution of 1,10-phenanthroline monohydrate (2 mmol) was added to the filtrate. Within 15 s, a granular precipitate formed. The reaction mixture was allowed to cool, and the product was collected by filtration, washed with water, and dried under vacuum. The europium compound (melting point = 242 to 244°C) was recrystallized from acetonitrile and the terbium compound (m. p. = 247 to 248°C) from ethyl acetate. Melby and coworkers [44] also prepared the europium compound following a similar procedure, but recrystallized the product from 95% ethanol (m. p. = 247 to 249°C). The compounds thus obtained were characterized by elemental analyses and emission spectroscopy. The authors proposed that the central ion is eight-coordinate in each compound [44, 46].

The europium adduct was also prepared by extracting an aqueous solution containing europium chloride, thenoyltrifluoroacetone, and 1,10-phenanthroline with benzene, separating the benzene layer, and adding petroleum ether to the extractant solution to effect precipitation of the product. The resulting precipitate was collected by filtration, washed with petroleum ether, and dried in air. The melting point of the compound thus obtained was 236 to 238°C [82].

Kononenko and coworkers [85] prepared a series of adducts (M = Y, La, Nd, Sm, Eu, Dy) by mixing aqueous solutions of the respective rare earth chlorides containing urotropine with a solution of thenoyltrifluoroacetone and 1,10-phenanthroline in ethanol or acetone. In each case, after one to two hours, the resulting precipitate was collected by filtration, washed with water, and dried in air. The melting points (°C) of the compounds were: La, 228 to 229; Nd, 236 to 237; Sm, 241 to 243; Eu, 238 to 240; Dy, 243 to 244; Y, 246 to 248°C. The complexes of europium and samarium fluoresced under irradiation by ultraviolet light [85, 86].

The scandium compound precipitated from a mixture of scandium chloride and 1,10-phenanthroline in 40% (v/v) ethanol/water (pH = 3.5 with acetate buffer). The product was collected by filtration, washed several times with small portions of 40% ethanol, and dried in air. The fine, white, crystalline powder melted at 213°C. The compound is moderately soluble in benzene, acetone, toluene, xylene, ether, cyclohexanone, ethyl acetate, and chloroform; sparingly soluble in isopentyl alcohol, cyclohexanol, and isobutyl alcohol; and almost insoluble in carbon tetrachloride and ethanol [84].

With 1,2-Di(4-pyridyl)ethane $C_5H_4NCH_2CH_2C_5H_4N$ ($=C_{12}H_{12}N_2$)

$Nd(C_8H_4F_3O_2S)_3 \cdot C_{12}H_{12}N_2 \cdot H_2O$. The compound was prepared by dissolving equivalent amounts of the tris chelate and 1,2-di(4-pyridyl) ethane in ethanol. The crystal and molecular structures of the resulting light blue crystals were determined by single-crystal X-ray analysis (R = 0.108 for 1600 reflections). The compound crystallized in the monoclinic space group $P2_1/c\text{-}C_{2h}^5$ (No. 14) with four molecules in a unit cell of dimensions a = 19.03, b = 10.33, c = 22.39 Å, and β = 94.3°. The neodymium ion is coordinated to six oxygen atoms from the thenoyltrifluoroacetonate groups, one oxygen atom from the water molecule, and one nitrogen atom of the Lewis base. The mean Nd-O (thenoyltrifluoroacetone) bond distance is 2.44 Å (range 2.398 to 2.492 Å), and the Nd-O (water) and Nd-N interatomic bond distances are 2.442 and 2.62 Å, respectively. The coordination polyhedron was described as a distorted square antiprism with two chelated groups spanning s edges and one chelate group spanning one l edge (**Fig. 5-28**). The s edges of one square

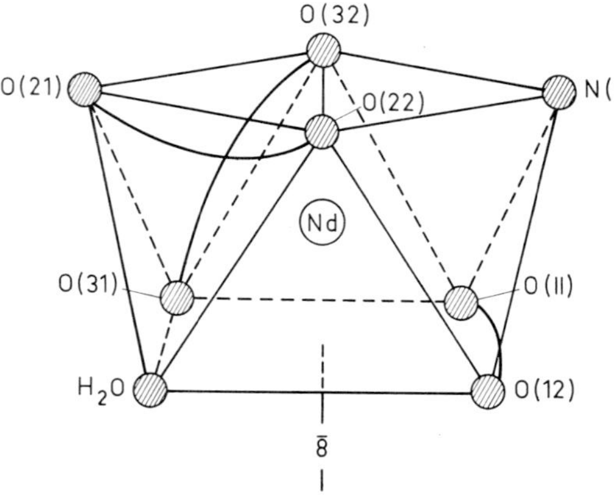

Fig. 5-28

Coordination polyhedron for $Nd(C_8H_4F_3O_2S)_3 \cdot C_{12}H_{12}N_2 \cdot H_2O$.

face of the antiprism range from 2.78 to 3.09 Å and those of the other square face from 2.76 to 3.00 Å, while the l edges range from 2.86 to 3.39 Å. The mean value of the angle between the metal-ligand bond and the C_4 symmetry axis is 57.1°. The chelate rings are approximately planar, as are the thienyl rings. Only one of the two nitrogen atoms of the 1,2-di(4-pyridyl)ethane group is coordinated to the neodymium ion. The other nitrogen atom is hydrogen bonded to a water molecule that is coordinated to another neodymium ion (nitrogen-oxygen bond distance = 2.67 Å) [87].

With 2,2':6',2"-Terpyridine (N N N) (= $C_{15}H_{11}N_3$)

$Eu(C_8H_4F_3O_2S)_3 \cdot C_{15}H_{11}N_3$. The compound was prepared by adding an aqueous solution of an europium salt (2 mmol), dropwise and with stirring, to a solution containing thenoyltrifluoroacetone (6 mmol), terpyridine (2 mmol), and 6 ml of 1 N sodium (or potassium) hydroxide in hot 95% ethanol. The mixture was allowed to cool, and the product was collected and recrystallized from 95% ethanol (m. p. = 247 to 251°C). The compound was characterized by elemental analysis and emission spectroscopy [44].

With Other N-Donor Ligands

The pyridine (No. 1), quinoline (No. 3), 4,4'-bipyridine (No. 4) and diphenylguanidine (No. 5) adducts listed in Table 5/53 were prepared by adding 0.1 mmol of the base to a solution of the tris neodymium chelate in 3 to 5 ml of benzene, followed by addition of a tenfold amount of petroleum ether to effect precipitation of the product. The pyridine adduct was also obtained by dissolving the tris chelate in pyridine and allowing the excess amine to evaporate at room temperature. The aniline adduct (No. 2) was prepared by extracting an aqueous solution containing neodymium chloride, thenoyltrifluoroacetone, and aniline with benzene, separating the benzene layer, and adding petroleum ether to the extractant solution to effect precipitation of the product. The solid was collected by filtration, washed with petroleum ether, and dried in air [82].

Table 5/53
Ternary Complexes of Neodymium with Thenoyltrifluoroacetone and Neutral Nitrogen Donors [82].

no.	neutral nitrogen donor	formula of complex	melting point in °C
1	pyridine (= C_5H_5N = py)	$Nd(C_8H_4F_3O_2S)_3 \cdot 2\,py$	180 to 182
2	aniline (= C_6H_7N)	$Nd(C_8H_4F_3O_2S)_3 \cdot 2\,C_6H_7N$	120 to 122
3	quinoline (= C_9H_7N)	$Nd(C_8H_4F_3O_2S)_3 \cdot 2\,C_9H_7N$	180 to 182
4	4,4'-bipyridine (= $C_{10}H_8N_2$)	$Nd(C_8H_4F_3O_2S)_3 \cdot C_{10}H_8N_2$	187 to 188
5	diphenylguanidine (= $C_{13}H_{13}N_3$)	$Nd(C_8H_4F_3O_2S)_3 \cdot C_{13}H_{13}N_3$	160 to 165

5.1.21.3.2 With Oxygen Donors

With γ-Picoline N-Oxide $CH_3C_5H_4NO$ (= 4-Methylpyridine 1-Oxide = C_6H_7NO)

$Eu(C_8H_4F_3O_2S)_3 \cdot 2\,C_6H_7NO$. The compound was prepared by adding an aqueous europium chloride (2 mmol) solution, dropwise and with stirring, to a hot solution containing thenoyltrifluoroacetate (6 mmol), picoline N-oxide (2 mmol), and 6 ml of 1 N sodium (or potassium) hydroxide in 95% ethanol. The solution was allowed to cool, and the product was collected and recrystallized from 95% ethanol (m. p. = 234 to 236°C). The compound was characterized by elemental analysis and emission spectroscopy [44].

With Triphenylphosphine Oxide $(C_6H_5)_3PO$ (= $C_{18}H_{15}OP$)

$M(C_8H_4F_3O_2S)_3 \cdot 2(C_6H_5)_3PO$. Ferraro, Healy [26], in a study of the synergistic effect of thenoyltrifluoroacetone and organophosphorus compounds on extraction of rare earths from aqueous media, isolated the triphenylphosphine oxide adducts (M = Nd, Sc) according to the following procedure: Two volumes of a 0.025 M rare earth nitrate solution were added to one volume of a hexane solution of 0.10 molar thenoyltrifluoroacetone and 0.05 M triphenylphosphine oxide. (In the case of neodymium, alkali was added to raise the pH of the aqueous phase in order to effect extraction of the metal ion into the organic phase.) The mixture was shaken for 20 min, then the aqueous phase was removed, and the hexane solution allowed to evaporate. When about nine-tenths of the hexane had evaporated, the resulting precipitate was collected by filtration and washed with a small portion of water and hexane. Recrystallization was effected from hexane, cyclohexane, or benzene. The neodymium compound was isolated as a solid (m. p. = 285°C), whereas the scandium compound was obtained as an oil. Infrared studies (KBr discs or Nujol mulls) revealed that the ν(P=O) stretching mode shifted to lower wave numbers in the spectra of the complexes (1150 and 1172 cm^{-1} for scandium and neodymium, respectively), relative to the free ligand value (1195 cm^{-1}) [26].

Melby and coworkers [44] effected synthesis of the europium compound according to the procedure described on p. 223 for preparation of the analogous picoline N-oxide adduct. The compound thus obtained melted at 251 to 253°C and was characterized further by elemental analysis and emission spectroscopy.

The crystal and molecular structures of the neodymium compound were determined by single-crystal X-ray analysis (R = 0.084 for 5505 independent reflections). The complex crystallizes in the triclinic space group $P\bar{1}$-C_i^1 (No. 2) with two molecules in a unit cell of dimensions a = 23.64, b = 12.15, c = 11.19 Å, α = 109.4°, β = 104.2°, and γ = 90.8° (D_{calc} = 1.55 g/cm^3, D_{obs} = 1.54 g/cm^3). The molecule is monomeric, and the neodymium ion is eight-coordinate, surrounded by six diketonate oxygen atoms and two oxygen atoms from the triphenylphosphine oxide groups, arranged at the vertices of a dodecahedron. The neodymium-oxygen (thenoyltrifluoroacetonate) bond distances range from 2.404 to 2.501 Å, whereas the neodymium-oxygen (triphenylphosphine oxide) distances are 2.399 and 2.423 Å [88].

$M(C_8H_4F_3O_2S)_2NO_3 \cdot 2(C_6H_5)_3PO$. The mixed nitrate-thenoyltrifluoroacetonate salts (M = Eu, Tb) were isolated while attempting to prepare the tris chelate adducts according to the procedure described on p. 223 for preparation of the picoline N-oxide adducts, except that europium nitrate was used in place of the chloride salt. The terbium compound melted at 232 to 234°C. The adducts were characterized by elemental analyses and emission spectroscopy [44].

With Other Organophosphine Oxides (= L)

$Eu(C_8H_4F_3O_2S)_3 \cdot nL$ (n = 1 or 2). **Formation in Solution.** The equilibrium constants, β_n, corresponding to the formation of the adducts in cyclohexane were determined by solvent extraction of europium(III) from an aqueous chloride solution (I = 0.1) at 24 ± 2°C into a mixture of thenoyltrifluoroacetone and the alkylphosphine oxide [4]. The NMR spectra of the adducts are discussed in [89].

phosphine oxide		compound	β_n *)
$(C_8H_{17})_3PO$	$(=C_{24}H_{51}OP)$	$Eu(C_8H_4F_3O_2S)_3 \cdot 2C_{24}H_{51}OP$	2.59×10^{13}
$(C_6H_{13})_2POCH_2PO(C_6H_{13})_2$	$(=C_{25}H_{54}O_2P_2)$	$Eu(C_8H_4F_3O_2S)_3 \cdot C_{25}H_{54}O_2P_2$	7.09×10^{10}
$(C_6H_{13})_2PO(CH_2)_3PO(C_6H_{13})_2$	$(=C_{27}H_{58}O_2P_2)$	$Eu(C_8H_4F_3O_2S)_3 \cdot C_{27}H_{58}O_2P_2$	1.40×10^{11}
$(C_6H_{13})_2PO(CH_2)_4PO(C_6H_{13})_2$	$(=C_{28}H_{60}O_2P_2)$	$Eu(C_8H_4F_3O_2S)_3 \cdot C_{28}H_{60}O_2P_2$	5.53×10^{11}

*) $\beta_n = [Eu(C_8H_4F_3O_2S)_3 \cdot nL] \cdot [Eu(C_8H_4F_3O_2S)_3]^{-1} \cdot [L]^{-n}$, in cyclohexane.

Isolated Compounds. The compounds $M(C_8H_4F_3O_2S)_3 \cdot 2(C_8H_{17})_3PO$ with M = Sc, Nd were isolated in the form of oils according to the procedure described above for preparation of the analogous triphenylphosphine oxide adducts [15, 26]. A slight shift of the ν(P=O) stretching mode to lower wave numbers in the spectra of the adducts (1145 and 1140 cm^{-1}) for scandium and neodymium, respectively, relative to the free ligand value (1152 cm^{-1}) was attributed to ligation of the phosphine oxide molecule through the phosphoryl group [26]. The NMR spectrum of the scandium compound is discussed in [89].

With Tributyl Phosphate $(C_4H_9O)_3PO$ (= $C_{12}H_{27}O_4P$)

$M(C_8H_4F_3O_2S)_3 \cdot n(C_4H_9O)_3PO$ (n = 1 or 2). **Formation in Solution.** The equilibrium constants (Table 5/54) corresponding to formation of tributylphosphate adducts of the tris(thenoyltrifluoroacetonato) rare earth chelates were determined in a variety of organic solvents by a solvent extraction method [8, 35, 40]. The values determined at 25°C for the scandium, lanthanum, europium, and lutetium adducts in carbon tetrachloride [31] are given below.

M	Sc^{3+}	La^{3+}	Eu^{3+}	Lu^{3+}
lg β_1	3.44	4.83	5.15	5.69
lg β_2	*)	9.33	8.89	6.67

*) No evidence for formation of the bis adduct.

Table 5/54

Equilibrium Constants Corresponding to Formation of Tributyl Phosphate Adducts of Tris(thenoyltrifluoroacetonato) Rare Earth Chelates $M(C_8H_4F_3O_2S)_3 \cdot n(C_4H_9O)_3PO$*) (n = 1, 2).

solvent		Eu^{3+}			Sc^{3+} [40]
	t in °C	lg β_1	lg β_2	Ref.	lg β_1
hexane	25	5.87	10.78	[35]	4.61
heptane	25	6.27	11.14	[35]	4.42
cyclohexane	25	6.08	10.96	[35]	4.52
cyclohexane	30	5.68	10.28	[8]	
methylene chloride	25	3.32	5.24	[35]	2.94
chloroform	25	3.40	5.20	[35]	1.89
chloroform	30	3.23	4.94	[8]	
carbon tetrachloride	25	5.05	8.40	[35]	3.38
bromoform	25	3.66	5.62	[35]	2.16
benzene	25	4.70	8.00	[35]	2.92
benzene	30	4.27	7.13	[8]	
toluene	25	4.84	7.98	[35]	3.10
chlorobenzene	25	4.48	7.26	[35]	2.70
chlorobenzene	30	4.19	6.86	[8]	
o-dichlorobenzene	25	4.30	7.20	[35]	2.76
o-dichlorobenzene	30	4.36	6.87	[8]	
1,2-dichloroethane	30	3.16	5.25	[8]	
isopropylbenzene	25	4.98	8.56	[35]	

*) $\beta_n = [M(C_8H_4F_3O_2S)_3 \cdot n(C_4H_9O)_3PO] \cdot [M(C_8H_4F_3O_2S)_3]^{-1} \cdot [(C_4H_9O)_3PO]^{-n}$.

The dependence of the concentration based equilibrium constants on the solvent was accounted for in terms of the activity coefficients for the species involved in each solvent [35, 40]. The equilibrium constants in terms of activities were evaluated at lg K_1 = 2.95 for the scandium adduct [40], and lg K_1 = 4.39 and lg K_2 = 2.98 for the europium adduct [35] in the various solvents.

The thermodynamic parameters ΔH in kcal · mol^{-1}, ΔS in cal · mol^{-1} · K^{-1} (Table 5/55) associated with the formation of the $Eu(C_8H_4F_3O_2S)_3 \cdot n(C_4H_9O)_3PO$ (n = 1, 2) adducts in various solvents were found to vary with the dielectric constant of the medium. These variations were suggestive of an electrostatic interaction between the metal chelate and tributyl phosphate [8].

Table 5/55

Enthalpy and Entropy Changes Corresponding to Formation of the $Eu(C_8H_4F_3O_2S)_3 \cdot n(C_4H_9O)_3PO$ (n = 1, 2) Adducts.

solvent	dielectric constant	$-\Delta H_1$ *)	$-\Delta H_2$ *)	ΔS_1	ΔS_2
cyclohexane	2.1	2.58±0.31	11.90±0.88	17.4±1.0	−18.2±2.9
chloroform	4.81	2.45±0.77	5.24±1.37	6.6±2.6	−9.5±4.5
1,2-dichloroethane	10.36	5.82±0.60	4.53±1.25	−4.8±2.0	−5.4±4.1
benzene	2.28	5.73±0.91	8.58±0.97	0.6±3.0	−15.3±3.3
chlorobenzene	5.62	6.24±1.02	9.27±0.24	−1.5±3.4	−18.4±0.8
1,2-dichlorobenzene	9.93	7.09±0.54	6.86±0.29	−3.5±1.9	−11.2±1.0

*) Determined from the temperature coefficient of the equilibrium constant.

Isolated Compounds $M(C_8H_4F_3O_2S)_3 \cdot 2(C_4H_9O)_3PO$. Sinha and coworkers [79] isolated the europium compound according to procedures described by Halverson, Brinen, and Leto [90]. In one method, the tris chelate was mixed with a dilute hydrocarbon solution of tributyl phosphate (mole ratio of tributyl phosphate to metal chelate in the range 2:1 to 4:1). In a second procedure, an aqueous europium nitrate solution was extracted with a solution of thenoyltrifluoroacetone and tributyl phosphate in ether, and the product was recovered from the ether solution. The emission spectrum of the compound in 3:1 (v/v) ethanol/methanol at 77 K revealed lines at 16235, 16342 cm^{-1} ($^5D_0 \rightarrow {}^7F_2$), 16867, 16937, 16976 ($^5D_0 \rightarrow {}^7F_1$), and 17283, 17341 cm^{-1} ($^5D_0 \rightarrow {}^7F_0$). The appearance of two bands in the region of the $^5D_0 \rightarrow {}^7F_0$ transition is indicative of the presence of two independent species or dissociation of the adduct leading to an equilibrium mixture in solution [79].

Ferraro and Healy [26] isolated a scandium compound, which they formulated as the bis adduct, according to the procedure described on p. 224 for preparation of triphenylphosphine oxide adducts. The compound was isolated in the form of an oil. The NMR spectrum of the adduct is discussed in [89].

With Methyl Isobutyl Ketone $(CH_3)_2CHCH_2COCH_3$ (= $C_6H_{12}O$) and Dibutyl Sulfoxide $(C_4H_9)_2SO$ (= $C_8H_{18}OS$)

$M(C_8H_4F_3O_2S)_3 \cdot nL$ (n = 1, 2). Equilibrium constants, corresponding to the reactions described by the equations $M(C_8H_4F_3O_2S)_3 + nL \rightleftharpoons M(C_8H_4F_3O_2S)_3 \cdot nL$ (n = 1, 2) in carbon tetrachloride, were determined at 25°C by a solvent extraction method. The authors noted that $Sc(C_8H_4F_3O_2S)_3$ adds one molecule of dibutyl sulfoxide but forms only a weak complex with methyl isobutyl ketone. In general, the adducts derived from the lanthanum, europium, and lutetium chelates were more stable than those derived from the scandium chelate [31].

metal ion	methyl isobutyl ketone		dibutyl sulfoxide	
	lg β_1	lg β_2	lg β_1	lg β_2
Sc^{3+}	*)	**)	3.30	*)
La^{3+}	2.0	2.9	5.09	8.58
Eu^{3+}	1.8	2.5	5.21	*)
Lu^{3+}	1.7	**)	—	—

*) Weak interaction. — **) No evidence for formation of the adduct.

With N-Butylacetanilide $C_6H_5N(C_4H_9)C(O)CH_3$ (= $C_{12}H_{17}NO$)

$Nd(C_8H_4F_3O_2S)_3 \cdot 2C_{12}H_{17}NO$. The compound was isolated in the form of an oil according to the procedure described on p. 224 for preparation of the triphenylphosphine oxide adducts of the tris thenoyltrifluoroacetonate chelates [15].

With Derivatives of 1,2-Dihydro-3H-pyrazol-3-one

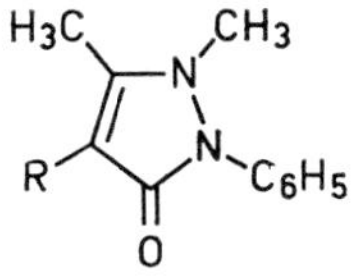

R = H (= Antipyrine = $C_{11}H_{12}N_2O$); R = $N(CH_3)_2$ (= 4-Dimethylaminoantipyrine = Pyrimidone = $C_{13}H_{17}N_3O$); R = CH_2-$C_{11}H_{11}N_2O$ (= Di-(4-antipyrinyl)methane = $C_{23}H_{24}N_4O_2$)

Each complex of approximate formula $\mathbf{M(C_8H_4F_3O_2S)_3 \cdot L}$ (M = Y, La, Pr, Nd, Sm, Eu, Tb, Er for L = $C_{11}H_{12}N_2O$, M = Sm, Eu, Tb, Dy for L = $C_{13}H_{17}N_3O$, and $C_{23}H_{24}N_4O_2$) was prepared by adding a solution of 0.04 mol thenoyltrifluoroacetone and 0.02 mol of the antipyrine ligand in 3 ml of ethanol to an aqueous solution of 0.01 mol of the rare earth chloride. After further addition of 2 ml of 40% hexamethylenetetramine (or urotropine) and water up to a total volume of 10 ml, an oily material separated from the mother liquor. This was dissolved in 5 ml of benzene and separated again by addition of 10 ml of petroleum ether.

It is shown that adduct formation of the antipyrine ligands increases the absorption intensities of the corresponding chelates $M(C_8H_4F_3O_2S)_3$ in water or benzene and modifies the structure of their luminescence spectra [98]. The IR spectrum of the lanthanum compound $La(C_8H_4F_3O_2S)_3 \cdot C_{23}H_{24}N_4O_2$ was studied and on this basis a coordination number of eight in these complexes was assumed [99].

5.1.21.4 Tetrakis Chelates of Type M'[M($C_8H_4F_3O_2S)_4$]

Ammonium Salt $NH_4[Pr(C_8H_4F_3O_2S)_4]$. The complex was prepared by mixing stoichiometric amounts of anhydrous praseodymium chloride and thenoyltrifluoroacetone in 95% ethanol and neutralizing the resulting mixture with aqueous ammonia in ethanol. The solution was allowed to evaporate slowly under a dry nitrogen atmosphere. The crystalline product thus obtained was dried under vacuum [43].

The crystal and molecular structures of the compound were determined by single-crystal X-ray analysis (R = 0.111 for 2969 reflections). The salt crystallizes in space group $P\bar{1}$-C_i^1 (No. 2) of the triclinic system, with two molecules in a unit cell of dimensions a = 10.10, b = 17.90, c = 12.14 Å, α = 78.57°, β = 103.97°, and γ = 86.95° (D_{obs} = 1.65 g/cm³; D_{calc} = 1.67 g/cm³). The crystal consists of discrete monomeric $Pr(C_8H_4F_3O_2S)_4^-$ ions, possibly bound together through weak water and ammonium hydrogen bonds. The coordination sphere around the praseodymium ion consists of eight essentially equidistant (2.42 to 2.49 Å) oxygen atoms, located at the vertices of a dodecahedron bisdiphenoid) (see **Fig.** 5-**29a** and **b**). Each ligand bridges adjacent vertices A-B of the two orthogonal trapezoids (each designated BAAB), along edges g of the dodecahedron—see Fig. 5-29. The angles Θ_A and Θ_B which the metal-A and metal-B bonds make with the four-fold symmetry axis are 41.4° and 65.7°, respectively. The essentially coplanar ligands all show folding about the oxygen-oxygen line in the six-membered chelate rings (dihedral angle ranges from 10.5 to 23.6°). Since one ligand has its terminal groups (thienyl and CF_3) reversed relative to the other three ligands (**Fig.** 5-**30**), the overall idealized (i. e., neglecting the ring folding and nonplanarity) symmetry including the total ligand is the minimum trivial symmetry of C_1. Considering only the eight-coordinated oxygen atoms, the symmetry is very nearly S_4 (and not D_{2d}). The displacements, normal to the trapezoidal plane, amount to 0.06 Å [43].

Molar magnetic susceptibilities, measured from 68 to 297 K, obeyed Curie's law ($\chi = c \cdot T^{-1}$ with c = 1.607) and yielded a moment of 3.59 μ_B, which is in agreement with the Van Vleck calculated value [43].

Dimethylammonium Salt $(CH_3)_2NH_2[Eu(C_8H_4F_3O_2S)_4]$. The compound was obtained in crystalline form by dissolving dimethylamine, thenoyltrifluoroacetone, and europium chloride in ethanol (mole ratio = 4:4:1). The crystalline precipitate was collected, washed, and air dried. The melting point of the compound was 194 to 195°C [49].

Fig. 5-29

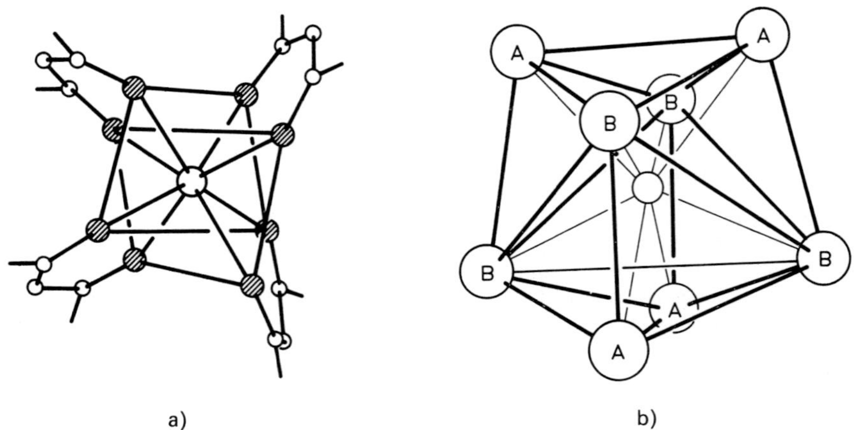

a) b)

Bisphenoidal arrangement of the inner coordination sphere of $NH_4[Pr(C_8H_4F_3O_2S)_4]$ a), and the idealized $\bar{4}2m$ dodecahedron b).

Fig. 5-30 Fig. 5-31

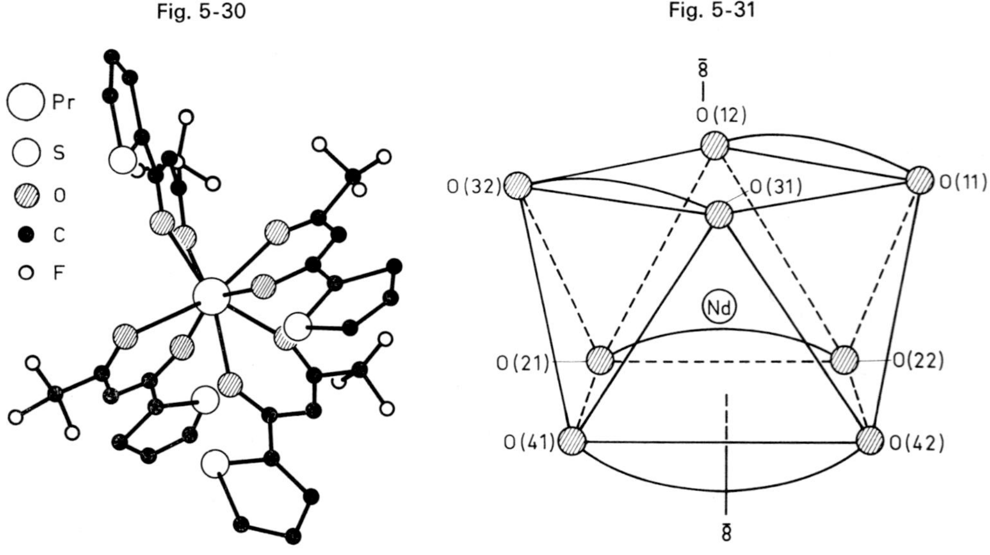

Molecular structure of the $[\alpha\text{-}Pr(C_8H_4F_3O_2S)_4]^-$ ion.

The coordination polyhedron of the tetrakis-(thenoyltrifluoroacetonato)neodymate(III) ion.

Triethylammonium Salts $(C_2H_5)_3NH[M(C_8H_4F_3O_2S)_4]$. Each compound (M = La, Nd, Eu, Tb) was prepared by adding a rare earth salt (4 mmol) in 20 ml of water to a solution, heated at reflux, of thenoyltrifluoroacetone (18 mmol) and triethylamine (18 mmol) in 95% ethanol. The solution was filtered while hot. In some instances the chelate began separating as an oil during the initial phase of cooling, in which case, crystallization was effected either by gently swirling the cloudy solution or scratching of the inner surface of the flask with a glass rod. The mixture was allowed to stand overnight at room temperature to afford analytically pure crystalline chelate. The melting points (°C) of the compounds thus obtained are: La, 135; Nd, 135; Eu, 133; Tb, 158 [44].

Tetraethylammonium Salt $(C_2H_5)_4N[Ce(C_8H_4F_3O_2S)_4]$. The cerium(III) complex was prepared under a pure nitrogen atmosphere using deoxygenated solvents according to the following procedure: A solution of cerium(III) chloride heptahydrate (2 mmol) and 2,2-dimethoxypropane (10 cm^3) in absolute ethanol (15 cm^3) was added to a mixture of thenoyltrifluoroacetone (8 mmol), sodium hydroxide (8 mmol), and tetraethylammonium chloride (2 mmol) in absolute ethanol (25 cm^3). The solvent was removed under reduced pressure, the residue was extracted with ethyl acetate (30 cm^3), and the suspended solid was removed by filtration. The solution was concentrated under vacuum to approximately 10 cm^3. Crystallization of the product was induced by slow addition of n-hexane. The red-orange colored solid was collected, washed with n-hexane, and dried under vacuum. The compound was characterized by elemental analysis and its electronic absorption spectrum, both in acetonitrile (21300 cm^{-1} (sh); 22700 cm^{-1} (sh), 29600 cm^{-1}; $\varepsilon = 65000\ dm^3 \cdot mol^{-1} \cdot cm^{-1}$; 37400 cm^{-1}, $\varepsilon = 26000\ dm^3 \cdot mol^{-1} \cdot cm^{-1}$) and in the solid state [45].

Tetrapropylammonium Salt $(C_3H_7)_4N[Eu(C_8H_4F_3O_2S)_4]$. The compound was prepared by adding 10% aqueous tetrapropylammonium hydroxide (8 mmol) to a solution of thenoyltrifluoroacetone (8 mmol) and europium chloride (2 mmol) in 75 ml of hot ethanol. On cooling, the solution deposited brightly fluorescent, coarse, light orange crystals. The precipitate was collected, and additional product was obtained by adding water to the filtrate (yield 74%). The complex was purified by recrystallization from ethanol (melting point = 188 to 189°C). The compound was characterized by elemental analysis and emission spectroscopy [46].

Tetrabutylammonium Salts $(C_4H_9)_4N[M(C_8H_4F_3O_2S)_4]$. Each compound (M = La to Lu except for Pm) was prepared according to a procedure similar to that described for preparation of the ammonium salt (see p. 227). Slow evaporation of the ethanolic solutions of these complexes resulted in two crystalline modifications, designated α and β. Solutions of the neodymium and europium complexes produced crystals of both modifications (dimorphism), requiring physical separation of the crystal types. Separation was accomplished under a stereomicroscope [47].

X-ray analysis revealed that the α-modification of the chelates crystallized in space group C2/c-C_{2h}^6 (No. 15) with four molecules per unit cell, whereas the β-modification crystallized in space group P2/n-C_{2h}^4 (No. 13) with two molecules per unit cell. The compounds within each group are isomorphous. The unit cell dimensions of all compounds are recorded in Table 5/56 together with their densities. The praseodymium (representative of the β-modifications) and samarium (representative of the α-modifications) chelates were chosen for single-crystal X-ray analysis (R = 0.118 and 0.080 for approximately 2500 reflections, recorded for the praseodymium and samarium compounds, respectively) [48]. In both structures, the central ion is eight-coordinate, surrounded by eight essentially equidistant oxygen atoms from the thenoyltrifluoroacetate groups at the vertices of a square antiprism. Disorder was observed in the thienyl rings of the samarium chelate [48].

Tetrahexylammonium Salt $(C_6H_{13})_4N[Eu(C_8H_4F_3O_2S)_4]$. The compound was prepared according to the procedure described for the preparation of the tetrapropylammonium salt of the tetrakis europium chelate. The complex was recrystallized from ethanol (melting point = 170 to 172°C), and characterized by elemental analysis and emission spectroscopy [46].

Pyridinium Salts $C_5H_5NH[M(C_8H_4F_3O_2S)_4]$. The neodymium [50] and terbium [44] chelates were prepared according to the procedure described for preparation of the triethylammonium salts of the thenoyltrifluoroacetato chelates (see p. 228). The melting point of the terbium chelate was 193°C [44]. The europium compound was obtained by extraction with benzene from an aqueous solution containing the rare earth chloride, thenoyltrifluoroacetone, and pyridine. The benzene layer was separated and petroleum ether added to the extract solution to effect precipitation of the product. The precipitate was collected by filtration, washed with petroleum ether, and dried in air. The melting point of the compound thus obtained was 150 to 152°C [82].

Table 5/56

Unit Cell Dimensions and Densities of the α- and β-Modifications of the Crystalline $(C_4H_9)_4N[M(C_8H_4F_3O_2S)_4]$ Compounds [47].

α-modification						β-modification					
M	density in g/cm^{-3}	a in Å	b in Å	c in Å	β	M	density in g/cm^{-3}	a in Å	b in Å	c in Å	β
La	1.506	19.17	18.15	16.74	109.8°	Pr	1.472	22.32	10.45	12.23	96.7°
Ce	1.516	19.21	18.08	16.59	109.4°	Nd	1.476	22.47	10.52	11.99	95.8°
Nd	1.523	19.16	18.15	16.57	109.1°	Eu	1.490	22.13	10.52	12.19	97.0°
Sm	1.551	19.14	18.10	16.54	109.3°	Tb	1.505	22.63	10.64	12.02	96.3°
Eu	1.540	19.19	18.03	16.61	109.3°	Dy	1.517	22.50	10.55	11.95	96.0°
Gd	1.552	19.19	18.10	16.54	109.3°	Ho	1.516	22.56	10.67	11.96	96.7°
						Er	1.517	21.78	10.58	12.08	96.4°
						Tm	1.532	22.34	10.52	11.83	96.0°
						Yb	1.532	22.56	10.60	11.86	94.0°
						Lu	1.540	21.75	10.49	12.19	96.0°

The crystal and molecular structures of the praseodymium compound were determined by single-crystal X-ray analysis (R = 0.124 for 3180 reflections). This compound crystallizes in space group $P2_1/c\text{-}C^5_{2h}$ (No. 14) with four molecules in a unit cell of dimensions a = 10.93, b = 22.25, c = 17.77 Å, β = 97.4° (D_{calc} = 1.72 g/cm^{-3}, D_{obs} = 1.71 g/cm^{-3}). The neodymium ion is coordinated to eight oxygen atoms of the four thenoyltrifluoroacetonato groups with a mean metal-oxygen bond distance of 2.44 Å (range = 2.283 to 2.511 Å). The coordination polyhedron was described as a distorted square antiprism in contrast to the dodecahedral structure reported for the $Pr(C_8H_4F_3O_2S)_4^-$ ion as the ammonium salt (see p. 228). As depicted in **Fig. 5-31**, p. 228, the s edges of the square antiprism are spanned by the chelate rings. The distance of the neodymium ion from the two square faces is the same (1.40 Å), but the angle between the two planes is 5.4°. The mean angle that the neodymium-oxygen bonds make with the four-fold symmetry axis is 55.0°. The six-membered chelate rings all show folding about the oxygen-oxygen line with a dihedral angle in the range of 17.7° to 18.8° [50].

2-Aminopyridinium Salt $NH_2C_5H_4NH[Eu(C_8H_4F_3O_2S)_4]$. The compound was obtained by extraction with benzene of an aqueous solution containing the rare earth chloride, thenoyltrifluoroacetone, and 2-aminopyridine according to the procedure described on p. 229 for the pyridinium compound (melting point = 188 to 190°C) [82].

4-Methylpyridinium Salts $CH_3C_5H_4NH[M(C_8H_4F_3O_2S)_4]$. The stabilities of the γ-picolinium salts of the lanthanum and yttrium chelates in basic acetone and basic dichloromethane were studied by NMR spectroscopy [52, 53]. The authors proposed that the tetrakis chelates dissociate in basic acetone to form the tris chelates, but do not dissociate in basic chloroform [52]. The infrared spectrum (KBr disc) was recorded for the yttrium complex [53].

2,4,6-Trimethylpyridinium Salts $(CH_3)_3C_5H_2NH[M(C_8H_4F_3O_2S)_4]$. The europium [44, 46] and terbium [46] chelates were prepared according to a procedure similar to that described for preparation of the triethylammonium salts of the thenoyltrifluoroacetate ions (see p. 228). The melting points (°C) of the compounds thus obtained are: Eu, 158 [44]; Eu, 159 to 160 [46]; Tb, 155 to 157 [46]. The compounds were characterized further by elemental analyses and emission spectroscopy [44, 46]. The neodymium and europium compounds were obtained by extraction with benzene from aqueous solutions containing the rare earth chloride, thenoyltrifluoroacetone, and 2,4,6-trimethylpyridine (= collidine) according to the procedure described on p. 229 for the pyridinium compound. The melting points (°C) of the compounds were: Nd, 155 to 156; Eu, 158 to 160 [82].

Piperidinium Salts $C_5H_{11}NH[M(C_8H_4F_3O_2S)_4]$. The europium compound was prepared by adding an ethanolic solution (10 ml) of europium chloride (2 mmol), dropwise, to a solution of thenoyltrifluoroacetone (8 mmol) and piperidine (8 mmol) in about 50 ml of ethanol. The solution was reduced in volume by evaporation under reduced pressure, using the minimum of heat, until crystallization commenced. After cooling the reaction mixture in an ice/salt mixture, the well-formed crystals were collected by filtration and air dried [54]. The europium and neodymium compounds were prepared by extraction into benzene according to the procedure described on p. 229 for the pyridinium salt. The melting points (°C) of the compounds thus obtained were: Nd, 182 to 184; Eu, 190 to 192 [82].

The emission spectrum of the europium chelate was recorded for samples dispersed in styrene and polystyrene [55] and in solution (3:1 ethanol/methanol and acetonitrile) [56].

Quinolinium Salt $C_9H_7NH[Nd(C_8H_4F_3O_2S)_4]$. The compound was obtained by extraction with benzene from an aqueous solution containing neodymium chloride, thenoyltrifluoroacetone, and quinoline according to the procedure described on p. 229 for the pyridinium compound (melting point = 165 to 168°C).

Isoquinolinium Salts $C_9H_7NH[M(C_8H_4F_3O_2S)_4]$. The europium chelate was prepared by allowing europium chloride, thenoyltrifluoroacetone, and isoquinoline to react in hot ethanol. The mixture was allowed to cool to room temperature, and the crystals thus formed were collected, washed with water, and dried under vacuum. The product was recrystallized from carbon tetrachloride (melting point = 170 to 171°C) [46].

The crystal and molecular structures of the cerium complex (method of preparation not presented) were determined by single-crystal X-ray analysis (R = 0.128 for 5325 reflections). The compound crystallizes in the monoclinic space group $P2_1/c\text{-}C_{2h}^5$ (No. 14) with four molecules in a unit cell of dimensions a = 22.76, b = 10.85, c = 20.13 Å, β = 112.58°. The cerium ion is coordinated by eight oxygen atoms from the four thenoyltrifluoroacetate groups, with a mean metal-oxygen bond distance of 2.470 Å (range 2.425 to 2.563 Å). The value of 2.563 Å, which is significantly longer than the mean metal-oxygen bond distance, was attributed to a hydrogen bond O···H-N (N-O bond distance = 2.87 Å) between the ligated oxygen atom and the NH^+ hydrogen atom of the isoquinolinium group, which effectively weakens the metal-oxygen interactions. The coordination polyhedron was described as a distorted D_{2d} dodecahedron with the bidentate ligands spanning the m edges (see **Fig. 5-32**) of the two trapezoids (the angle between the two trapezoidal planes is 89.8° compared with the idealized angle of 90°). Two of the chelate rings are planar, whereas the other two are folded about the oxygen-oxygen line, with a dihedral angle of about 18° [57].

Fig. 5-32

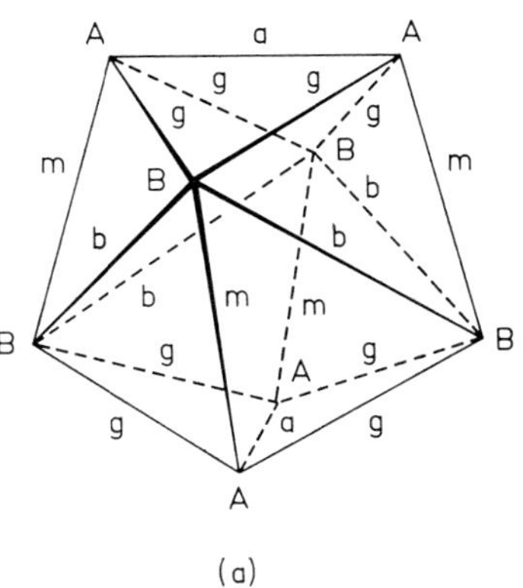

(a)

7 24 38 49 10 21 52 35

(b)

The D_{2d} dodecahedron (a) with oxygen sites labeled in $Ce(C_8H_4F_3O_2S)_4^-$ (b)
(For atom numbering scheme, see [57]).

Diphenylguanidinium Salt $(C_{13}H_{13}N_3H)[Nd(C_8H_4F_3O_2S)_4]$ and 4,4'-Bipyridinium Salt $(C_{10}H_8N_2H)[Nd(C_8H_4F_3O_2S)_4]$. The compounds were obtained by extraction with benzene from an aqueous solution containing neodymium chloride, thenoyltrifluoroacetone, and diphenylguanidine or 4,4'-bipyridine according to the procedure described on p. 229 for the pyridinium salt (melting point = 188 to 190°C and 190 to 193°C, respectively) [82].

5.1.21.5 Cerium(IV) Compound $Ce(C_8H_4F_3O_2S)_4$

The cerium(IV) chelate was prepared by adding an ethanolic solution of thenoyltrifluoroacetone to a 50% aqueous ethanolic solution of ceric ammonium nitrate. The resulting precipitate was collected by filtration, washed with successive portions of water and ethanol, and finally dried in a desiccator at room temperature. The compound thus obtained melted at 190°C [91]. The compound was also prepared by adding a solution of thenoyltrifluoroacetone in ethanol to an aqueous solution of ammonium cerium(IV) sulfate tetrahydrate. The product was extracted with toluene or benzene from the crude precipitate, and allowed to crystallize from the extractant solution. The compound was recrystallized from benzene and dried under vacuum [92].

Infrared spectral data (spectrum obtained for samples dispersed in Nujol and hexachlorobutadiene mulls [92] and KBr discs [91]) are tabulated and discussed in [91] and [92]. The ν(Ce-O) stretching mode was assigned to a band at 240 cm^{-1} [92]. The electronic absorption spectrum of the red-brown compound in acetonitrile revealed bands at 22500 (sh), 29900 (73200), 36000 (27000), and 47000 cm^{-1} (sh) (absorption coefficient, ε in $dm^3 \cdot mol^{-1} \cdot cm^{-1}$, given in parentheses) [45].

Preliminary single-crystal X-ray analysis revealed that the compound crystallizes in the orthorhombic system, in either space group $P2_12_12\text{-}D_2^3$ (No. 18) or $Pma2\text{-}C_{2v}^4$ (No. 28). There are four molecules in a unit cell of dimensions a = 17.532, b = 20.719, and c = 10.710 Å (D_{calc} = 1.749 g/cm^3, D_{obs} = 1.76 g/cm^3) [91].

5.1.21.6 Mixed Complexes Derived from Thenoyltrifluoroacetone and Other Ligands (HL)

Complexes with HL = ethylene diamine-N,N,N',N'-tetraacetic acid and N'-(2-hydroxyethyl)-ethylenediamine-N,N,N'-triacetic acid are described in "Rare Earth" D 1, pp. 236/7 and 234, respectively.

Acetato-Thenoyltrifluoroacetato Compound

$Eu(C_8H_4F_3O_2S)_2(C_2H_3O_2)$. The mixed ligand complex was prepared by adding, with vigorous shaking, an aqueous solution of sodium acetate (0.03 mol) and europium chloride (0.005 mol) to a solution of thenoyltrifluoroacetone (0.0165 mol) in absolute ethanol. The reaction mixture, containing a precipitate, was heated on a water bath for 15 to 20 min. After standing at room temperature overnight, the precipitate was collected, washed with absolute ethanol, and air dried. The emission spectrum of the compound in 3:1 (v/v) ethanol/methanol at 77 K revealed the following peaks in the regions of the $^5D_0 \rightarrow {}^7F_J$ (J = 0, 1, 2) transitions: $^5D_0 \rightarrow {}^7F_2$ = 16216, 16329 cm^{-1}; $^5D_0 \rightarrow {}^7F_1$ = 16866, 16896, 16963, 16984 cm^{-1}; and $^5D_0 \rightarrow {}^7F_0$ = 17327 cm^{-1} [79].

Tetrakis Chelate Derived from Thenoyltrifluoroacetone and Acetylacetone ($= C_5H_8O_2$ = Hacac)

$M(C_8H_4F_3O_2S)_3Hacac$. Each chelate (M = Nd, Sm, Gd, Dy, Er) was prepared by extracting an aqueous solution (25 ml) of the rare earth chloride (3 g) with a solution of thenoyltrifluoroacetone (9 g) and acetylacetone (4 g) in about 40 ml of benzene. Dilute aqueous ammonia was added dropwise, with stirring, to the resulting mixture until a final pH of 3.5 was reached in the aqueous phase. Upon further stirring, the organic phase thickened, and the product left was an oil which was purified by holding at elevated temperatures under vacuum (2 Torr). The mixed chelate species appeared stable under vacuum up to a temperature of 86°C. The amount of acetylacetone present in the compounds was determined by titrating samples dissolved in pyridine using tetrabutylammonium hydroxide dissolved in 1:9 (v/v) methanol/benzene as the titrant. A small absorption band at 1715 cm^{-1} in the infrared spectra of the complexes was assigned to the ν(C=O) stretching mode of the undissociated acetylacetonate molecule [93].

Mixed Complex Derived from Thenoyltrifluoroacetone and 1,1,1-Trifluoro-4-mercapto-4-(2-thienyl)-3-buten-2-one

$C_4H_3S-C(SH)=CH-C(=O)-CF_3$ (= $C_8H_5F_3OS_2$)

$Pr(C_8H_4F_3O_2S)_2(C_8H_4F_3OS_2)$. A warm solution of praseodymium chloride (0.25 mmol) in 95% ethanol was added to an ethanolic solution of thenoyltrifluoroacetone (0.01 mol) previously neutralized with alcoholic ammonia to pH 7.0. No precipitation occurred. The solution was cooled to −70°C, and hydrogen sulfide was pumped into the reaction mixture for 20 min. The system was allowed to come to room temperature, and evaporation of a sample yielded small crystals. Larger crystals were obtained by allowing the solution to evaporate slowly in a refrigerator over a period of 4 months. Pink crystals were removed and dried under vacuum. Mass spectral evidence was presented for the formulation of the mixed ligand chelate [94].

Preliminary single-crystal X-ray studies revealed that the compound formulated as the dihydrate crystallizes in the monoclinic system, space group $P2_1/c-C_{2h}^5$ (No. 14) with four molecules in a unit cell of dimensions a = 14.294, b = 13.258, c = 20.006 Å, and β = 133°35′ [80]. Infrared studies indicated formation of both a praseodymium-sulfur bond (νPr-S = 430 cm^{-1}) with ligation of the mercapto ligand and Pr-O bonds (νPr-O = 610 cm^{-1}) [80, 94].

References to 5.1.21:

[1] N. Ikeda, K. Kimura, H. Asai, N. Oshima (Radioisotopes [Tokyo] **19** [1970] 1/6). — [2] Yu. A. Zolotov, N. V. Shakhova, I. P. Alimarin (Zh. Analit. Khim. **23** [1968] 1321/6; J. Anal. Chem. [USSR] **23** [1968] 1164/8). — [3] T. Sekine, A. Koizumi, M. Sakairi (Bull. Chem. Soc. Japan **39** [1966] 2681/4). — [4] M. A. Carey, C. V. Banks (J. Inorg. Nucl. Chem. **31** [1969] 533/50). — [5] R. A. Bolomey, L. Wish (J. Am. Chem. Soc. **72** [1950] 4483/6).

[6] J. N. Mathur, S. A. Pai, P. K. Khopkar, M. S. Subramanian (J. Inorg. Nucl. Chem. **39** [1977] 653/7). — [7] L. Genov, I. L. Dukov (Monatsh. Chem. **104** [1973] 750/7). — [8] S. A. Pai, J. N. Mathur, P. K. Khopkar, M. S. Subramanian (J. Inorg. Nucl. Chem. **39** [1977] 1209/11). — [9] F. Kassierer, A. S. Kertes (Proc. 3rd Symp. Coord. Chem., Debrecen, Hung., 1970, p. 119/32). — [10] L. Genov, I. Dukov, G. Kassabov (Acta Chim. Acad. Sci. Hung. **95** [1977] 361/5).

[11] B. Kuznik, L. Genov, G. Georgiev (Monatsh. Chem. **106** [1975] 1543/51). — [12] P. K. Khopkar, J. N. Mathur (J. Inorg. Nucl. Chem. **39** [1977] 2063/7). — [13] T. Mitsuji, M. Sakurada, K. Nomura (Nara Kyoiku Daigaku Kiyo Shizen Kagaku **22** [1973] 31/42 from C.A. **81** [1974] No. 17373). — [14] V. T. Mishchenko, R. S. Lauer, N. P. Efryushina, N. S. Poluektov (Zh. Analit. Khim. **20** [1965] 1073/81; J. Anal. Chem. [USSR] **20** [1965] 1129/37). — [15] T. V. Healy, D. F. Peppard, G. W. Mason (J. Inorg. Nucl. Chem. **24** [1962] 1429/48).

[16] N. Suzuki, K. Akiba, H. Asano (Anal. Chim. Acta **52** [1970] 115/22). — [17] N. Suzuki, K. Akiba, T. Kanno, T. Wakabayashi, K. Takaizumi (J. Inorg. Nucl. Chem. **30** [1968] 3047/55). — [18] T. Sekine (Acta Chem. Scand. **19** [1965] 1476/82). — [19] T. Sekine, M. Sakairi, F. Shimada, Y. Hasegawa (Bull. Chem. Soc. Japan **38** [1965] 847/9). — [20] P. Tanipanichskul, J. Foos, A. Iuchenko, R. Guillaumont (J. Inorg. Nucl. Chem. **38** [1976] 315/8).

[21] T. Wakabayashi, K. Takaizumi, K. Seto, N. Suzuki, K. Akiba (Bull. Chem. Soc. Japan **41** [1968] 1854/8). — [22] S. Oki, T. Omori, T. Wakahayashi, N. Suzuki (J. Inorg. Nucl. Chem. **27** [1965] 1141/50). — [23] F. W. Cornish (CR-891 [1952] from C.A. **1952** 11001). — [24] T. V. Healy (J. Inorg. Nucl. Chem. **19** [1961] 314/27). — [25] T. V. Healy (J. Inorg. Nucl. Chem. **19** [1961] 328/39).

[26] J. R. Ferraro, T. V. Healy (J. Inorg. Nucl. Chem. **24** [1962] 1463/74). — [27] H. Irving, D. N. Edgington (J. Inorg. Nucl. Chem. **21** [1961] 169/80). — [28] P. G. Manning (Can. J. Chem. **41** [1963] 658/66). — [29] T. Sekine, D. Dyrssen (J. Inorg. Nucl. Chem. **29** [1967] 1457/73). — [30] T. Sekine, D. Dyrssen (J. Inorg. Nucl. Chem. **29** [1967] 1475/80).

[31] T. Sekine, D. Dyrssen (J. Inorg. Nucl. Chem. **29** [1967] 1481/7). — [32] T. Sekine, M. Ono (Bull. Chem. Soc. Japan **38** [1965] 2087/94). — [33] T. Taketatsu, C. V. Banks (Anal. Chem. **38** [1966] 1524/8). — [34] T. K. Keenan, J. F. Suttle (J. Am. Chem. Soc. **76** [1954] 2184/5). — [35] K. Akiba (J. Inorg. Nucl. Chem. **35** [1973] 2525/35).

[36] L. I. Kononenko, M. A. Tishchenko, R. A. Vitkun, V. N. Drobyazko (Ukr. Khim. Zh. **37** [1971] 829/34; Soviet Progr. Chem. **37** No. 8 [1971] 78/82). — [37] T. Honjyo (Bull. Chem. Soc. Japan **42** [1969] 995/9). — [38] L. I. Kononenko, R. A. Vitkun (Zh. Neorgan. Khim. **13** [1968] 3328/32; Russ. J. Inorg. Chem. **13** [1968] 1715/8). — [39] T. Taketatsu, N. Toriumi (Bull. Chem. Soc. Japan **41** [1968] 1275). — [40] K. Akiba, T. Ishikawa, N. Suzuki (J. Inorg. Nucl. Chem. **33** [1971] 4161/70).

[41] T. Sekine, D. Dyrssen (J. Inorg. Nucl. Chem. **29** [1967] 1489/98). — [42] R. W. Taft, E. H. Cook (J. Am. Chem. Soc. **81** [1959] 46/54). — [43] R. A. Lalancette, M. Cefola, W. C. Hamilton, S. J. LaPlace (Inorg. Chem. **6** [1967] 2127/34). — [44] L. R. Melby, N. J. Rose, E. Abramson, J. C. Caris (J. Am. Chem. Soc. **86** [1964] 5117/25). — [45] M. Ciampolini, F. Mani, N. Nardi (J. Chem. Soc. Dalton Trans. **1977** 1325/8).

[46] H. Bauer, J. Blanc, D. L. Ross (J. Am. Chem. Soc. **86** [1964] 5125/31). — [47] R. T. Criasia, M. Cefola (J. Inorg. Nucl. Chem. **37** [1975] 1814/5). — [48] R. T. Criasia (Fordham Univ. 1973, p. 1/199; Diss. Abstr. Intern. B **34** [1974] 3687/8). — [49] E. J. Schimitschek, R. B. Nehrich, J. A. Trias (J. Chem. Phys. **42** [1965] 788/90). — [50] J. G. Leipoldt, L. D. C. Bok, S. S. Basson, A. E. Laubscher, J. S. Van Vollenhoven (J. Inorg. Nucl. Chem. **39** [1977] 301/3).

[51] L. I. Kononenko, R. A. Vitkun (Zh. Neorgan. Khim. **15** [1970] 1345/50; Russ. J. Inorg. Chem. **15** [1970] 690/3). — [52] N. Kono, Y. Yano, K. Nagashima (Bunseki Kagaku **25** [1976] 649/52 from C.A. **86** [1977] No. 34857). — [53] Y. Yano, N. Kono (Bunseki Kagaku **24** [1975] 371/7 from C.A. **84** [1976] No. 128391). — [54] S. J. Lyle, A. D. Witts (Inorg. Chim. Acta **5** [1971] 481/4). — [55] E. Z. Georgadze, V. I. Kapanadze, R. N. Kukharskii, V. S. Chagulov (Soobshch. Akad. Nauk Gruz. SSR **79** [1975] 581/4 from C.A. **84** [1976] No. 51665).

[56] L. D. Derkacheva, A. D. Kudryavtseva, G. V. Peregudov, A. I. Sokolovskaya (Zh. Prikl. Spektrosk. **6** [1967] 346/9; J. Appl. Spectrosc. [USSR] **6** [1967] 233/5 from C.A. **67** [1967] No. 48795). — [57] A. T. McPhail, Pui-Suen Wong Tschang (J. Chem. Soc. Dalton Trans. **1974** 1165/71). — [58] D. Purushottam, V. Ramachandra Rao, Bh. S. V. Raghava Rao (Anal. Chim. Acta **33** [1965] 182/97). — [59] D. Purushottam, Bh. S. V. Raghava Rao (Indian J. Chem. **4** [1966] 109/10). — [60] S. Lis, S. Siekierski (Chem. Anal. [Warsaw] **17** [1972] 979/87).

[61] S. Lis, S. Siekierski (Inst. Nucl. Res. Warsaw Rept. No. 1247/V/C [1970] from C.A. **76** [1972] No. 20826). — [62] S. Lis (Proc. 3rd Anal. Chem. Conf., Budapest 1970, p. 433/41 from C.A. **74** [1971] No. 134285). — [63] R. G. Charles, R. C. Ohlmann (J. Inorg. Nucl. Chem. **27** [1965] 255/9). — [64] E. Uhlemann, F. Dietze (Z. Anorg. Allgem. Chem. **386** [1971] 329/34). — [65] J. G. White (Inorg. Chim. Acta **16** [1976] 159/62).

[66] P. C. Mehta, S. P. Tandon (Z. Naturforsch. **26a** [1971] 759/62). — [67] G. S. Shephard, D. A. Thornton (Helv. Chim. Acta **54** [1971] 2212/21). — [68] S. Lis (Inst. Nucl. Res. Warsaw Rept. No. 1430/V/C [1972] from C.A. **79** [1974] No. 36786). — [69] K. I. Gur'ev (Nekotorye Vopr. Nelinein. Opt. Teor. Spektrosk. Kvant. Khim. [1972] 32/40 from C.A. **79** [1973] No. 141234). — [70] P. C. Mehta, S. S. L. Surana, S. P. Tandon (Can. J. Spectrosc. **18** [1973] 55/60).

[71] G. A. Domrachev, V. P. Ippolitova (Tr. Khim. Khim. Tekhnol. **1966** No. 2, p. 227/40 from C.A. **68** [1968] No. 73665). — [72] Kh. A. Ainitdinov, O. L. Lebedev, A. V. Michurina (Opt. Spektroskopiya **18** [1965] 532/3; Opt. Spectrosc. [USSR] **18** [1965] 303/4). — [73] N. Filipescu, M. R. Kagan, N. McAvoy, F. A. Serafin (Nature **196** [1962] 467/8). — [74] S. P. Tandon, P. C. Mehta (J. Chem. Phys. **52** [1970] 4313/5). — [75] P. C. Mehta, S. P. Swami (J. Chem. Phys. **53** [1970] 414/7).

[76] T. Isobe, S. Misumi (Bull. Chem. Soc. Japan **47** [1974] 281/4). — [77] S. P. Tandon, R. C. Govil (Spectrosc. Letters **4** [1971] 73/7). — [78] M. Das, S. E. Livingstone (Australian J. Chem. **28** [1975] 1513/6). — [79] V. J. Rao, D. R. Rao, A. P. B. Sinha (Indian J. Chem. **8** [1970] 270/4). — [80] N. G. Dowling (Fordham Univ. 1971, p. 1/77; Diss. Abstr. Intern. B **32** [1971] 2092).

[81] A. S. Kertes, E. F. Kassierer (Inorg. Chem. **11** [1972] 2108/11). — [82] E. V. Melent'eva, L. I. Kononenko, N. S. Poluektov (Ukr. Khim. Zh. **32** [1966] 1147/53; Soviet Progr. Chem. **32** [1966] 867/71). — [83] J. G. Leipoldt, L. D. C. Bok, S. S. Basson, A. E. Laubscher (J. Inorg. Nucl. Chem. **38** [1976] 1477/9). — [84] L. I. Kononenko, I. G. Orlova, N. S. Poluektov (Ukr. Khim. Zh. **32** [1966] 627/31; Soviet Progr. Chem. **32** [1966] 482/5). — [85] L. I. Kononenko, M. A. Tishchenko, R. A. Vitkun, N. S. Poluektov (Zh. Neorgan. Khim. **10** [1965] 2465/70; Russ. J. Inorg. Chem. **10** [1965] 1341/4).

[86] E. A. Bozhevol'nov, A. G. Stepanova, L. G. Totskaya (Zh. Analit. Khim. **32** [1977] 2254/9; J. Anal. Chem. [USSR] **32** [1977] 1791/5). — [87] J. G. Leipoldt, L. D. C. Bok, S. S. Basson, J. S. Van Vollenhoven, A. E. Laubscher (J. Inorg. Nucl. Chem. **38** [1976] 2241/4). — [88] J. G. Leipoldt, L. D. C. Bok, A. E. Laubscher, S. S. Basson (J. Inorg. Nucl. Chem. **37** [1975] 2477/80). — [89] J. R. Ferraro (J. Inorg. Nucl. Chem. **26** [1964] 225/30). — [90] F. Halverson, J. S. Brinen, J. R. Leto (J. Chem. Phys. **41** [1964] 2752/60).

[91] Y. Baskin, N. S. Krishna Prasad (J. Inorg. Nucl. Chem. **25** [1963] 1011/9). — [92] T. Yoshimura, C. Miyake, S. Imoto (Bull. Chem. Soc. Japan **46** [1973] 2096/101). — [93] C. Woo, W. F. Wagner, D. E. Sands (J. Inorg. Nucl. Chem. **34** [1972] 307/12). — [94] N. Dowling, M. Cefola, J. G. White (J. Inorg. Nucl. Chem. **41** [1979] 106/8). — [95] H. G. Brittain (J. Chem. Soc. Dalton Trans. **1979** 1187/91).

[96] S. M. Shilov, I. M. Batyaev (Zh. Neorgan. Khim. **25** [1980] 409/11; Russ. J. Inorg. Chem. [USSR] **25** [1980] 223/4). — [97] V. S. Khomenko, T. A. Rasshinina, V. M. Suboch (Vestsi Akad. Navuk Belarussk.SSR Ser. Khim. Navuk **1979** No. 3, pp. 36/40 from C.A. **91** [1979] No. 100799). — [98] L. I. Kononenko, M. A. Tishchenko (Zh. Analit. Khim. **24** [1969] 1823/7; J. Anal. Chem. [USSR] **24** [1969] 1479/82). — [99] M. A. Tishchenko, G. I. Gerasimenko, Yu. N. Anisimov, N. S. Poluektov (Dokl. Akad. Nauk SSSR **250** [1980] 122/5; Dokl. Chem. Proc. Acad. Sci. USSR **250** [1980] 16/9).

5.1.22 Complexes with Other Thienyl Derivatives of 1,3-Diketones

5.1.22.1 With Diketones of Type

$C_4H_3S{-}C(=O){-}CH_2{-}C(=O){-}R$ R = CF_2OCF_3 (= $C_9H_5F_5O_3S$) R = (perfluoro-2-methyltetrahydrofuran-3-yl ring: F, F, F, F, CF_2, O, F, F) (= $C_{12}H_5F_9O_3S$)

$Eu(C_9H_4F_5O_3S)_n^{3-n}$ (n = 1 to 4) and $Eu(C_{12}H_4F_9O_3S)_n^{3-n}$ (n = 1 to 3) Complexes

Formation constants for the rare earth chelates were determined potentiometrically (glass electrode) at 20°C in 80% (v/v) methanol/water, I = 0.15 NaCl. For the $Eu(C_9H_4O_3F_5)_n^{3-n}$ complex, lg K_1 = 6.46, lg K_2 = 5.72, lg K_3 = 4.86, lg K_4 = 3.42 and for the $Eu(C_{12}H_4O_3F_9)_n^{3-n}$ complex, lg K_1 = 6.03, lg K_2 = 5.62, and lg K_3 = 4.00, Yu. A. Fialkov, P. A. Yufa, A. G. Goryushko, N. K. Davidenko, L. M. Yagupol'skii (Zh. Org. Khim. **11** [1975] 1066/9; J. Org. Chem. [USSR] **11** [1975] 1054/7).

5.1.22.2 With 1-(3-Benzo[b]thienyl)-4,4,4-trifluoro-1,3-butanedione

$C_8H_5S{-}C(=O){-}CH_2{-}C(=O){-}CF_3$ (= $C_{12}H_7F_3O_2S$)

$Ce(C_{12}H_6F_3O_2S)_4$

The reagent reacts with cerium(IV) in slightly acidic media, yielding an orange-red complex which is extractable into benzene. The extracted complex exhibited a maximum at 424 nm in the electronic absorption spectrum, with a molar extinction coefficient of 5.51×10^3 l · mol^{-1} · cm^{-1}. The composition of the complex in solution was found to be 4:1 by a mole ratio method and confirmed by a Job's plot, W. J. Holland, A. E. Veel, J. Gerard (Mikrochim. Acta **1970** 297/300).

5.1.22.3 With 1,3-Bis(2-thienyl)-1,3-propanedione

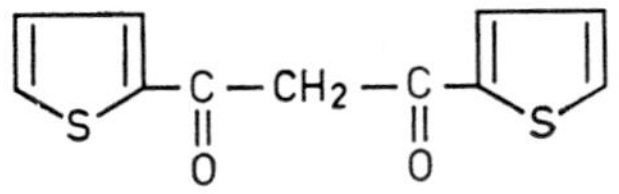

(= Dithenoylmethane = $C_{11}H_8O_2S_2$)

$M(C_{11}H_7O_2S_2)_3$

The europium, gadolinium, and terbium chelates were prepared by treating ethanolic solutions containing the respective rare earth chlorides and dithenoylmethane with piperidine [1, 2]. The holimum compound was obtained by adding a saturated solution of ethanolic ammonia to a solution containing dithenoylmethane and the hydrated rare earth nitrate in aqueous ethanol. The gadolinium chelate was characterized by its absorption (λ_{max} = 375 nm) and phosphorescence (λ_{max} = 513 nm) spectra [1], the europium and terbium compounds by their fluorescence spectra [2], and the holmium chelate by its infrared and electronic absorption spectra [3].

References to 5.1.22.3:

[1] W. F. Sager, N. Filipescu, F. A. Serafin (J. Phys. Chem. **69** [1965] 1092/100). — [2] N. Filipescu, W. F. Sager, F. A. Serafin (J. Phys. Chem. **68** [1964] 3324/46). — [3] D. Purushottam, V. Ramachandra Rao (Sci. Cult. [Calcutta] **36** [1970] 484).

5.1.23 Complexes with Selenophenyl Derivatives of 1,3-Diketones

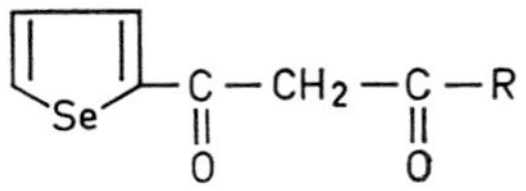

5.1.23.1 With 1-(2-Selenophenyl)-1,3-butanedione
($R=CH_3$; Selenophenoylacetone = $C_8H_8O_2Se$)
and Its 4,4,4-Trifluoro Derivative ($R=CF_3 = C_8H_5F_3O_2Se$)

$Nd(C_8H_7O_2Se)_n^{3-n}$ and $Nd(C_8H_4F_3O_2Se)_n^{3-n}$ Complexes (n = 1 to 3)

Formation in Solution. Formation constants for rare earth complexes derived from selenophenoylacetone and selenophenoyltrifluoroacetone were determined by a solvent extraction method at 25°C in aqueous media, I = 0.1 [1, 3].

complex	lg K_1	lg K_2	lg K_3	lg β_3	Ref.
$Nd(C_8H_7O_2Se)_n^{3-n}$	5.60	5.10	4.20	14.90	[1]
$Eu(C_8H_7O_2Se)_n^{3-n}$	6.24	6.05	5.59	17.88	[3]
$Nd(C_8H_4F_3O_2Se)_n^{3-n}$	5.00	4.55	3.50	13.05	[1]

Isolated Compounds. Byrke and coworkers [2] prepared a neodymium complex of composition $Nd(C_8H_7O_2Se)_3 \cdot H_2O$, by adding a saturated solution of ethanolic ammonia to a solution of neodymium nitrate and the ligand in aqueous ethanol. The compound was characterized by its infrared spectrum. Lebedev and coworkers [4] effected preparation of $Eu(C_8H_7O_2Se)_3$ and $Eu(C_8H_4F_3O_2Se)_3$ chelates according to a procedure described in [5]. The compounds were characterized by their emission spectra.

5.1.23.2 With 1-Phenyl-3-(2-selenophenyl)-1,3-propanedione
($R = C_6H_5$ = Selenophenoylbenzoylmethane = $C_{13}H_{10}O_2Se$)

$Eu(C_{13}H_9O_2S)_n^{3-n}$. The formation constants for the europium complexes (n = 1 to 3) in the aqueous phase were determined by a solvent extraction method at 25°C, I=0.1 (KNO_3): $\beta_1 = 3.15 \times 10^5$, $\beta_2 = 1.09 \times 10^{11}$, $\beta_3 = 1.19 \times 10^6$ [3].

$M(C_{13}H_9O_2S)_3 \cdot 2H_2O$. Byrke and coworkers [2] effected synthesis of the yttrium, praseodymium, and erbium compounds by adding saturated solutions of ethanolic ammonia to solutions of the respective hydrated rare earth nitrate and ligand in aqueous ethanol. The compounds were characterized by elemental analyses and their infrared spectra. Lebedev and coworkers [4] prepared the europium chelate according to the procedure described by Sevchenko, Trofimov [5] for the preparation of europium tris(benzoylacetonato) complexes.

5.1.23.3 With Other Selenophenyl Derivatives

(2-selenophenyl)–C(=O)–CH2–C(=O)–R (=HL)

no.	R =	formula of ligand	no.	R =	formula of ligand
1	$-C_2H_5$	$C_9H_{10}O_2Se$	6	3-pyridyl	$C_{12}H_9NO_2Se$
2	$-C_3H_7$	$C_{10}H_{12}O_2Se$			
3	p-tolyl ($-C_6H_4-CH_3$)	$C_{14}H_{12}O_2Se$	7	4-pyridyl	$C_{12}H_9NO_2Se$
4	2-furyl	$C_{11}H_8O_3Se$	8	2-thienyl	$C_{11}H_8O_2SSe$
5	2-pyridyl	$C_{12}H_9NO_2Se$	9	2-selenophenyl	$C_{11}H_8O_2Se_2$

Lebedev and coworkers [4] prepared a series of europium tris chelates (ligands see preceding table) of general composition, EuL_3, according to a procedure described by Sevchenko, Trofimov [5] for the preparation of europium tris(benzoylacetonato) complexes. The emission spectra of the compounds (crystalline samples at 77 K and propyl alcohol solutions) revealed three groups of bands, corresponding to the $^5D_0 \rightarrow {}^7F_0$ (5790 Å), $^5D_0 \rightarrow {}^7F_1$ (5880 to 5980 Å), and $^5D_0 \rightarrow {}^7F_2$ (6105 to 6235 Å) transitions. Faint transitions from $^5D_1$ to ^7F_n (n = 0 to 3) were also observed. The number of lines and band widths of particular transitions were dependent on the nature of the ligand [4].

For equilibrium constants of cerium(III) complexes with ligand 5 of preceding table, see [6].

References to 5.1.23:

[1] V. M. Peshkova, I. P. Efimov, N. N. Magdesieva (Zh. Analit. Khim. **21** [1966] 499/501; Russ. J. Anal. Chem. **21** [1966] 446/8). — [2] A. I. Byrke, N. N. Magdesieva, L. I. Martynenko, V. I. Spitsyn (Zh. Neorgan. Khim. **12** [1967] 666/71; Russ. J. Inorg. Chem. **12** [1967] 348/51). — [3] A. S. Berlyand, A. I. Byrke, L. I. Martynenko (Zh. Neorgan. Khim. **13** [1968] 2106/10; Russ. J. Inorg. Chem. **13** [1968] 1089/91). — [4] O. L. Lebedev, N. N. Magdesieva, A. V. Michurina, Kh. A. Ainitdinov, Yu. K. Yur'ev (Zh. Strukt. Khim. **7** [1966] 521/5; J. Struct. Chem. [USSR] **7** [1966] 493/6). — [5] A. N. Sevchenko, A. K. Trofimov (Zh. Eksperim. Teor. Fiz. **21** [1951] 220/9; C.A. **1952** 2403).

[6] I. P. Efimov, L. S. Voronets, L. G. Makarova, V. M. Peshkova (Vestn. Mosk. Univ. Khim. **24** No. 4 [1969] 121/3; Mosc. Univ. Chem. Bull. **24** No. 4 [1969] 83/4).

5.1.24 With Formyl- and Oxoalkyl Derivatives of Cyclohexanone

5.1.24.1 With 2-Formylcyclohexanone (= 2-(Hydroxymethylene)cyclohexanone = $C_7H_{10}O_2$) and Its Alkyl Derivatives

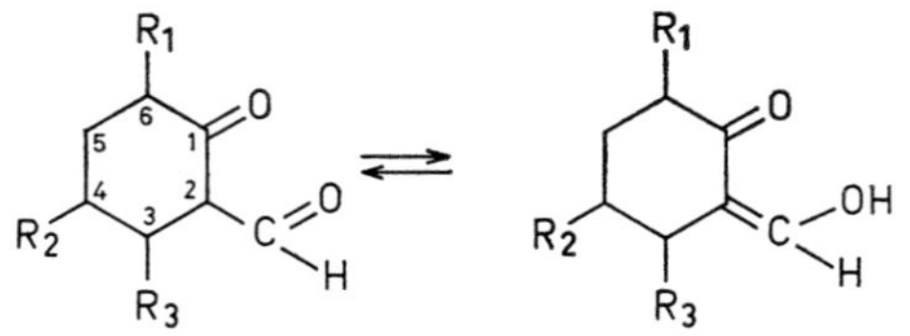

Each compound that is listed in Table 5/57 was prepared by adding an aqueous solution (15 ml) of the rare earth acetate (0.0017 mol) to a suspension of 0.0025 mol of bis(2-formyl)cyclohexanonato zinc(II) or its alkylated derivatives in 100 ml of acetone. (The zinc complex was prepared by adding an aqueous-alcoholic solution of zinc acetate (0.01 mol) to a solution of the ligand in the form of its potassium salt in aqueous ethanol. The resulting precipitate was collected by filtration, washed with portions of ether and water, and dried under vacuum.) The mixture was stirred for 2 h then extracted with petroleum ether. The extractant solution was dried with molecular sieves, the ether was removed, and the residual product was dried under vacuum.

The complexes were obtained as yellow powders, readily soluble in common organic solvents. They are hygroscopic, decomposing in air after a long exposure. The infrared spectra of the chelates (mineral oil mulls) revealed a strong, characteristic band in the region 1610 to 1620 cm^{-1}, corresponding to the stretching vibration of the chelated carbonyl group. The ultraviolet absorption spectra of the compounds in chloroform and alcohol were characterized by a strong band at 305 nm (lg ε = 4.4), which was assigned to a $\pi\rightarrow\pi^*$ transition [1]. The IR data indicate, that complexation with the metal ions distortes the original geometry of the ligand. Force constants of the metal-O bonds are given in [2].

Table 5/57
Rare Earth Complexes Derived from Alkyl Derivatives of 2-Formylcyclohexanone.

no.	R_1 =	R_2 =	R_3 =	formula of ligand	formula of complex	ν(C=O) in cm^{-1}
1	H	H	H	$C_7H_{10}O_2$	$Pr(C_7H_9O_2)_3$	—
2	H	H	H	$C_7H_{10}O_2$	$Eu(C_7H_9O_2)_3$	1600
3	CH_3	H	H	$C_8H_{12}O_2$	$Eu(C_8H_{11}O_2)_3$	1620
4	n-C_4H_9	H	H	$C_{11}H_{18}O_2$	$Eu(C_{11}H_{17}O_2)_3$	1620
5	H	tert-C_4H_9	H	$C_{11}H_{18}O_2$	$Eu(C_{11}H_{17}O_2)_3$	1610
6	iso-C_3H_7	H	CH_3	$C_{11}H_{18}O_2$	$Eu(C_{11}H_{17}O_2)_3$	1600

5.1.24.2 With 2-Butyrylcyclohexanone (= 2-(1-Oxobutyl)-cyclohexanone = $C_{10}H_{16}O_2$)

$M(C_{10}H_{15}O_2)_n^{3-n}$ Complexes (n = 1 to 3). **Formation in Solution**

Formation constants for 2-butyrylcyclohexanonato rare earth chelates were determined potentiometrically (glass electrode) at 30°C in 3:1 (v/v) acetone/water, I = 0.1 ($NaClO_4$) [1].

metal ion	Y^{3+}	La^{3+}	Pr^{3+}	Nd^{3+}	Sm^{3+}	Eu^{3+}	Gd^{3+}	Dy^{3+}	Er^{3+}	Yb^{3+}
lg K_1	9.13	8.52	9.20	9.43	9.69	9.92	9.42	9.88	10.72	10.81
lg K_2	8.02	7.33	8.53	8.53	8.79	8.92	8.78	9.02	8.95	9.76
lg K_3			8.02	8.59	8.44	8.09	8.45	8.63	9.03	9.61
lg β_3			25.75	26.55	26.92	26.93	26.65	27.53	26.70	30.18

Isolated Compounds $M(C_{10}H_{15}O_2)_2(OH) \cdot H_2O$

Each compound (M = La, Pr, Nd, Sm, Eu, Gd) was prepared by adding aqueous ammonia, dropwise, with vigorous stirring, to an alcoholic solution of the rare earth nitrate and ligand until slight precipitation or turbidity appeared. The solution was filtered, the filtrate evaporated to dryness, and the products obtained as residues, were analyzed [3].

5.1.24.3 With 2-Isovalerylcyclohexanone
(= 2-(3-Methyl-1-oxobutyl)-cyclohexanone = $C_{11}H_{18}O_2$)

$M(C_{11}H_{17}O_2)_n^{3-n}$ Complexes (n = 1 to 3). **Formation in Solution**

Formation constants for 2-isovalerylcyclohexanonato rare earth chelates were determined potentiometrically (glass electrode) at 30°C in 3:1 (v/v) acetone/water, I = 0.1 ($NaClO_4$). The value of lg K_1 increases with decreasing ionic radius up to europium, decreases at gadolinium, then increases for the heavier members of the series [4].

metal ion	Y^{3+}	La^{3+}	Pr^{3+}	Nd^{3+}	Sm^{3+}	Eu^{3+}	Gd^{3+}	Dy^{3+}	Er^{3+}	Yb^{3+}
lg K_1	9.56	8.95	9.43	9.54	9.63	9.74	9.49	9.43	10.04	10.18
lg K_2	8.60	7.49	8.51	8.55	9.08	9.11	8.93	9.01	9.79	9.34
lg K_3			7.62	7.93	8.17	8.76	8.09	8.30	8.41	8.90

5.1.24.4 With 2-Butyryl-6-methylcyclohexanone
(= 6-Methyl-2-(1-oxobutyl)cyclohexanone = $C_{11}H_{18}O_2$)

$M(C_{11}H_{17}O_2)_n^{3-n}$ Ions (n = 1, 2). **Formation in Solution**

Formation constants for the 1:1 and 2:1 (ligand : metal) rare earth chelates derived from 2-butyryl-6-methylcyclohexanone were determined potentiometrically (glass electrode) at 30°C in 3:1 (v/v) acetone/water I = 0.1 ($NaClO_4$). The value of lg K_1 increases with decreasing ionic radius up to europium, decreases at gadolinium, then increases across the heavier end of the lanthanide series [4].

metal ion	Y^{3+}	La^{3+}	Pr^{3+}	Nd^{3+}	Sm^{3+}	Eu^{3+}	Gd^{3+}	Dy^{3+}	Er^{3+}	Yb^{3+}
lg K_1	10.58	10.31	10.40	10.69	10.75	10.58	10.32	10.62	10.70	10.74
lg K_2	9.90	9.58	9.30	10.08	10.26	10.24	9.94	10.07	10.11	9.84

References to 5.1.24:

[1] V. M. Potapov, V. G. Bakhmutskaya, I. G. Il'ina, E. G. Rukhadze, G. I. Vinnik (Zh. Obshch. Khim. **46** [1976] 2117/21; J. Gen. Chem. [USSR] **46** [1976] 2037/40). — [2] E. G. Rukhadze, M. A. Salimov, V. G. Bakhmutskaya, V. M. Potapov (Str. Svoistra Primen. Beta Diketonatov Metal. Mater 3rd Vses. Semin., Moscow 1977 [1978], pp. 144/50 from C.A. **91** [1979] No. 174321). — [3] N. K. Dutt, S. Sanyal (J. Inorg. Nucl. Chem. **34** [1972] 651/5). — [4] N. K. Dutt, S. Sanyal, U. U. M. Sharma (J. Inorg. Nucl. Chem. **34** [1972] 2261/4).

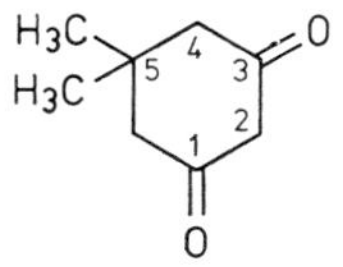

5.1.25 With 5,5-Dimethyl-1,3-cyclohexanedione
(= Dimedone = $C_8H_{12}O_2$)

$M(C_8H_{11}O_2)_n^{3-n}$ Ions. (n = 1, 2). **Formation in Solution**

Formation constants for the 1:1 and 2:1 rare earth chelates with dimedone were determined potentiometrically (glass electrode) at 30°C in 3:1 (v/v) acetone/water, I = 0.1. The values of lg K_1 and lg K_2 were considered to be low compared to the basicity of the ligand (pKa = 7.27). This observation and the small ratio of the two successive formation constants were attributed to monodentate ligation of dimedone [1].

metal ion . .	Y^{3+}	La^{3+}	Pr^{3+}	Nd^{3+}	Sm^{3+}
lg K_1	2.78	2.48	2.64	2.70	2.86
lg K_2	2.59	2.33	2.55	2.40	2.59

Isolated Compounds $Sm(C_8H_{11}O_2)_n(C_3H_6O)_{3-n}$ (n = 1 to 3).

The 2-propanolates were prepared by treating samarium isopropoxide with a stoichiometric amount of dimedone [2].

References to 5.1.25:

[1] N. K. Dutt, U. U. M. Sarma (J. Inorg. Nucl. Chem. **37** [1975] 606/7). — [2] M. Hasan, S. N. Misra, R. N. Kapoor (Vijnana Parishad Anusandhan Patrika **13** [1970] 31/6 from C.A. **74** [1971] No. 150591).

5.1.26 Complexes With Optically Active β-Diketones

Introduction. Chiral tris(β-diketonato)europium(III) chelates have been synthesized and tested for their effectiveness as NMR shift reagents for the direct determination of enantiomeric compositions [1 to 4]. Judged by the dual criteria of generality of application as reagents for determining enantiomeric compositions as well as ease of preparation, Whitesides and coworkers [1] concluded that the most useful of these chiral shift reagents are tris(*d,d*-dicampholylmethanato)europium(III), tris(3-trifluoroacetyl-*d*-nopinonato)europium(III), and tris(3-trifluoroacetyl-*d*-camphorato) europium(III) (see below). The first compound was found to be the most effective reagent for the resolution of enantiotropic resonances, whereas the latter two compounds are considerably less effective, but are more easily prepared [1]. Whitesides and coworkers [1] noted that it was not possible to predict in any detail the influence of a particular shift reagent on a mixture of enantiomers and that several chiral shift reagents generally have to be tested, varying sample concentration and temperature, in order to achieve useful resolution among the signals of enantiotropic protons in a complex. Fraser, Petit, Saunders [5] demonstrated the effectiveness of tris(3-heptafluorooxobutyl-(*d*)-camphorato)europium(III) for separating NMR signals of the enantiomers in racemic mixtures of alcohols, sulfoxides, an expoxide, and an aldehyde.

5.1.26.1 With 3-Trifluoroacetyl-*d*-camphor

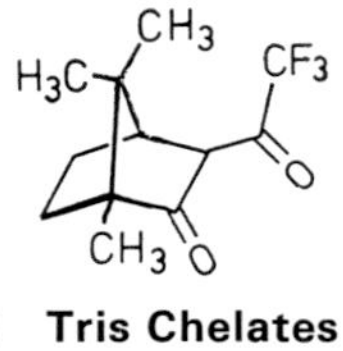

3-Trifluoromethylhydroxymethylene-*d*-camphor (= $C_{12}H_{15}F_3O_2$)

5.1.26.1.1 Tris Chelates $M(C_{12}H_{14}F_3O_2)\cdot nH_2O$ (n = 0, 1)

Goering, Eikenberry, Koermer [4] effected synthesis of the anhydrous europium chelate by treating an aqueous ethanolic solution of europium chloride and 3-trifluoroacetyl-*d*-camphor with base. The chelate was isolated as a resinous precipitate by adding water to the reaction mixture, extracting with pentane, washing the pentane layer with water, and then allowing the organic layer to evaporate to dryness. The residual product was dehydrated under vacuum. The chiral compound was shown to be useful as an NMR shift reagent for direct determination of enantiomeric composition.

Schurig [7] effected preparation of a series of chelates (M = Pr, Sm, Eu, Dy, Ho, Yb, n = 0) by the quantitative homogeneous exchange reaction of rare earth nitrates with the barium salt of 3-trifluoromethylhydroxymethylene-*d*-camphor in ethanol, e.g., $2\,M(NO_3)_3\cdot nH_2O + 3\,Ba(C_{12}H_{14}F_3O_2)_2 \rightleftharpoons 2\,M(C_{12}H_{14}F_3O_2)_3 + 3\,Ba(NO_3)_2 + 2n\,H_2O$, according to the following procedure: An ethanolic solution (25 ml) of the rare earth nitrate hydrate (2 mmol) was added, with stirring, to a warm solution of $Ba(C_{12}H_{14}F_3O_2)_2$ in 95% ethanol, whereby barium nitrate precipitated. (The preparation of $Ba(C_{12}H_{14}F_3O_2)_2$ is reported in [8].) The reaction mixture was stirred for 15 min and then filtered, and the solvent removed under reduced pressure with a rotary evaporator. The residue was extracted with warm n-hexane, the extractant solution filtered until clear, and the solvent subsequently removed under reduced pressure. The vacuum was maintained for 90 min while the sample was heated at 90°C. The resulting glassy chelates were dried under vacuum over phosphorus(V) oxide. The samples were purified further by sublimation at 180 to 200°C under vacuum (4×10^{-3} Torr) [7].

Kalashnikov and coworkers [9, 10] obtained a series of monohydrated chelates (M = Y, La to Lu except for Pm) and the anhydrous scandium chelate according to a similar procedure as described above, except that acetone was used as the solvent medium for the exchange reactions with $Ba(C_{12}H_{14}F_3O_2)_2$. The compounds thus obtained are insoluble in water and readily soluble in alcohols, ethers, esters, ketones, hydrocarbons, and amines [9,10]. The ν(OH) stretching mode due to water was observed in the region 3200 to 3500 cm^{-1} of the infrared spectra of the chelates [9]. The chelates decomposed in the temperature range 190 to 220°C [6]. The scandium chelate melted at 141°C [10]. The heat of vaporization of the latter compound is 31.30 kcal/mol [10].

Infrared spectra for the series of chelates are reported in [11]. Brittain, Richardson [12] investigated the total emission, and circularly polarized emission spectra of the europium chelate, in powder form, at liquid nitrogen temperature, and in a variety of pure solvent and mixed solvent systems at room temperature. The detailed features of the spectra were related to structural characteristics of the chelate system and to the nature of chelate-solvent interactions. The emission anisotropy factor, g_{em}, was found to be a sensitive probe for studying chelate-solvent adduct formation and for deducing information about the relative coordinating strengths of various solvent molecules [12]. Circularly polarized emission studies of the europium chelate dissolved in pure dimethyl sulfoxide and in 2:1 1-phenylethylamine (optically active): dimethyl sulfoxide solvent mixtures were also reported. The solvent induced circularly polarized emission intensities and the magnitudes of the observed emission anisotropy factors (g_{em}) suggested stereospecifity and chiral discrimination in the interaction between the chiral chelate system and the optically active solvent molecules [13].

5.1.26.1.2 Lewis Base Adducts of the Tris Chelates

With Primary Amines RNH_2

$Eu(C_{12}H_{14}F_3O_2)_3 \cdot 2RNH_2$. Formation in Solution. The formation constants $K_n = [M(C_{12}H_{14}F_3O_2)_3 \cdot nRNH_2] \cdot [M(C_{12}H_{14}F_3O_2)_3 \cdot (n-1)RNH_2]^{-1} \cdot [RNH_2]^{-1}$ (n = 1, 2), were determined spectrophotometrically (emission titration) for several adducts of the tris chelate of europium with primary amines:

substrate	formula	$K_1 \cdot 10^{-1}$ *)	$K_2 \cdot 10^{-3}$ *)	$\beta_2 \cdot 10^{-4}$ **)
propylamine	$C_3H_7NH_2$	6.68	1.35	9.01
isopropylamine	$C_3H_7NH_2$	5.88	1.03	6.05
butylamine	$C_4H_9NH_2$	6.20	1.36	8.43
sec-butylamine	$C_4H_9NH_2$	5.47	1.07	5.83
tert-butylamine	$C_4H_9NH_2$	4.43	1.09	4.82

*) The units of K_1 and K_2 are $l \cdot mol^{-1}$, and each is associated with an error of ±0.05. — **) The units of β_2 are $l^2 \cdot mol^{-2}$, and each constant carries an error of ±0.1.

Both linear amines have values of K_2 that are essentially 1.35×10^3, whereas the branched amines have values of K_2 that average to 1.06×10^3. This grouping was attributed to the greater steric hindrance provided by the branched alkyl groups, relative to the linear groups toward binding the second substrate molecule. The binding of the first molecule of substrate also displayed a dependence on the steric nature of the amine, with nonbranched amine molecules binding the tris chelate with the greatest ease, those with one branch finding greater difficulty, and that with the tert-butyl group demonstrating the greatest difficulty in binding the chelate [14].

With Dimethylformamide ($= C_3H_7NO = DMF$)

$\mathbf{Pr_2(C_{12}H_{14}F_3O_2)_6(DMF)_3}$. The adduct was prepared by saturating hot dimethylformamide with the unsolvated tris chelate and allowing the solution to cool slowly to room temperature. Slow isothermal evaporation of a hexane solution of the adduct yielded air stable, light green crystals of the product.

The crystal and molecular structures of the adduct were determined by single-crystal X-ray analysis ($R = 0.0709$ for 8779 reflections). The compound crystallizes in the monoclinic system, space group $P2_1$-C_2^2 (No. 4) in a unit cell with dimensions $a = 17.930$, $b = 19.403$, $c = 13.049$ Å, and $\beta = 91°5'$ ($\rho_{calc} = 1.45$ g/cm^3, $\rho_{obs} = 1.44$ g/cm^3). The oxygen atoms of the dimethylformamide ligands are equally shared by the two praseodymium atoms and act as bridges between the two $Pr(C_{12}H_{14}F_3O_2)_3$ moieties. Each praseodymium atom is nine-coordinate, with the oxygen atoms occupying the vertices of a distorted monocapped square antiprism. The average Pr-O (diketonate) and Pr-O (DMF) bond distances are 2.46 and 2.60 Å, respectively, and the Pr-Pr distance is 4.078 Å. The average Pr-O (DMF) -Pr bond angle is 103.6°. No short interligand approach distances were observed in the structure, indicating that steric crowding is not a serious problem [15].

With Trialkylphosphates $(RO)_3PO$ ($R = CH_3$, C_2H_5, C_4H_9, C_6H_5)

Luminescence titrations were used to study the adduct formation between $Eu(C_{11}H_{15}O_2F_3)_3$ and the phosphate esters. The formation constants of the adducts (lg $K = 1.861$, 2.130, and 2.270, respectively) were calculated from the enhancement of the $^5D_0$-$^7F_2$ Eu^{III} transition [16].

5.1.26.2 With 3-(2,2,3,3,4,4,4-Heptafluoro-1-oxobutyl)-*d*-camphor

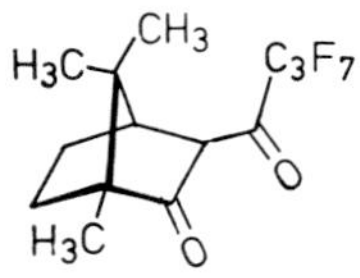

H3C CH3 C3F7 H3C O O

($= C_{14}H_{15}F_7O_2$)

$\mathbf{Eu(C_{14}H_{14}F_7O_2)_3}$ was prepared according to the procedure described by K. J. Eisentraut, R. E. Sievers (J. Am. Chem. Soc. **87** [1965] 5254/6) for preparation of the tris(2,2,6,6-tetramethyl-3,5-heptanedionato)rare earth(III) chelates (see p. 125) [5]. The formation of adducts between $Eu(C_{14}H_{14}F_7O_2)_3$ and trialkyl- or triphenyl phosphates $(RO)_3PO$ with $R = CH_3$, C_2H_5, C_4H_9, C_6H_5 was studied by luminescence titrations. From the enhancement of the $^5D_0$-$^7F_2$ Eu^{III} transition the following formation constants were calculated: lg $K = 2.881$ (for $R = CH_3$), 3.246 (for $R = C_2H_5$), 3.212 (for $R = C_4H_9$), 2.720 (for $R = C_5H_6$), respectively [16].

5.1.26.3 With 3-(2,2-Dimethyl-1-oxopropyl)-*d*-camphor

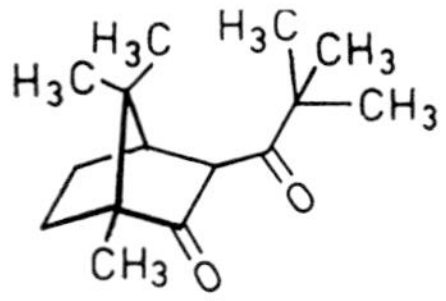

H3C H3C H3C CH3 CH3 CH3 O O

($= C_{15}H_{24}O_2$)

$\mathbf{M(C_{15}H_{23}O_2)_3}$. The europium compound was prepared by rapidly adding, with vigorous stirring, a solution (15 ml) of europium chloride hexahydrate (2.74 mmol) in 50% aqueous ethanol to a solution obtained by mixing a 95% ethanolic solution (10 ml) of 3-(2,2-dimethyl-1-oxopropyl)-

d-camphor (8.20 mmol) with a solution of sodium hydroxide (10 mmol) in 15 ml of 50% aqueous ethanol. The reaction was carried out under a nitrogen atmosphere. The product immediately precipitated as a light yellow solid. The reaction mixture was stirred for 30 min and then diluted with 30 ml of ice water. The precipitate was collected by filtration, dried under vacuum (0.1 Torr) at room temperature, and then broken into a fine powder. The substance was purified by dissolving it in absolute ethanol, separating the insoluble solids by centrifugation, and effecting precipitation of the product by adding ice water to the resulting solution. The precipitate was collected by filtration, dried under vacuum (0.1 Torr) at room temperature for 12 h, and then further purified by fractional molecular distillation at 134 to 140°C at 0.01 Torr [1, 2]. The yellow-colored solid melted at 131 to 134°C (discolored at 108 to 112°C) and exhibited an optical rotation of $[\alpha]_D^{15} = +38.2°$ [1]. The praseodymium chelate was obtained as a light green powder following the same procedure [1]. The latter compound softened at 145 to 150°C and became a clear liquid at 230 to 243°C with $[\alpha]_D^{25} = +46.0°$ [1].

5.1.26.4 With 3-(*d*-Fencholyl)-*d*-camphor

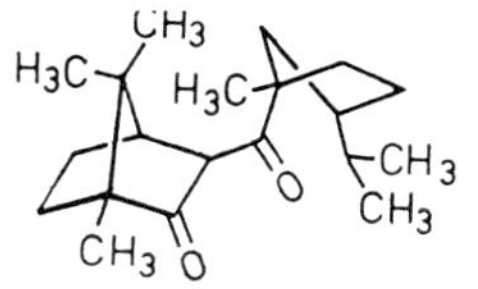

$(= C_{20}H_{32}O_2)$

$Eu(C_{20}H_{31}O_2)_3$ was prepared from 2.5 g (8.2 mmol) of 3-(*d*-fencholyl)-*d*-camphor according to the procedure described for preparation of the 3-(2,2-dimethyl-1-oxopropyl)-*d*-camphorato chelates [1, 3]. Purification by molecular distillation (150 to 200°C at 0.01 Torr) yielded yellow needles that could be powdered. The compound melted at 113 to 115°C (softened at 80 to 90°C) and exhibited an optical rotation of $[\alpha]_D^{25} = +72.4°$ [1].

5.1.26.5 With 3-(*l*-Fencholyl)-*d*-camphor

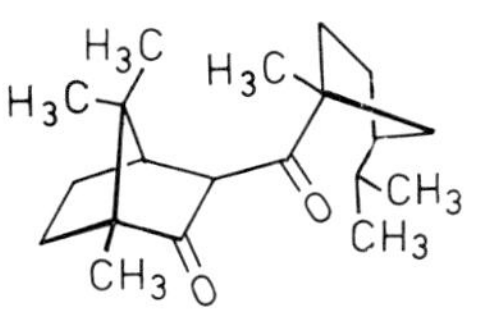

$(= C_{20}H_{32}O_2)$

$Eu(C_{20}H_{31}O_2)_3$ was prepared from 2.5 g (8.2 mmol) of 3-(*l*-fencholyl)-*d*-camphor according to the procedure described for preparation of the 3-(2,2-dimethyl-1-oxopropyl)-*d*-camphorato chelates [2, 3]. The product was obtained as a yellow glass which melted at 112 to 114°C (softened at 80°C) and exhibited an optical rotation of $[\alpha]_D^{25} = 57.4°$ [1].

5.1.26.6 With *d*-Campholyltrifluoroacetone

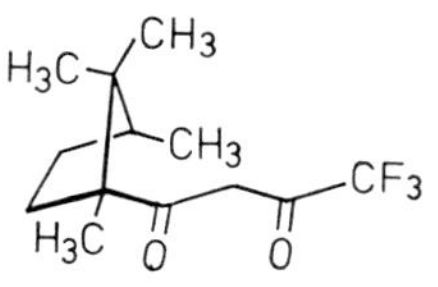

$(= C_{13}H_{19}F_3O_2)$

$Eu(C_{13}H_{18}F_3O_2)_3$ was prepared from 2.64 g (10.0 mmol) of *d*-campholyltrifluoroacetylmethane according to the procedure described for preparation of the 3-trifluoroacetyl-*d*-nopinato chelate. The crude product was purified by fractional molecular distillation (170 to 180°C at 0.004 Torr) to yield the product in the form of an amorphous yellow solid (m.p. ≈ 100°C), $[\alpha]_D^{24} = +36.8°$ [1].

5.1.26.7 With *l*-Fencholyltrifluoroacetone

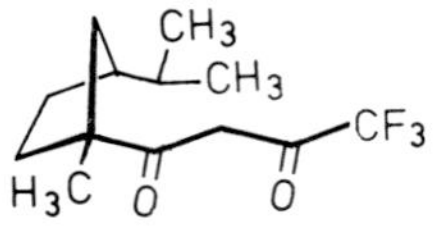

$(=C_{13}H_{19}F_3O_2)$

$Eu(C_{13}H_{18}F_3O_2)_3$ was prepared from 1.32 g (5.0 mmol) of *l*-fencholyltrifluoroacetone using the procedure described for preparation of tris(3-trifluoroacetyl-*d*-nopinato)europium(III). The product was obtained as a viscous yellow glass that softened below ambient temperature ($[\alpha]_D^{24}$ = −9.5°). Attempts to purify the compound by thin layer chromatography, molecular distillation, or sublimation resulted in decomposition of the chelate [1].

5.1.26.8 With *d,d*-Dicampholylmethane

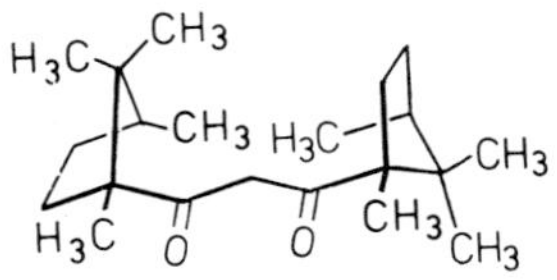

$(=C_{21}H_{36}O_2)$

$M(C_{21}H_{35}O_2)_3$. The europium chelate was prepared by adding a methanolic solution of europium chloride hexahydrate (0.27 mol) to a methanolic solution of *d,d*-dicampholylmethane (0.081 mol) and sodium methoxide (0.081 mol). The reaction was carried out under a nitrogen atmosphere. The resulting suspension, containing a cream-white precipitate, was stirred vigorously at 35 to 40°C for 2 h, cooled to 0°C, and filtered, yielding brittle beige lumps and a cream-colored amorphous solid. The product was dissolved in pentane, the solution filtered, and the solvent allowed to evaporate. The residue was heated at 100°C under vacuum (0.1 Torr) for 36 h to yield a white powder (m.p. = 222.0 to 227.5°C, $[\alpha]_D^{25}$ = +28.6°). The praseodymium (m.p. = ≈80°C, $[\alpha]_D^{24}$ = +54.5°) and holmium (m.p. = ≈250°C, $[\alpha]_D^{24}$ = +40.2°) compounds were similarly obtained as a mint green solid and white amorphous solid, respectively. Samples for analysis were purified by fractional molecular distillation (190 to 210°C at 0.003 Torr for M = Pr and 250 to 270°C at 0.005 Torr for M = Ho) [1].

5.1.26.9 With *d,l*-Dicampholylmethane

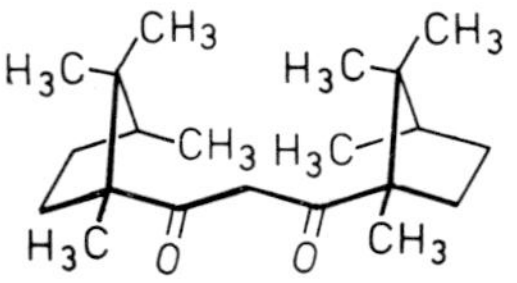

$(=C_{21}H_{36}O_2)$

$Eu(C_{21}H_{35}O_2)_3$ was prepared from 0.55 g (1.73 mmol) of *d,l*-dicampholylmethane and 0.21 g (0.58 mmol) of europium chloride hexahydrate according to the procedure described for preparation of the analogous *d,d*-dicampholylmethanato chelates. The compound melted at 222.5 to 225.0°C ($[\alpha]_D^{25}$ = 0.0) [1].

5.1.26.10 With *l,l*-Difencholylmethane

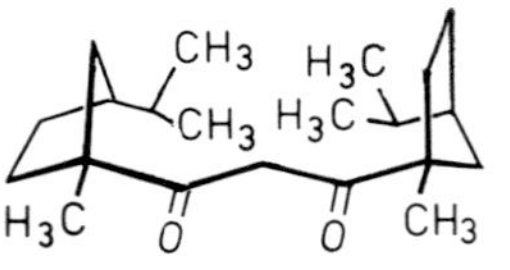

$(=C_{21}H_{36}O_2)$

$Eu(C_{21}H_{35}O_2)_3$ was prepared from 2.63 g (8.24 mmol) of *l,l*-difencholylmethane using a procedure similar to that described for preparation of the *d,d*-dicampholylmethanato chelates but with the following modifications: After stirring the reaction mixture was diluted with 20 ml of ice water and filtered at 0°C. The gummy residue that was collected was dried at 20°C under vacuum (0.1 Torr) for 6 h and then extracted with ether. The insoluble residue was removed by centrifugation. After removing the ether at reduced pressure, the resulting yellow, oily crude product was purified by fractional distillation at 200 to 250°C under vacuum (0.01 Torr) to give a light yellow glass that softened below 25°C. An analytical sample was further purified by fractional molecular distillation (180 to 250°C at 0.01 Torr). The optical rotation, $[\alpha]_D^{25}$, of the compound was −18.8° [1].

5.1.26.11 With *d,l*-Difencholylmethane

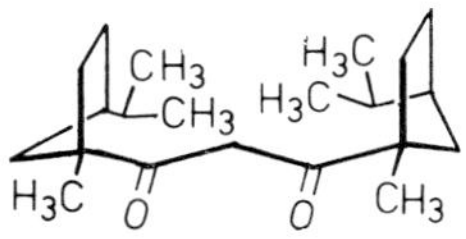

($=C_{21}H_{36}O_2$)

$Eu(C_{21}H_{35}O_2)_3$ was prepared using a procedure analogous to that described for the *l,l*-difencholylmethanato chelate. The product also had physical properties similar to those of the *l,l*-difencholylmethanato chelate ($[\alpha]_D^{25} = -5.2°$) [1].

5.1.26.12 With *d*-Campholyl-*d*-fencholylmethane

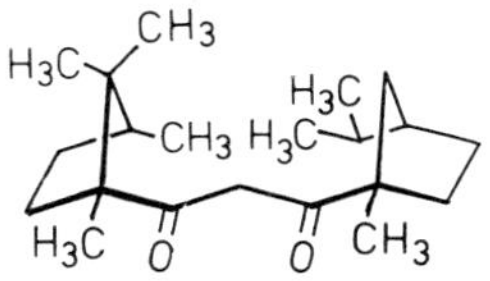

($=C_{21}H_{36}O_2$)

$Eu(C_{21}H_{35}O_2)_3$ was prepared according to the procedure described for preparation of the analogous *l,l*-difencholylmethanato chelate [1, 2]. The product was obtained as a viscous yellow glass ($[\alpha]_D^{25} = +33.6°$) [1].

5.1.26.13 With *d*-Campholyl-*l*-fencholylmethane

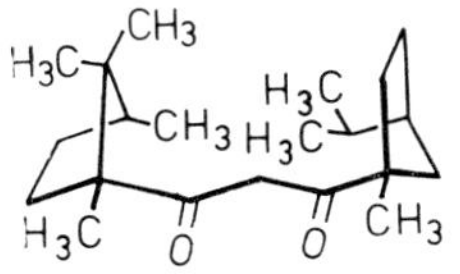

($=C_{21}H_{36}O_2$)

$Eu(C_{21}H_{35}O_2)_3$ was prepared from 0.874 g (2.73 mol) of *d*-campholyl-*l*-fencholylmethane using a procedure similar to that described on p. 244 for preparation of the *d,d*-dicampholylmethanato chelates, with the following modifications: After the reaction mixture was stirred for 1 h, 200 ml of water was added, and the product was extracted from the resulting oily mixture with pentane. The organic layer was separated and washed with water, and the solvent was removed. The residue was heated at 100°C under vacuum (0.1 Torr) for 36 h, yielding a viscous yellow glass ($[\alpha]_D^{25} = +27.8°$). A pure analytical sample was obtained by fractional molecular distillation (200 to 250°C at 0.01 Torr) [1].

5.1.26.14 With *l*-Campholyl-*l*-fencholylmethane

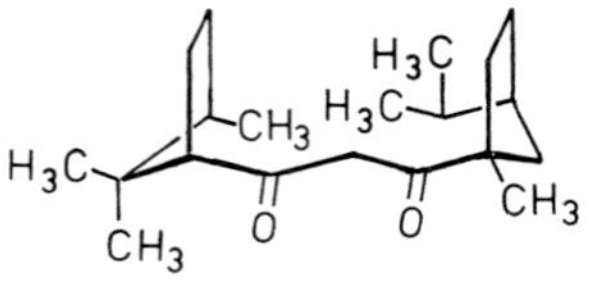

($=C_{21}H_{36}O_2$)

$Eu(C_{21}H_{35}O_2)_3$ was obtained in the form of a viscous yellow glass according to the procedure described for preparation of the analogous *d*-campholyl-*l*-fencholylmethanato chelate ($[\alpha]_D^{20}$ = −39.2°) [1].

5.1.26.15 With 2-Formyl-*d*- and 2-Formyl-*l*-menthone

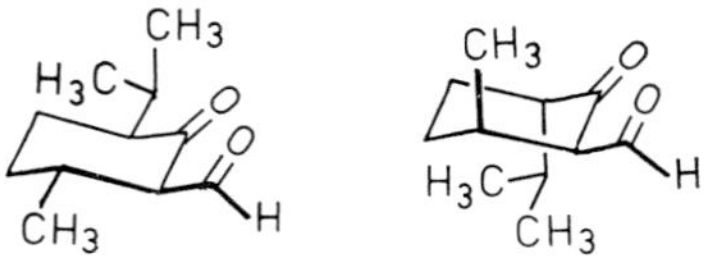

($=C_{11}H_{18}O_2$)

$Eu(C_{11}H_{17}O_2)_3$. The synthesis of tris(2-formylmenthonato)europium(III) was effected by the exchange reaction between bis(2-formylmenthonato)zinc(II) and europium acetate (solvent not specified) [4].

5.1.26.16 With 2-Trifluoroacetyl-*l*-menthone

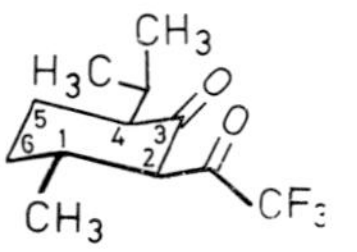

($=C_{12}H_{17}F_3O_2$)

$Eu(C_{12}H_{16}F_3O_2)_3$ was prepared from 2.40 g (10.0 mmol) of 2-trifluoroacetyl-*l*-menthone according to the procedure described below for preparation of the tris(3-trifluoro-acetyl-*d*-nopinato)-europium(III) chelate. The product was isolated as a yellow glass that softened below ambient temperature ($[\alpha]_D^{24}$ = −131°). Attempts to purify the compound by thin layer chromatography, molecular distillation, or sublimation resulted in decomposition of the chelate [1].

5.1.26.17 With 3-Trifluoroacetyl-*d*-nopinone

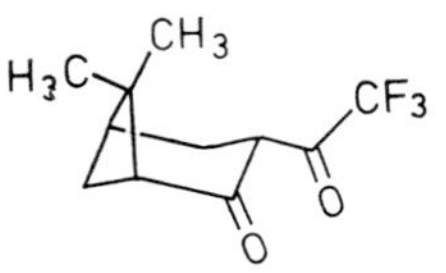

($=C_{11}H_{13}F_3O_2$)

$Eu(C_{11}H_{12}F_3O_2)_3$ was prepared from 17.5 g (0.075 mol) of 3-trifluoroacetyl-*d*-nopinone according to the procedure described for preparation of the *d,d*-dicampholylmethanato chelates, with the following modifications. After stirring the reaction mixture for 1 h, 200 ml of water was added, and the product was extracted from the resulting oily mixture with three 150-ml portions of pentane. The combined organic layers were washed with water, the solvent evaporated, and the residue dried under vacuum (0.1 Torr) at 100°C for 36 h. The bright yellow amorphous solid obtained melted at approximately 80 to 100°C and exhibited an optical rotation of $[\alpha]_D^{24}$ = −67.2°. Attempts to purify the compound by thin layer chromatography, molecular distillation, or sublimation resulted in decomposition of the chelate [1].

5.1.26.18 With 3-Trifluoroacetyl-*d*-(8-lanosten-3-one)

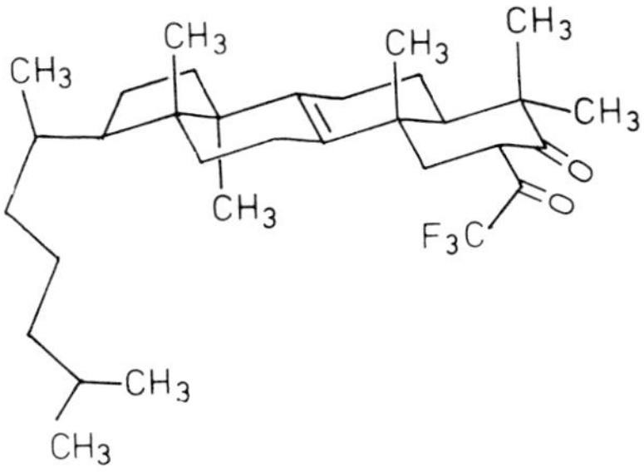

(= $C_{32}H_{49}F_3O_2$)

Eu($C_{32}H_{48}F_3O_2$)$_3$ was prepared by the exchange reaction between the corresponding barium(II) complex and europium chloride hexahydrate in ethanol. The reaction mixture was heated to 70°C and vigorously stirred for 1 h to give a pale yellow solution and a white precipitate. The mixture was cooled to 0°C, filtered to remove the precipitated barium chloride, diluted with water, and extracted with several portions of pentane. The combined pentane extracts were washed once with water. The solvent was removed, and the residue was heated at 100°C under vacuum (0.01 Torr) for 24 h. The bright yellow solid obtained melted at ≈ 220°C [1]. The barium chelate used in the exchange reaction was prepared by shaking an aqueous barium chloride solution with an ether solution of 2-trifluoroacetyl-*d*-dihydrolanosterone and then storing the mixture overnight at 0°C. The resulting precipitate was collected by filtration and washed successively with portions of water and methanol [1].

References to 5.1.26:

[1] M. D. McCreary, D. W. Lewis, D. L. Wernick, G. M. Whitesides (J. Am. Chem. Soc. **96** [1974] 1038/54). — [2] G. M. Whitesides, D. W. Lewis (J. Am. Chem. Soc. **92** [1970] 6979/80). — [3] G. M. Whitesides, D. W. Lewis (J. Am. Chem. Soc. **93** [1971] 5914/6). — [4] H. L. Goering, J. N. Eikenberry, G. S. Koermer (J. Am. Chem. Soc. **93** [1971] 5913/4). — [5] R. R. Fraser, M. A. Petit, J. K. Saunders (Chem. Commun. **1971** 1450/1).

[6] V. M. Potapov, V. G. Bakhmutskaya, I. G. Il'ina, G. I. Vinnik, E. G. Rukhadze (Zh. Obshch. Khim. **45** [1975] 2104/5; J. Gen. Chem. [USSR] **45** [1975] 2071). — [7] V. Schurig (Tetrahedron Letters [1972] 3297/300). — [8] V. Schurig (Inorg. Chem. **11** [1972] 736/8). — [9] S. M. Kalashnikov, L. K. Tarasov, V. I. Spitsyn (Zh. Neorgan. Khim. **20** [1975] 3132/3; Russ. J. Inorg. Chem. **20** [1975] 1733/4). — [10] L. K. Tarasov, S. M. Kalashnikov, V. I. Spitsyn (Zh. Neorgan. Khim. **21** [1976] 93/5; Russ. J. Inorg. Chem. **21** [1976] 49/51).

[11] A. N. Kurskii, Yu. A. Pentin, L. K. Tarasov, V. I. Spitsyn (Str. Svoistva Primen. Beta-Diketonatov Metal. Mater. 2nd Vses. Semin., Moscow 1976 [1978], pp. 28/34 from C.A. **89** [1978] No. 206724). — [12] H. G. Brittain, F. S. Richardson (J. Am. Chem. Soc. **98** [1976] 5858/63). — [13] H. G. Brittain, F. S. Richardson (J. Am. Chem. Soc. **99** [1977] 65/70). — [14] H. G. Brittain (J. Am. Chem. Soc. **101** [1979] 1733/6). — [15] J. A. Cunningham, R. E. Sievers (J. Am. Chem. Soc. **97** [1975] 1586/8).

[16] H. G. Brittain (Inorg. Chem. **19** [1980] 640/3).

5.2 Complexes with Other Diketones and Polyketones

5.2.1 With Tribenzoylmethane ($(C_6H_5CO)_3CH$ (= $C_{22}H_{16}O_3$))

Eu($C_{22}H_{15}O_3$)$_3$ · C_2H_5OH · 1.5 H_2O. The compound was prepared by adding 30 ml of sodium hydroxide solution (50% aqueous ethanol), dropwise, with vigorous stirring and over a period of 5 h, to a solution (400 ml) of tribenzoylmethane (6.6 mmol) and europium nitrate (2.0 mmol) in 75% (v/v) ethanol/water. The alcohol was then removed under suction and the crude product collected by filtration. The solid was extracted with a small volume of hot chloroform, unreacted tribenzoylmethane was separated from solution by effecting its precipitation with ethanol, and the product was then precipitated from the ethanolic solution with petroleum ether (boiling range 40 to 60°C). The product was recrystallized from absolute ethanol and analyzed.

The complex melted with decomposition at 127 to 135°C. Differential scanning calorimetric analysis of the chelate revealed a thermogram with a broad peak centered at 110°C corresponding to the loss of solvent prior to melting at 127 to 135°C. The complex was nonvolatile. It was fairly soluble in most common solvents, but the solutions lacked stability. The substance emitted bright red fluorescence, both in the solid state and chloroform solution, when irradiated with ultraviolet light (340 nm). The solid europium complex lost its characteristic fluorescence upon removal of the solvent molecules, suggesting that the latter are coordinated to the metal ion. Recrystallization of the product from benzene exposed to air resulted in decomposition, with formation of a product corresponding approximately to $Eu(C_{22}H_{15}O_3)(C_7H_5O_2)_2$, where $C_7H_5O_2^-$ is the benzoate ion. Tribenzoylmethane and dibenzoylmethane were observed as products in the benzene mother liquor following decomposition. Recrystallization of $Eu(C_{22}H_{15}O_3)_3 \cdot C_2H_5OH \cdot 1.5H_2O$ from benzene under a nitrogen atmosphere yielded the original product, with perhaps slight changes in the degree and nature of solvation, M. Ismail, S. J. Lyle, J. E. Newbery (J. Inorg. Nucl. Chem. **31** [1969] 2091/3).

5.2.2 With 2-Acetyl-1,3-indandione

C−CO−CH3 ($=C_{11}H_8O_3$)

$[M(C_{11}H_7O_3)_n]^{3-n}$ Complexes (n = 1 to 3). **Formation in Solution**

The stepwise stability constants of lanthanide complexes in 60 vol% aqueous acetone at 22°C and ionic strength 0.1 ($NaClO_4$) were determined by pH potentiometric titrations.

M	lg K_1	lg K_2	lg K_3
La	3.54	3.16	2.78
Ce	3.67	3.29	2.91
Pr	3.76	3.39	3.01
Nd	3.86	3.48	3.08
Sm	4.06	3.68	3.30
Eu	3.91	3.53	3.15
Gd	3.84	3.46	3.08

M	lg K_1	lg K_2	lg K_3
Tb	3.93	3.55	3.17
Dy	3.97	3.59	3.21
Ho	3.98	3.60	3.22
Er	3.91	3.53	3.15
Tm	3.87	3.49	3.11
Yb	3.86	3.48	3.08

These complexes are less stable than the corresponding β-diketonates, I. I. Zheltvai, E. V. Melent'eva, M. A. Tishchenko (Zh. Neorgan. Khim. **24** [1979] 1214/8; Russ. J. Inorg. Chem. **24** [1979] 675/8).

5.2.3 With Squaric Acid

($=C_4H_2O_4$)

Stability constants and thermodynamic parameters determined by pH potentiometric titrations and calorimetry (ΔH) for the formation of squarate complexes of the rare earth ions are listed in Table 5/58. The values of the stability constants are substantially lower than those of kojate, acetylacetonate, and tropolonate complexes, in agreement with their relative basicities. Extraction of several lanthanide squarate compounds from an acidic (pH = 1.1) aqueous solution into a benzene solution containing dinonylnaphthalenesulfonic acid was used to determine the effect of ionic strength on the stability constants of the 1:1 squarate complexes. The values reported in Table 5/58 decrease with increasing ionic strength as expected and correspond to the uninegatively charged anion complex [2]. The differences between these values and the values reported in Choppin's paper, [1, 3] which are larger by one or two orders of magnitude, are not discussed.

Table 5/58

Stability Constants and Thermodynamic Parameters for Rare Earth Squarato Complexes at 25°C and Ionic Strength I = 0.10 M ($NaClO_4$) [1].

a) Obtained from pH Potentiometric Titrations*):

M	β_1	$-\Delta G_1$	ΔH_2	ΔS_1	$\beta_2 \times 10^{-3}$	$-\Delta G_2$	ΔH_2	ΔS_2
Y	533	3.72	2.37	20.4	17.6	5.79	3.33	30.6
La	516	3.70	1.51	17.6	—	—	—	—
Ce	525	3.71	1.78	18.4	—	—	—	—
Pr	533	3.72	1.89	18.8	—	—	—	—
Nd	543	3.73	1.99	19.3	—	—	—	—
Sm	643	3.83	2.02	19.6	12.1	5.57	3.04	28.9
Eu	699	3.88	2.00	19.7	13.2	5.62	4.31	33.3
Gd	735	3.91	2.01	19.9	14.6	5.68	3.79	31.8
Tb	748	3.92	2.18	20.4	16.1	5.74	3.44	30.8
Dy	735	3.91	2.18	20.4	17.6	5.79	4.08	33.1
Ho	711	3.89	2.14	20.2	18.5	5.82	3.83	30.9
Er	653	3.84	2.42	21.0	17.9	5.80	4.57	34.8
Tm	601	3.79	2.33	20.5	16.4	5.75	4.02	32.8
Yb	543	3.73	2.42	20.6	14.1	5.66	3.83	31.8
Lu	474	3.65	2.47	20.5	10.8	5.50	3.90	31.5

*) Errors: $\beta_1 = \pm 10\%$; $\Delta G_1 = \pm 0.05$; $\Delta H_1 = \pm 0.10$; $T\Delta S_1 = \pm 0.10$; $\beta_2 = \pm 50\%$; $\Delta G_2 = \pm 0.20$; $\Delta H_2 = \pm 0.25$; $T\Delta S_2 = \pm 0.25$.

b) Calculated from Calorimetric Measurements**) [3].

M	β_1	$-\Delta G_1$	ΔH_2	ΔS_1	$\beta_2 \times 10^{-3}$	$-\Delta G_2$	ΔH_2	ΔS_2
Y	598	3.78	2.31	20.4	14.8	5.69	3.67	31.0
La	540	3.72	1.51	17.6	15.1	5.70	3.00	29.2
Ce	559	3.74	1.69	18.2	13.6	5.36	3.14	28.5
Pr	531	3.71	1.84	18.6	8.5	5.36	3.14	28.5
Nd	597	3.78	1.87	18.9	16.4	5.75	2.96	29.2
Sm	662	3.84	2.00	19.6	15.4	5.71	2.97	29.1
Eu	568	3.75	2.12	19.7	—	—	—	—
Gd	588	3.77	2.22	20.1	—	—	—	—
Tb	797	3.95	2.12	20.4	—	—	—	—
Dy	824	3.97	2.10	20.3	—	—	—	—
Ho	1044	4.11	2.06	20.7	—	—	—	—
Er	1027	4.10	2.11	20.8	—	—	—	—
Tm	448	3.61	2.61	20.8	14.8	5.69	3.24	29.0
Yb	488	3.66	2.53	20.8	9.8	5.78	3.27	30.0
Lu	448	3.61	2.56	20.7	11.9	5.56	3.48	30.0

**) Errors: $\beta_1 = \pm 10\%$; $\Delta G_1 = \pm 0.005$; $\Delta H_1 = \pm 0.005$; $T\Delta S_1 = \pm 0.05$; $\beta_2 = \pm 25\%$; $\Delta G_2 = \pm 0.13$; $\Delta H_2 = \pm 0.15$; $T\Delta S_2 = \pm 0.15$.

$pKa_1 + pKa_2 = 4.15$; ΔG and ΔH in kcal/mol; ΔS in $cal \cdot mol^{-1} \cdot K^{-1}$.

References to 5.2.3:

[1] E. Orebaugh, G. R. Choppin (Inorg. Chem. **17** [1978] 2300/2). — [2] A. T. Kandil, N. Souka, K. Farah (J. Radioanal. Chem. **49** [1979] 13/20). — [3] E. Orebaugh, G. R. Choppin (J. Coord. Chem. **5** [1975/76] 123/8).

5.2.4 With 3-Hydroxy-4-methyl-3-cyclobutene-1,2-dione ($=C_5H_4O_3$)

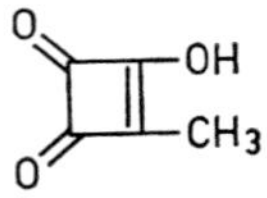

Stability constants and thermodynamic parameters for the 1:1 complex of europium(III) at ionic strength I = 1.0 M ($NaClO_4$) and pH = 2 have been reported: $K_1 = 6.1 \pm 0.4$ (2°C); 5.2 ± 0.5 (10°C); 4.5 ± 0.5 (25°C); 3.6 ± 1.0 (37.5°C); 2.7 ± 0.5 (51°C); $\Delta G = -0.89 \pm 0.06$ kcal/mol; $\Delta H = -2.7 \pm 0.4$ kcal/mol; and $\Delta S = -6.4 \pm 0.6$ cal · $mol^{-1} \cdot K^{-1}$. The values were determined by solvent extraction techniques using bis(2-ethyl)hexylphosphoric acid and radioactive Eu^{3+} as a tracer. The entropy data are typical of outer-sphere complex formation. The structurally similar squarate (and tropolonate) complexes appear to be inner-sphere species, S. S. Yun, G. R. Choppin (J. Inorg. Nucl. Chem. **38** [1976] 332/3).

5.2.5 Croconic Acid ($=C_5H_2O_5$)

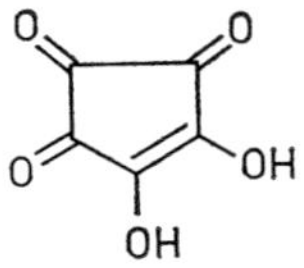

Stability constants and thermodynamic parameters determined by pH potentiometric titrations and calorimetry (ΔH) for the formation of croconate complexes of the rare earth ions are listed in Table 5/59.

Table 5/59
Stability Constants and Thermodynamic Parameters for Rare Earth Croconato Complexes*).

M	$\beta_1 \times 10^{-3}$	$-\Delta G_1$	ΔH_1	ΔS_1	$\beta_2 \times 10^{-3}$	$-\Delta G_2$	ΔH_2	ΔS_2
Y	0.612	3.80	2.71	21.8	29.1	2.29	1.20	11.7
La	1.13	4.17	0.97	17.2	31.3	1.96	1.08	10.3
Ce	1.26	4.23	0.82	16.9	24.9	1.77	1.05	9.5
Pr	1.64	4.39	0.72	17.1	28.5	1.69	1.28	10.0
Nd	1.70	4.41	0.63	16.9	(27.1)	(1.63)	(0.77)	(6.0)
Sm	1.21	4.23	0.86	17.1	(6.8)	(1.00)	(−0.81)	(−2.3)
Eu	1.48	4.21	1.25	18.3	(15.0)	(1.48)	(0.00)	(0.8)
Gd	0.96	4.07	1.73	19.5	(16.2)	(1.67)	(0.58)	(7.5)
Tb	0.878	4.01	2.39	21.5	85.0	2.71	0.44	10.6
Dy	0.759	3.93	2.57	21.8	44.0	2.40	0.85	10.9
Ho	0.791	3.95	2.62	22.1	48.0	2.43	1.12	11.9
Tm	0.819	3.97	2.64	22.2	39.0	2.29	1.68	13.3
Yb	0.845	3.99	2.61	22.2	37.5	2.25	1.88	13.8
Lu	0.823	3.98	2.58	22.0	28.0	2.09	2.37	15.0

T = 25.0°C and 0.1 M ($NaClO_4$); ΔG and ΔH in kcal/mol; ΔS in cal · $mol^{-1} \cdot K^{-1}$.

*) Errors: $\beta_1 = 10\%$; $\beta_2 = 25\%$; $\Delta G_1 = \pm 0.005$; $\Delta H_1 = 0.005$; $\Delta S_1 = \pm 0.2$; $\Delta G_2 = \pm 0.15$; $\Delta H_2 = \pm 0.17$; $\Delta S_2 = \pm 0.4$.

Reference to 5.2.5:
G. R. Choppin, E. Orebaugh (Inorg. Chem. **17** [1978] 2300/2).

5.2.6 With Rhodizonic Acid ($=C_6H_2O_6$)

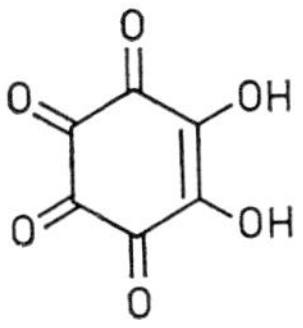

Complex formation of rhodizonic acid with Sc, Y, La, Nd, and Yb ions was observed by following the activity of ^{90}Y on cation and anion exchange columns in the presence and absence of sodium rhodizonate at pH 6.5 to 7.0. The occurrence of precipitation at molar ratios of 2:1 to 1:2.5 yttrium to rhodizonic acid was taken as evidence for the formation of a neutral complex, and the presence of red-brown solutions at higher mole ratios (>2.5) was taken as being indicative of the formation of anionic complexes. No complexes were actually isolated, and no stability constant measurements were made.

Reference to 5.2.6:
A. A. K. Al-Mahdi, T. Schonfeld (Mikrochim. Acta **1962** 254/64).

6 Complexes with Quinones

John H. Forsberg
Department of Chemistry, Monsanto Hall, Saint Louis New University
Saint Louis, Missouri, USA

6.1 With Semiquinone Radical Anions

Complexing of various semiquinone radical anions with diamagnetic rare earth ions has been studied by electron spin resonance.

The ESR spectrum of an aqueous solution of the ortho-semiquinone radical (generated by bubbling air or oxygen through an alkaline solution of catechol) in the presence of excess yttrium(III) ion revealed 18 lines. The spectrum of the *o*-semiquinone radical in the absence of the metal ion, revealed 9 lines, indicating that coupling occurred between the ligand electron spin and yttrium nuclear spin ($I = {}^1/_2$, for ^{89}Y) in the complex, from which a coupling constant $a_Y = 0.65$ Gauss was obtained. With $^{139}La(III)$ ($I = {}^7/_2$) and the *o*-semiquinone radical in dimethyl sulfoxide, 52 of the theoretical 72 lines were resolved, with $a_{La} = 2.07$ Gauss. The observation of hyperfine splitting indicates some degree of covalent bonding between the ligand and metal ion. The smaller value found for the yttrium coupling constant compared with the lanthanum coupling constant was considered consistent with the smaller nuclear magnetic moment of yttrium. The composition of the complex in solution was not discussed, D. R. Eaton (Inorg. Chem. **3** [1964] 1268/71).

6.2 With Derivatives of 1,2-Benzoquinone

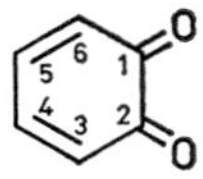

Razuvaev, Abakumov, Klimov [1] observed that 3,5-di-tert-butyl-1,2-benzoquinone, ($= C_{14}H_{20}O_2$), and its 6-nitro derivative ($= C_{14}H_{19}NO_4$), reacted with diamagnetic rare earth chlorides (M = Y, La, Lu) at room temperature in both solvating solvents (tetrahydrofuran, 1,2-dimethoxyethane, diglyme, ethanol) and in benzene. The ESR spectra of the lanthanum and lutetium complexes derived from both ligands (compositions not defined) were similar and characterized by hyperfine structure due to interaction of one unpaired electron with one proton (4-position) and one metal nucleus ($I = {}^7/_2$ for both ^{139}La and ^{175}Lu). For the 3,5-di-tert-butyl-1,2-benzoquinone complexes, $a_H = a_M = 3.2$ and 5.5 Oe for M = La and Lu, respectively, whereas for the 6-nitro derivative, $a_H = a_M = 3.0$ and 5.1 Oe for M = La and Lu, respectively (data for alcoholic solutions). The spectra contained nine hyperfine components, with the intensity ratio 1:2:2:2:2:2:2:2:1. The absence of coupling with a hydrogen atom in the 6-position was attributed to low unpaired electron spin density in that position. The spectra of the yttrium complexes in benzene or tetrahydrofuran, on the other hand, revealed only a simple doublet, due to coupling with 4-H, and no additional components due to coupling with yttrium ($I = {}^1/_2$). The latter observation was attributed to the small magnetic moment ($-0.136\,\mu_B$) of the ^{89}Y nucleus [1].

The triplet character observed in EPR spectra of the 3,5-di-tert-butyl-1,2-benzoquinone-rare earth complexes (M = Sc, Y, La, Lu; compositions not defined) was attributed to the protons of the aromatic ring. Depending on the $a_M:a_H$ ratio (values of a_H and a_M are recorded below), 24, 19, and 10 components were observed, respectively, in the lanthanum, scandium, and lutetium systems. Data for the scandium and lanthanum complexes were obtained for alcoholic solutions, whereas data for the yttrium and lutetium systems were obtained for tetrahydrofuran solutions.

M^+	Sc	Y	La	Lu
a_H in Oe	3.8	3.7	3.7	5.3
a_M in Oe	2.5	—	3.1	5.3

The absence of splitting in the spectrum of the yttrium complex due to the metal ion was attributed to the small magnetic moment of the ^{89}Y nucleus [2].

Lobanov, Abakumov, and Razuvaev [3] isolated compounds of composition $M(C_{14}H_{20}O_2)_3$ (M = La, Gd), containing the semiquinone ligand derived from 3,5-di-tert-butyl-1,2-benzoquinone, by allowing the rare earth chloride to react with a solution of the quinone in the presence of disodium 3,5-di-tert-butyl-1,2-pyrocatecholate in tetrahydrofuran. The compounds were considered to be complexes derived from quasiradical ligands.

3,4,5,6-Tetrachloro-1,2-benzoquinone (= chloranil = $C_6Cl_4O_2$), also reacted with rare earth chlorides (M = Y, La, Lu) in a variety of solvents at ambient temperatures to yield paramagnetic complexes (compositions not defined). The ESR spectra of the lanthanum and lutetium ($I = ^7/_2$) complexes in diethyl ether revealed 8 lines, corresponding to a coupling between the unpaired electron and metal nucleus. The spectrum of the yttrium complex revealed a singlet, indicating that the magnetic moment of ^{89}Y ($-0.136\,\mu_B$) is not sufficiently large to allow resolution of the hyperfine lines. The singlet character of the yttrium-chloranil spectrum confirms the interpretations of the ESR spectra discussed above for the benzoquinone-rare earth system [4].

References to 6.2:

[1] G. A. Razuvaev, G. A. Abakumov, E. S. Klimov (Dokl. Akad. Nauk SSSR **201** [1971] 624/7; Dokl. Chem. Proc. Acad. Sci. USSR **196/201** [1971] 968/70). — [2] G. A. Abakumov, E. S. Klimov, V. V. Ershov, I. S. Belostotskaya (Izv. Akad. Nauk SSSR Ser. Khim. **1975** 927/30; Bull. Acad. Sci. USSR Div. Chem. Sci. **1975** 841/3). — [3] A. V. Lobanov, G. A. Abakumov, G. A. Razuvaev (Dokl. Akad. Nauk SSSR **235** [1977] 824/7 from C.A. **87** [1977] No. 126509). — [4] G. A. Abakumov, E. S. Klimov (Izv. Akad. Nauk SSSR Ser. Khim. **1972** 1199/201; Bull. Acad. Sci. USSR Div. Chem. Sci. **1972** 1157/9).

6.3 With Derivatives of 1,4-Naphthoquinone

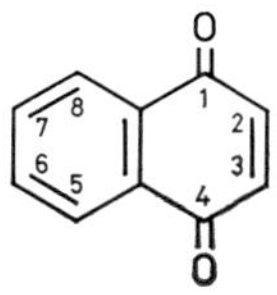

O
8 1
7 2
6 3
5 4
O

6.3.1 With 2-Hydroxy-1,4-naphthoquinone (= Lawson = $C_{10}H_6O_3$)

$M(C_{10}H_5O_3)_3 \cdot n\,H_2O$ (n = 0, 1). Each compound (n = 0 for M = La, Pr, Nd, Sm; n = 1 for M = Ce, Gd, Dy) was prepared by adding, with stirring, an alcoholic (80%) solution of a rare earth chloride to an alcoholic solution of the ligand (1:3 mol ratio). Precipitation of the product was effected by adding an aqueous ammonia (1:20 v/v) solution dropwise. The resulting suspension was stirred for 4 h and then allowed to cool in ice for 2 h. The precipitate was collected by filtration, washed repeatedly with alcohol, and dried under vacuum.

The ν(C=O) stretching mode of the noncoordinated carbonyl group appeared in the region 1625 to 1640 cm^{-1} in the infrared spectra of the chelates (Nujol mulls), whereas the ν(C=O) mode of the coordinated carbonyl group was assigned to a band near 1655 cm^{-1}. The ν(C-O) stretching absorption of the hydroxo group was observed as a strong band at 1215 cm^{-1} in the spectrum of the free ligand, but appeared as a shoulder in the region 1220 to 1222 cm^{-1} in the spectra of the chelates, an observation interpreted in terms of ligation of the hydroxo oxygen atom. The absence of a band near 3200 cm^{-1} due to the phenolic hydroxo group confirmed loss of the hydroxo hydrogen atom upon chelation. The hydrates revealed ν(O-H) stretching modes due to water molecules in the region 3100 to 3600 cm^{-1}.

The physical properties of the chelates are summarized in Table 6/1. The magnetic moments of the chelates corresponded to the free ion values, indicating that the 4f electrons do not participate significantly in bond formation [1].

Table 6/1
Physical Properties of the Tris(2-hydroxy-1,4-naphthoquinonato) Rare Earth(III) Chelates $M(C_{10}H_5O_3)_3 \cdot nH_2O$.

M	n	color	decomposition temp. in °C	magnetic moment in μ_B *)
La	0	yellow orange	258	diamagnetic
Ce	1	dark orange	292	3.20
Pr	0	orange	295	3.40
Nd	0	yellow orange	285	3.59
Sm	0	brownish orange	268	2.24
Gd	1	yellow orange	302	7.93
Dy	1	brown	309	10.47

*) Magnetic susceptibilities were measured at ambient temperatures and corrected for diamagnetic contributions of the ligands.

References to 6.3.1:

[1] S. B. Padhye, C. R. Joshi, B. A. Kulkarni (J. Inorg. Nucl. Chem. **39** [1977] 1289/90).

6.4 With Derivatives of Anthraquinone

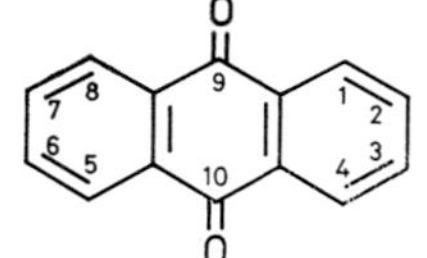

(=9,10-Dioxo-9,10-dihydroanthracene)

6.4.1 With 1,2,4-Trihydroxyanthraquinone (= Purpurin = $C_{14}H_8O_5$)

Rare earth chlorides (M = La, Ce, Pr, Nd, Sm) interact in aqueous media with 1,2,4-trihydroxyanthraquinone, forming pink-colored, 1:2 (metal:ligand) chelates. The wavelengths of maximum absorbance for the chromogenic reagent and its complexes are 480 and 520 nm, respectively. The logarithm for the apparent stability constant, as determined spectrophotometrically for each complex, is 9.1 ± 0.2 [1].

6.4.2 With 1,2,5,8-Tetrahydroxyanthraquinone (= Quinalizarin = $C_{14}H_8O_6$)

The instability constants for the 1:1 chelates (charges were omitted from the chelate formulation) were determined spectrophotometrically in 20% (v/v) methanol/water at pH 6.8 (λ_{max} = 560 nm), and the values are recorded below [2].

metal ion	La^{3+}	Pr^{3+}	Nd^{3+}	Gd^{3+}	Tb^{3+}
$K_1 \times 10^4$	0.93	1.19	1.47	1.66	1.56

Srivastava, Banerji [3] and Tataev, Bagdasarov [4] interpreted their spectrophotometric data in terms of formation of 1:2 (metal:ligand) chelates for M = La [3] and Pr [4]. Maximum absorbance of the lanthanum chelate was observed at 510 to 520 nm and pH 6.8 to 8.1 (lg β_2 = 10.1 at 30°C in 50% ethanol at pH 6.8, I = 0.1) [3]. Maximum absorbance for the 1:2 praseodymium chelate was observed at 584 nm in neutral media (pH 7), but at 658 nm at pH 8 to 10 (instability constant of the complex was determined as 3.02×10^{-12}) [4].

6.4.3 With 1-Amino-4-hydroxyanthraquinone (= $C_{14}H_9NO_3$)

Spectrophotometric studies revealed that 1-amino-4-hydroxyanthraquinone forms reddish-violet colored 1:1 chelates (degree of ligand ionization not defined) with the tripositive ions of lanthanum, cerium, praseodymium, neodymium, and samarium in methanol. The color of the solutions remained unchanged for 3 or 4 days. The stability constants, determined spectrophotometrically at 30°C, and absorption maxima of the chelates are given below. Range of value reflects two different methods used to calculatete the stability constants [5].

metal ion	La^{3+}	Ce^{3+}	Pr^{3+}	Nd^{3+}	Sm^{3+}
lg K	5.16 to 5.26	5.25 to 5.33	5.36 to 5.42	5.40 to 5.45	5.25 to 5.46
λ_{max} in nm*)	605	610	605	610	605

*) Free ligand absorbs at 530 nm.

6.4.4 With 1,2-Dihydroxy-3-nitroanthraquinone (= Nitroalizarin = $C_{14}H_7NO_6$)

Spectrophotometric studies indicated that nitroalizarin forms pink colored 1:2 (metal:ligand) chelates with rare earth ions (M = La, Ce, Pr, Nd, Sm, Gd, Dy) [6, 7]. The logarithms of the formation constants for the complexes in 95% ethanol were determined to be of the order of 11.0±0.3 [6]. The wavelength of maximum absorbance was between 520 and 550 nm [6, 7].

$Y(C_{14}H_5NO_6)Cl \cdot 3H_2O$ was prepared by allowing yttrium chloride to react with the free ligand in an aqueous medium. Infrared data were interpreted in terms of bidentate ligation via the C_1-OH group and the oxygen atom of the C_9-CO group. Thermogravimetric analysis indicated coordination by water molecules, whereas the molar conductance in dimethyl sulfoxide revealed nonelectrolyte behavior [8].

6.4.5 With Sodium 9,10-Dioxo-9,10-dihydro-2-anthracenesulfonate (= $C_{14}H_7NAO_5S$)

$Eu(C_{14}H_7O_5S)_3 \cdot 3H_2O$ was prepared by mixing warm, aqueous solutions containing stoichiometric amounts of europium chloride and the sodium salt of the ligand. The resulting precipitate was collected by filtration, washed with alcohol, and dried under vacuum. The compounds are thermally stable, decomposing without melting at ≈520°C. The compound is very soluble in dimethylformamide but insoluble in common solvents such as alcohol, ether, benzene and acetone.

The absorption spectrum of the complex in dimethylformamide revealed a strong band at 2900 to 3900 Å. The fluorescence spectrum revealed weak bands at 17270 cm^{-1} ($^5D_0 \rightarrow {}^7F_0$), 17140 cm^{-1} ($^5D_1 \rightarrow {}^7F_3$), and 15385 cm^{-1} ($^5D_0 \rightarrow {}^7F_3$) and intense bands at 17020, 16920, 16820 cm^{-1} ($^5D_0 \rightarrow {}^7F_1$), and 16350, 16300, 16245, 16140, 16075 cm^{-1} ($^5D_0 \rightarrow {}^7F_2$) [9].

6.4.6 With Sodium 3,4-Dihydroxy-9,10-dioxo-9,10-dihydro-2-anthracenesulfonate (= Alizarin Red S = $C_{14}H_7NaO_7S$)

$M(C_{14}H_5O_7S)_n^{3-3n}$ Species (n = 1, 2)

Formation constants for the 1:1 and 1:2 (metal:ligand) chelates derived from the ligand in its trinegatively charged form were determined potentiometrically (glass electrode) at 30°C in aqueous media, I = 0.05 ($NaClO_4$). The lanthanum(III) ion formed only the 1:1 chelate [10]. Proton dissociation constants of the ligand are (30°C, I = 0.1 $NaClO_4$): lg K_1 = 4.60, lg K_2 = 11.20 [10].

metal ion	Y^{3+}	La^{3+}	Pr^{3+}	Nd^{3+}
lg K_1	12.68	11.77	12.56	12.73
lg K_2	11.14	—	8.90	21.63

$M(C_{14}H_6O_7S)_2^-$ Ions

The formation of each 1:2 (metal:ligand) alizarin red S rare earth chelate (M = Sc, Y, Pr, Nd) derived from the ligand in its dinegatively charged form was established spectrophotometrically [11 to 13]. The stability constants, wavelengths of maximum absorbance, and the solution pH values at which the measurements were made are recorded below. The anionic natures of the 1:2 chelates were confirmed by electrophoresis experiments [11, 13].

metal ion	t in °C	pH	λ_{max} in nm	lg β*)	Ref.
Sc^{3+}	25	4.0	500	9.0 to 9.4	[11]
Y^{3+}	25	4.0	510	9.5 to 9.6	[11]
Pr^{3+}	—	4.5	530	9.21	[12]
Nd^{3+}	—	4.5	530	9.34	[12]
Gd^{3+}	25	4.0	520	8.6	[13]
Tb^{3+}	25	4.0	530	9.0 to 9.2	[13]
Yb^{3+}	25	4.0	530	8.6 to 8.7	[13]
Lu^{3+}	25	4.0	530	9.1 to 9.2	[13]

*) Range of values reflects different methods of calculation.

For reactions of rare earth elements with alizarin red S in the presence of ammonia and amines, see also [16 to 18].

6.4.7 With 3-[Di(carboxymethyl)aminomethyl]-1,2-dihydroxyanthraquinone (= Alizarin Complexon = $C_{19}H_{15}NO_8$)

Complex formation by yttrium(III) with alizarin-complexone in aqueous media in the pH range 2 to 5 was investigated spectrophotometrically and by pH measurement and potentiometric titrations. The data were interpreted in terms of formation of two complexes, formulated as $Y(C_{19}H_{13}NO_8)^+$ and $Y(C_{19}H_{12}NO_8)$ at pH 3 to 5 (lg K = 31.31 and 28.03, respectively, at 22°C, I = 0.1 KNO_3). The acid dissociation constants of the $Y(C_{19}H_{13}NO_8)^+$ and $Y(C_{19}H_{12}NO_8)$ species were pK = 3.28 and 16.84, respectively [14].

6.4.8 With Carminic Acid (= $C_{22}H_{20}O_{13}$)

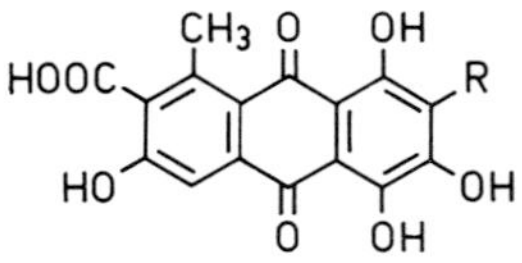

(R = D-glucopyranosil = $C_6H_{11}O_5$)

Rare earth chlorides (M = Y, La, Gd, Lu) react with carminic acid in aqueous media (pH 5.0 to 5.5) to form violet-red 1:2 (metal:ligand) chelates of composition $M(C_{22}H_{18}O_{13})_2(OH)^{2-}$. A solution of the chromogenic reagent has a yellow-orange color under these conditions. The solutions of the cerium subgroup complexes remained clear for a long period of time, whereas solutions of complexes of the yttrium subgroup yielded precipitates after 1 to 2 h. The wavelengths of maximum absorbance are 532, 535, and 572 nm for solutions of the lanthanum, gadolinium, and lutetium complexes, respectively. The instability constants of the chelates, recorded below, were determined spectrophotometrically, I = 0.1 [15].

metal ion	Y^{3+}	La^{3+}	Gd^{3+}	Lu^{3+}
$K \cdot 10^9$	1.95	3.72	1.905	5.36

References to 6.4:

[1] J. Ahmad, N. Ahmad, S. M. F. Rahman (J. Indian Chem. Soc. **44** [1967] 485/6). — [2] S. Zielinski, L. Lomozik (Prace Kom. Mat. Przyr. Poznan. Tow. Przyj. Nauk Prace Chem. **12** No. 5 [1971] 315/23 from C.A. **76** [1972] No. 50794). — [3] K. C. Srivastava, S. K. Banerji (J. Prakt.

Chem. [4] **38** [1968] 327/38). — [4] O. A. Tataev, K. N. Bagdasarov (Elektrokhim. Opt. Metody Anal. **1963** 212/6 from C.A. **61** [1964] 4957). — [5] A. K. Jain, V. P. Aggarwala, P. Chand, S. P. Garg (Talanta **19** [1972] 1481/2).

[6] J. Ahmad, N. Ahmad, S. M. F. Rahman (J. Indian Chem. Soc. **44** [1967] 444/7). — [7] S. M. F. Rahman, N. Ahmad, J. Ahmad (J. Indian Chem. Soc. **46** [1969] 615/7). — [8] S. M. F. Rahman, N. Ahmad, V. Kumar (Indian J. Chem. **12** [1974] 899/900). — [9] B. Blanzat, J. Loriers (Compt. Rend. C **264** [1967] 4/7). — [10] S. D. Makhijani, S. P. Sangal (J. Indian Chem. Soc. **52** [1975] 788/90).

[11] K. N. Munshi, S. N. Sinha, S. P. Sangal, A. K. Dey (Microchem. J. **7** [1963] 473/84). — [12] L. L. Soni, G. Viswanath (J. Indian Chem. Soc. **48** [1971] 799/802). — [13] S. P. Sangal (J. Prakt. Chem. [4] **36** [1967] 126/37). — [14] I. G. Prisyagina, G. S. Tereshin (Zh. Neorgan. Khim. **20** [1975] 66/71; Russ. J. Inorg. Chem. **20** [1975] 36/9). — [15] N. S. Poluektov, R. S. Lauer, M. A. Sandu (Zh. Anal. Khim. **25** [1970] 2118/24; J. Anal. Chem. [USSR] **25** [1970] 1821/6).

[16] L. S. Serdyuk, G. P. Fedorova (Zh. Neorgan. Khim. **4** [1959] 88/96; C.A. **1959** 12817/8). — [17] L. S. Serdyuk, G. P. Fedorova (Ukr. Khim. Zh. **27** [1961] 252/6; C.A. **1961** 21947). — [18] L. S. Serdyuk, V. F. Silich (Ukr. Khim. Zh. **29** [1963] 848/54; C.A. **60** [1964] 2325).

6.5 Complexes with Triphenylmethane Dyes and Analogous Compounds

6.5.1 With Aluminon

NH$_4$OOC, HO, COONH$_4$, O, C, COONH$_4$, OH

(=Triammonium Aurintricarboxylate = $(NH_4)_3C_{22}H_{11}O_9$)

Formation constants for 1:1 Aluminon-rare earth chelates of composition $M(C_{22}H_{10}O_9)^-$ were determined spectrophotometrically at 25°C in aqueous media, pH = 6.0 [1, 2]. The liberation of a hydrogen ion upon ligation prompted the proposal that chelation occurs through a phenolic oxygen atom and an oxygen atom of the adjacent carboxylate group. The complexes revealed maximum absorbance at 540 nm. The pH range of stability for each chelate is given in Table 6/2 [1]. For potentiometric studies on chelate formation of Aluminon with cerium(IV), see [3]. Absorption and luminescence spectra of rare earth metal complexes with aurintricarboxylic acid and some amines are reported in [4].

Table 6/2

Formation Constants and pH Range of Stability for 1:1 complexes with Aluminon, $M(C_{22}H_{10}O_9)^-$, at 25°C and pH 6.0 [1].

metal ion	pH range of stability	lg K*)	metal ion	pH range of stability	lg K*)
Pr^{3+}	4.5 to 8.5	4.2 to 4.4	Dy^{3+}	4.5 to 8.0	4.7 to 4.9
Nd^{3+}	4.5 to 8.0	4.4	Ho^{3+}	4.5 to 8.0	4.9 to 5.0
Sm^{3+}	4.5 to 8.5	4.5	Er^{3+}	4.5 to 8.0	4.9 to 5.1
Eu^{3+}	4.5 to 8.0	4.5 to 4.6	Tm^{3+}	4.5 to 8.0	5.0 to 5.2
Gd^{3+}	4.5 to 8.0	4.6 to 4.8	Yb^{3+}	4.5 to 8.5	5.1 to 5.3
Tb^{3+}	4.5 to 8.5	4.6 to 4.8	Lu^{3+}	4.5 to 8.5	5.2 to 5.5

*) Range of values reflects different methods of calculation.

References to 6.5.1:

[1] S. P. Sangal (J. Prakt. Chem. **36** [1967] 126/37). — [2] S. P. Sangal, A. K. Dey (Microchem. J. **12** [1967] 168/76). — [3] B. K. Avinashi, S. K. Banerji (Indian J. Chem. **10** [1972] 213/4). — [4] A. Janowski, J. Rzeszotarska, N. Sadlej (J. Lumin. **5** [1972] 453/60 from C.A. **78** [1973] No. 49995).

6.5.2 With Eriochrome Azurol B

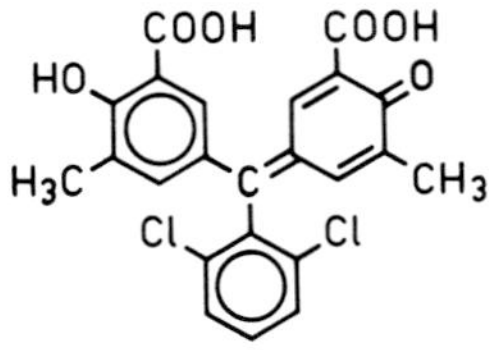

($= C_{23}H_{16}Cl_2O_6$)

Spectrophotometric studies indicated that Eriochrome Azurol B interacted with yttrium(III) forming a 1:2 (metal:ligand) complex of composition $Y(C_{23}H_{13}Cl_2O_6)_2^{3-}$ at pH 6.0 ($\lambda_{max} = 550$ nm). The logarithm of the overall thermodynamic stability constant ($I = 0$) for the chelate, determined spectrophotometrically at 27°C, was 9.6 ± 0.2 (lg $\beta_2 = 9.2 \pm 0.2$ at $I = 0.1$ M $NaClO_4$) [1]. The chelation of Sc^{3+}, Y^{3+}, and some lanthanides with Eriochrome Azurol B in the presence and absence of cetyl trimethylammonium bromide was studied in detail by Vekhande, Munshi [2].

References to 6.5.2:

[1] S. P. Pande, K. N. Munshi (J. Indian Chem. Soc. **50** [1973] 649/50). — [2] C. Vekhande, K. N. Munshi (Microchem. J. **23** [1978] 28/41; C.A. **89** [1978] No. 16175).

6.5.3 With Chromal Blue G

(=Disodium 2''-chloro-4'-hydroxy-5,5''-dimethyl-4''-nitro fuchsone-3,3'-dicarboxylate $= C_{23}H_{14}ClNNa_2O_8$)

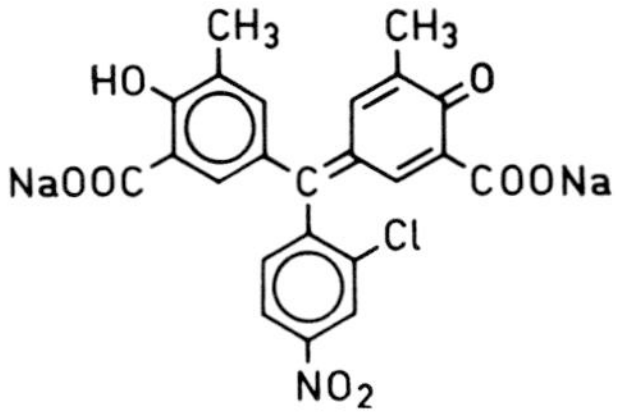

Spectrophotometric studies revealed that scandium chloride reacted with Chromal Blue G in an aqueous acetate buffer to form a 1:2 (metal:ligand) chelate (degree of ligand ionization not defined). The maximum absorbance of the complex was observed at 590 nm in acidic media, but above pH 7.0, the position of the band maximum shifted toward longer wavelength. The formation constant, determined spectrophotometrically at pH 6.0 and 25°C, was estimated to be 2.8×10^{10}, K. Uesugi (Bull. Chem. Soc. Japan **42** [1969] 2051/4).

6.5.4 With Pyrocatechol Violet

(=3,3′,4′-Trihydroxyfuchsone-2″-sulfonic Acid = $C_{19}H_{14}O_7S$)

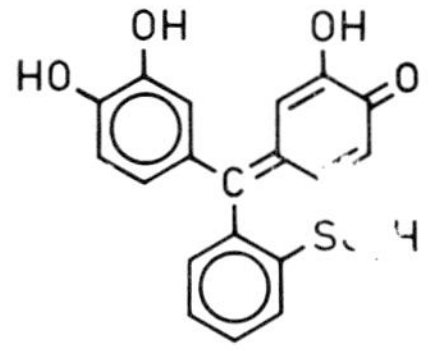

The stepwise formation constants for the 1:1 and 2:1 (ligand:metal) Pyrocatechol Violet-rare earth chelates were determined potentiometrically (glass electrode) at 25°C in aqueous media, I = 0.1 ($NaClO_4$) [1, 2]. Although the degree of ligand ionization in the metal chelates was not defined in either of the studies, three stepwise proton-ligand formation constants were reported in [1] (lg K_1 = 9.50, lg K_2 = 7.50, lg K_3 = 2.70), whereas only two values (lg K_2 = 9.17, lg K_2 = 5.44) were reported in [2]. Europium(III) and gadolinium(III) formed blue 1:1 complexes with Pyrocatechol Violet above pH 6 (λ_{max} = 660 nm at pH 7.2). The dissociation constant (determined spectrophotometrically) reported for both complexes was 2.3×10^{-6} [3]. Kharlamova and coworkers [4], in a spectrophotometric study of the reaction of cerium and yttrium subgroup rare earth salts with the chromogenic reagent, concluded that only a single complex with a 1:1 component ratio was formed in both neutral and weakly alkaline media [4]. Spectrophotometric studies of Shaulina, Kirillov, Churakova [5] showed that complexation occurs by the reaction of $M(OH)^{2+}$ with H_2L^{2-} (H_4L = Pyrocatechol Violet) to give 1:1 complexes $Ln(OH)(H_2L)$. For the reaction of Y, Ce, and La with Pyrocatechol Violet in the presence of H_3BO_3 see [6].

metal ion	La^{3+} [1]	Y^{3+} [1]	Y^{3+} [2]
lg K_1	14.44*), 18.50**)	19.40*), 15.85**)	6.81***)
lg K_2	4.30*), 7.89**)	8.80*), 12.35***)	

*) By interpolation at half $\bar{n}$ values.—**) By interpolation at various $\bar{n}$ values.—***) Mean value obtained by interpolation at half $\bar{n}$ values and various $\bar{n}$ values.

References to 6.5:

[1] R. L. Kushwaha, C. K. Verma, Om Prakash, S. P. Mushran (J. Indian Chem. Soc. **54** [1977] 285/8). — [2] S. P. Pande, K. N. Munshi (J. Indian Chem. Soc. **51** [1974] 339/40). — [3] T. R. Babaeva, E. A. Bashirov, M. K. Akhmedli (Azerb. Khim. Zh. **1966** 122/6 from C.A. **67** [1976] No. 39819). — [4] L. N. Kharlamova, E. G. Bol'shakova, K. I. Gur'ev, R. K. Chernova (Zh. Neorgan. Khim. **21** [1976] 2035/9; Russ. J. Inorg. Chem. **21** [1976] 1121/3). — [5] L. P. Shaulina, A. I. Kirillov, G. N. Churakova (Zh. Neorgan. Khim. **25** [1980] 951/3 from C.A. **92** [1980] No. 204357).

[6] L. S. Serdyuk, U. F. Silich (Zh. Anal. Khim. **18** [1963] 166/71; C.A. **58** [1963] 11955).

6.5.5 With Eriochrome Cyanine R (= Solochrome Cyanine R)

(= Trisodium 4′-hydroxy-5,5′-dimethyl-2″-sulfonato-3,3′-fuchsonedicarboxylate = $C_{23}H_{15}Na_3O_9S$)

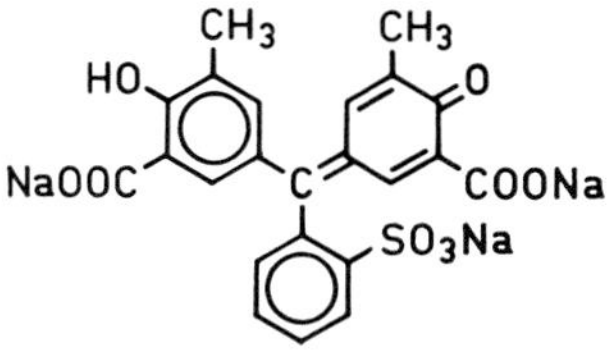

Tataev and coworkers [1] reported values for the stepwise acid dissociation constants of eriochrome cyanine R (pK_1 = −1.8, pK_2 = 3.0, pK_3 = 4.8, pK_4 = 12.5) and indicated that the metal:ligand mole ratio of the rare earth chelates is dependent on solution pH and concentration of reagents

[1]. Jog and coworkers [2] determined the stability constants for the 1:1 chelates spectrophotometrically at 25°C and pH = 5.8 (M = lg K_1): La, 5.17; Pr, 5.65; Nd, 5.20. They noted that the λ_{max} for the 1:1 chelates was dependent on pH: for pH 1 to 2.0, λ_{max} = 500 nm; for pH 2 to 4.5, λ_{max} = 540 nm; for pH 5 to 8.3, λ_{max} = 500 nm; and for pH > 9, λ_{max} = 440 nm [2].

Munshi and coworkers [3, 4] proposed formation of 1:2 (metal:ligand) chelates over the pH range 5.5 to 8.0. The conditional stability constants, determined spectrophotometrically at pH = 5.0 to 6.5 and 25°C, and the absorption maxima for the red colored 1:2 chelates are recorded below [3, 4]. Additional stability constant data are found in [5].

Spectrophotometric investigations of the complex formation between La^{3+} and Eriochrome Cyanine R show three complexes at pH 5.3 to 5.5 (λ_{max} = 460 nm), 6.2 to 6.5 (λ_{max} = 490 nm), and 8.2 to 9.0 (λ_{max} = 545 nm) with a shoulder at 570 to 580 nm. At pH > 9.5, the lanthanum compound precipitates as a violet flocculation. Stability constants at 25°C and I = 0.2 ($NaClO_4$) are 4.9×10^7, 7.0×10^7, and 1.0×10^4, respectively [6].

metal ion	λ_{max} in nm	lg β_2
Sc^{3+}	535	
Y^{3+}	530	
La^{3+}	480	
Pr^{3+}	490	10.6
Nd^{3+}	500	9.6
Sm^{3+}	510	9.6
Gd^{3+}	520	9.8
Ho	520	9.8
Tm^{3+}	530	9.4
Yb^{3+}	530	9.1
Lu^{3+}	530	9.2

Spectrophotometric studies indicated that Solochrome Cyanine R interacted with yttrium(III) in aqueous solution, forming a 1:1 chelate of composition $Y(C_{23}H_{14}O_9S)^-$ at pH 6.0, (λ_{max} = 530 nm). The logarithm of the thermodynamic stability constant for the complex (I = 0), determined spectrophotometrically at 27°C, was 5.0 ± 0.2 (lg K = 4.4 ± 0.2 at I = 0.1 M $NaClO_4$) [7]. A study of the increased sensitivity of Solochrome Cyanine R for the micro determination of scandium, yttrium, and some lanthanides caused by the presence of cetyltrimethylammonium bromide is reported in [8].

References to 6.5.5:

[1] O. A. Tataev, K. N. Bagdasarov, S. A. Akhmedov, Kh. A. Mirzaeva, L. G. Anisimova, S. D. Abdurakhmanova, R. R. Abdullaev (Primen. Org. Reagentov Anal. Khim. **1974** 30/8 from C.A. **86** [1977] No. 100297). — [2] M. S. Jog, A. T. Choudhary, S. P. Sangal (Chem. Era **12** [1976] 381/4). — [3] K. N. Munshi, A. K. Dey (Rev. Chim. Minerale **5** [1968] 619/28). — [4] K. N. Munshi, S. C. Srivastava, A. K. Dey (J. Indian Chem. Soc. **45** [1968] 817/20). — [5] R. K. Chernova, E. G. Bol'shakova, E. G. Kulapina (Izv. Vysshikh Uchebn. Zavedenii Khim. Khim. Tekhnol. **18** [1975] 884/8 from C.A. **83** [1975] No. 137738).

[6] J. F. C. Boodts, W. Saffioti (Anais Acad. Brasil. Cienc. **51** [1979] 97/101). — [7] S. P. Pande, K. N. Munshi (J. Indian Chem. Soc. **50** [1973] 649/50). — [8] C. Vekhande, K. N. Munshi (Indian J. Chem. A **16** [1978] 1079/82 from C.A. **91** [1979] No. 32360).

6.5.6 With Chrome Azurol S

(= Trisodium Salt of 2'',6''-Dichloro-4-hydroxy-5,5'-dimethyl-3''-sulfo-3,3'-fuchsonedicarboxylic Acid = $C_{23}H_{13}Cl_2Na_3O_9S$)

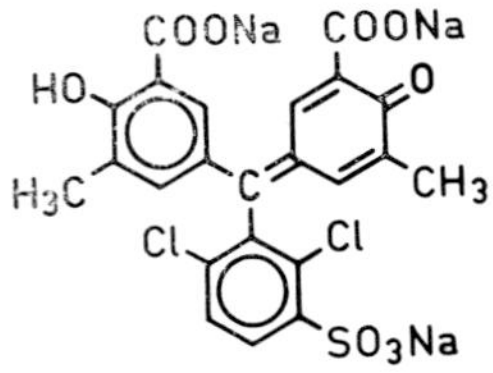

$M(C_{23}H_{13}Cl_2O_9S)_n^{3-3n}$ Species (n = 1, 2)

Formation constants for the 1:1 Chrome Azurol S chelates of scandium, yttrium, and lanthanum were determined spectrophotometrically at 25°C in aqueous media (ionic strength could not be maintained as the addition of an inert electrolyte effected precipitation of the chelate) [1, 5, 6].

metal ion	λ_{max} in nm	pH	lg K_1	Ref.
Sc^{3+}	460	5.0	5.5	[1]
Y^{3+}	490	6.0	4.3	[1]
La^{3+}	460	6.0	4.8	[1]
Sm^{3+}	490	5.0 to 6.5	4.8	[6]
Eu^{3+}	540	4.5 to 8.5	4.2	[6]
Eu^{3+}	560	7.0		[5]
Gd^{3+}	510	5.0 to 8.5	4.4	[6]
Tb^{3+}	500	4.5 to 8.0	4.7	[6]
Dy^{3+}	530	5.0 to 7.5	4.2	[6]
Ho^{3+}	550	5.5 to 8.5	4.3	[6]

The conditional overall formation constants, β_2, for the 2:1 (ligand : metal) chelates of scandium and yttrium were determined spectrophotometrically as 9.5×10^{11} (pH 5.6) [2] and 1.5×10^{10} (pH 7.6) [3], respectively. Spacu, Plostinaru [4] postulated the formation of 1:2 (metal : ligand) chelates for M = La, Ce, Dy, and Lu in aqueous media at 4.7 to 5 on the basis of spectrophotometric studies (λ_{max} = 490 to 510 nm). The dissociation constant of the lutetium chelate was determined by Job's method as 4×10^{-11} [4]. For complexes of Sc^{3+}, Y^{3+}, and La^{3+} with Chrome Azurol S and 1,10-phenanthroline, see [7].

References to 6.5.6:

[1] S. N. Sinha, S. P. Sangal, A. K. Dey (J. Indian Chem. Soc. **44** [1967] 203/7). — [2] R. Ishida, N. Hasegawa (Bull. Chem. Soc. Japan **40** [1967] 1153/8). — [3] R. Ishida (Bunseki Kagaku **15** [1966] 829/34 from C.A. **66** [1967] No. 91395). — [4] S. P. Spacu, S. Plostinaru (Roumaine Chim. **12** [1967] 383/7). — [5] R. W. Cattrall, S. J. E. Slater (Microchem. J. **16** [1971] 602/9).

[6] S. P. Sangal (J. Prakt. Chem. **36** [1967] 126/37). — [7] L. I. Ganago, L. A. Alinovskaya (Zh. Anal. Khim. **28** [1973] 494/9).

6.5.7 With Phthalexon S

(= 3,3'-Bis[bis(carboxymethyl)amino]phenolsulfophthalein = $C_{27}H_{24}N_2O_{13}S$, $R_1 = R_2 = H$)

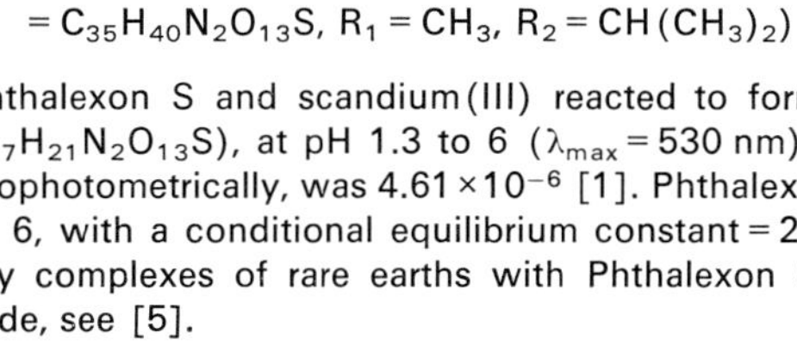

and Thymolphthalexon S (= 3,3'-Bis[bis(carboxymethyl)amino]thymolsulfophthalein = $C_{35}H_{40}N_2O_{13}S$, $R_1 = CH_3$, $R_2 = CH(CH_3)_2$)

Phthalexon S and scandium(III) reacted to form a bright pink 1:1 complex of composition $Sc(C_{27}H_{21}N_2O_{13}S)$, at pH 1.3 to 6 ($\lambda_{max}$ = 530 nm) [1, 2]. The dissociation constant, determined spectrophotometrically, was 4.61×10^{-6} [1]. Phthalexon S also formed a 1:1 chelate with yttrium(III) at pH 6, with a conditional equilibrium constant = 2.78×10^4 [3]. For the formation of binary and ternary complexes of rare earths with Phthalexon S in the presence of cetyltrimethylammonium bromide, see [5].

Thymolphthalexon S and scandium(III) formed a blue-violet, 1:2 (metal : ligand) chelate (λ_{max} = 590 nm) in aqueous solution at pH 2.3 to 3.8 [2, 4]. The apparent formation constant of the complex, determined spectrophotometrically, was 2.73×10^{11} [4]. The chromogenic reagent formed a 1:1 chelate with yttrium(III) at pH 7 with a conditional equilibrium constant, $\beta = 1.41 \times 10^4$, and a 1:2 chelate at pH 6, with $\beta = 6.00 \times 10^8$ [3].

References to 6.5.7:

[1] A. I. Cherkesov, N. M. Alykov, M. A. Karib'yants (Ftaleksony **1970** 151/6 from C.A. **76** [1972] No. 121155). — [2] M. A. Karib'yants, A. I. Cherkesov, N. M. Alykov (Izv. Vysshikh Uchebn. Zavedenii Khim. Khim. Tekhnol. **16** [1973] 523/6 from C.A. **79** [1974] No. 35661). — [3] N. M. Alykov, A. V. Grunin, A. I. Krasnov (Ftaleksony **1970** 128/33 from C.A. **76** [1972] No. 80558). — [4] N. M. Alykov, A. I. Cherkesov, M. A. Karib'yants (Ftaleksony **1970** 157/64 from C.A. **76** [1972] No. 135350). — [5] A. I. Kirillov, L. P. Shaulina, G. N. Koroleva, N. S. Poluektov (Zh. Analit. Khim. **32** [1977] 2154/8; J. Anal. Chem. [USSR] **32** [1977] 1714/7).

6.5.8 With Xylenol Orange

(=3,3'-Bis[bis(carboxymethyl)aminomethyl]cresolsulfophthalein = $C_{30}H_{30}N_2O_{13}$)

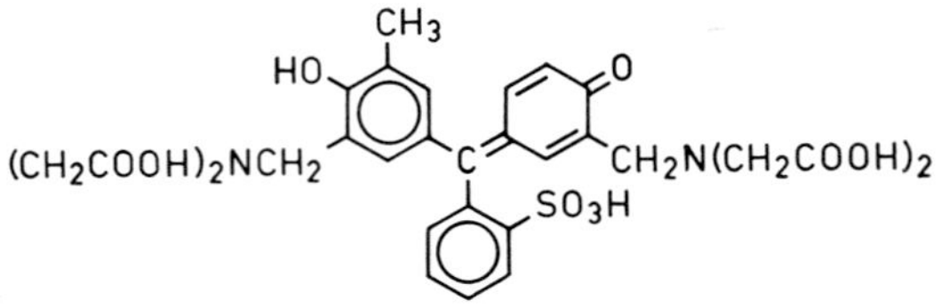

$M(C_{30}H_{30-n}N_2O_{13})^{3-n}$ Ions (n = 3 to 6) and $M(C_{30}H_{24}N_2O_{13})(OH)^{4-}$ Species

The existence in aqueous media of several forms of 1:1 Xylenol Orange rare earth chelates, the compositions of which differ mainly in the degree of ligand ionization, has been proposed [1 to 13]. Representative formation constant data for the various species are recorded in Table 6/3. The potentiometric data, obtained over the pH range 3 to 11, are not in agreement with the spectrophotometric data, which were obtained over the more limited pH range of 3 to 5. The values obtained potentiometrically revealed an increasing trend in stability for the complexes of lanthanum(III) through lutetium(III), with the values for the yttrium(III) chelates lying between those of dysprosium(III) and holmium(III) [6]. Proton dissociation constants for the complexes are recorded in [2].

The rates of complex formation were measured at 18°C by the temperature jump method (temperature jump amounted to about 7°C) in aqueous media, I = 0.7 (KNO_3), pH = 3.90 [14 to 16]. The values for the rate constants of complex formation and dissociation are recorded in [14, 15].

Rates for the ligand exchange reactions between the Xylenol Orange chelates and the ethylenediaminetetraacetate ion, $M(C_{30}H_{30-n}N_2O_{13})^{3-n} + edta^{4-} \rightleftharpoons M(edta)^- + (C_{30}H_{30-n}N_2O_{13})^{-n}$, were measured by the stopped flow method in the pH range 4.2 to 5.7 (in this pH range, Xylenol Orange exists predominantly in the forms with n = 2 to 4). The values for the rate constants at 25°C and activation energies for the exchange reactions are recorded in [16 to 18]. The reactions are first order in the Xylenol Orange complex and independent of the concentration of ethylenediaminetetraacetate ion. In the first stage of the mechanism, as proposed, the denticity of Xylenol Orange decreases as a result of ligation of water molecules, $MX + n\,H_2O \underset{k_{-1}}{\overset{k_1}{\rightleftharpoons}} M(H_2O)_nX'$, where X and X' represent Xylenol Orange with a greater and lesser denticity, respectively (charges omitted). The resulting complex then dissociates according to the scheme, $M(H_2O)_nX' \underset{k_{-2}}{\overset{k_2}{\rightleftharpoons}} M(III) + X$, where M(III) represents the solvated rare earth ion and X represents the various forms of the Xylenol Orange anion. Finally, the metal ion reacts with the ethylenediaminetetraacetate ion to complete the reaction. The entropies of activation for the exchange processes are generally negative (0 to −60 cal · mol · K^{-1}), an observation attributed to the attachment of n water molecules preceding the ligand dissociation [16 to 18]. For differential kinetic analysis of three-component mixture, see [19]. Complexes with Xylenol Orange and diphenylguanidine or Bromopyrogallol Red are studied in [20] and [21], respectively.

Table 6/3
Formation Constants for Rare Earth Complexes Derived from Xylenol Orange.

metal ion	method*)	temp. in °C	medium ($NaClO_4$)	lg K_1: $M(C_{30}H_{27}N_2O_{13})$	$M(C_{30}H_{26}N_2O_{13})^-$	$M(C_{30}H_{25}N_2O_{13})^{2-}$	$M(C_{30}H_{24}N_2O_{13})^{3-}$	$M(C_{30}H_{24}N_2O_{13})(OH)^{4-}$	Ref.
Sc^{3+}	sp	20	0.2		12.00				[1]
Y^{3+}	sp	20	0.2			12.81			[1]
Y^{3+}	gl	30	0.1	5.30	7.70	10.24	11.96	15.37	[2]
La^{3+}	sp	20	0.2			11.67			[1]
La^{3+}	sp	20	0.1			12.90			[3]
Ce^{3+}	gl	30	0.1**)	4.35	6.01	7.83	9.00	11.75	[2]
Pr^{3+}	gl	30	0.1	4.59	6.27	8.16	9.37	12.22	[2]
Nd^{3+}	gl	30	0.1	4.74	6.50	8.47	9.79	12.80	[2]
Sm^{3+}	gl	30	0.1	4.89	6.88	8.90	10.38	13.46	[2]
Eu^{3+}	gl	30	0.1	4.91	6.92	9.03	10.50	13.65	[2]
Gd^{3+}	gl	30	0.1	5.06	7.20	9.45	11.05	14.26	[2]
Tb^{3+}	gl	30	0.1	5.12	7.35	9.72	11.33	14.67	[2]
Dy^{3+}	gl	30	0.1	5.22	7.54	9.98	11.63	15.00	[2]
Ho^{3+}	gl	30	0.1	5.38	7.88	10.58	12.39	15.96	[2]
Er^{3+}	gl	30	0.1	5.45	8.11	10.94	12.93	16.60	[2]
Tm^{3+}	gl	30	0.1	5.54	8.32	11.21	13.27	17.04	[2]
Yb^{3+}	gl	30	0.1	5.64	8.54	11.56	13.74	17.60	[2]
Lu^{3+}	gl	30	0.1	5.74	8.80	11.86	14.11	18.35	[1]
Lu^{3+}	sp	20	0.2			14.09			

*) Abbreviations: gl = determined potentiometrically using a glass electrode, sp = spectrophotometric determination; **) medium (NaCl).

References to 6.5.8:

[1] V. N. Kumok, V. V. Serebrennikov (Zh. Neorgan. Khim. **11** [1966] 90/2; Russ. J. Inorg. Chem. **11** [1966] 47/9). — [2] R. Nayan, A. K. Dey (J. Indian Chem. Soc. **54** [1977] 192/9). — [3] V. I. Kornev, L. V. Kardapolova (Zh. Neorgan. Khim. **22** [1977] 1405/8; Russ. J. Inorg. Chem. **22** [1977] 764/6). — [4] K. N. Munshi, A. K. Dey (Mikrochim. Acta **1968** 1059/65). — [5] L. S. Serdyuk, V. S. Smirnova (Zh. Analit. Khim. **19** [1964] 451/6; Russ. J. Anal. Chem. **19** [1964] 413/7).

[6] A. K. Dey, R. Nayan (Proc. 12th Rare Earth Res. Conf., Vail, Colo., 1976, Vol. 1, p. 192/201). — [7] L. G. Anisimova, L. D. Shapovalova (Sb. Nauchn. Soobshch. Dagestan. Univ. Kafedra Khim. **1969** 39/41 from C.A. **77** [1972] No. 28522). — [8] M. Z. Yampol'skii, E. R. Oskotskaya (Uch. Zap. Kursk. Gos. Pedagog. Inst. **23** [1966] 244/55 from C.A. **67** [1967] No. 6226). — [9] E. T. Beschetnova (Sb. Nauchn. Soobshch. Dagestan. Univ. Kafedra Khim. **1969** 66/70 from C.A. **77** [1972] No. 69662). — [10] L. D. Fomenko, A. E. Okun, M. Z. Yampol'skii (Uch. Zap. Kursk. Gos. Pedagog. Inst. **23** [1966] 216/20 from C.A. **67** [1967] No. 104847).

[11] V. I. Kornev, G. I. Manasheva (Koord. Khim. **4** [1978] 1539/43 from C.A. **89** [1979] No. 221818). — [12] O. V. Kon'kova (Zh. Analit. Khim. **19** [1964] 73/8; Russ. J. Anal. Chem. **19** [1964] 63/7). — [13] A. K. Dey, R. Nayan (Acta Ciencia Indica **2** [1976] 104/10 from C.A. **86** [1977] No. 61057). — [14] L. I. Budarin, E. P. Zhuchenko, K. B. Yatsimirskii (Zh. Neorgan. Khim. **19** [1974] 1170/4; Russ. J. Inorg. Chem. **19** [1974] 639/41). — [15] K. B. Yatsimirskii, L. I. Budarin (Dokl. Akad. Nauk SSSR **180** [1968] 1383/4; Dokl. Chem. Proc. Acad. Sci. USSR **178/181** [1968] 581/2).

[16] K. B. Yatsimirskii, L. I. Budarin (Dokl. Akad. Nauk SSSR **170** [1966] 1107/9; Dokl. Chem. Proc. Acad. Sci. USSR **166/171** [1966] 991/2). — [17] E. P. Zhuchenko, L. I. Budarin, K. B. Yatsimirskii (Teor. Eksperim. Khim. **5** [1969] 507/10; Theor. Exptl. Chem. [USSR] **5** [1969] 329/31). — [18] L. I. Budarin, K. B. Yatsimirskii (Teor. Eksperim. Khim. **4** [1968] 474/9; Theor. Exptl. Chem. [USSR] **4** [1968] 305/9). — [19] K. B. Yatsimirskii, A. G. Khachatryan, L. I. Budarin (Dokl. Akad. Nauk SSSR **211** [1973] 1139/41 from C.A. **80** [1974] No. 22322). — [20] E. T. Beschetnova, L. N. Malinovskaya, A. P. Golovina, N. B. Zorov (Zh. Analit. Khim. **27** [1972] 2152/5 from C.A. **78** [1973] No. 52138).

[21] A. M. Ayubova, F. B. Imamverdieva (Azerb. Khim. Zh. No. 6 [1978] 116/20 from C.A. **91** [1979] No. 101431).

6.5.9 With Glycinecresol Red

(= 3,3'-Bis(carboxymethylaminomethyl)cresolsulfophthalein = $C_{27}H_{28}N_2O_9S$)

The instability constants of the 1:3 (metal : ligand) Glycinecresol Red chelates of scandium and yttrium were determined spectrophotometrically at 4.73×10^{-20} ($\lambda_{max} = 490$ nm at pH 5.0) [1] and 3.94×10^{-18} ($\lambda_{max} = 540$ nm at pH 5 to 6), respectively [2]. The violet color of the yttrium chelate develops immediately at pH 6.5 and is stable for several days over a wide temperature range [2].

References to 6.5.9:

[1] O. A. Tataev, E. T. Beschetnova, E. A. Yarysheva, V. K. Guseinov (Zh. Analit. Khim. **24** [1969] 255/7; Russ. J. Anal. Chem. **24** [1969] 152/4). — [2] E. T. Beschetnova O. A. Tataev (Sb. Nauchn. Soobshch. Dagestan. Univ. Kafedra Khim. **1969** 26/8 from C.A. **76** [1972] No. 41667).

6.5.10 With para-Xylenolphthalexon S

(= 3,3'-Bis[bis(carboxymethyl)aminomethyl]xylenolsulfophthalein = $C_{33}H_{36}N_2O_{13}S$)

Spectrophotometric measurements revealed that ytterbium chloride reacts with p-Xylenolphthalexon S in aqueous solution with formation of two complex compounds. A 1:1 complex, $Yb(C_{33}H_{32}N_2O_{13}S)^-$, with $\lambda_{max} = 610$ nm ($\varepsilon = 2.40 \times 10^4$ $l \cdot mol^{-1} \cdot cm^{-1}$), predominates in an excess of the metal ion (pH range 3.5 to 4.5) and a 1:2 complex, $Yb(C_{33}H_{32}N_2O_{13}S)_2(OH)^{6-}$, with $\lambda_{max} = 580$ nm ($\varepsilon = 2.50 \times 10^4$ $l \cdot mol^{-1} \cdot cm^{-1}$), predominates in an excess of the reagent (pH range 6.7 to 7.2). The instability constants (pK values) of the two species, determined spectrophotometrically, I = 0.1 (KCl), were 14.92 and 24.68, respectively. A dimeric structure was proposed for the 1:1 chelate. Molecular structures showing the sites of ligation for both the 1:1 and 1:2 chelates are depicted in the original literature, A. I. Kirillov, G. N. Koroleva, N. S. Poluektov (Zh. Neorgan. Khim. **20** [1975] 3228/33; Russ. J. Inorg. Chem. **20** [1975] 1784/7).

6.5.11 With Methylthymol Blue

(= Tetrasodium Salt of 3,3'-Bis[bis(carboxymethyl)aminomethyl]thymolsulfophthalein = $C_{37}H_{40}N_2Na_4O_{13}S$)

Several forms of the Methylthymol Blue rare earth chelates in aqueous media have been proposed [1 to 11]. Zielinski and Lomozik [1 to 3] reported that the chelate composition depended on the concentration of reactants in solution, and the molar composition, pH, and ionic strength of the solutions. They proposed the existence of three species (I = 0.4 M KCl) with compositions $M(C_{37}H_{40}N_2O_{13}S)^-$ (pH ca. 6 to 7), $M(C_{37}H_{39}N_2O_{13}S)^{2-}$ (pH > 8), and $M_2(C_{37}H_{40}N_2O_{13}S)^{2+}$ (pH ca. 3.3 to 5.3) for M = Y, La, Pr, Nd, Sm, Gd [2]. In each case, the number of protons released on formation of the chelate was determined by titration with an aqueous solution of 0.001 M sodium hydroxide [1]. A fourth species of composition $M_2(C_{37}H_{40}N_2O_{13}S)_3^{6-}$ (M = Y, La, Pr, Nd, Sm, Gd) was identified in solutions (I = 0.4 M KCl) having pH > 6.1 and containing close to the stoichiometric ratio (2:3) of reactants. Reducing the ionic strength of the latter solutions to 0.1 (KCl) brought about the disappearance of the 2:3 species [3]. Instability and stability constants for the complexes are recorded in Table 6/4.

Mal'kova and Fateeva [4] interpreted their spectrophotometric data, obtained at 20°C for weakly acidic solutions maintained at constant ionic strength, I = 0.4 (KCl), in terms of a mixture of $M(C_{37}H_{40}N_2O_{13}S)^-$ and $M_2(C_{37}H_{40}N_2O_{13}S)(OH)_2$ (M = La, Nd, Sm, Gd, Ho, Er) in equilibrium. Equilibrium constants which defined the systems are recorded in Table 6/4 [4]. Similar results are reported in [5]. At slightly higher pH, the existence of 1:2 (metal:ligand) chelates of composition $M(C_{37}H_{38}N_2O_{13}S)_2^{9-}$ was proposed [6]. Equilibrium constants corresponding to dissociation of the latter species are also recorded in Table 6/4 [4, 6].

Table 6/4
Instability or Stability Constants for Methylthymol Blue Rare Earth Chelates.

metal ion	instability constants*) [1] $M(C_{37}H_{39}N_2O_{13}S)^{2-}$	$M(C_{37}H_{40}N_2O_{13}S)^-$	$M_2(C_{37}H_{40}N_2O_{13}S)^{2+}$	stability constants**) [3] $M_2(C_{37}H_{40}N_2O_{13}S)_3^{6-}$
Y^{3+}	1.09×10^{-6}	0.59×10^{-5}	3.7×10^{-8}	3.53×10^{17}
La^{3+}	1.99×10^{-5}	0.93×10^{-4}	2.2×10^{-8}	2.66×10^{18}
Pr^{3+}	1.04×10^{-5}	0.40×10^{-4}	1.6×10^{-8}	1.74×10^{18}
Nd^{3+}	8.09×10^{-6}	0.35×10^{-4}	8.9×10^{-9}	2.22×10^{18}
Sm^{3+}	3.92×10^{-6}	0.13×10^{-4}	4.7×10^{-9}	4.79×10^{20}
Gd^{3+}	2.76×10^{-6}	0.85×10^{-5}	2.3×10^{-9}	1.36×10^{20}

*) Determined spectrophotometrically at pH 6.5, I = 0.4 (KCl).—**) Determined spectrophotometrically at pH 6.90, except for M = La, which was determined at pH 8.00, I = 0.4 (KCl).

Based on spectrophotometric and pH titration data, Tereshin and coworkers [7] proposed formation of the yttrium chelates, $Y_2(C_{37}H_{39}N_2O_{13}S)_2^{4-}$ and $Y_2(C_{37}H_{38}N_2O_{13}S)_2^{6-}$ at pH 7 and 10, respectively. The value of the equilibrium constant (K) corresponding to complete dissociation of the $Y_2(C_{37}H_{39}N_2O_{13}S)_2^{4-}$ species, determined spectrophotometrically at pH 6.96, was 1.2×10^{-50} ($K=[Y^{3+}]^2\cdot[(C_{37}H_{38}N_2O_{13}S^{6-}]^2\cdot[H^+]^2\cdot[Y_2(C_{37}H_{39}N_2O_{13}S)_2^{4-}]^{-1}$, whereas the acid dissociation constants of the chelate had values $pK_1=8.0$ and $pK_2=9.5$. For the $Y_2(C_{37}H_{38}N_2O_{13}S)_2^{6-}$ species, pK = 37 (K = instability constant) [7]. Budensinsky and Antonescu [10] interpreted spectrophotometric data obtained for the lanthanum system in terms of a chelate of composition $La_2(C_{37}H_{38}N_2O_{13}S)_2^{6-}$ at pH 5.54, with an overall stability constant, lg β = 39.8. They accounted for their data in the alkaline region (pH range 10.00 to 11.81) in terms of the $La_2(C_{37}H_{38}N_2O_{13}S)_2(OH)_2^{8-}$ species, with lg β = 23.2 ($\beta=[La_2(C_{37}H_{38}N_2O_{13}S)_2(OH)_2^{8-}]\cdot[La^{3+}]^{-2}\cdot[C_{37}H_{38}N_2O_{13}S]^{-2}\cdot[OH]^{-2}$).

The rates of formation of the $M(C_{37}H_{38}N_2O_{13}S)^{3-}$ species, formed according to the equation, $M_2(C_{37}H_{38}N_2O_{13}S)+C_{37}H_{41}N_2O_{13}S^{3-}\rightleftharpoons 2\,M(C_{37}H_{38}N_2O_{13}S)^{3-}+3\,H^+$ (M = Nd, Sm, Eu, Gd, Ho, Er, Tm, Lu), were determined spectrophotometrically. When the reactants were mixed under the conditions selected for the kinetic studies, the solutions instantaneously acquired the blue coloration characteristic of the $M_2(C_{37}H_{38}N_2O_{13}S)$ complex. The solution then gradually turned green,

Table 6/5
Equilibrium Constants Corresponding to Dissociation of Methylthymol Blue Rare Earth Complexes, $M(C_{37}H_{40}N_2O_{13}S)^-$, $M_2(C_{37}H_{40}N_2O_{13}S)(OH)_2$, and $M(C_{37}H_{38}N_2O_{13}S)_2^{9-}$.

metal ion	$M(C_{37}H_{40}N_2O_{13}S)^-$ [4] pH range	K_1*)	$M_2(C_{37}H_{40}N_2O_{13}S)(OH)_2$ [4] pH range	K_2**)	$M(C_{37}H_{38}N_2O_{13}S)_2^{9-}$ [6] pH range	$-\lg\beta_2$***)
La^{3+}		1.1×10^{-6}		6.3×10^{-7}	6.60 to 7.05	20.0
Nd^{3+}	5.10 to 5.98	1.0×10^{-5}	5.10 to 5.98	1.0×10^{-3}	6.00 to 7.03	14.0
Sm^{3+}	5.10 to 6.07	1.6×10^{-5}	5.10 to 6.07	1.3×10^{-3}	6.00 to 7.00	10.0
Gd^{3+}	5.10 to 5.95	2.5×10^{-5}	5.10 to 5.95	3.2×10^{-3}	5.95 to 7.05	11.0
Ho^{3+}	4.47 to 5.64	3.2×10^{-3}	4.47 to 5.64	1.0×10^{-1}	5.95 to 7.00	1.0
Er^{3+}	3.95 to 5.10	1.0×10^{-3}	3.95 to 5.10	3.2×10^{-3}	6.00 to 7.00	0.0

*) $K_1=[M(C_{37}H_{40}N_2O_{13}S)^-]\cdot[H^+]^2\cdot[(C_{37}H_{42}N_2O_{13}S)^{2-}]^{-1}\cdot[M^{3+}]^{-1}$, determined spectrophotometrically at 20°C, I = 0.4 (KCl).— **) $K_2=[M_2(C_{37}H_{40}N_2O_{13}S)(OH)_2]\cdot[H^+]^2\cdot[M(C_{37}H_{40}N_2O_{13}S)^-]^{-1}\cdot[M^{3+}]^{-1}$, determined spectrophotometrically at 20°C, I = 0.4 (KCl).— ***) For M = La to Gd, $\beta_2=[M(C_{37}H_{38}N_2O_{13}S)_2^{9-}]\cdot[H^+]^4\cdot[(C_{37}H_{40}N_2O_{13}S)^{4-}]^{-2}\cdot[M^{3+}]^{-1}$, and for M = Ho and Er, $\beta_2=[M(C_{37}H_{38}N_2O_{13}S)_2^{9-}]\cdot[H^+]^2\cdot[(C_{37}H_{39}N_2O_{13}S)^{5-}]^{-2}\cdot[M^{3+}]^{-1}$. Values of β_2 were determined spectrophotometrically at 20°C, I = 1.0 (KCl).

corresponding to accumulation of the $M(C_{37}H_{38}N_2O_{13}S)^{3-}$ ion. The rate constants, activation energies, and entropies for the reactions indicated above, are recorded in Table 6/6. Each reaction was second order, first order in both the Methylthymol Blue anion, $C_{37}H_{41}N_2O_{13}S^{3-}$ and the dinuclear complex, $M_2(C_{37}H_{38}N_2O_{13}S)$. The effective rate constants decreased as the pH was raised, suggesting a preequilibrium involving protonation of the reacting species: $M_2(C_{37}H_{38}N_2O_{13}S) + H^+ \rightleftharpoons M_2(C_{37}H_{39}N_2O_{13}S)^+$. The rate determining step, as was proposed, involved interaction of the protonated form of the complex with the anionic ligand, $C_{37}H_{41}N_2O_{13}S^{3-}$. The effective rate constant for the formation of the $Gd(C_{37}H_{38}N_2O_{13}S)^{3-}$ complex was shown to be independent of the concentration of ammonium acetate introduced as the buffer solution [12].

Table 6/6

Rate Constants and the Activation Energies and Entropies for the Formation of the 1:1 Methylthymol Blue Rare Earth Complexes of Composition $M(C_{37}H_{38}N_2O_{13}S)^{3-}$.

metal ion	rate constant*) k in $l \cdot mol^{-1} \cdot s^{-1}$	activation energy in $kcal \cdot mol^{-1}$	$\Delta S^{\ddagger}$ $cal \cdot mol^{-1} \cdot K^{-1}$
Nd^{3+}	67 ± 9	7.6 ± 0.2	−25 ± 1
Sm^{3+}	37 ± 5	2.8 ± 0.8	−42 ± 8
Eu^{3+}	44 ± 9	14 ± 2	−36 ± 4
Gd^{3+}	50 ± 25	10 ± 1	−18 ± 2
Ho^{3+}	17 ± 3	11 ± 2	−20 ± 3
Er^{3+}	13 ± 4	12 ± 2	−18 ± 3
Tm^{3+}	6.3 ± 3	10 ± 1	−21 ± 2
Lu^{3+}	5.6 ± 0.4	9.0 ± 3.4	−24 ± 7

*) Determined spectrophotometrically at 10°C, I = 0.5 ($NH_4C_2H_3O_2$, $NaClO_4$).

The decomposition of the 1:1 chelates, $Ln(MTBX)_n$ with Ln = Sm, Gd, Er, Tm, Lu where MTB is used in the formula to denote several forms of the metal chelate, $Ln(C_{37}H_{38+n}N_2O_{13}S)^{(6-n)-}$ coexisting in equilibrium, was studied spectrophotometrically and described according to the equation $2\,Ln(MTB) \rightleftharpoons Ln_2(C_{37}H_{38}N_2O_{13}S) + C_{37}H_{41}N_2O_{13}S^{3-} \rightleftharpoons C_{37}H_{40}N_2O_{13}S^{4-} + H^+$. The reaction was found to be first order with respect to the concentration of the complex, Ln(MTB), and mixed order with respect to the hydrogen ion concentration. A proposed mechanism for the decomposition involved two parallel paths for protonation, with subsequent dissociation of the resulting monoprotonated and diprotonated complexes. The rate constants for the reaction along each pathway, determined at 15°C, I = 0.5 ($NH_4C_2H_3O_2$, $NaClO_4$) are recorded in the original literature [13].

Kinetic studies of the formation and dissociation of a thulium-Methylthymol Blue complex and the substitution of Methylthymol Blue by ethylenediaminetetraacetate ion in the erbium chelate are described in references [14] and [15], respectively. For ion associates of Sc^{3+}, Y^{3+}, and La^{3+} with Methylthymol Blue and diphenylguanidine, see [16].

References to 6.5.11:

[1] S. Zelin'ski, L. Lomozik (Zh. Neorgan. Khim. **21** [1976] 692/6; Russ. J. Inorg. Chem. **21** [1976] 372/5). — [2] S. Zelin'ski, L. Lomozik (Zh. Neorgan. Khim. **22** [1977] 928/34; Russ. J. Inorg. Chem. **22** [1977] 514/7). — [3] S. Zelin'ski, L. Lomozik (Zh. Neorgan. Khim. **22** [1977] 1491/5; Russ. J. Inorg. Chem. **22** [1977] 812/4). — [4] T. V. Mal'kova, N. A. Fateeva (Zh. Neorgan. Khim. **13** [1968] 2094/101; Russ. J. Inorg. Chem. **13** [1968] 1082/6). — [5] T. V. Mal'kova, N. A. Fateeva, K. B. Yatsimirskii (Zh. Neorgan. Khim. **12** [1967] 915/21; Russ. J. Inorg. Chem. **12** [1967] 481/5).

[6] T. V. Mal'kova, N. A. Fateeva, K. B. Yatsimirskii (Zh. Neorgan. Khim. **12** [1967] 3342/7; Russ. J. Inorg. Chem. **12** [1967] 1769/71). — [7] G. S. Tereshin, A. R. Rubinshtein, I. V. Tananaev (Zh. Analit. Khim. **20** [1965] 1082/92; Russ. J. Anal. Chem. **20** [1965] 1138/46). — [8] L. S. Serdyuk, V. S. Smirnaya (Zh. Analit. Khim. **20** [1965] 161/4; Russ. J. Anal. Chem. **20** [1965] 146/8). — [9] M. K. Akhmedli, D. G. Gambarov (Zh. Analit. Khim. **22** [1967] 1183/5; Russ. J. Anal. Chem. **22** [1967] 997/9). — [10] B. Budesinsky, E. Antonescu (Collection Czech. Chem. Commun. **28** [1963] 3264/70).

[11] S. N. Drozdova (Uch. Zap. Kafedry Obshch. Khim. Kursk. Gos. Pedagog. Inst. **35** [1967] 154/61 from C.A. **71** [1969] No. 45460). — [12] T. V. Mal'kova, L. V. Krasukhina (Zh. Neorgan. Khim. **20** [1975] 2656/63; Russ. J. Inorg. Chem. **20** [1975] 1470/4). — [13] T. V. Mal'kova, L. V. Krasukhina (Zh. Neorgan. Khim. **20** [1975] 3234/9; Russ. J. Inorg. Chem. **20** [1975] 1787/90). — [14] T. V. Mal'kova, L. V. Krasukhina (Izv. Vysshikh Uchebn. Zavedenii Khim. Khim. Tekhnol. **16** [1973] 1326/30 from C.A. **79** [1974] No. 149938). — [15] O. I. Davydova, L. V. Krasukhina, T. V. Mal'kova (Tr. Ivanov. Khim. Tekhnol. Inst. **1973** No. 15, pp. 208/14 from C.A. **81** [1974] No. 127242).

[16] E. T. Beschetnova, A. P. Golobina, N. B. Zorov (Zh. Analit. Khim. **28** [1973] 1943/6; J. Anal. Chem. [USSR] **28** [1973] 1725/7).

6.5.12 With Glycinethymol Blue

(= 3,3'-Bis(carboxymethylaminomethyl)thymolsulfophthalein = $C_{33}H_{40}N_2O_9S$)

$Sc(C_{33}H_{36}N_2O_9S)^-$ Ion. The instability constant of the Glycinethymol Blue scandium chelate was determined spectrophotometrically in aqueous media, I = 0.05 (KCl): $K = 2.60 \times 10^{-4}$ to 3.52×10^{-4}. The anionic nature of the chelate was established by electromigration studies [1].

Ternary complexes of rare earth elements with Glycinethymol Blue and cetyltrimethylammonium bromide are investigated by Kirillov et al. [2].

References to 6.5.12:

[1] S. Zielinski, L. Lomozki, J. Andrzejewska (Zh. Neorgan. Khim. **20** [1975] 329/35; Russ. J. Inorg. Chem. **20** [1975] 182/6). — [2] A. I. Kirillov, O. P. Makarenko, N. S. Poluektov, R. S. Lauer, L. P. Shaulina (Zh. Analit. Khim. **28** [1973] 1618/21).

6.5.13 With Fluorescein and Its Derivatives

Fluorescein = 9-(2-Carboxyphenyl)-6-hydroxy-3H-xanthene-3-one (= $C_{20}H_{12}O_5$)
Gallein = 4,5'-Dihydroxyfluorescein (= $C_{20}H_{12}O_7$)
Dibromogallein = 3,3'-Dibromo-4,'5'-dihydroxyfluorescein (= $C_{20}H_{10}Br_2O_7$)

Rare earth fluorescein chelates of various compositions were obtained according to the following procedures:

$La(C_{20}H_{11}O_5)(OH)_2 \cdot 2H_2O$ was precipitated at pH 6.7 on mixing a 0.05 M $La(NO_3)_3$ solution with a 0.05 M alcoholic solution of fluorescein. $Ce(C_{20}H_{11}O_5)(OH)_2 \cdot 2H_2O$ was precipitated in the pH range 6.7 to 6.9 according to the procedure used for preparation of the analogous lanthanum complex, except that 0.06 M solutions of each reagent were used. Precipitation of $Pr_7(C_{20}H_{10}O_5)_2)OH)_{17} \cdot 6H_2O$ was effected by mixing 0.04 M $Pr(NO_3)_3$ solution with a 0.01 M alcoholic solution of fluorescein at pH 6.8 and 60°C. The compound was collected and dried at 80°C. $Nd_5(C_{20}H_{10}O_5)(OH)_{13} \cdot 6H_2O$ was obtained when a 0.06 M $Nd(NO_3)_3$ solution was mixed with a 0.015 M alcoholic solution of fluorescein in the pH range 6.7 to 7.2. The precipitate was washed with an ethanol/water mixture and dried. $Sm_{15}(C_{20}H_{10}O_5)_4(OH)_{37} \cdot 18H_2O$ was precipitated in the pH range 6.9 to 7.2 according to a procedure similar to that described for preparation of the lanthanum complex, except that the concentration of the reactant solutions was 0.06 molar.

All compounds were insoluble in common organic solvents. The color and magnetic moment of each complex are recorded below. The magnetic moments in μ_B at room temperature are in agreement with the values calculated from the Van Vleck formula, but at other temperatures, an anomalous behavior was observed which was attributed to either the polymeric nature of the complexes and/or to ligand field effects. Electrical conductivity data obtained for the solid compounds suggested that these complexes have semi-conducting properties and may be expected to be antiferromagnetic. The solid state conductivity generally decreased with an increase in the atomic number of the rare earth. Infrared data for both the complexes and free ligand were tabulated and interpreted in terms of bidentate ligation via the carboxy and hydroxyl groups. Bands at ≈ 565 cm^{-1} were assigned to a $\nu(M\text{-}O)$ stretching mode. A polymeric molecular structure with hydroxy bridges was proposed [1].

compound	color	μ_B	temp. in K
$La(C_{20}H_{11}O_5)(OH)_2 \cdot 2H_2O$	reddish brown	diamagnetic	
$Ce(C_{20}H_{11}O_5)(OH)_2 \cdot 2H_2O$	dark chocolate	2.48	299.6
$Pr_7(C_{20}H_{10}O_5)_2(OH)_{17} \cdot 6H_2O$	dull red	3.42	302
$Nd_5(C_{20}H_{10}O_5)(OH)_{13} \cdot 6H_2O$	red	3.64	287
$Sm_{15}(C_{20}H_{10}O_5)(OH)_{37} \cdot 18H_2O$	dark red	1.64	302

For the determination of dysprosium with gallein, see [2], for that of scandium with dibromogallein, see [3].

References to 6.5.13:

[1] R. A. Bhobe (J. Indian Chem. Soc. **51** [1974] 846/9). — [2] V. C. Shastri, A. P. Joshi (J. Indian Chem. Soc. **56** [1979] 192/3). — [3] V. P. Antonovich, G. I. Ibragimov, E. M. Nevskaya, E. I. Shelikhina (Zh. Analit. Khim. **34** [1979] 81/6 from C. A. **90** [1979] No. 179489).

6.5.14 With 2-Hydroxyphenylfluorone

(= 2,6,7-Trihydroxy-9-(2-hydroxyphenyl)-3H-xanthene-3-one = $C_{19}H_{12}O_6$)

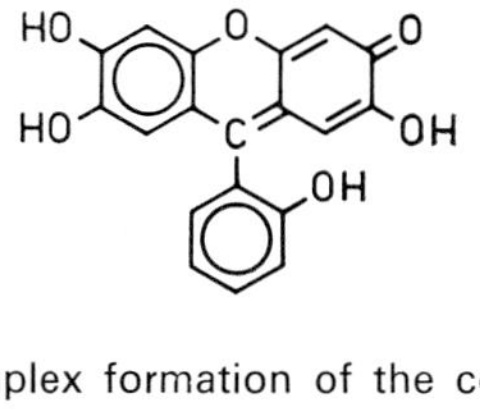

Complex formation of the cerium subgroup elements (M = La, Ce, Pr, Nd, Sm, Eu, Gd) with 2-hydroxyphenylfluorone was studied spectrophotometrically over a wide range of hydrogen ion concentrations in 20% (v/v) ethanol/water. Complex formation was observed at pH > 4.0 with maximum yields at pH 6.0 to 7.0 ($\lambda_{max} \approx 550$ nm). The data were interpreted in terms of formation of the 1:2 (metal:ligand) chelates (degree of ligand ionization not defined). The formation constants, recorded below, were determined at pH 6.0, I = 0.1 (KNO_3).

metal ion . .	La^{3+}	Ce^{3+}	Pr^{3+}	Nd^{3+}	Sm^{3+}	Eu^{3+}	Gd^{3+}
K*)	9.0×10^{6}	1.4×10^{11}	1.2×10^{11}	1.3×10^{11}	6.1×10^{11}	2.1×10^{12}	2.7×10^{12}

*) Complex formation reached equilibrium after 30 min at 20°C, except for the cerium system which reacted equilibrium at 50°C after 30 min, O. V. Sivanova, N. N. Gritsienko (Zh. Neorgan. Khim. **21** [1976] 1202/7; Russ. J. Inorg. Chem. **21** [1976] 654/7).

6.5.15 With Pyrogallol Red

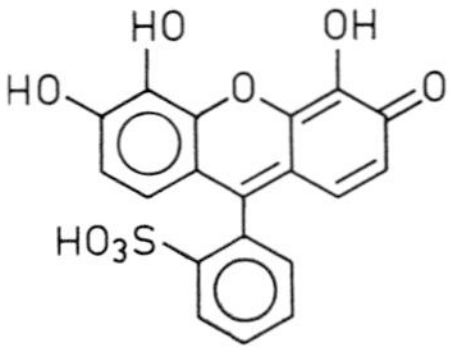

(= $C_{19}H_{12}O_8S$)

and Bromopyrogallol Red (= 3′,3″-Dibromosulfogallein = $C_{19}H_{10}Br_2O_8S$)

Formation constants, recorded below, for the 1:1 bromopyrogallol red rare earth chelates (degree of ligand ionization not defined) were determined spectrophotometrically at 25°C in aqueous media (acetate buffer). The stability constant (determined spectrophotometrically, pH = 6.2), and ligational enthalpy and entropy changes for the 1:1 yttrium chelate, reported by Srivastava [2], are 6.0, −3.5 kcal/mol^{-1}, and 15.53 cal·mol·K^{-1}, respectively. Formation constant data are also reported in [3] to [7]. The wavelengths (nm) of maximum absorbance for the 1:1 chelates are: M = Sc, 590; M = Y, Ho, Er, 580; M = Ce to Dy, 690; and M = La, 720 [1]. Based on spectrophotometric studies, Sivanova and coworkers [8] postulated the formation of both 1:1 and 2:1 chelates of lanthanum, praseodymium, and samarium. The absorption maxima of the 1:1 and 2:1 complexes, respectively, were assigned to bands at 575 and 695 nm for lanthanum, 570 and 685 nm for praseodymium, and 560 and 665 nm for samarium [8].

metal ion .	Sc^{3+}	Y^{3+}	La^{3+}	Ce^{3+}	Pr^{3+}	Nd^{3+}	Sm^{3+}	Eu^{3+}	Gd^{3+}	Tb^{3+}	Dy^{3+}	Ho^{3+}	Er^{3+}
lg K_1 . . .	5.4	5.9	5.3	5.2	5.1	5.1	5.0	4.5	5.2	4.3	4.2	4.2	4.2

A quantum chemical study of the structure of the ligand and of the rare earth complexes (MO-LCAO method) shows, that coupling of the rare earths to bromopyrogallol red takes place both through the o-dihydroxy group, resulting in a long wave absorbance maximum, and through the o-hydroxyquinone group, with absorption in the short-wave region [9].

For ternary complexes with pyrogallol red and diphenylguanidine and their use for the extraction-spectrophotometric determination of rare earth elements in mineral waters, see [10] for those with bromopyrogallol red and Xylenol Orange, see [11].

References to 6.5.15:

[1] M. S. Jog, S. C. Pande, V. L. Shah, S. P. Sangal (J. Indian Chem. Soc. **51** [1974] 908/9). — [2] K. C. Srivastava (J. Chinese Chem. Soc. [Taipei] **23** [1976] 229/32). — [3] S. C. Pande, S. P. Sangal (Chem. Era **11** [1975] 29/33 from C.A. **85** [1976] No. 86752). — [4] M. K. Akhmedli, D. G. Gambarov (Uch. Zap. Azerb. Gos. Univ. Ser. Fiz. Mat. Khim. Nauk **1967** No. 2, pp. 24/7 from C.A. **69** [1968] No. 101147). — [5] L. S. Serdyuk, U. F. Silich (Izv. Vysshikh Uchebn. Zavedenii Khim. Khim. Tekhnol. **9** [1966] 190/4 from C.A. **65** [1966] No. 9814).

[6] M. K. Akhmedli, A. M. Ayubova, T. R. Babaeva (Azerb. Khim. Zh. **1972**, No. 4, pp. 121/5 from C.A. **79** [1973] No. 73185). — [7] R. L. Kushwaha, C. K. Verma, Om Prakash, S. P. Mushran (J. Indian Chem. Soc. **54** [1977] 285/8). — [8] O. V. Sivanova, G. S. Ivankovich, N. G. Kachkovskaya (Issled. Obl. Khim. Redkozem. Elem. **1969** 21/2 from C.A. **76** [1972] No. 132187). — [9] O. V. Sivanova, V. N. Rodnikova, S. P. Mushtakova (Zh. Analit. Khim. **34** [1979] 2134/7 from C.A. **92** [1980] No. 223428 — [10] A. I. Kirillov, C. P. Makarenko, N. A. Vlasov (Izv. Vysshikh Uchebn. Zavedenii Khim. Khim. Tekhnol. **16** [1973] 1150/3 from C.A. **80** [1974] No. 85839). — [11] A. M. Ayubova, F. B. Imamverdieva (Azerb. Khim. Zh. **1978** No. 6, pp. 116/20; C.A. **91** [1979] No. 101431).

7 Complexes with Ethers and O-Heterocycles

Edward R. Birnbaum
Department of Chemistry, New Mexico State University
Las Cruces, New Mexico, USA

7.1 With Diethyl Ether H_5C_2-O-C_2H_5 (= $C_4H_{10}O$)

There are no reports of definite adducts of rare earth salts with diethyl ether. A few adducts of MR_3 compounds (R = alkyl) which have been reported, will be described in the respective section on organometallic rare earth compounds.

7.2 With 1,2-Diethoxyethane $(CH_3CH_2OCH_2)_2$ (= Ethylene Glycol Diethyl Ether = $C_6H_{14}O_2$)

The molar ratios of the precipitates, obtained by stirring the anhydrous chloride with the dry ether in the absence of air at 25°C and drying in vacuum over P_4O_{10} are listed below together with the solubilities of MCl_3 salts in ethylene glycol diethyl ether, E. M. Kirmse, K. J. Zwietasch (Z. Chem. [Leipzig] **7** [1967] 281).

M	Sc	Y	La	Pr	Nd
solubility in g/100 g solution .	1.22	—	—	—	—
M : Cl : ether . .	1:3.00:1.13	1:2.97:1.37	1:2.96:0.70	1:3.00:0.89	1:3.03:0.92

M	Sm	Eu	Gd	Tb	Dy
solubility in g/100 g solution .	—	0.25	0.33	0.22	0.25
M : Cl : ether . .	1:2.88:0.96	1:2.91:1.10	1:2.91:0.99	1:2.93:1.37	1:2.91:1.70

M	Ho	Er	Tm	Yb
solubility in g/100 g solution .	0.37	0.67	0.88	1.15
M : Cl : ether . .	1:2.97:1.82	1:2.93:1.86	1:2.97:2.00	1:3.03:0.94

7.3 With Tetrahydrofuran (= thf = C_4H_8O)

The tetrahydrofuran adduct of scandium chloride was obtained by adding the anhydrous salt to a large excess of the solvent with mild (50°C) heating. Pale pink plates of the tris adduct, $ScCl_3 \cdot 3$thf, were obtained [1, 2]. An adduct of scandium tetrahydroborate, $Sc(BH_4)_3 \cdot$ thf, was prepared by distilling dry tetrahydrofuran into a mixture of anhydrous scandium chloride and lithium tetrahydroborate (slight excess). After standing for some days, the mixture was filtered, the solvent was removed under vacuum, and the resultant white solid was purified by rapid sublimation at ≈80°C onto an ice-salt probe. The extremely moisture-sensitive sublimate contained no chlorine and appeared to be thermally stable at room temperature [3].

Adducts of tetrahydrofuran with several different salts of the lanthanides have been prepared. Each adduct of the chloride salts was prepared using a special apparatus to maintain a dry, oxygen-free atmosphere by extracting the anhydrous chloride salt into tetrahydrofuran which had been previously distilled from $LiAlH_4$. After cooling the extract and allowing to stand for 4 h at room temperature, the solution was filtered and the crystals dried for 25 to 40 min at 0.5 Torr. Formulations corresponding to $MCl_3 \cdot 1.5thf$ (M = La, Ce); $MCl_3 \cdot 2thf$ (M = Pr, Nd); $MCl_3 \cdot 3.5thf$ (M = Y, Sm to Tm); and $MCl_3 \cdot 3thf$ (M = Yb, Lu) were reported [4]. A similar method reported earlier yielded nonstoichiometric adducts and in general fewer moles of tetrahydrofuran per mole of chloride salt [2, 5]. Washing the adducts isolated from tetrahydrofuran solution with ether or benzene alters the mole ratio of the Eu, Tb, and Dy chloride salts from 3.5 to 2.0 [4]. The bromide salts were prepared in a similar way, resulting in adducts with the formulations: $MBr_3 \cdot 4thf$ (M = La to Pr); $MBr_3 \cdot 3.5thf$ (M = Y, Nd to Er); and $MBr_3 \cdot 3thf$ (M = Tm, Yb, Lu) [6]. The tetrahydrofuran adducts of the cyanide salts were prepared by adding a solution of LiCN in tetrahydrofuran to the bromide thf adduct, resulting in precipitation of the bis adducts, $M(CN)_3 \cdot 2thf$ where M = Y, Pr to Lu [7]. Adducts of the thiocyanate salts can be isolated by treating the anhydrous chloride salts with KSCN in thf or a mixture of thf and benzene. The resulting formulations are: $M(SCN)_3 \cdot 3thf \cdot C_6H_6$ (M = Yb) and $M(SCN)_3 \cdot 4thf$ (M = Y, La, Ce, Pr, Nd, Sm, Gd, Dy, Ho, and Er) [8]. For adducts of rare earth halogenides with thf, see also [13].

Reactions of the chloride salts with $NaBH_4$ in tetrahydrofuran are reported to yield bis adducts with the formulation $M(BH_4)_3 \cdot 2thf$ where M = La and Nd [9]. Tris tetrahydrofuran adducts of lanthanide borohydride salts have been prepared by two methods [10]. The first involves the reaction of aluminium borohydride with the rare earth methoxide salts. Crystals of $M(BH_4)_3 \cdot 3thf$ (M = Y, Er, Gd) were obtained from tetrahydrofuran solution. The products were washed with cyclohexane and benzene and then vacuum sublimed. The second method involves the reaction of diborane with the rare earth methoxide salts in a solution of tetrahydrofuran at room temperature. The excess of thf and $B(OCH_3)_3$ are pumped off and the products washed with fresh thf and vacuum filtered. The tris adducts for all the lanthanides and yttrium are reportedly prepared by these two methods, but only the Y, Er, and Gd salts are discussed [10]. For adducts of tetrahydroborates with thf, see also [14 to 18].

The X-ray crystal structure determination of the scandium chloride tris adduct places the monoclinic crystals in space group $P2^1/c\text{-}C^5_{2h}$ (No. 14) with Z = 4 and cell parameters a = 8.890(4), b = 12.842(6), c = 15.485(6) Å, and β = 92.24(5)°. The volume of the unit cell is 1767 Å^3 with a calculated density of 1.38 g/ml and a measured density of 1.37 g/ml. The final R-factor was 0.077 for 1227 reflections. The molecular structure (**Fig. 7-1**) shows octahedral coordination about the scandium ion. The Sc-Cl distances are 2.406(4), 2.420(4), and 2.415(4) Å for Cl_1-Cl_3, respectively, and the Sc-O distances are 2.236(8), 2.147(7), and 2.164(7) Å, for O_1-O_3, respectively. The major deviation in the structure from octahedral can be seen from the fact that the Cl_1-Sc-ligand angles all fall in the range 92.5° to 96.7° [1].

Fig. 7-1

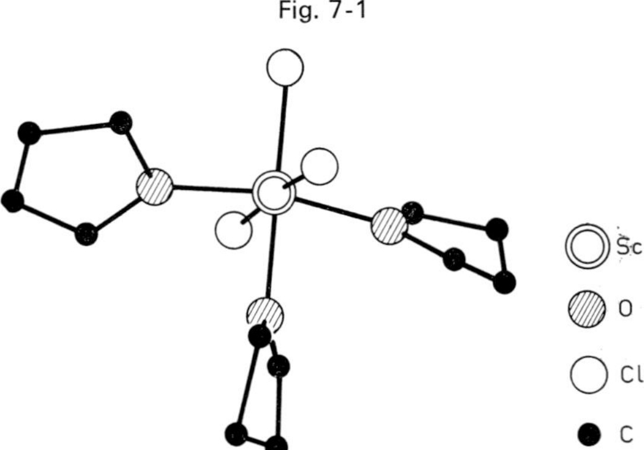

Coordination polyhedron of $ScCl_3 \cdot 3thf$.

X-ray crystal structure determinations of the tris tetrahydrofuran adducts of yttrium and the lanthanide borohydrides are reportedly in progress. The cell dimensions of the orthorhombic crystals —space group $Pbcn\text{-}D^{14}_{2h}$ (No. 60)—are a = 9.28, b = 14.52, c = 14.53 Å for the Gd derivative and a = 9.26, b = 14.49, c = 14.43 Å for the Y derivative. There are four molecules in the unit cell in both cases, and the metal ions are at sites of twofold symmetry. The La derivative appears to be triclinic with space group $P1\text{-}C^1_1$ (No. 1) [10].

The infrared spectrum of the $Sc(BH_4)_3 \cdot thf$ adduct, recorded as Nujol and hexachlorobutadiene mulls, is reported, and a band at 450 cm^{-1} is tentatively assigned to the Sc-O stretching mode. A molecular ion, $[Sc(BH_4)_3 \cdot thf]^+$, at m/e = 159 to 162, is found in the mass spectrum of the adduct, as well as peaks at m/e = 142 to 149, corresponding to $[Sc(BH_4)_2 \cdot thf]^+$, and at m/e = 128 to 134, corresponding to $[Sc(BH_4) \cdot thf]^+$ [3]. The complete optical spectra of $Er(BH_4)_3 \cdot 3thf$, as well as those of mixed crystals containing La, Gd, and Y, are reported. The magnetic susceptibility of the Er derivative follows the Curie law over the range from 310 to 4.2 K, μ_{eff} = 8.83 B.M. [10]. At low temperatures the Pbcn crystals undergo a phase transition to a structure with nonequivalent lanthanide ion sites, as indicated by optical spectra of the mixed Y, Er, and Gd adducts. Vibrational intensity is observed at 1.6 and 77 K [10].

All the compounds described above are air and moisture sensitive. The adducts can in general be decomposed thermally, resulting in the precipitation of the corresponding anhydrous salts. Occasionally intermediate decomposition steps can be ascertained, as shown for the bis adducts of the lanthanide chloride salts (M = Pr, Nd), where $MCl_3 \cdot 0.5thf$ adducts are obtained between 60 and 80°C and for the 3.5 adducts (M = Y, Dy, Ho, Er) where $MCl_3 \cdot thf$ adducts are obtained [11]. Similar results were obtained for the bromide adducts, where $MBr_3 \cdot thf$ was obtained for M = La, Ce, and Pr, where $MBr_3 \cdot 0.5thf$ was obtained for M = Sm [12]. The thiocyanate adducts are reported to decompose at ≈150°C [8].

References to 7.3:

[1] J. L. Atwood, K. D. Smith (J. Chem. Soc. Dalton Trans. **1974** 921/3). — [2] S. Herzog, K. Gustav, E. Krueger, H. Oberender, R. Schuster (Z. Chem. [Leipzig] **3** [1963] 428/9). — [3] J. H. Morris, W. E. Smith (J. Chem. Soc. Chem. Commun. **1970** 245). — [4] K. Rossmanith (Monatsh. Chem. **100** [1969] 1484/8). — [5] K. Rossmanith, C. Auer-Welsbach (Monatsh. Chem. **96** [1965] 602/5).

[6] K. Rossmanith (Monatsh. Chem. **97** [1966] 1357/64). — [7] K. Rossmanith (Monatsh. Chem. **97** [1966] 1698/712). — [8] K. Rossmanith (Monatsh. Chem. **93** [1962] 1121/7). — [9] U. Mirsaidov, T. G. Rotenberg, T. N. Dymova (Dokl. Akad. Nauk Tadzh.SSR **19** No. 2 [1976] 30/3 from C.A. **85** [1976] No. 136374). — [10] E. R. Bernstein, K. M. Chen (Chem. Phys. **10** [1975] 215/28).

[11] K. Rossmanith, C. Auer-Welsbach (Monatsh. Chem. **96** [1965] 606/13). — [12] K. Rossmanith, H. Blaha (Monatsh. Chem. **98** [1967] 1060/5). — [13] G. B. Deacon, A. J. Koplick (Inorg. Nucl. Chem. Letters **15** [1979] 263/5). — [14] U. Mirsaidov, A. Rakhimova, T. N. Dymova (Zh. Neorgan. Khim. **23** [1978] 3326/9). — [15] U. Mirsaidov, A. Kurbonbekov, Kh. Sh. Dzhuraev (Zh. Neorgan. Khim. **24** [1979] 2855/6).

[16] U. Mirsaidov, T. G. Rotenberg, Ya. Samiev (Zh. Neorgan. Khim. **24** [1979] 1995/6). — [17] U. Mirsaidov, A. Rakhimova (Izv. Akad. Nauk Tadzh. SSR Otd. Fiz. Mat. Geol. Khim. Nauk **1979** No. 2, pp. 100/3). — [18] O. V. Kravchenko, S. E. Kravchenko, V. B. Polyakova, K. N. Semenenko (Koord. Khim. **6** [1980] 76/80).

7.4 With Dioxane (= 1,4-Dioxane = $C_4H_8O_2$)

Dioxane adducts of scandium chloride were prepared in two ways. The first involved adding 150 mg of the salt to 10 ml of dioxane. The salt slowly dissolved after heating under reflux for a half-hour. Fine crystals precipitated, which were drawn off and dried in vacuo, resulting in the bis adduct, $ScCl_3 \cdot 2C_4H_8O_2$ [1]. The second method, which produced the hygroscopic tris adduct, $ScCl_3 \cdot 3C_4H_8O_2$, involved taking the previous filtrate and adding hexane until cloudiness occurred. After standing for one or two hours in a closed vessel, colorless crystals appeared, which were separated and washed with hexane [1]. The dioxane adduct of scandium perchlorate was prepared in the same manner as the lanthanide perchlorate adducts [2]. The adduct of scandium thiocyanate was prepared in the same manner as samarium thiocyanate using a 1:3 molar ratio, except that the solution was evaporated to dryness and stored over P_4O_{10} until constant mass was reached [3]. The formulations reported for these two compounds are $Sc(ClO_4)_3 \cdot 7H_2O \cdot 4C_4H_8O_2$ and $Sc(SCN)_3 \cdot 2C_4H_8O_2$.

Adducts of the lighter lanthanide nitrate salts were obtained by complete evaporation under vacuum of solutions of the hydrated nitrate salts dissolved in warm dioxane, which after storage in a desiccator over calcium chloride, had the formulation $M(NO_3)_3 \cdot 2H_2O \cdot C_4H_8O_2$ (M = La and Ce) and $M(NO_3)_3 \cdot 2H_2O \cdot 2C_4H_8O_2$ (M = Pr, Nd, and Sm). Adducts of the heavier lanthanide nitrate salts were prepared in a similar fashion except that the dioxane solutions were only partially evaporated. The remaining viscous solutions were cooled and the walls of the containers scratched to start crystallization or seeded with crystals of the yttrium compound. The crystals were collected and maintained over calcium chloride, yielding the formulation $M(NO_3)_3 \cdot 3H_2O \cdot 2.5C_4H_8O_2$ (M = Y, Gd, Tb, Dy, Ho, Er, Tm, and Yb) [4].

Adducts of the lanthanide chloride salts were prepared by allowing the anhydrous chlorides to react with an excess of dioxane (an exothermic reaction). After allowing the mixture to stand for 24 h in a closed flask, the product was collected in a dry box and dried in vacuo over sodium hydroxide [5]. Compounds with the formulations $MCl_3 \cdot C_4H_8O_2$ (M = La, Ce, Pr, Nd, Sm, Eu, and Gd) and $MCl_3 \cdot 2C_4H_8O_2$ (M = Y, Dy, and Er) were obtained. Adducts of the perchlorate salts were prepared by heating the lanthanide oxides with an appropriate amount of 70% perchloric acid (including a few drops of perhydrol) on a water bath. The solutions were evaporated to near dryness, the residues ground with excess dioxane, and the resulting crystals recrystallized from hot dioxane. All compounds were stored in a vacuum desiccator over calcium chloride [5, 6]. Compounds with the formulations $M(ClO_4)_3 \cdot 9H_2O \cdot 4C_4H_8O_2$ (M = Y, La to Gd, and Er) [6] and $M(ClO_4)_3 \cdot 9H_2O \cdot 4C_4H_8O_2$ (M = Tb, Dy, Ho, Tm, Yb, and Lu) [7] were obtained.

Adducts of the lanthanide thiocyanate salts were prepared in the same fashion as those of the lighter lanthanide nitrate adducts, resulting in compounds with the formulation $M(SCN)_3 \cdot 2H_2O \cdot 4C_4H_8O_2$ (M = La to Nd) [8]. A second preparation of the dioxane adduct of samarium thiocyanate involved the reaction of ethanolic solutions of $Sm(ClO_4)_3$ (dried at 80 to 110°C for 3 to 4 h) and KSCN in a 1:3 ratio to produce $Sm(SCN)_3$ by precipitation. The thiocyanate salt was dissolved in a 50% (v/v) ethanol/dioxane solution, and after standing over phosphorus(V) oxide, crystals separated from the mother liquor. These were then dried over P_4O_{10}. The resulting compound had the formulation $Sm(SCN)_3 \cdot 7C_4H_8O_2$. If a 1:2 molar ratio of $Sm(ClO_4)_3$ to KSCN was used, the divalent samarium adduct, $Sm(SCN)_2 \cdot 5C_4H_8O_2$ was reported to form [9].

The adducts of the lanthanide chloride salts are extremely hygroscopic and unstable, slowly losing dioxane at room temperature. The X-ray powder patterns were obtained for these adducts. The two groups of compounds, $MCl_3 \cdot C_4H_8O_2$ and $MCl_3 \cdot 2C_4H_8O_2$, appear to form two isomorphous series [5]. The perchlorate adducts, $M(ClO_4) \cdot 9H_2O \cdot 4C_4H_8O_2$, also form two isomorphous series. The first series included Y and La to Dy, whereas the second includes Ho to Lu [7]. Unit cell dimensions and volumes from X-ray diffraction data have been reported for several of the dioxane adducts of the perchlorate salts (Table 7/1) [3]. Also included in Table 7/1 are the densities and refractive indices of the crystals. The space groups possible for these compounds are Pmmm-$D_{2h}^1 = V_h^1$ (No. 47), P2mm-C_{2v}^1 (No. 25), and P222-$D_2^1 = V^1$ (No. 16). Due to the hygroscopicity of these compounds no good single crystals could be grown [10]. X-ray powder diffraction data for the lanthanide thiocyanate adducts $M(SCN)_3 \cdot 2H_2O \cdot 4C_4H_8O_2$ are consistent with a single isomorphous series for M = La to Nd [8]. X-ray powder diffraction data of the scandium perchlorate adduct are also available [2].

Table 7/1
Physical Data for Dioxane Adducts of $M(ClO_4)_3$ Salts, $M(ClO_4)_3 \cdot 9H_2O \cdot 4C_4H_8O_2$.

M	unit cell parameters			densities			refractive indices		
	a_0 in Å	b_0 in Å	c_0 in Å	V in Å^3	d_{obs}*)	d_{calc}	N_x^+	N_y^+	N_z^+
La	16.07	17.87	19.50	5.600	1.69	1.69	1.454	1.457	1.464
Ce	16.05	17.70	19.39	5.508	—	1.72	1.455	1.458	1.465
Pr	15.83	17.76	19.35	5.440	1.71	1.75	1.453	1.456	1.464
Nd	15.81	17.76	19.31	5.422	—	1.76	1.458	1.461	1.469
Sm	15.85	17.56	19.24	5.355	1.70	1.79	1.460	1.464	1.473

*) Density obtained by flotation, in g/cm^3.

The infrared spectra of the nitrate adducts, recorded in Nujol, are typical of a weak interaction with dioxane, although shifts of the intense 875 and 1122 cm^{-1} bands to 850 and 1080 cm^{-1} for La, 855 and 1080 cm^{-1} for Ce, 865 and 1110 cm^{-1} for Nd, 870 and 1110 cm^{-1} for Dy, and 862 and 1110 cm^{-1} for Tm were observed. The nitrate ion appears to be coordinated in these adducts as judged by the splitting of the 1390 cm^{-1} band of the ionic nitrate [4]. The infrared spectra reported for the adducts of the lanthanide chloride salts, using Nujol mulls between NaCl plates, show strong water bands due to the hygroscopicity of these compounds. Several ring stretching modes are reported, but no interpretation is made [5]. The infrared spectra of the lanthanide perchlorate adducts are also discussed. The lack of shift in the 875 and 1122 cm^{-1} peaks is characteristic of a weak interaction between dioxane and the metal ion [7]. These spectra were also recorded as Nujol mulls. The infrared spectra of the adducts of the lanthanide thiocyanates show the strong ν_{CN} stretching vibration split into two peaks at 2083 and 2049 cm^{-1}. Water bands are also observed. The strong 874 cm^{-1} band in pure dioxane is shifted to 865 cm^{-1} in the adducts. Several other weak bands appear in the spectra [8]. The infrared spectrum of the scandium perchlorate adduct also shows little change from the pure dioxane spectrum [2], as it is the case for the scandium thiocyanate adduct [3].

The $Sm(SCN)_3 \cdot 7C_4H_8O_2$ adduct loses the solvate molecules between 48 to 85°C and decomposes completely between 195 to 400°C [9]. Thermogravimetric curves are reported for the scandium thiocyanate adduct. A stable mono adduct is reported with a decomposition temperature of 120 ± 1°C, whereas the bis adduct decomposes at 65°C [3].—The lanthanide perchlorate adducts are very soluble in water, methanol, ethanol, acetone, and ethylacetate. They are insoluble in chloroform, carbon tetrachloride, and benzene. These adducts could also be recrystallized from dioxane [7]. The scandium thiocyanate adducts are soluble in alcohols, dioxane, pyridine, and acetone but insoluble in carbon tetrachloride, chloroform, and ether [3]. Both the divalent and trivalent samarium adducts, $Sm(SCN)_2 \cdot 5C_4H_8O_2$ and $Sm(SCN)_3 \cdot 7C_4H_8O_2$, are soluble in water and various organic solvents but insoluble in carbon tetrachloride, chloroform, and benzene [9].

References to 7.4:

[1] V. H. Funk, B. Kohler (Z. Anorg. Allgem. Chem. **325** [1963] 67/71). — [2] M. Perrier, E. Giesbrecht, W. G. R. deCamargo, G. Vicentini (Chem. Ber. **95** [1962] 257/9). — [3] L. N. Komissarova, T. M. Sas, A. A. Krasnoyarskaya, V. A. Gagarina (Zh. Neorgan. Khim. **15** [1970] 2952/9; Russ. J. Inorg. Chem. **15** [1970] 1537/41). — [4] G. Vicentini, M. Perrier, E. Giesbrecht (J. Inorg. Nucl. Chem. **26** [1964] 2207/10). — [5] M. Perrier, E. Giesbrecht, G. Vicentini (Anais Assoc. Brasil. Quim. **25** [1966] 39/42).

[6] G. Vicentini, M. Perrier, E. Giesbrecht (Chem. Ber. **94** [1961] 1153/5). — [7] G. Vicentini, J. V. Valarelli, M. Perrier, E. Giesbrecht (J. Inorg. Nucl. Chem. **24** [1962] 1351/3). — [8] G. Vicentini, R. Forneris, M. Perrier (Anais Acad. Brasil. Cienc. **38** [1966] 261/4). — [9] A. M. Golub, A. N. Borshch (Ukr. Khim. Zh. **32** [1966] 923/5; Soviet Progr. Chem. **32** [1966] 685/6). — [10] W. G. R. deCarmargo, J. V. Valarelli (Acta Cryst. **16** [1963] 321/2).

7.5 Complexes with Macrocyclic Ethers

General Literature:

J. F. Desreux, La Complexation des Lanthanides par les Macrocycles, Bull. Classe Sci. Acad. Roy. Belg. [5] **64** [1978] 814/39.

7.5.1 With Crown Ethers

7.5.1.1 With 12-Crown-4 (= 1, 4, 7, 10-Tetraoxacyclododecane = $C_8H_{16}O_4$)

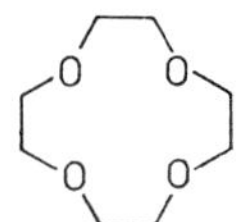

Nitrates. The addition of an anhydrous solution of $Nd(NO_3)_3$ in acetonitrile leads to the synthesis of a 1:1 complex [1]. For preparation of the complexes $M(NO_3)_3 \cdot C_8H_{16}O_4$ with M = La to Lu, a solution of 5 mmol of 12-crown-4 in 50 ml CH_3CN was added dropwise to a solution of 5 mmol of the respective hydrated lanthanide nitrate in 50 ml CH_3CN. The resulting mixture was stirred at 60°C for 24 h. The polycrystalline complexes were filtered off after cooling, washed with CH_2Cl_2 and dried 24 h over P_4O_{10}. Anhydrous complexes were obtained for Ln = Nd to Lu. For Ln = La to Pr monohydrated complexes were isolated, which could be further dried under high vacuum (48 h at 10^{-6} Torr). X-ray powder diagrams of the complexes with Ln = Nd to Lu show that all these complexes are isomorphous. The vibrational spectra indicate the presence of bidentate nonequivalent NO_3 groups, but they cannot rule out the presence of monodentate nitrate groups. The Lu and Nd complexes are stable at ≦270°C but decompose completely at 280 and 340°C, respectively. The M-O bond lengths are longer for M = La than for M = Er [2].

Perchlorates. Complexes of composition $M(ClO_4)_3 \cdot 2C_8H_{16}O_4$ with M = La, Pr, Eu, Ho, Er, Tm, Yb were prepared by dropwise addition with vigorous stirring of a 5 ml acetonitrile solution containing 1 mmol of anhydrous lanthanide salt to a twofold excess of 12-crown-4, dissolved in 1 ml acetonitrile. The precipitate formed nearly immediately, and the suspension was refluxed for 1 h. After cooling, the crystals were filtered, washed with cold acetonitrile and held overnight in vacuum at 80°C [1].

Mixed complex with 12-crown-4 and 15-crown-5, see p. 277.

Reference to 7.5.1:

[1] J. F. Desreux, G. Duyckaerts (Inorg. Chim. Acta **35** [1979] L313/L315). — [2] J.-C. G. Bünzli, D. Wessner (Inorg. Chim. Acta **44** [1980] L55/L58).

7.5.1.2 With 15-Crown-5 (=1, 4, 7, 10, 13-Pentaoxacyclopentadecane = $C_{10}H_{20}O_5$)

Nitrates. Adducts of 15-crown-5 with the lighter lanthanide nitrates were prepared by mixing solutions of the partially dehydrated (0.5 to 1.5 mol H_2O) lanthanide nitrate salts with solutions of the ligand in acetonitrile. After partial evaporation of the solvent, each crystalline complex was collected and dried for 3 to 4 days either in a desiccator with P_4O_{10} (for M = La to Eu) or in a high vacuum (M = Gd) at 5°C/10^{-5} Torr. In each case, the 1:1 complex was formed, $M(NO_3)_3 \cdot C_{10}H_{20}O_5$ with M = La to Gd, and only the gadolinium compound was hygroscopic [2]. The infrared spectra of the compounds taken as Nujol mulls or KBr pellets are typified by a −43 and −18 cm^{-1} shift in the ν(COC) stretching vibrations at 1120 and 1092 cm^{-1}, respectively. Nonsystematic shifts of ±10 to 30 cm^{-1} in the ring stretching and deformation modes occur in the 510 to 990 cm^{-1} region [2]. The nitrate groups appear to be coordinated since six bands appear in the nitrate region of the spectrum [1]. The magnetic moments in Bohr Magnetons are as follows: Ce, 2.26; Pr, 3.26; Nd, 3.28; Sm, 1.42; Eu, 3.25; Gd, 7.89. Errors in the last digit vary from ±0.01 to 0.03μ_B. The lanthanum adduct has a molar susceptibility of -0.35×10^{-6} cgs unit. All measurements were made at 21°C [2]. The thermogravimetric study of complexes $M(NO_3)_3 \cdot C_{10}H_{20}O_5 \cdot nH_2O$ indicate that 4:3 complexes $[M(NO_3)_3]_4 \cdot 3C_{10}H_{20}O_5$ are formed for M = Gd to Lu [3].

Chlorides. A hygroscopic adduct of 15-crown-5 with praseodymium chloride was prepared by the dropwise addition of a solution of 5 mmol of the ligand in 20 to 35 ml of a 1:3 CH_3OH/CH_3CN mixture to a solution of 5 mmol of the hydrated salt in an equal volume of solvent. The reaction mixture was stirred at 60°C for 5 to 20 h. After cooling, the complex was filtered off and dried in a desiccator for 3 to 4 days over P_4O_{10}, resulting in the formulation $PrCl_3 \cdot C_{10}H_{20}O_5$ [1]. The Raman spectrum of a powdered sample was obtained, using an Ar laser as the excitation source. The infrared spectrum as Nujol mulls is presented, although no detailed analysis is made [1]. The magnetic moment of the 1:1 adduct is reported to be 3.49(2) μ_B, compared to 3.57 μ_B for the anhydrous salt at 21±1°C [1].

Perchlorate. A hygroscopic adduct of 15-crown-5 with praseodymium perchlorate has been prepared using the same method (10 mmol of ligand) as for the chloride salts, except that acetonitrile was the solvent used and an additional day of drying time at 50°C and 10^{-2} Torr was required. In the case of the perchlorate, only the bis adduct was obtained, $Pr(ClO_4)_3 \cdot 2C_{10}H_{20}O_5$. In the infrared spectrum (Nujol mull) a band at ≈865 cm^{-1} is obtained, and a Raman band at 874 cm^{-1} is also present. The perchlorate appears to be ionic from the infrared spectrum. The magnetic moment is reported to be 3.37(2) μ_B at 21±1°C [1].

Thiocyanates. Two adducts of praseodymium thiocyanate were obtained by the procedure used for the perchlorate salts, $Pr(SCN)_3 \cdot C_{10}H_{20}O_5$ and $Pr(SCN)_3 \cdot C_{10}H_{20}O_5 \cdot H_2O$. The anhydrous adduct is hygroscopic, whereas the other is not. The infrared spectra of these adducts are consistent with N-bonded thiocyanate groups. The infrared and Raman spectra of these adducts are also reported. The magnetic moment is reported to be 3.48(2) μ_B at 21±1°C [1].

For spectroscopic properties of thiocyanates, nitrates, and perchlorates, see also [4].

With 12-Crown-4 ($=C_8H_{16}O_4$) and 15-Crown-5 ($=C_{10}H_{20}O_5$)

A mixed hygroscopic adduct of 12-crown-4, and 15-crown-5 with praseodymium perchlorate has been reported, $Pr(ClO_4)_3 \cdot C_8H_{16}O_4 \cdot C_{10}H_{20}O_5$. This adduct was prepared in a similar fashion to the 15-crown-5 adduct of $Pr(ClO_4)_3$. The infrared spectrum in Nujol and the Raman powder spectrum are mentioned briefly. The perchlorate ions appear to be ionic from the lack of splitting of the bands attributed to these ions in the infrared spectrum. The magnetic moment of this compound is reported to be 3.46±0.02 μ_B at 21±1°C [1].

References to 7.5.1.2:

[1] J.-C. G. Bünzli, D. Wessner, H. T. T. Oanh (Inorg. Chim. Acta **32** [1979] L33/L36). — [2] J.-C. G. Bünzli, D. Wessner (Helv. Chim. Acta **61** [1978] 1454/61). — [3] J.-C. G. Bünzli, D. Wessner, P. Tissot (Experientia Suppl. No. 37 [1979] 44/8). — [4] S. M. de B. Costa, M. M. Queimado, J. J. R. Frausto da Silva (J. Photochem. **12** [1980] 31/9).

7.5.1.3 With 18-Crown-6 (=1,4,7,10,13,16-Hexaoxacyclooctadecane $= C_{12}H_{24}O_6$)

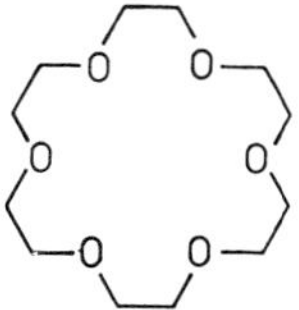

Nitrates. Adducts of the lighter lanthanide salts with 18-crown-6 were prepared in the same fashion as the 15-crown-5 adducts, resulting in $M(NO_3)_3 \cdot C_{12}H_{24}O_6$ for M = La to Pr [2]. For the ions Nd to Gd, this procedure yielded $[M(NO_3)_3]_4 \cdot 3C_{12}H_{24}O_6$. These latter compounds could be recrystallized from ethanol unchanged. In the case of the neodymium compound, partial solvent evaporation yielded a second crystalline fraction corresponding to the 1:1 compound, $Nd(NO_3)_3 \cdot C_{12}H_{24}O_6$ [2]. The heats of formation in anhydrous acetonitrile were determined by using an especially constructed calorimeter. The heats and entropies of formation were calculated [6]. A thermogravimetric analysis was also reported [4].

The $(M(NO_3)_3)_4 \cdot 3C_{12}H_{24}O_6$ adducts of M = La to Pr could be prepared from the 1:1 adducts by heating the solid compounds at 225°C/2 × 10^{-2} Torr for 24 h for La, 190°C/2 × 10^{-2} Torr for Ce, and 170°C/720 Torr for 16 h for Pr. None of the adducts, except for Gd, is hygroscopic. The possibility that the $[M(NO_3)_3]_4 \cdot 3C_{12}H_{24}O_6$ compounds are a mixture of $M(NO_3)_3$ with $M(NO_3)_3 \cdot C_{12}H_{24}O_6$ is unlikely since these compounds can be recrystallized from ethanol, and they exhibit unique X-ray powder diagrams when compared to the simple nitrate salts. In the infrared spectra of the 1:1 compounds, taken as Nujol mulls or KBr pellets, there is a −29 cm^{-1} shift in the ν(COC) stretching mode at 1107 cm^{-1}, and a −41 cm^{-1} shift for the 4:3 compounds [2]. The nitrate groups appear to be coordinated since six bands appear in the nitrate region of the spectra [1]. The magnetic moments of the compounds in μ_B are as follows: for $M(NO_3)_3 \cdot C_{12}H_{24}O_6$: M = Ce, 2.30 ± 0.02; M = Pr, 3.28 ± 0.02; M = Nd, 3.30 ± 0.02 and for $[M(NO_3)_3]_4 \cdot 3C_{12}H_{24}O_6$: M = Ce, 2.25 ± 0.02; M = Pr, 3.29 ± 0.02; M = Nd, 3.27 ± 0.02; M = Sm, 1.44±0.01; M = Eu, 3.24±0.02; M = Gd, 7.90±0.03. The lanthanum compounds both had a diamagnetic susceptibility of -0.35×10^{-6} cgs unit. All measurements were made at 21°C [2]. The crystals of $Nd(NO_3)_3 \cdot C_{12}H_{24}O_6$ are orthorhombic, space group Pbca-D_{2h}^{15} (No. 61), with a = 12.232(2), b = 15.622(3), and c = 21.799(3) Å. Density D(exp) = 1.891, D(calc) = 1.896 g/cm³ for Z = 8. The neodymium atom is coordinated to 3 bidentate NO_3 groups and to 6 O atoms of the ether with the macrocycle in a distorted boat conformation [5].

Chloride. A 1:1 hygroscopic adduct of praseodymium chloride with the macrocyclic ether, 18-crown-6 was prepared in the same manner as the adduct of that salt with 15-crown-5. It has the formulation $PrCl_3 \cdot C_{12}H_{24}O_6$ [1]. Part of the infrared spectrum is reported, and some mention of the Raman spectrum is made, using the same conditions as for the 15-crown-5 compound. The magnetic moment for the compound is reported to be 3.48±0.02 μ_B at 21 ± 1°C [1].

Perchlorate. A hygroscopic adduct of 18-crown-6 with praseodymium perchlorate has also been prepared, using the same method as for the 15-crown-5 compound. In this case the 1:1 compound was formed, $Pr(ClO_4)_3 \cdot C_{12}H_{24}O_6$ [1]. At least one coordinated perchlorate group appears to be present from the infrared spectrum taken as a Nujol mull. The 1:1 adduct has a magnetic moment of 3.45±0.02 μ_B at 21 ± 1°C.

Thiocyanate. A nonhygroscopic adduct of 18-crown-6 with praseodymium thiocyanate was prepared by the same procedure as for the 15-crown-5 adduct, resulting in $Pr(NCS)_3 \cdot C_{12}H_{24}O_6$ [1]. The infrared spectrum is reported and suggests that the thiocyanate groups are N-bonded. The 1:1 adduct has a magnetic moment of 3.41 ± 0.02 μ_B at 21 ± 1°C [1].

β-Diketonates. Using NMR line-width techniques, the kinetics of the exchange behavior of 18-crown-6 with the tris(6,6,7,7,8,8,8-heptafluoro-2,2-dimethyl-3,5-octanedionato)salt of europium was determined. A temperature study over the range −65 to +50°C suggested that the mechanism may be dissociative, and a plot of lg k vs. 1/T yielded a value of 9.0 kcal/mol for the activation energy and a value of 4 cal · mol⁻¹ · K⁻¹ for the entropy of activation. The association constant of 18-crown-6 with the Eu and Pr salt was between 16 and 21 l/mol and remained unchanged over the temperature range from +30 to −20°C, yielding a ΔH of complexation ≈0 kcal/mol and a ΔS of ≈6 cal · mol⁻¹ · K⁻¹ [3].

References to 7.5.1.3:

[1] J.-C. G. Bünzli, D. Wessner, H. T. T. Oanh (Inorg. Chim. Acta **32** [1979] L33/L36). — [2] J.-C. G. Bünzli, D. Wessner (Helv. Chim. Acta **61** [1978] 1454/61). — [3] V. A. Bidzilya (Teor. Eksperim. Khim. **14** [1978] 708/13; Theor. Exptl. Chem. [USSR] **14** [1978] 553/7). — [4] J.-C. G. Bünzli, D. Wessner, P. Tissot (Experientia Suppl. No. 37 [1979] 44/8). — [5] J.-C. G. Bünzli, B. Klein, D. Wessner (Inorg. Chim. Acta **44** [1980] L147/L149).

[6] B. Ammann, J.-C. G. Bünzli (Experientia Suppl. No. **37** [1979] 49/53).

7.5.1.4 With Dicyclohexyl-18-crown-6 (= cis, syn, cis-2,5,8,15,18,21-Hexaoxatricyclo[20.4.0.0^{9,14}] hexacosane = $C_{20}H_{36}O_6$)

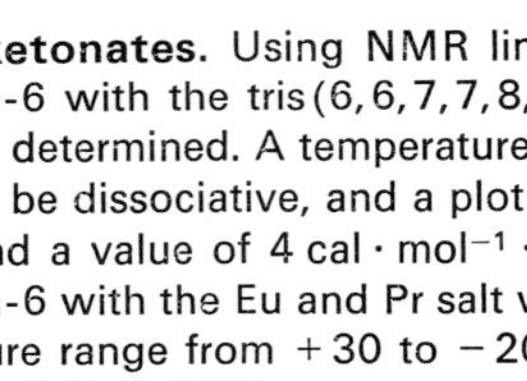

Complexes of the ligand dicyclohexyl-18-crown-6 with lanthanide perchlorate and nitrate salts were prepared by adding the ligand to solutions of the salts, resulting in the formulations $MX_3 \cdot C_{20}H_{36}O_6$. The solvent and the particular lanthanide ions used were not reported [1]. Complexes of dicyclohexyl-18-crown-6 with lanthanide nitrate, chloride, and thiocyanate salts were prepared by adding 1 mmol of the crystalline polyether to 1 mmol of the cation in acetonitrile (nitrates and thiocyanates) or anhydrous ethanol (chlorides), resulting in the formulations $M(NO_3)_3 \cdot C_{20}H_{36}O_6$ (M = La, Nd), $Eu(NO_3)_3 \cdot C_{20}H_{36}O_6 \cdot 2H_2O$, $MCl_3 \cdot C_{20}H_{36}O_6$ (M = La, Nd), $EuCl_3 \cdot C_{20}H_{36}O_6 \cdot H_2O$, $M(SCN)_3 \cdot C_{20}H_{36}O_6$ (M = La, Nd, Eu, and Tb). The complexes of the perchlorate salts were prepared by adding 1 mmol of the crystalline polyether to each perchlorate salt, previously heated to 200°C under reduced pressure ($\approx 10^{-3}$ Torr) in a vacuum thermoanalyzer, and dissolved in anhydrous acetonitrile. After the solutions were heated moderately, microcrystalline products formed from the cooled solutions, with the formulations $M(ClO_4)_3 \cdot C_{20}H_{36}O_6$ (M = La, Nd). To obtain the compounds $Eu(ClO_4)_3 \cdot C_{20}H_{36}O_6 \cdot 2H_2O$ and $Ho(NO_3)_3 \cdot C_{20}H_{36}O_6 \cdot 4H_2O$, it was necessary to concentrate the solutions containing rigorously stoichiometric proportions of $EuCl_3$ and polyether and to add successively small amounts of carbon tetrachloride or anhydrous diethyl ether until turbidity settled. The compounds precipitated slowly from cooled solutions by prolonged stirring or on standing overnight. The complexes were washed with anhydrous ethanol and dried in vacuo at room temperature. Temperatures of incipient thermal decomposition are 270°C for the anhydrous nitrate and chloride complexes, and 5 to 10°C less for the hydrated europium complexes. The thiocyanate complexes decompose at slightly lower temperatures (La = 270°C; Nd = 275°C; Eu = 265°C; Tb = 255°C) whereas the perchlorate complexes decompose at higher temperatures (La = 300°C; Nd = 295°C; Eu = 285°C). Conductivity data at 25°C in $\Omega^{-1} \cdot cm^2 \cdot mol^{-1}$: nitrates in acetonitrile; 90 (La), 95 (Nd), 75 (Eu): chlorides in acetonitrile/ethanol mixture (5:1); 75 (La), 72 (Nd): perchlorates in acetonitrile; 220 (La), 225 (Nd), 245 (Eu): thiocyanates in acetonitrile; 38 (La), 35 (Nd), 38 (Eu), 32 (Tb) [3].

The crystal structure of the lanthanum nitrate derivative was reported, the crystals being monoclinic and falling in the space group Cc-C_s^4 (No. 9) [2]. The cell parameters are a = 11.440, b = 22.995, c = 11.892 Å, and β = 118.74°, with 4 molecules to the unit cell and a cell volume of 2743 Å^3. The calculated density was 1.689 g/cm^3. After 2209 significant reflections the R index was 0.108. The molecular structure shown in **Fig. 7-2** corresponds to a 12-coordinate lanthanide ion with the six ether oxygen atoms and six oxygen atoms from the three bidentate nitrate groups directly bound to the lanthanum ion. There is a twofold pseudo-axis normal to the polyether ring. The six ether oxygen atoms are not coplanar, with the two diametrically opposed atoms displaced 0.81 Å below the plane of the other four, equalizing the distance between the twelve oxygen atoms. The La-O distances fall within 2.63 to 2.71 Å (La-O nitrate) and 2.61 to 2.92 Å (La-O ether) [2]. The proton (^{1}H) NMR spectra

Fig. 7-2

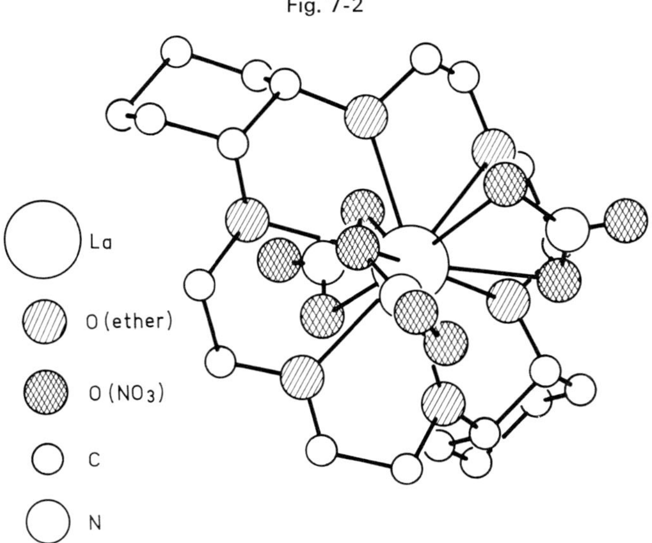

The coordination geometry for $La(NO_3)_3 \cdot C_{20}H_{36}O_6$.

of the La^{3+}, Sm^{3+}, and Pr^{3+} perchlorates with this ligand are reported in a $CD_3CN/CDCl_3$ 1:1 mixed solvent [1]. The effects of added water (2 mol/mol complex) and tributyl phosphate are also shown. The shifts are large, the praseodymium spectrum extending from ≈ +5 to −15 ppm. The spectra are in the slow exchange limit. An attempt to calculate the shifts using the nonaxial pseudocontact shift equation is made, resulting in the axial molar susceptibility term, $\chi_Z - \bar{\chi} = -542 \times 10^6$ cgs unit and a nonaxial term, $\chi_X - \chi_Y = 927 \times 10^6$ cgs unit, for the perchlorate-crown ether spectrum. Similar fits were made for the spectra where water and tributylphosphate were added, resulting in a change in $\chi_Z - \bar{\chi}$ to -144×10^6 and 72×10^6 cgs unit, respectively, and a change in $\chi_X - \chi_Y$ to 843×10^6 and 1300×10^6 cgs unit, respectively. The ^{13}C spectra of the praseodymium adduct was also reported [1].

References to 7.5.1.4:

[1] M. E. Harman, F. A. Hart, M. B. Hursthouse, G. P. Moss, P. R. Raithby (J. Chem. Soc. Chem. Commun. **1976** 396/7). — [2] G. A. Catton, M. E. Harman, F. A. Hart, G. E. Hawkes, G. P. Moss (J. Chem. Soc. Dalton Trans. **1978** 181/4). — [3] A. Seminara, A. Musemeci (Inorg. Chim. Acta **39** [1980] 9/17).

7.5.1.5 With Benzo-15-crown-5 (= 2,3,5,6,8,9,11,12-Octahydrobenzo-1,4,7,10,13-pentaoxacyclopentadecin = $C_{14}H_{20}O_5$)

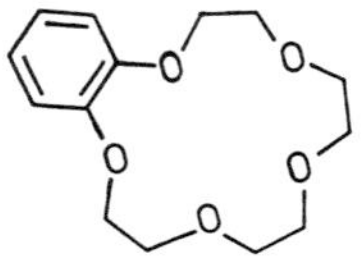

7.5.1.5.1 Scandium Compounds

The compound $ScCl_3 \cdot 2C_{14}H_{20}O_5 \cdot H_2O$ was prepared (under nitrogen) by adding a solution of $ScCl_3$ (1 mmol/50 ml) in acetonitrile to a solution of the ligand in the same solvent (1 mmol/150 ml). The volume was reduced to ≈100 ml under reduced pressure, and ≈200 ml of ether was added to the clear solution. The resulting white precipitate was isolated by suction filtration and dried under vacuum over P_4O_{10} at room temperature for 24 h. The thiocyanate compound, $Sc(NCS)_3 \cdot C_{14}H_{20}O_5 \cdot 1.5C_4H_8O \cdot 2H_2O$ (C_4H_8O = tetrahydrofuran), was prepared similarly except a scandiumthiocyanate-ether adduct was used as the starting salt, tetrahydrofuran was used as the solvent, and ≈100 ml of n-hexane was used to precipitate the adduct [4]. Attempts to dry these adducts above 70°C resulted in the formation of glassy amorphous solids of uncertain compositions. Infrared spectra of these compounds, obtained as Nujol mulls between CsI plates, were devoid of bands that could be assigned to Sc-O stretching vibrations. The Sc-N vibration was assigned to the 343 cm^{-1} band in the thiocyanate compound. Both ionic and coordinated anions are suggested to be present based on the infrared data. The ν(CO) stretching vibrations appear at 1257, 1239, and 1180 cm^{-1} in the chloride compound and 1251 and 1210 cm^{-1} in the thiocyanate compound, compared to 1252 and 1228 cm^{-1} in the free ligand. In attempts to obtain the mass spectra of these compounds, the highest m/e fragment peak was the polyether ligand rather than the compound, consistent with the relative ease of decomposition of these compounds [4].

The NMR spectra of these compounds in $^2H_6$-acetone have two principal resonances corresponding to the aromatic and methylene resonances which occur at 3.18 and 6.12/6.38 τ, respectively. The methylene peak is split into two peaks, with the lower τ value associated with the methylene protons adjacent to the benzene ring. The chloride compound has these resonances at 3.08 and 6.02/6.32 τ, respectively, whereas the thiocyanate compound has these resonances at 3.08 and 6.00/6.31 τ, respectively. The conductivities in acetonitrile of the chloride compound are typical of a 1:2 electrolyte (Λ_M = ≈200 $\Omega^{-1} \cdot cm^2 \cdot mol^{-1}$) while the thiocyanate compound appears to be a 1:1 electrolyte (Λ_M = ≈30 $\Omega^{-1} \cdot cm^2 \cdot mol^{-1}$) [4]. At low concentrations ($>10^{-4}$ M), extensive anion dissociation appears to be taking place, as judged by a sharp rise in the conductivity plot with decreasing concentration.

7.5.1.5.2 Lanthanide Compounds

Nitrates. Complexes of benzo-15-crown-5 with the lanthanide nitrates were prepared by mixing solutions of the hydrated lanthanide nitrate salts in anhydrous acetonitrile (1 mm/5 ml) with equimolar proportions of a solution of the ligand in the same solvent. From the clear solutions, on heating to 50 to 60°C with stirring, the complexes gradually precipitated as crystalline solids. The precipitates were filtered, washed with acetonitrile, and dried in vacuo [1]. Presumably all the lanthanide derivatives were prepared in this fashion, although analytical data are presented only for $M(NO_3)_3 \cdot C_{14}H_{20}O_5$ with M = Ce and Er [1]. A second report of the preparation of these complexes differs from the first report in that solvated complexes were obtained for the heavier lanthanide ions [2]. Complexes free from solvent were obtained by stirring solutions of the hydrated lanthanide salts in acetone (7 mmol/10 ml) with solutions of the ligand (7 to 10 mmol/10 ml) in the same solvent at room temperature. Crystalline precipitates formed immediately. After standing for 1 h, these were removed by filtration, washed with two 10-ml portions of acetone, and dried at 20°C (35 Torr) for 1 to 2 h. Additional products were obtained by concentrating the filtrates under vacuum, adding hexane, and cooling to 0°C. Traces of water and/or acetone can be removed by drying at 130 to 150°C, (0.5 Torr) for 18 h, resulting in the formulation $M(NO_3)_3 \cdot C_{14}H_{20}O_5$, where M = La, Ce, Pr and Nd [2]. Solvated complexes of the heavier lanthanides were prepared in a similar fashion. However, since no precipitates were obtained in these cases, the volumes were reduced by distillation of the acetone to $^1/_3$ of the original volumes. After the dropwise addition of hexane to the solutions until permanent turbidity resulted, the solutions were cooled to room temperature and allowed to stand overnight. The resulting crystalline precipitates were washed with two 10 ml portions of hexane and dried at 25°C (35 Torr) for 1 to 2 h. Additional products could be obtained by cooling the filtrates to −78°C. The solvated complexes had the composition, $M(NO_3)_3 \cdot C_{14}H_{20}O_5 \cdot 3H_2O \cdot (CH_3)_2CO$ with M = Sm, Eu, Gd, Tb, Dy, Ho, Er, Tm, Yb, and Lu. The water and acetone could be removed by drying at 70 to 90°C (0.1 to 0.5 Torr) for 48 to 72 h. However, the unsolvated complexes are glassy, amorphous solids which do not analyze for the pure compounds, $M(NO_3)_3 \cdot C_{14}H_{20}O_5$. Table 7/2 gives the decomposition temperatures of the complexes. Several thermoanalytical curves for these complexes are also reported [2].

The infrared spectra of these adducts, taken as Nujol or Fluorolube mulls between NaCl plates, show shifts of 20 to 40 cm^{-1} to lower energy of the $\nu(CO)$ bands at ≈1225 to 1260 cm^{-1} and 1120 to 1140 cm^{-1} [2]. The nitrate ions appear to be complexed and essentially unchanged from the original spectra of the hydrated salt. In the solvated complexes, bands at 1695 to 1698 cm^{-1} and 1640 to 1635 cm^{-1} appear, attributable to the $\nu(CO)$ of acetone and $\delta(OH)$ of water, respectively. The small shift (≈15 to 20 cm^{-1}) for the $\nu(CO)$ of the acetone molecule suggests that this ligand is only very weakly bound to the metal ion, or hydrogen bonded in the lattice [2].

The compounds are practically insoluble in benzene and carbon tetrachloride, and moderately soluble in acetonitrile and other polar solvents [1]. Conductivity data for millimolar solutions in acetonitrile are typical for nonelectrolytes ($\Lambda_M = 20\,\Omega^{-1} \cdot cm^2 \cdot mol^{-1}$) [1, 2]. Nuclear magnetic resonance data for these compounds were obtainable only for the solvated derivatives which were sufficiently soluble in $^2H_6$-acetone. However, no detail is given due to the large line widths of the peaks. The complexes did function as shift reagents, as judged by their effects on the spectrum of 4-vinylpyridine in acetone [2].

Table 7/2

Decomposition Temperatures (in °C) for Benzo-15-crown-5 Complexes of the Lanthanide Nitrate Salts.

$M(NO_3)_3 \cdot C_{14}H_{20}O_5$	T_{dec}	$M(NO_3)_3 \cdot C_{14}H_{20}O_5 \cdot 3H_2O \cdot (CH_3)_2CO$	T_{dec}	$M(NO_3)_3 \cdot C_{14}H_{20}O_5 \cdot 3H_2O \cdot (CH_3)_2CO$	T_{dec}
M = La	280 to 285	M = Sm	85 to 100	M = Ho	120 to 123
Ce	235 to 245	Eu	104 to 108	Er	138 to 141
Pr	275 to 276	Gd	105 to 110	Tm	136 to 140
Nd	270	Tb	118 to 121	Yb	138 to 142
Sm	232 to 235	Dy	118 to 123	Lu	126 to 136

Chlorides. The complexes $NdCl_3 \cdot C_{14}H_{20}O_5 \cdot 2H_2O$ and $EuCl_3 \cdot C_{14}H_{20}O_5 \cdot 4H_2O$ were obtained by adding 1 mmol of the crystalline polyether to 1 mmol of the chloride in anhydrous ethanol. The neodymium compound slowly precipitates as a crystalline powder on heating the solution. To obtain the compound $EuCl_3 \cdot C_{14}H_{20}O_5 \cdot 4H_2O$, it is necessary to concentrate the solution containing rigorously stoichiometric proportions of $EuCl_3$ and polyether and to add successively small amounts of carbon tetrachloride or anhydrous diethyl ether until turbidity settles; the compound precipitates slowly from cooled solutions by prolonged stirring or on standing overnight. The complexes are washed with anhydrous ethanol and dried in vacuo at room temperature. Temperature of incipient thermal decomposition for the neodymium complex: 230°C; for the europium complex: 225°C. Conductivity data for the chlorides in acetonitrile/ethanol mixture (5:1) at 25±0.1°C: 33 and 39 $\Omega^{-1} \cdot cm^2 \cdot mol^{-1}$, respectively [5].

Thiocyanates. Complexes of benzo-15-crown-5 with the lanthanide thiocyanate salts were prepared in the same manner as for the nitrate salts, using the hydrated thiocyanate salts as the starting materials [1]. Presumably, all the lanthanide derivatives were prepared in this fashion, although analytical data for $M(SCN)_3 \cdot C_{14}H_{20}O_5 \cdot H_2O$ are presented only for M = La and Yb. The conductivity data for these compounds in millimolar acetonitrile solution are typical of nonelectrolytes (Λ_M = 10 to 20 $\Omega^{-1} \cdot cm^2 \cdot mol^{-1}$) [1]. The thiocyanate compounds are practically insoluble in benzene and carbon tetrachloride and moderately soluble in acetonitrile and other polar solvents. Several thermoanalytical curves for these compounds are reported.

Table 7/3 lists the temperatures at which decomposition begins (T_{dec}), the calculated "activation energy" parameters (E_A^*), and the enthalpies of decomposition (ΔH_{dec}) in kcal/mol [3].

Table 7/3
Thermodynamic Parameters for the Decomposition of $M(SCN)_3 \cdot C_{14}H_{20}O_5 \cdot H_2O$ Compounds.

	in N_2 atmosphere			reduced pressure		
M	T_{dec} in °C	E_A^*	ΔH_{dec}	T_{dec} in °C	E_A^*	ΔH_{dec}
La	230	65	22	220	98	26
Ce	220	40	33	212	54	17
Pr	226	55	15	210	49	23
Nd	222	97	19	210	48	22
Sm	218	67	40	224	40	30
Eu	218	46	36	200	52	41
Gd	220	40	34	216	56	44
Tb	218	44	38	204	22	26
Dy	218	48	52	215	19	24
Ho	220	59	64	219	13	24
Er	218	38	44	204	20	18
Tm	218	34	37	170	29	22
Yb	230	36	35	202	32	23
Lu	230	37	40	200	38	25

*) Heating rate was 4°C/min.

References to 7.5.1.5:

[1] A. Cassol, A. Seminara, G. DePaoli (Inorg. Nucl. Chem. Letters **9** [1973] 1163/8). — [2] R. B. King, P. R. Heckley (J. Am. Chem. Soc. **96** [1974] 3118/23). — [3] A. Seminara, S. Gurrieri, G. Siracusa, A. Cassol (Thermochim. Acta **12** [1975] 173/8). — [4] D. J. Olszanski, G. A. Melson (Inorg. Chim. Acta **26** [1978] 263/9). — [5] A. Seminara, A. Musumeci (Inorg. Chim. Acta **39** [1980] 9/17).

7.5.1.6 With Dibenzo-18-crown-6 (=6,7,9,10,17,18,20,21-Octahydrodibenzo[b,k]-1,4,7,10,13,16-hexaoxacyclooctadecin = $C_{20}H_{24}O_6$)

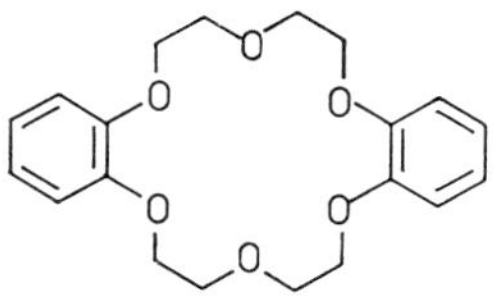

7.5.1.6.1 Scandium Compounds

The compound $ScCl_3 \cdot C_{20}H_{24}O_6 \cdot 1.5\,C_4H_8O$ (C_4H_8O = tetrahydrofuran) was prepared (under nitrogen) by adding a solution of $ScCl_3$ (1 mmol/150 ml) in tetrahydrofuran to a solution of the ligand in the same solvent (1 mmol/100 ml). The volume was reduced to ≈100 ml using reduced pressure, and ≈200 ml of hexane was added to the clear solution. The resulting white precipitate was isolated by suction filtration and dried over P_4O_{10} in vacuum for 24 h at room temperature. If acetonitrile was used as the solvent, the analogous acetonitrile adduct was prepared, $ScCl_3 \cdot C_{20}H_{24}O_6 \cdot 1.5\,CH_3CN$ [5]. The bis adduct of $Sc(SCN)_3$ was prepared in a similar fashion using the ether adduct of $Sc(SCN)_3$ as the starting material to produce $Sc(SCN)_3 \cdot 2\,C_{20}H_{24}O_6 \cdot 3\,C_4H_8O$ [5]. Attempts to dry these adducts above 70°C resulted in the formation of glassy, amorphous solids of uncertain compositions. Infrared spectra of these compounds, obtained as Nujol mulls between CsI plates, were devoid of bands that could be assigned to Sc-O stretching vibrations. The Sc-Cl stretching vibrations were found at 330 cm^{-1} for the acetonitrile adduct and at 356, 333, and 330 cm^{-1} for the tetrahydrofuran adduct. An Sc-N vibrational mode was assigned to the band at 320 cm^{-1} for the thiocyanate adduct. Both ionic and coordinated anions are suggested to be present based on the infrared data. The ν(CO) stretching vibrations appear at 1257 and 1230 cm^{-1} in the chloride tetrahydrofuran compound, at 1251, 1239, and 1223 cm^{-1} in the chloride acetonitrile compound, and at 1251, 1232, and 1217 cm^{-1} in the thiocyanate compound, compared to 1257 and 1230 cm^{-1} in the free ligand [5]. In attempts to obtain the mass spectra of these compounds, the highest m/e fragment peak was the polyether ligand, consistent with the relative ease of decomposition of these compounds.

The NMR spectra of these compounds in $^2H_6$-acetone have two principal resonances corresponding to the aromatic and methylene resonances which occur at 3.17 and 6.00 τ, respectively, for the free ligand. The chloride tetrahydrofuran compound has three resonances at 3.12 and 5.95 τ, whereas the chloride acetonitrile compound has them at 3.07 and 5.90 τ, and the thiocyanate compound has them at 3.10 and 5.90 τ. The conductivities of these compounds are typical of 1:1 electrolytes for the chloride salts and of a 1:2 electrolyte for the thiocyanate salt. At low concentrations (> 10^{-4} molar), extensive anion dissociation appears to be taking place, as judged by a sharp rise in the conductivity plot with decreasing concentrations [5]. Typical Λ_M values at mmolar concentrations in acetonitrile are ≈100 $\Omega^{-1} \cdot cm^2 \cdot mol^{-1}$ for the chloride tetrahydrofuran compound, ≈20 $\Omega^{-1} \cdot cm^2 \cdot mol^{-1}$ for the chloride acetonitrile compound, and ≈200 $\Omega^{-1} \cdot cm^2 \cdot mol^{-1}$ for the thiocyanate adduct.

7.5.1.6.2 Lanthanide Compounds

Nitrates. Complexes of dibenzo-18-crown-6 with the lanthanide nitrates were prepared using the same procedure as for benzo-15-crown-5 described earlier [1]. Presumably, all the lanthanide derivatives of the 1:1 complex were prepared, $M(NO_3)_3 \cdot C_{20}H_{24}O_6$, but analyses for only the Pr and Tm compounds are presented [1]. The complexes of the lighter lanthanide salts were also prepared by boiling, under reflux, mixtures of the hydrated lanthanide nitrate salts (2 mmol) and the ligand (3 mmol) in 35 to 55 ml of acetonitrile for 30 min with vigorous stirring. After cooling to room temperature, each precipitate was removed by filtration, washed with two 10 ml portions of dichloromethane, and dried at 25°C (35 Torr) for 1 to 2 h. Additional product was recovered from the filtrate by adding excess dichloromethane or diethyl ether and by cooling. Traces of water and/or acetonitrile could be removed by drying at 130 to 150°C (0.1 Torr) for ≈16 h, yielding $M(NO_3)_3 \cdot C_{20}H_{24}O_6$, where M = La, Ce, Pr, and Nd [2]. The other lanthanide ions yielded either nonstoichiometric compounds or mixtures of the ether and salt. Decomposition temperatures reported for the lighter lanthanide compounds are: La, 280 to 285°C; Ce, 220 to 225°C; Pr, 270 to 275°C; and Nd, 247 to 248°C [2]. Several thermoanalytical curves for these complexes are also reported [3].

The infrared spectra of these compounds were obtained as Nujol or Fluorolube mulls. The ν(CO) bands at ≈1225 to 1260 cm^{-1} and 1120 to 1140 cm^{-1} show shifts of 20 to 40 cm^{-1} to lower wave numbers upon complexation. The nitrate ions appear to be complexed and the spectra unchanged from the original hydrated salt spectra. The compounds are practically insoluble in acetonitrile and other polar solvents [1]. Conductivity data for millimolar solutions in acetonitrile are typical of nonelectrolytes (10 to 20 $\Omega^{-1} \cdot cm^2 \cdot mol^{-1}$) [1, 2]. The dibenzo-18-crown-6 macrocyclic polyether could also be used to separate various lanthanides from each other [2]. The fluorescence spectra of the europium chloride and nitrate derivatives are reported in glycerol, water, and alcohol solutions [6].

Chlorides. The $PrCl_3 \cdot C_{20}H_{24}O_6$ and $SmCl_3 \cdot C_{20}H_{24}O_6 \cdot 2H_2O$ complexes were prepared by adding 1 mmol of the crystalline polyether to 1 mmol of the chloride in anhydrous ethanol. Temperature of incipient thermal decomposition for the praseodymium complex: 260°C; for the samarium complex: 250°C. Conductivity data for the chlorides in acetonitrile/ethanol mixture (5:1) at 25 ± 0.1°C: 48 and 46 $\Omega^{-1} \cdot cm^2 \cdot mol^{-1}$, respectively [7].

Perchlorates. The complexes $M(ClO_4)_3 \cdot C_{20}H_{24}O_6$ (M = La, Nd) were prepared by using the same procedure as for benzo-15-crown-5 described earlier. Temperature of incipient thermal decomposition for the lanthanum complex: 300°C; for the neodymium complex: 305°C. Conductivity data for the perchlorates in acetonitrile at 25 ± 0.1°C: 220 and 225 $\Omega^{-1} \cdot cm^2 \cdot mol^{-1}$, respectively [7].

Adducts of the lanthanide perchlorates with dibenzo-18-crown-6 were also prepared by adding the polyether (2.2 mmol) to a solution of each anhydrous perchlorate (2 mmol) dissolved in anhydrous acetonitrile (15 ml) at room temperature. The crystals which separated in a few minutes were removed by filtration, washed with an acetonitrile/dichloromethane mixture, and dried in vacuo (0.2 Torr) at room temperature. By careful exclusion of water, compounds with the formulations $M(ClO_4)_3 \cdot C_{20}H_{24}O_6$ with M = La, Ce, Pr, Nd, Sm, Eu, Dy, and Ho, and $M(ClO_4)_3 \cdot C_{20}H_{24}O_6 \cdot H_2O$ with M = Er and Yb were obtained. The anhydrous derivatives transform to the trihydrates when exposed to air for a few hours [8]. The crystals of $Sm(ClO_4)_3 \cdot C_{20}H_{24}O_6$ are orthorhombic, space group Fdd2 with the cell dimensions a = 37.130(15), b = 12.987(5), and c = 11.312(5) Å, D(calc) = 1.97 g/cm³ for Z = 8. The samarium atom is coordinated to the six oxygen atoms of the polyether molecule and four oxygen atoms from the perchlorate ions, one ion acting in a bidentate fashion. There is a twofold axis through the Sm atom and the Sm-O distances vary between 2.36(1) and 2.64(1) Å [8].

Thiocyanates. Complexes of dibenzo-18-crown-6 with the lanthanide thiocyanate salts were prepared in the same manner as for the nitrate salts, using the hydrated thiocyanate salts as the starting materials [1]. Presumably, all the lanthanide derivatives were prepared in this fashion, although analytical data for $M(SCN)_3 \cdot C_{20}H_{24}O_6$ are presented only for M = Nd and Dy [1]. The thiocyanate compounds are practically insoluble in benzene and carbon tetrachloride and moderately soluble in acetonitrile and other polar solvents. The conductivity data for these compounds in millimolar acetonitrile solution are typical of nonelectrolytes (Λ_M = 10 to 20 $\Omega^{-1} \cdot cm^2 \cdot mol^{-1}$) [1]. Several thermoanalytical curves for these compounds are also reported [3]. Table 7/4 lists the temperatures at which decomposition begins, the calculated "activation energy" parameters (E_A^*) and the enthalpies of decomposition (ΔH_{dec}) in kcal/mol [3].

Table 7/4

Thermodynamic Parameters for the Decomposition*) of $M(SCN)_3 \cdot C_{20}H_{24}O_6$ Compounds.

M	in N_2 atmosphere			reduced pressure		
	T_{dec} in °C	E_A^*	ΔH_{dec}	T_{dec} in °C	E_A^*	ΔH_{dec}
La	270	41	75	216	53	82
Ce	270	134	134	250	41	68
Pr	248	80	85	250	33	37

Table 7/4 (Continued)

M	in N_2 atmosphere			reduced pressure		
	T_{dec} in °C	E_A^*	ΔH_{dec}	T_{dec} in °C	E_A^*	ΔH_{dec}
Nd	222	34	41	174	29	23
Sm	238	154	154	236	47	35
Eu	238	68	73	224	50	58
Gd	232	64	51	242	77	63
Tb	256	92	89	242	40	39
Dy	240	80	75	230	38	43
Ho	236	87	80	210	44	50
Er	230	73	98	220	46	54
Tm	260	171	183	234	52	58
Yb	224	42	57	212	55	63
Lu	235	53	67	228	59	66

*) Heating rate was 4°C/min.

β-Diketonates. The association constant of dibenzo-18-crown-6 with the 6,6,7,7,8,8,8-heptafluoro-2,2-dimethyl-3,5-octanedionato complexes of europium and praseodymium was reported to be between 16 and 21 l/mol and to remain unchanged over the temperature range from +30 to −20°C, yielding a ΔH of complexation of ≈0 kcal/mol and a ΔS of ≈6 cal · mol^{-1} · K^{-1} [4]. The unraveling of the kinetics of this interaction using NMR line width techniques was made impossible by a suggested conformational change which preceded the actual exchange process.

References to 7.5.1.6:

[1] A. Cassol, A. Seminara, G. DePaoli (Inorg. Nucl. Chem. Letters **9** [1973] 1163/8). — [2] R. B. King, P. R. Heckley (J. Am. Chem. Soc. **96** [1974] 3118/23). — [3] S. Gurrieri, A. Seminara, G. Siracusa, A. Cassol (Thermochim. Acta **11** [1975] 433/9). — [4] V. A. Bidzilya (Teor. Eksperim. Khim. **14** [1978] 708/13; Theor. Exptl. Chem. [USSR] **14** [1978] 553/7). — [5] D. J. Olszanski, G. A. Melson (Inorg. Chim. Acta **26** [1978] 263/9).

[6] L. G. Koreneva, V. A. Barabanova, V. F. Zolin, S. L. Davydova (Koord. Khim. **4** [1978] 193/9; Soviet J. Coord. Chem. **4** [1978] 143/8). — [7] A. Seminara, A. Musemeci (Inorg. Chim. Acta **39** [1980] 9/17). — [8] M. Ciampolini, N. Nardi, R. Cini, S. Mangani, P. Orioli (J. Chem. Soc. Dalton Trans. **1979** 1983/6).

7.5.1.7 With Dibenzo-30-crown-10

(= $C_{28}H_{40}O_{10}$)

Hygroscopic complexes of the lanthanide perchlorates with dibenzo-30-crown-10 were prepared by adding a solution of each anhydrous (La, Ce, Pr, Nd, Sm, and Eu) or nearly anhydrous (Dy, Ho, Er, and Yb) perchlorate dissolved in anhydrous acetonitrile to the polyether. The crystals, which separated in a few minutes, were removed by filtration, washed with an acetonitrile/dichloromethane mixture, and dried in vacuo at room temperature. The complexes isolated have the formulations: $M(ClO_4) \cdot C_{28}H_{40}O_{10}$ (M = La, Ce, Pr, Nd, Sm, Eu), $Dy(ClO_4)_3 \cdot C_{28}H_{40}O_{10} \cdot 2H_2O \cdot 2CH_3CN$, $M(ClO_4)_3 \cdot C_{28}H_{40}O_{10} \cdot 2H_2O \cdot 1.5CH_3CN$ (M = Ho, Er), $Ho(ClO_4)_3 \cdot C_{28}H_{40}O_{10} \cdot H_2O \cdot 2CH_3CN$, $Yb(ClO_4)_3 \cdot C_{28}H_{40}O_{10} \cdot 2CH_3CN$. The molar conductivities in acetonitrile at 25°C at mmol concentrations vary from 340 to 373 $\Omega^{-1} \cdot cm^2 \cdot mol^{-1}$, M. Ciampolini, N. Nardi (Inorg. Chim. Acta **32** [1979] L9/L11).

7.6 Complexes with Furanone Derivatives

7.6.1 With Ascorbic Acid

(= 3,4-Dihydroxy-5-(1,2-dihydroxyethyl)-5H-furan-2-one = $C_6H_8O_6$ = H_2L)

Complexes of ascorbic acid with the yttrium and lanthanide ions can be prepared by boiling solutions of the chloride salts with the acid at pH 5.5 to 6.3 using a 1:3 molar ratio in a total volume of 20 to 25 ml. The complexes which precipitate on boiling are formulated as oxo compounds $MO(C_6H_7O_6) \cdot 2H_2O$ with M = Y, La, Nd, and Er [1]. Using the absorption spectrum of the cation, 1:1 complexes identified in aqueous solution have dissociation constants of 0.235 for praseodymium, 0.27 for neodymium, 0.229 for holmium and 0.206 for erbium [1]. The effects of ascorbate on the absorption spectra of several of the lanthanides have been reported [2]. Small shifts in the positions of the peaks are observed as well as an approximate doubling of the extinction coefficients. Using these data, stability constants for complexes of several of the metals with the singly (HL^-) and doubly ionized acid (L^{2-}) are reported in Table 7/5 [2, 3].

Table 7/5
Instability Constants of Rare Earth Ascorbato Complexes [2].

M	pK(MHA)	pK(MA)	M	pK(MHA)	pK(MA)	M	pK(MHA)	pK(MA)
Sc	1.3	—	Nd b)	—	8.65	Ho d)	1.0	6.4
Y	1.8	—	Sm	1.2	—	Er e)	0.5	5.9
La	2.5	—	Eu c)	0.8	—	Er b)	—	9.03
Ce	0.9	—	Gd	0.1	—	Yb	1.1	—
Pr	0.8	—	Tb	1.3	—	Lu	1.3	—
Nd a)	1.2	6.8	Dy	0.6	—			

a) λ_{max} in nm = 582 to 583, ε in $l \cdot mol^{-1} \cdot cm^{-1}$ = 18 at pH = 2 to 4, ε = 2.7 at pH = 6.0.— b) I = 0.3 M $NaClO_4$ [3].— c) λ_{max} = 380, ε = 300 at pH = 3 to 4.— d) λ_{max} = 451, ε = 22 at pH = 2 to 4, ε = 112 at pH = 6.— e) λ_{max} = 378, ε = 47 at pH = 2 to 4, ε = 314 at pH = 5.9.

References to 7.6.1:

[1] R. S. Lauer, N. P. Efryushchina, N. S. Poluektov (Zh. Neorgan. Khim. **11** [1966] 1883/6; Russ. J. Inorg. Chem. **11** [1966] 1006/8). — [2] K. P. Stolyarov, I. A. Amantova (Talanta **14** [1967] 1237/44). — [3] N. S. Poluektov, N. P. Efryushina (Ukr. Khim. Zh. **36** [1970] 164/9; Soviet Progr. Chem. **36** [1970] 51/7).

7.7 Complexes with Pyranone Derivatives

7.7.1 With Dehydroacetic Acid

(= 3-Acetyl-4-hydroxy-6-methyl-2-pyrone = $C_8H_8O_4$)

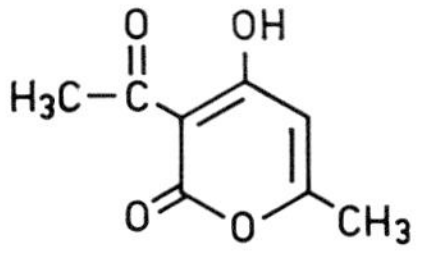

The stability constants of the dehydroacetic acid complexes of the rare earth ions have been determined by pH potentiometric titrations in a 50% (v/v) aqueous dioxane mixture:

M	lg K_1	lg K_2	lg β_2
Sc	5.07	4.49	9.56
Y	4.80	3.98	8.78
La	4.05	3.18	7.23
Ce	4.13	3.28	7.41
Pr	4.27	3.41	7.68
Nd	4.34	3.52	7.86
Sm	4.48	3.72	8.20
Eu	4.62	3.82	8.44
Gd	4.50	3.73	8.23
Tb	4.63	3.82	8.45
Dy	4.72	3.96	8.68
Ho	4.84	3.99	8.83
Er	4.96	3.96	8.92
Tm	5.02	3.87	8.90
Yb	5.13	3.72	8.85
Lu	4.92	3.58	8.50

All values were obtained in 50% (v/v) aqueous dioxane and 0.01 M $NaClO_4$ at 35°C with a standard deviation of 0.02 to 0.03 units, G. C. S. Manku (Australian J. Chem. **24** [1971] 925/34).

7.7.2 With 2,6-Dimethyl-4-pyrone (= 2,6-Dimethyl-4H-pyran-4-one = $C_7H_8O_2$)

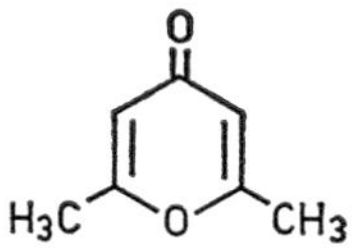

7.7.2.1 Scandium Compounds

Adducts of 2,6-dimethyl-4-pyrone of several scandium salts have been prepared [3]. Each hydrated perchlorate, nitrate, chloride, and bromide salt (0.01 mol) was dissolved in 30 ml of dry triethyl orthoformate and heated for 30 min at 50°C after which the ligand, (1.06 mol) dissolved in 30 ml of absolute ethanol, was added. After 1 h, the fine crystalline product was filtered off, washed with alcohol and ether, and dried in vacuo over anhydrous calcium chloride. If precipitation did not occur immediately, the solution was partially evaporated in a vacuum until crystallization began. For the scandium iodide adduct, a solution of the iodide salt in ethanol, prepared from mixing $Sc(ClO_4)_3 \cdot 7H_2O$ and KI in ethanol, was used as the starting material. The scandium thiocyanate adducts were prepared similarly, except that a tris adduct was obtained using a 1:4 molar ratio and a hexakis adduct was obtained using a 1:8 molar ratio. Table 7/6 lists the adducts isolated along with their melting points and other physical properties. The differential thermal analyses data show the strong exothermic effects due to explosive decomposition of the scandium perchlorate and nitrate adducts. Thermogravimetric analyses suggest that the $ScCl_3 \cdot 4C_7H_8O_2$ adduct loses one $C_7H_8O_2$ molecule over the 150 to 250°C range, whereas the $ScBr_3 \cdot 6C_7H_8O_2$ adduct loses two pyrone molecules at 250°C and a third molecule of pyrone at 390°C. The iodide adduct does not appear to form an intermediate product. The $Sc(SCN)_3 \cdot 6C_7H_8O_2$ adduct loses three pyrone molecules over the very broad 150 to 360°C range [3]. The infrared spectra, taken as KBr pellets, show a splitting of the free ligand ν(CO) stretching vibration at 1608 cm^{-1} and a shift of ≈70 cm^{-1} to lower wave numbers, indicative of coordination of the ligand, and the presence of some uncoordinated ligand. The spectra of the perchlorate salts show no splitting of the 620 and 1085 cm^{-1} bands, indicative of only ionic perchlorate groups. The spectrum of the nitrate adduct, however, has bands at 625, 1040, 1380, and 830 cm^{-1} typical of coordinated, C_{2v} nitrate groups [3].

The conductivity and molecular weight measurements for these scandium adducts are shown in Table 7/6. The cryoscopic measurements in nitrobenzene are indicative of extensive dissociation of these adducts, suggesting that none of the adducts is stable in solution [3].

Table 7/6
Physical Properties for 2,6-Dimethyl-4-Pyrone (= L) Adducts $ScX_3 \cdot nL$.

adduct	melting point in °C	ν(Sc-O)	ν(Sc-X)	ν(CO)	molar conductance*) Λ_m**)			molecular weight***)	
					H_2O	DMF	nitrobenzene	calc	found
$Sc(ClO_4)_3 \cdot 6L$	193±2*)	360	310	1533	393	229	59.0	1088	280
$ScCl_3 \cdot 4L$	—	360	310	1545, 1610	397	86	2.42	648	—
$ScBr_3 \cdot 6L$	97±2	360	300	1535, 1610	397	188	11.5	1170	—
$ScI_3 \cdot 6L$	102±1	363	—	1535, 1608	381	189	16.05	1170	150
$Sc(NO_3)_3 \cdot 4L$	129±1*)	358	290, 320	1542	268	148	1.66	728	241
$Sc(SCN)_3 \cdot 6L$	114±1	360	288, 305, 320	1535, 1610	448	220	4.65	964	197
$Sc(SCN)_3 \cdot 3L$	175±1	360	290, 306, 320	1535	390	208	4.90	591	280

*) Decomposition explosively at 350°C (perchlorate) and 180°C (nitrate). — **) Λ_M in $\Omega^{-1} \cdot cm^2 \cdot mol^{-1}$ at 25°C, concentration 1 mmol. (Typical values for 1:3 electrolyte in water, 370 to 380; in dimethylformamide, 200 to 250; in nitrobenzene, 60 to 90. Typical values for 1:2 electrolyte in dimethylformamide, 135 to 175; in nitrobenzene, 40 to 60. Typical values for 1:1 electrolyte in dimethylformamide, 70 to 80; in nitrobenzene, 20 to 30.) — ***) Cryoscopic measurements in g/mol. Compounds omitted were insufficiently soluble.

7.7.2.2 Lanthanide Compounds

Adducts of 2,6-dimethyl-4-pyrone with lanthanide chlorides were prepared by treating saturated ethanolic solutions of the anhydrous lanthanide trichloride salts with the ligand. The complexes of the heavier lanthanides crystallized immediately, whereas those of the lighter lanthanides crystallized after partial removal of the solvent. The crystals were collected and dried in vacuo over $CaCl_2$ at room temperature. Free ligand present in the products could be washed out with warm benzene. Molar ratios of ligand to MCl_3 of 1:1 to 8:1 produced the tris adducts $MCl_3 \cdot 3C_7H_8O_2$ with M = Y and Ce to Lu. The lanthanum tris adduct could only be obtained pure by using molar ratios less than 3. Molar ratios of 3 or greater produced mixtures of the bis and tris adducts for no apparent reason. The melting ranges of the tris adducts are given below.

M	m.p. in °C
Y	240 to 255
La	>350
Ce	290 (d)
Pr	235 (d)
Nd	231 to 235
Sm	213 to 218
Eu	220 to 230
Gd	215 to 218

M	m.p. in °C
Tb	240 to 250 (d)
Dy	260 (d)
Ho	250 to 265 (d)
Er	240 to 245
Tm	250 to 260
Yb	235 (d)
Lu	250 to 260

The bis lanthanum adduct did not melt below 350°C [1]. The octakis adducts of lanthanum [1], neodymium [1], and samarium perchlorate [2], $M(ClO_4)_3 \cdot 8\,C_7H_8O_2$ (M = La, Nd, Sm), as well as the heptakis adducts, $M(ClO_4)_3 \cdot 7\,C_7H_8O_2$ (M = Eu, Gd, Er, Yb) [2], were isolated by this method, using the anhydrous perchlorate salt as the starting material. The chloride adducts are highly hygroscopic [1] whereas the perchlorate adducts are not [2].

The single-crystal X-ray structure of the lanthanum perchlorate adduct, $La(ClO_4)_3 \cdot 8\,C_7H_8O_2$, has been reported [4]. The crystals are monoclinic, space group $P2_1/c\text{-}C_{2h}^5$ (No. 14), with the cell dimensions a = 12.543(4), b = 15.956(4), c = 32.914(4) Å, and β = 99°20'(10'). There are four molecules to the unit cell, and the calculated density is 1.26 g/cm³. The final R index was 0.072. The coordination polyhedron around the lanthanum ion is a severely distorted square antiprism with internal angles of the two squares ranging between 85° and 92°. **Fig. 7-3** shows the overall molecular structure, and Table 7/7 lists bond distances and angles for the coordination polyhedron.

Fig. 7-3

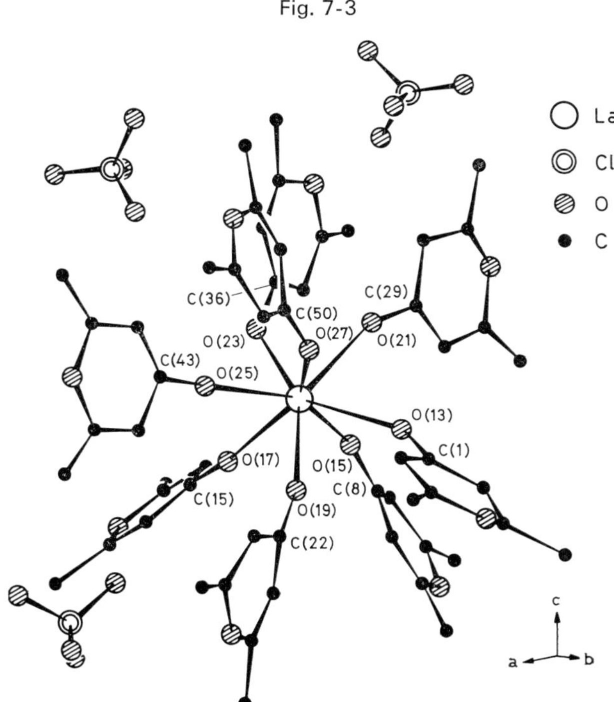

Overall molecular structure of $La(ClO_4)_3 \cdot 8\,C_7H_8O_2$.

The crystal and molecular structures of two additional 2,6-dimethyl-4-pyrone complexes have been reported, $Er(ClO_4)_3 \cdot 7\,C_7H_8O_2$ and $LaCl_3 \cdot H_2O \cdot C_7H_8O_2$ [5]. The erbium complex falls in space group $P\bar{1}\text{-}C_i^1$ (No. 2) with the cell dimensions a = 14.288(4), b = 14.495(4), c = 14.346(4) Å, α = 103°46'(3'), β = 92°52'(3'), γ = 94°56'(3') with 2 formula units in the unit cell. After 7123 reflections the final weighted R factor was 0.056. The erbium coordination polyhedron is a distorted pentagonal bipyramid with Er-O distances of 2.209, 2.256 Å (apical), and 2.2256, 2.300, 2.305, 2.325, 2.327 Å (equatorial) with σ = 0.006 Å. The angle between the apical oxygen atoms is 171.1 (2)° and between the contiguous equatorial oxygen atoms is 69° to 75°. The lanthanum compound falls in space group $P2_1/m\text{-}C_{2h}^2$ (No. 11) with cell dimensions a = 7.255(3), b = 9.152(3), c = 9.781 (3) Å, β = 103°49'(3') and with 2 formula units to the unit cell. The nonhydrogen atoms were refined to R = 0.078. The lanthanum coordination polyhedron is also a slightly distorted pentagonal bipyramid containing a mirror plane. The La-Cl (apical) distance is 2.72 Å. The La-Cl (equatorial) distances are 2.91 (twice) and 2.92 Å (twice). The La-O($C_7H_8O_2$) (apical) and La-O (water) (equatorial) distances are 2.34 and 2.51 Å, respectively. The polyhedra form infinite chains along the b axis sharing Cl-Cl edges [5].

Table 7/7
Bond Distances and Angles for the Coordination Polyhedron in Octakis (2,6-Dimethyl-4-Pyrone) Lanthanum(III) Perchlorate.

distances	Å	angles*)	deg
La-O(13)	2.56(2)	La-O(13)-C(1)	142.7
La-O(15)	2.53(2)	La-O(15)-C(8)	137.7
La-O(17)	2.48(3)	La-O(17)-C(15)	161.7
La-O(19)	2.37(2)	La-O(19)-C(22)	164.3
La-O(21)	2.51(2)	La-O(21)-C(29)	147.4
La-O(23)	2.50(2)	La-O(23)-C(36)	137.1
La-O(25)	2.35(2)	La-O(25)-C(43)	169.1
La-O(27)	2.52(2)	La-O(27)-C(50)	138.8

*) E. S. D. < 2.6°.

The infrared spectra, obtained as Nujol mulls between polyethylene films, show the ν(CO) stretching mode of the pyrone displaced from 1600 cm^{-1} to ≈1550 cm^{-1} in all the adducts, consistent with coordination of the carbonylic oxygen atom. The spectra in the perchlorate region show no evidence of splitting of the ionic ClO_4^- group band at 1080 cm^{-1} [1, 2]. The solubilities of the chloride complexes in ethanol decrease markedly from lanthanum to lutetium. The conductivities of the chloride adducts are typical of nonelectrolyte species [1]. The perchlorate adducts are not sufficiently soluble to obtain conductivity data [2].

References to 7.7.2:

[1] C. Castellani Bisi (J. Inorg. Nucl. Chem. **32** [1970] 2899/903). — [2] C. Castellani Bisi, M. Cola, A. Perotti, V. Tazzoli (J. Inorg. Nucl. Chem. **34** [1972] 3259/60). — [3] F. Kutek, B. Dusek (Zh. Neorgan. Khim. **19** [1974] 2360/5; Russ. J. Inorg. Chem. **19** [1974] 1289/92). — [4] C. Castellani Bisi, A. Della Giusta, A. Coda, V. Tazzoli (Cryst. Struct. Commun. **3** [1974] 381/6). — [5] C. Castellani Bisi, M. Gorio, E. Cannillo, A. Coda, V. Tazzoli (Acta Cryst. A **31** [1975], Suppl., p. 134).

7.7.3 With 3-Hydroxy-2-methyl-4-pyrone (= Maltol = $C_6H_6O_3$)

Several complexes of the lanthanides with maltol have been prepared by adding 6 ml of the ligand (6 mmol) dissolved in ethanol to an aqueous solution (20 ml) of each lanthanide nitrate. The pH of each solution was raised by the dropwise addition of sodium hydroxide (≈3 mol) with constant stirring until the solution became turbid. At this stage, the solution was filtered and the filtrate adjusted to pH 6.5. A solid separated immediately, and the mixture was kept on a water bath for ≈30 min. Each solid complex was washed with hot alcohol containing a small amount of water to remove unreacted ligand. Drying in air resulted in $La(C_6H_5O_3)_3$ and $M(C_6H_5O_3)_3 \cdot H_2O$ with M = Y, Pr, Nd, Sm, Gd, Dy, and Yb [1]. Thermogravimetric and differential thermal analyses data showed an endothermic peak at 80°C for the praseodymium complex corresponding to the loss of a water molecule. However, weight loss continued up to 110°C, and decomposition occurred above

300°C (310°C for the praseodymium and 390°C for the neodymium compound) [1]. The infrared spectra of the complexes in Nujol show the ν(C=C) stretching mode at 1615 and 1555 cm^{-1} in the free ligand shifted to 1570 and 1505 cm^{-1} in the samarium complex. The ν(CO) stretching mode shifts ≈ 50 cm^{-1} upon coordination. The phenolic OH band at ≈ 3200 cm^{-1} is missing in the complex, and a band at 3400 cm^{-1} is present for the hydrated compounds [1].

The stability constants of maltol complexes with the ions of the rare earths have been determined by potentiometric titration with NaOH and are listed in Table 7/8.

Table 7/8
Stability Constants for Lanthanide Maltol Complexes.*)

M	lg K_1	lg K_2	lg K_3	lg β_3	M	lg K_1	lg K_2	lg K_3	lg β_3
Y	6.70	5.39	4.88	15.29	Eu	6.72	5.31	3.80	15.83
La	5.76	4.41	3.17	13.34	Gd	6.58	5.33	3.95	15.86
Pr	6.13	4.83	3.55	14.51	Dy	6.88	5.61	4.12	16.61
Nd	6.22	4.92	3.54	14.68	Er	6.98	5.68	4.21	16.87
Sm	6.51	5.19	3.73	15.43	Yb	7.06	5.83	4.39	17.28

*) T = 30 ± 0.5°C, I = 0.1 M ($NaClO_4$); pKa = 8.62.

References to 7.7.3:

[1] N. K. Dutt, U. U. M. Sharma (J. Inorg. Nucl. Chem. **37** [1975] 1801/2). — [2] N. K. Dutt, U. V. M. Sharma (J. Inorg. Nucl. Chem. **32** [1970] 1035/8).

7.7.4 With Kojic Acid (= 5-Hydroxy-2 hydroxymethyl-4-pyrone = $C_6H_6O_4$)

HO, O, O, CH_2OH

Complex Formation in Solution. The stability constants of kojate species are listed in Table 7/9 as well as the available thermodynamic parameters [1 to 5]. All values reported were obtained by potentiometric pH titrations of solutions of the acid and metal salt with sodium hydroxide or by calorimetry (ΔH measurements). The stabilities of these complexes appear to be determined by the entropy changes. The increased stability at the lower ionic strength (0.1 M compared to 2.0 M) is again due to an increased entropy term. The possibility of several different hydrated complex ions existing in solution is suggested as the reason for such a small separation of the lg K_1 values into two groups, as occurs for most stability constants of the lanthanide series [5].

Isolated Compounds. Each compound of kojic acid with the lanthanide ions has been prepared by the addition of an aqueous solution of kojic acid (4 mol) to an aqueous solution of the lanthanide acetate (1 mol). The solutions were kept stirred by scratching the walls of the containers. Precipitates formed immediately for the compounds with M = La to Nd but only after concentration for the remaining ions studied. The reaction mixtures were allowed to stand for 2 h at room temperature, after which the precipitates were removed by filtration, washed successively with water, ethanol and ether, and dried in vacuo at room temperature. The compounds isolated have the composition $M(C_6H_5O_4)_3 \cdot 2H_2O$ with M = La to Sm, Gd [6] and Dy, Ho, and Y [7]. Water of hydration was lost at 120°C [7]. Similar syntheses starting with metal salts other than acetates have been reported where sodium acetate was used to buffer the solution of kojic acid and salt. Each mixture was heated on a steam bath and allowed to crystallize. The product was washed with water and dried at 130°C. Anhydrous derivatives resulted, $M(C_6H_5O_4)_3$ with M = Nd and Sm [8]. The neodymium and samarium compounds turned brown at 275°C. In the infrared spectra of these adducts, obtained as KBr pellets, the ν(CO) stretching mode of the pyrone ring was assigned to bands in the range 1590 to 1640 cm^{-1}, shifted from 1680 and 1620 cm^{-1} in the free ligand. A ν(M-O) stretching mode was assigned to bands in the range 390 to 425 cm^{-1} (see Table 7/10) where the magnetic moments of these compounds are also reported [6, 7].

Table 7/9
Stability Constants and Thermodynamic Parameters for Lanthanide Kojato Complexes.

M	lg K_1 [a]	lg K_2 [a]	lg K_1 [b]	lg K_2 [b]	lg K_3 [b]	lg K_1 [c]	lg K_2 [c]	lg K_3 [c]	lg K_1 [d]
Y	5.43	5.38	6.18	5.19	4.15	6.00	5.04	3.64	—
La	5.14	4.13	5.35	4.21	3.31	5.16	3.90	3.15	5.30±0.08
Ce	—	—	—	—	—	5.56	4.36	3.50	—
Pr	5.18	4.58	5.77	4.77	3.89	5.60	4.40	3.58	5.69±0.05
Nd	—	—	5.80	4.83	4.03	5.68	4.48	3.62	5.75±0.05
Sm	5.38	5.20	6.03	5.02	3.94	6.10	5.15	3.86	6.00±0.04
Eu	5.35	5.10	6.15	5.10	4.03	—	—	—	6.12±0.03
Gd	5.49	5.17	6.09	5.12	4.05	5.92	5.02	3.80	6.09±0.03
Tb	—	—	—	—	—	—	—	—	6.25±0.02
Dy	5.74	5.39	6.39	5.47	4.35	6.26	5.18	3.82	6.34±0.02
Ho	—	—	—	—	—	—	—	—	6.35±0.02
Er	5.72	5.60	6.31	5.46	4.40	—	—	—	6.39±0.02
Tm	6.00	5.55	—	—	—	—	—	—	6.46±0.02
Yb	—	—	6.53	5.70	4.82	—	—	—	6.53±0.02
Lu	6.24	5.60	—	—	—	—	—	—	6.50±0.02

M	$-\Delta G_1$ [d]	$-\Delta H_1$ [d]	ΔS_1 [d]	lg K_1 [e]	$-\Delta G_1$ [e]	$-\Delta H_1$ [e]	ΔS_1 [e]
Y	—	—	—	5.84±0.03	7.97±0.04	1.04±0.09	23.2±0.3
La	7.22±0.11	0.57±0.27	22.3±1.0	5.11±0.02	6.97±0.03	1.37±0.07	18.8±0.3
Ce	—	—	—	5.28±0.03	7.20±0.04	1.52±0.08	19.1±0.3
Pr	7.76±0.07	0.90±0.12	23.0±0.5	5.41±0.02	7.38±0.03	1.44±0.05	19.9±0.2
Nd	7.84±0.07	0.55±0.19	24.4±0.7	5.42±0.03	7.39±0.04	1.43±0.11	20.0±0.4
Sm	8.19±0.05	0.51±0.23	25.8±0.8	5.66±0.03	7.73±0.04	1.19±0.15	21.9±0.5
Eu	8.35±0.04	0.34±0.20	26.9±0.7	5.72±0.03	7.80±0.04	1.22±0.09	22.1±0.3
Gd	8.30±0.04	0.34±0.14	26.7±0.5	5.72±0.03	7.80±0.04	1.08±0.10	22.5±0.4
Tb	8.52±0.02	0.28±0.16	27.6±0.5	5.88±0.03	8.02±0.04	1.07±0.12	23.3±0.4
Dy	8.65±0.02	0.36±0.17	27.9±0.6	5.97±0.03	8.14±0.04	1.12±0.10	23.5±0.4
Ho	8.66±0.02	0.46±0.11	27.5±0.4	5.94±0.03	8.10±0.04	1.08±0.11	23.5±0.4
Er	8.71±0.02	0.20±0.12	28.5±0.4	6.00±0.03	8.18±0.04	1.02±0.09	24.0±0.3
Tm	8.81±0.02	0.30±0.13	28.6±0.5	6.01±0.04	8.20±0.05	1.01±0.08	24.1±0.3
Yb	8.90±0.02	0.60±0.10	27.9±0.3	6.09±0.04	8.30±0.05	0.75±0.11	25.3±0.4
Lu	8.86±0.02	0.19±0.11	29.1±0.4	6.00±0.05	8.19±0.06	0.46±0.16	25.9±0.6

[a] Reference [1], T = 26°C, I = 2.00 M ($NaClO_4$).—[b] Reference [2], T = 30 ± 0.5°C, I = 0.1 M ($NaClO_4$); pKa = 7.80.—[c] Reference [3], T = 25°C, I = 0.1 M (KCl); pKa = 7.68.—[d] Reference [5], T = 25°C, I = 0.1 M ($NaClO_4$); pKa = 7.61; ΔG and ΔH values in kcal/mol, ΔS in cal · mol^{-1} · K^{-1}; $-\Delta G_1$, $-\Delta H_1$, and ΔS_1 for the pronation of the kojate anion are 10.38 kcal/mol, 3.47 kcal/mol, and 23.2 cal · mol^{-1} · K^{-1}, respectively.—[e] Reference [5], T = 25°C, I = 2.00 M ($NaClO_4$).

A third method of preparation, producing monohydrates, $M(C_6H_5O_4)_3 \cdot H_2O$ with M = Y, La, Pr, Nd, Sm, Gd, Dy and Yb involved mixing solutions of the ligand in ethanol (3 mmol/3 ml) with aqueous solutions of the rare earth nitrates. In each reaction, the pH was raised by the dropwise addition of sodium hydroxide (≈3 M) with constant stirring until the solution became turbid. At this stage the solution was filtered and the filtrate adjusted to pH 6.5. The solid separated immediately. The mixture was kept on a water bath for ≈30 min. The solid complexes obtained were washed with hot alcohol containing a small amount of water to remove unreacted ligand and then dried in air [9]. Thermogravimetric and differential thermal analyses show an endothermic change over the range 50 to 120°C corresponding to the loss of water. Decomposition occurs over the range 250 to 600°C without any plateaus. Infrared spectra in Nujol mulls show the ν(C=C) bands at 1625 and 1575 cm^{-1} in the free ligand shifted to 1560 and 1515 cm^{-1} in the samarium complex. The ν(CO) band at ≈1650 cm^{-1} shifts ≈35 cm^{-1} upon coordination. The phenolic OH band at ≈3200 cm^{-1} is absent in the complexes, and a band at ≈3400 cm^{-1} confirms the presence of water [6]. A small but different red shift is observed for each peak as well as an intensification of each peak.

The kojate salts of the lanthanides behave as nonelectrolytes in nitromethane solution, as judged by their conductivities (see Table 7/10). The electronic absorption spectra of the praseodymium and neodymium compounds in dimethylformamide have been discussed [1].

Table 7/10
Physical Properties of Rare Earth Kojates.

M	ν(M-O)*) in cm^{-1}	dec temp. in °C	μ_{eff} in μ_B	$\chi'_m \times 10^6$**)	Λ_M***)	Ref.
Y	400 (m)	250 to 295	—	—	25	[7]
La	396 (m)	295 to 340	—	—	24	[6]
Ce	410 (w)	293 to 330	2.33	2020	22	[6, 10]
Pr	400 (w)	270 to 325	3.69	5907	16	[6, 10]
Nd	405 (w, sh)	290 to 326	3.46	4908	20	[6, 10]
Sm	425 (w)	280 to 315	1.53	987	26	[6, 10]
Gd	400 to 415 (w, b)	270 to 310	7.90	26390	24	[6, 10]
Dy	390 (m)	260 to 290	10.44	—	23	[7]
Ho	390 (m)	270 to 295	10.15	—	29	[7]

*) KBr pellets, m = medium, w = weak, sh = shoulder, b = broad. — **) Molar susceptibility, χ'_m, in cgs units at ≈303 K corrected for dimagnetism. Magnetic moments calculated from the Curie Law, assuming $\mu_{eff} = 2.84\ \chi'_m$ — ***) Molar conductance (Λ_M in $\Omega^{-1} \cdot cm^2 \cdot mol^{-1}$) in ≈$10^{-3}$ M nitromethane. The oscillator strengths of the "hypersensitive" transitions of the neodymium and holmium complexes containing 1 and 2 moles of kojate bond to the metal ion have also been reported [11].

References to 7.7.4:

[1] H. Yoneda, G. R. Choppin, J. L. Bear, J. V. Quagliano (Inorg. Chem. **3** [1964] 1642/4). — [2] N. K. Dutt, S. Sanyal, U. U. M. Sharma (J. Inorg. Nucl. Chem. **34** [1972] 2261/4). — [3] R. C. Agarwala, S. P. Gupta (Acta Ciencia Indica **1** [1974] 31/2). — [4] K. P. Stolyarov, I. A. Amantova (Vestn. Leningr. Univ. Fiz. Khim. **1971** No. 4, pp. 117/22 from C.A. **76** [1972] No. 118044). — [5] R. Stampfli, G. R. Choppin (J. Coord. Chem. **1** [1971] 173/8).

[6] R. C. Agarwala, S. P. Gupta, D. K. Rastogi (J. Inorg. Nucl. Chem. **36** [1974] 208/11). — [7] R. C. Agarwala, S. P. Gupta (Current Sci. [India] **43** [1974] 263/5). — [8] C. Musante (Gazz. Chim. Ital. **79** [1949] 679/83). — [9] N. K. Dutt, U. U. M. Sarma (J. Inorg. Nucl. Chem. **37** [1975] 1801/2). — [10] R. C. Agarwala, S. P. Gupta (Current Sci. [India] **44** [1975] 3/4).

[11] G. R. Choppin, R. L. Fellows (J. Coord. Chem. **3** [1973/74] 209/15).

7.7.5 With Chlorokojic Acid (= 5-Hydroxy-2-chloromethyl-4-pyrone = $C_6H_5ClO_3$)

Formation constants of chlorokojic acid with the ions of the rare earth series have been determined by potentiometric titration at T = 30 ± 0.5°C and I = 0.1 M ($NaClO_4$). The lg K_1 values gradually decrease with increasing atomic number, with a break at gadolinium [1].

M	lg K_1	lg K_2	M	lg K_1	lg K_2
Y	6.00	5.32	Eu	5.98	5.21
La	5.28	4.48	Gd	5.98	5.14
Pr	5.70	4.82	Dy	6.13	5.38
Nd	5.73	4.92	Er	6.19	5.44
Sm	5.83	5.08	Yb	6.28	5.69

The complexes have been prepared by the third method described for the preparation of kojic acid complexes [2]. Anhydrous compounds were obtained, $M(C_6H_4O_3Cl)_3$ with M = Y, La, Pr, Nd, Sm, Gd, Dy, and Yb. The infrared spectra taken in Nujol show the ν(C=C) bands at 1610 and 1580 cm^{-1} shifted to 1585 and 1520 cm^{-1} in the complex. The ν(CO) band at 1645 cm^{-1} is shifted to 1615 cm^{-1} in the complex, and the phenolic OH band at 3200 cm^{-1} is absent in the complex [2].

References to 7.7.5

[1] N. K. Dutt, S. Sanyal, U. U. M. Sharma (J. Inorg. Nucl. Chem. **34** [1972] 2261/4). — [2] N. K. Dutt, U. U. M. Sharma (J. Inorg. Nucl. Chem. **37** [1975] 1801/2).

7.8 Complexes with Derivatives of Coumarin

7.8.1 With 3-Acetyl-4-hydroxycoumarin (= $C_{11}H_8O_4$)

Formation constants for the 1:1 and 1:2 (metal:ligand) chelates were determined potentiometrically (glass electrode) at 35°C in 50% (v/v) aqueous dioxane, I = 0.1 ($NaClO_4$):

metal ion	lg K_1	lg K_2	lg β_2	metal ion	lg K_1	lg K_2	lg β_2
Sc^{3+}	4.32	3.41	7.73	Gd^{3+}	3.73	2.82	6.55
Y^{3+}	4.02	3.08	7.10	Tb^{3+}	3.88	2.93	6.81
La^{3+}	3.23	2.45	5.68	Dy^{3+}	3.98	3.04	7.02
Ce^{3+}	3.32	2.52	5.84	Ho^{3+}	4.09	3.08	7.17
Pr^{3+}	3.47	2.59	6.06	Er^{3+}	4.21	3.12	7.33
Nd^{3+}	3.64	2.70	6.34	Tm^{3+}	4.34	3.08	7.42
Sm^{3+}	3.78	2.84	6.62	Yb^{3+}	4.46	2.92	7.38
Eu^{3+}	3.92	2.97	6.89	Lu^{3+}	4.31	2.82	7.13

Both lg K_1 and lg K_2 increase monotonically for complexes of the lighter lanthanides (M = La to Eu), before decreasing at gadolinium(III). For the heavier members of the series, lg K_1 and lg K_2 show two distinct trends: (1) lg K_1 increases monotonically from gadolinium(III) to ytterbium(III), before decreasing at lutetium(III); (2) lg K_2 increases from gadolinium(III) to dysprosium(III), then remains effectively constant (within experimental error) through thulium(III), before decreasing at ytterbium(III). The second trend was attributed to increasing steric hindrance to ligation with decreasing ionic radius. The lg K_1 and lg K_2 values for the yttrium(III) complexes are among those for complexes of the heavier lanthanides [1].

7.8.2 With 4-Hydroxy-3-nitroso-coumarin (= $C_9H_5NO_4$)

Formation constants of the 4-hydroxy-3-nitrosocoumarin complexes have been determined by pH potentiometric titrations in a 50% (v/v) aqueous dioxane mixture at 21°C [2]:

M	lg K_1	lg K_2	lg β_2	M	lg K_1	lg K_2	lg β_2
Y	2.76	1.93	4.69	Dy	2.64	1.98	4.62
La	2.31	0.87	3.18	Ho	2.76	1.93	4.69
Ce	2.45	0.89	3.34	Er	2.84	1.93	4.77
Pr	2.50	0.81	3.31	Tm	2.90	1.93	4.83
Nd	2.52	1.23	3.75	Yb	2.96	1.94	4.90
Sm	2.58	1.60	4.18	Lu	3.07	1.96	5.03
Gd	2.62	1.93	4.55				

7.8.3 With 8-Amino-7-hydroxy-4-methylcoumarin (= $C_{10}H_9NO_3$)

$M(C_{10}H_8NO_3)_3 \cdot nH_2O$ (n = 8, 10). Each complex (n = 2 for M = La, Nd, Sm, Gd; n = 8 for M = Ce; n = 10 for M = Pr) was prepared by adding 75 ml of an 0.02 M aqueous solution of a rare earth(III) nitrate or acetate to 300 ml of a 0.02 M solution of the ligand in ethanol, and adjusting the pH of the reaction mixture to between 5.0 and 6.5 by the addition of sodium hydroxide. The precipitates were collected, recrystallized from ethanol, and dried under vacuum at room temperature [3, 4].

A comparison of the infrared spectra of the ligand and its complexes reveals the disappearance of absorption bands corresponding to the stretching (3442 cm^{-1}) and deformation (1334 cm^{-1}) modes of the phenolic group upon complexation, verifying the loss of the phenolic proton. Both the ν_{asym} (3368 cm^{-1}) and ν_{sym} (3302 cm^{-1}) NH stretching modes of the free ligand are shifted to lower wave numbers (≈3340 and 3288 cm^{-1}, respectively) upon complexation, suggesting ligation of the nitrogen atom of the amino group. The ν(M-O) and ν(M-N) stretching modes were assigned, respectively, to bands at 420 and 530 cm^{-1} (cerium complex) and 426 and 535 cm^{-1} (praseodymium complex) [3].

Physical properties of the compounds are summarized in Table 7/11. The effective magnetic moments of the cerium and praseodymium complexes are in agreement with the values obtained for the corresponding rare earth sulfates [3]. The molar conductivities for millimolar solutions of the compounds in water or nitromethane are indicative of nonelectrolyte behavior. Selected electronic absorption spectral data for both compounds were also reported [3, 4].

Table 7/11
Physical Properties of Complexes with 8-Amino-7-hydroxy-4-methylcoumarin, $M(C_{10}H_8NO_3)_3 \cdot nH_2O$.

M^{3+}	n	color	decomposition temp. in °C	Λ_M *) in $\Omega^{-1} \cdot cm^2 \cdot mol^{-1}$	magnetic moment in μ_B	Ref.
La^{3+}	2	light yellow	279 to 295	18		[2]
Ce^{3+}	8	dark green	310	23	2.18	[1]
Pr^{3+}	10	light green	320	25	3.53	[1]
Nd^{3+}	2	yellow	300 to 325	22		[2]
Sm^{3+}	2	yellowish brown	265 to 285	16		[2]
Gd^{3+}	2	greenish yellow	260 to 290	20		[2]

*) Values for the cerium(III) and praseodymium(III) chelates were determined in aqueous solutions, those for the remaining species were determined in nitromethane (temperature not specified).

7.8.4 With 7-Hydroxy-4-methylcoumarin-6-carboxylic Acid (= $C_{11}H_8O_5$)

$M(C_{11}H_7O_5)_3 \cdot 2H_2O$. Each complex (M = La, Nd, Sm, Gd) was prepared by mixing 300 ml of 0.02 M solution of the ligand in ethanol with 75 ml of an aqueous solution (0.02 M) of the rare earth acetate, then adjusting the pH of the reaction mixture to between 5.0 and 6.5 by the dropwise addition of sodium hydroxide solution. The complexes were isolated as crystalline solids.

A comparison of the infrared spectra of the ligand and its complexes reveals the disappearance of the absorption band associated with the $\nu(OH)$ stretching mode (2950 cm^{-1}) of the carboxyl group upon complexation, indicating loss of the carboxyl proton. The $\nu_{as}(COO^-)$ mode shifted to higher wave numbers (1715 to 1725 cm^{-1}) upon complexation relative to the free ligand value (1717 cm^{-1}), while the $\nu_s(COO^-)$ mode shifted to lower wave numbers (1574 to 1590 cm^{-1}) relative to the free ligand value (1688 cm^{-1}).

All the complexes behave as nonelectrolytes in nitromethane as revealed by molar conductance data (Λ_M = 15 to 21 $\Omega^{-1} \cdot cm^2 \cdot mol^{-1}$). The compounds decompose when heated in the ranges (in °C): La, 220 to 235; Nd, 235 to 250; Sm, 270 to 310; Gd, 290 to 325 [4].

References to 7.8:

[1] G. S. Manku (Australian J. Chem. **24** [1971] 925/34). — [2] G. S. Manku, R. D. Gupta, A. N. Bhat, B. D. Jain (J. Less-Common Metals **16** [1968] 343/50). — [3] D. K. Rastogi (Australian J. Chem. **25** [1972] 729/37). — [4] D. K. Rastogi, K. C. Sharma, M. P. Teotia (Indian J. Chem. **13** [1975] 1335/7).

7.9 Complexes with Hydroxyflavones

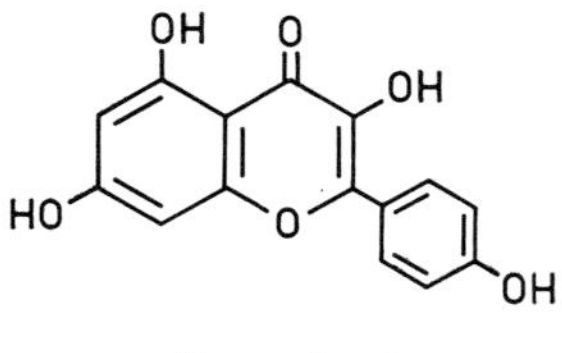

Kaempferol (3,5,7,4'-Tetrahydroxyflavone = $C_{15}H_{10}O_6$)

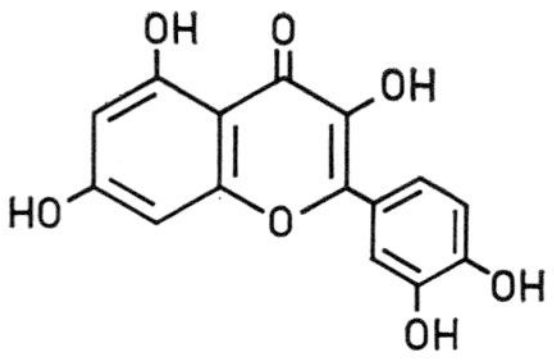

Quercetin (3,5,7,3',4'-Pentahydroxyflavone = $C_{15}H_{10}O_7$)

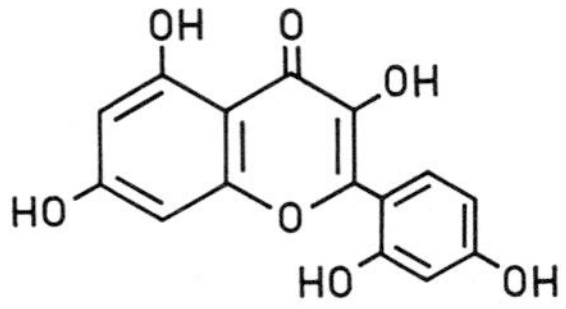

Morin (3,5,7,2',4'-Pentahydroxyflavone = $C_{15}H_{10}O_7$)

Kaempferol interacts with scandium(III) at pH 3 in 50% (v/v) ethanol/water medium to form a 1:1 chelate with an instability constant of 1.42×10^{-5}, as determined spectrophotometrically at ambient temperature (degree of ligand ionization not defined). The wavelength of maximum absorbance was 415 nm at pH 3.0. Above pH 4.0, the absorption maximum shifted to longer wavelengths, an observation attributed to formation of a hydroxo species [1].

Hamaguchi and coworkers [2] determined spectrophotometrically that quercetin reacts with scandium(III) at pH 4.4 to form a 1:1 chelate, λ_{max} = 435 nm. The instablility constant ($K = 2.7 \times 10^{-7}$) of the complex (degree of ligand ionization not defined) was determined spectrophotometrically at ambient temperature [2]. Nazarenko, Antonovich [3], in a spectrophotometric study conducted in more alkaline solutions at 20°C, I = 0.1 ($NaClO_4$), observed the formation of three different 1:1 quercetin-scandium chelates, two of which fluoresced on irradiation with ultraviolet light. Electromigration studies revealed that the complex formed in the pH range 5.5 to 6.0 was positively charged, whereas those formed over the pH ranges 6.0 to 6.5 and 7.6 to 8.8 were neutral. Each complex was derived from the uninegatively charged form of the ligand. The authors proposed that the fluorescent, positively charged complex, formed in the pH range 5.5 to 6.0, is composed of $Sc(OH)^{2+}$ ions ligated to the quercetin ions via the carbonyl and 7-hydroxy (with loss of the proton) oxygen atoms. The neutral, fluorescent compound, $Sc(OH)_2(C_{15}H_9O_7)$, formed at pH 6.0 to 6.5, is derived from $Sc(OH)_2^+$

ions, with the quercetin molecule ligated as described for the positively charged complex, $Sc(OH)(C_{15}H_9O_7)^+$. The nonfluorescent, neutral complex is also derived from $Sc(OH)_2^+$ ions, except that ligation via the carbonyl and 5-hydroxy (with loss of the proton) oxygen atoms was proposed [3].

Morin interacts with scandium to form two complexes: a positively charged, fluorescent species in the pH range 1.2 to 2.8, formulated as $Sc(OH)(C_{15}H_9O_7)^+$, and a neutral, nonfluorescent species in the pH range 6.0 to 7.6, formulated as $Sc(OH)_2(C_{15}H_9O_7)$. With analogies to the corresponding quercetin chelates, Nazarenko and Antonovich [3] proposed that morin is ligated via the carbonyl and 7-hydroxy (with loss of proton) oxygen atoms in the positively charged species, and via the carbonyl and 5-hydroxy (with loss of proton) oxygen atoms in the neutral species.

References to 7.9:

[1] B. S. Garg, K. C. Trikha, R. P. Singh (Anal. Chim. Acta **42** [1968] 343/6). — [2] H. Hamaguchi, R. Kuroda, R. Sugisita, N. Onuma, T. Shimizu (Anal. Chim. Acta **28** [1963] 61/7). — [3] V. A. Nazarenko, V. P. Antonovich (Zh. Analit. Khim. **22** [1967] 1812/7; Russ. J. Anal. Chem. **22** [1967] 1517/21).

7.10 Complex with Hematein

(=3,4,6a,10-Tetrahydroxy-6a,7-dihydro-6H-indeno[2.1-c]chromen-9-one = $C_{16}H_{12}O_6$)

HO OH OH O O OH

Yttrium(III) chloride in water reacted with hematein (formed by adding hydrogen peroxide to a solution of hematoxylin in water/alcohol) to form a violet colored, 1:2 (metal:ligand) chelate (λ_{max} = 560 nm, ε = 10830 $l \cdot mol^{-1} \cdot cm^{-1}$). The logarithm of the thermodynamic stability constant, determined spectrophotometrically at 27°C and pH 6.5, was 9.6 ± 0.2 (lg K = 8.7 to 9.35 at I = 0.1 ($NaClO_4$)). Electrophoresis and ion exchange experiments showed that the complex is anionic, suggesting the composition $Y(C_{16}H_{10}O_6)_2^-$. Bidentate ligation via the quinoid oxygen atom and the adjacent phenolic group was proposed, S. P. Pande, K. N. Munshi (J. Indian Chem. Soc. **49** [1972] 533/5).

Empirical Formula Index of Ligands

The ligands treated in this volume, excluding water, are arranged in the index according to the system of A. Hill (J. Am. Chem. Soc. **22** [1900] 478/94). In this system, the first criterion for the location of a ligand is the number of C atoms; the second, the number of hydrogen atoms. Then come the numbers of atoms of the other elements, set in alphabetical order.

The first column contains the empirical formulas of the ligands. The second column shows the names of the ligands. Additional ligands, occuring in mixed-ligand complexes, and adducts are placed as subheadings. Thus, the complex $M(C_5H_7O_2)_3 \cdot CH_3OH$ may be found under the main entry $C_5H_8O_2$ = 2,4-Pentanedione and the subheading "and CH_4O". But it is also listed under the main entry CH_4O and the subheading "and $C_5H_8O_2$". On the other hand, those additional ligands, which are described as main ligands in the volumes **D 1**, **D 2**, and **D 4**, occur only as subheadings in the formula index of this volume.—The last column lists the pertinent pages. The boldfaced **D 1** before the page numbers refers to those complex compounds, described in Rare Earth Elements D 1, in which the O-donor ligands of this volume occur as additional ligands.

Empirical formulas and names are based on the neutral ligand without regard of its state in the complex. The nomenclature corresponds to IUPAC rules. Customary trade names and trivial names are sometimes given to facilitate identification. Abbreviations for ligand names are listed only when they have been used in this volume.

C_6

C_8

C_{13}

C_{16} to C_{19}

Table of Conversion Factors

Following the notation in Landolt-Börnstein [7], values which have been fixed by convention are indicated by a bold-face last digit. The conversion factor between calorie and Joule that is given here is based on the thermochemical calorie, $cal_{th\,ch}$, and is defined as 4.184**0** J/cal. However, for the conversion of the ,,Internationale Tafelkalorie'', cal_{IT}, into Joule, the factor 4.1868 J/cal is to be used [1, p. 147]. For the conversion factor for the British thermal unit, the Steam Table Btu, BTU_{ST}, is used [1, p. 95].

Force	N	dyn	kp
1 N (Newton)	1	10^{5}	0.1019716
1 dyn	10^{-5}	1	1.019716×10^{-6}
1 kp	9.8066**5**	$9.8066\mathbf{5} \times 10^{5}$	1

Pressure	Pa	bar	kp/m²	at	atm	Torr	lb/in²
1 Pa (Pascal) = 1 N/m^2	1	10^{-5}	1.019716×10^{-1}	1.019716×10^{-5}	0.986923×10^{-5}	0.750062×10^{-2}	145.0378×10^{-6}
1 bar = 10^6 dyn/cm^2	10^{5}	1	10.19716×10^{3}	1.019716	0.986923	750.062	14.50378
1 kp/m^2 = 1 mm H_2O	9.8066**5**	$0.98066\mathbf{5} \times 10^{-4}$	1	10^{-4}	0.967841×10^{-4}	0.735559×10^{-1}	1.422335×10^{-3}
1 at = 1 kp/cm^2	$0.98066\mathbf{5} \times 10^{5}$	0.98066**5**	10^{4}	1	0.967841	735.559	14.22335
1 atm = 76**0** Torr	$1.0132\mathbf{5} \times 10^{5}$	1.0132**5**	1.033227×10^{4}	1.033227	1	760	14.69595
1 Torr = 1 mm Hg	133.3224	1.333224×10^{-3}	13.59510	1.359510×10^{-3}	1.315789×10^{-3}	1	19.33678×10^{-3}
1 lb/in^2 = 1 psi	6.89476×10^{3}	68.9476×10^{-3}	703.069	70.3069×10^{-3}	68.0460×10^{-3}	51.7149	1

Work, Energy, Heat	J	kWh	kcal	Btu	MeV
1 J (Joule) = 1 Ws = 1 Nm = 10^7 erg	1	2.778×10^{-7}	2.39006×10^{-4}	9.4781×10^{-4}	6.242×10^{12}
1 kWh	$\mathbf{3.6} \times 10^6$	1	860.4	3412.14	2.247×10^{19}
1 kcal	4184.**0**	1.1622×10^{-3}	1	3.96566	2.6117×10^{16}
1 Btu (British thermal unit)	1055.06	2.93071×10^{-4}	0.25164	1	6.5858×10^{15}
1 MeV	1.602×10^{-13}	4.450×10^{-20}	3.8289×10^{-17}	1.51840×10^{-16}	1

Power	kW	h. p. (PS)	kp m/s	kcal/s
1 kW = 10^{10} erg/s	1	1.35962	101.972	0.239006
1 h. p. (PS)	0.73550	1	**75**	0.17579
1 kp m/s	9.8066**5** $\times 10^{-3}$	0.01333	1	2.34384×10^{-3}
1 kcal/s	4.184**0**	5.6886	426.650	1

References:
[1] A. Sacklowski, Die neuen SI-Einheiten, Goldmann, München 1979. (Conversion tables in an appendix.)
[2] International Union of Pure and Applied Chemistry, Manual of Symbols and Terminology for Physicochemical Quantities and Units, Pergamon, London 1979; Pure Appl. Chem. **51** [1979] 1/41.
[3] The International System of Units (SI), National Bureau of Standards Specl. Publ. 330, [1972].
[4] H. Ebert, Physikalisches Taschenbuch, 5th Ed., Vieweg, Wiesbaden 1976.
[5] Kraftwerk Union Information, Technical and Economic Data on Power Engineering, Mülheim/Ruhr 1978.
[6] E. Padelt, H. Laporte, Einheiten und Größenarten der Naturwissenschaften, 3rd Ed., VEB Fachbuchverlag, Leipzig 1976.
[7] Landolt-Börnstein, 6. Aufl., II. Bd., 1. Tl., 1971, S. 1/14.